# INDUSTRIAL CATALYSIS AND SEPARATIONS

## Innovations for Process Intensification

# INDUSTRIAL CATALYSIS AND SEPARATIONS

## Innovations for Process Intensification

*Edited by*

**K. V. Raghavan, PhD, and B. M. Reddy, PhD**

Apple Academic Press

TORONTO NEW JERSEY

Apple Academic Press Inc.
3333 Mistwell Crescent
Oakville, ON L6L 0A2
Canada

Apple Academic Press Inc.
9 Spinnaker Way
Waretown, NJ 08758
USA


*Exclusive worldwide distribution by CRC Press, a member of Taylor & Francis Group*


Printed in the United States of America on acid-free paper

International Standard Book Number-13: 978-1-926895-96-3 (Hardcover)



**Library of Congress Control Number: 2014938654**

**Library and Archives Canada Cataloguing in Publication**

Industrial catalysis and separations: innovations for process intensification/edited by K.V. Raghavan, PhD, and B.M. Reddy, PhD.

Includes bibliographical references and index.
ISBN 978-1-926895-96-3 (bound)

1. Catalysis--Industrial applications. 2. Green chemistry. 3. Chemical processes. 4. Chemical engineering. 5. Sustainable engineering. I. Reddy, B. M., author, editor II. Raghavan, K. V., author, editor

TP156.C35I54 2014 660'.2995 C2014-902755-9

Apple Academic Press also publishes its books in a variety of electronic formats. Some content that appears in print may not be available in electronic format. For information about Apple Academic Press products, visit our website at **www.appleacademicpress.com** and the CRC Press website at **www.crcpress.com**

# ABOUT THE EDITORS

**K. V. Raghavan, PhD**

K. V. Raghavan, PhD, is the current Chairman of the Research Council of the Central Leather Research Institute (CLRI), Chennai, India. He is coordinating the working group deliberations on Low Carbon Energy Technologies of the CAETS, an international body of engineering academies. He is a Fellow of the Indian National Academy of Engineering (INAE), the Indian Institute of Chemical Engineers (IIChE), and the A.P. Akademy of Sciences (APAS), and is a Distinguished Fellow of the University of Grants Commission (UGC). Dr. Raghavan took up the Distinguished Professorship of the INAE at the Indian Institute of Chemical Technology, Hyderabad, in 2008. He became the Vice President (International Cooperation) of the INAE in 2011.

He has held various levels of scientific positions in three national laboratories of the Council of Scientific and Industrial Research (CSIR) of India. He was appointed as the Director of Central Leather Research Institute, Chennai in 1994, and in 1996 he took over the Directorship of the Indian Institute of Chemical Technology, Hyderabad. On successful completion of this tenure, he was appointed as the Chairman of Recruitment and Assessment Centre of DRDO, Ministry of Defence, Government of India in May 2004. In addition, Dr. Raghaven chairs scientific committees of public and private sector bodies on clean technologies, catalytic ethanol reforming, biorefining, and allied areas.

The recipient of six national awards, Dr. Raghavan has published more than 120 papers, filed 45 patents, and edited three books. His applied research efforts, covering a time span of over four decades, contributed to the successful development of more than 25 chemical processes with high industrial impact in bulk organics, specialities, oil field chemicals, and fluoroorganics. He has made significant contribution to the technological upgradation of the leather, agrochemical, and drugs/pharma sectors.

Chemical process development and design, reaction engineering, simulation and modeling, and chemical hazard analysis are his areas of specialization. His basic research contributions cover the simulation of complex reactions in fixed bed reactors, hydrodynamics of multiphase reaction systems, envirocatalysis for clean processing, zeolite catalysis for macromolecules, thermochemistry, and kinetics of charge transfer polymerization and modeling of chemical accidents. His current research activities are in process intensification of water gas shift reaction, catalytic CO2 decomposition, analysis of $CO_2$ capture technologies and characterization of the reactivity of ionic liquids.

He received his BTech from Osmania University in 1964 and a MS and PhD from the Indian Institute of Technology (IIT), Madras, India.

**B. M. Reddy, PhD**

Benjaram Mahipal Reddy, PhD, is a Chief Scientist and Head of the Division of Inorganic and Physical Chemistry at the CSIR-Indian Institute of Chemical Technology (IICT) in Hyderabad in India. IICT is one of the premier research institutes in India under the aegis of the Council of Scientific and Industrial Research (CSIR), New Delhi, India.

Dr. Reddy has published over 230 papers in peer-reviewed journals, several review articles, and book chapters, and he has eight patents to his credit. His papers have been cited more than 5600 times (h-index 41). He guided 25 PhD and 16 MTech/BTech students for their degrees. He is the recipient of CSIR Young Scientist Award and Catalysis Society of India Young Scientist Award in addition to several other awards, fellowships, and honors. Dr. Reddy is an elected fellow of the Indian National Academy of Engineering (FNAE); the National Academy of Sciences, India (FNASc); and the fellow of Andhra Pradesh Academy of Sciences (FAPAS).

After completing his BSc degree from Osmania University, Hyderabad, and MSc degree from Kakatiya University, Warangal, he joined IICT for his PhD degree and later moved to Germany for postdoctoral studies with Prof. Dr. Helmut Knözinger at Ludwig-Maximilians-Universität in Munich, and then to the USA with Prof. Eli Ruckenstein at the State University of New York, Buffalo, New York, and Texas A&M University, College Station, Texas. He has been a visiting scientist on various prestigious fellowships and exchange programs, namely, DAAD (Germany), CNRS (France), JSPS (Japan), Brain-Pool (South Korea), and SER (Switzerland), and worked at Ruhr University Bochum (Germany), Institute for Research on Catalysis (Lyon, France), AIST (Osaka, Japan), Inha University (Incheon, South Korea), Paul Scherrer Institute (Villigen, Switzerland), respectively. He has also visited Italy, South Africa, Hungary, China, Russia, and Australia as the Indian delegate member on science and technology activities.

His major research interests are solid superacids for organic synthesis, nanocomposite oxides, oxygen storage and release materials, carbon dioxide utilization, biomass conversion, nano-gold, and green technologies.

# CONTENTS

# LIST OF CONTRIBUTORS

**Suresh K. Bhargava**
Centre for Advanced Materials and Industrial Chemistry, School of Applied Sciences, RMIT University, Melbourne, Australia

**Damma Devaiah**
Inorganic and Physical Chemistry Division, CSIR-Indian Institute of Chemical Technology, Hyderabad – 500 007, India

**Venkata R. Doddi**
Organic and Biomolecular Chemistry Division, CSIR-Indian Institute of Chemical Technology, Hyderabad – 500 007, India

**Carlos A. Fernandez**
Pacific Northwest National Laboratory, 902 Battelle Blvd, Richland, WA 99352, USA

**Shiva Shanker Kaki**
Centre for Lipid Research, CSIR-Indian Institute of Chemical Technology, Uppal Road, Hyderabad – 500 007, India

**Narayana V. Kalevaru**
Leibniz-Institute for Catalysis, Albert-Einstein-Str. 29A, D-18059 Rostock, Germany

**Sudhir Kashyap**
Organic and Biomolecular Chemistry Division, CSIR-Indian Institute of Chemical Technology, Hyderabad – 500 007, India

**Palakodety R. Krishna**
Organic and Biomolecular Chemistry Division, CSIR-Indian Institute of Chemical Technology, Hyderabad – 500 007, India

**Jian Liu**
Energy and Environment Directorate, Pacific Northwest National Laboratory, 902 Battelle Blvd, Richland, WA 99352, USA

**D. Manjunath**
Membrane Separations Group, Chemical Engineering Division, CSIR-Indian Institute of Chemical Technology (IICT), Hyderabad - 500 007, India

**Andreas Martin**
Leibniz-Institute for Catalysis, Albert-Einstein-Str. 29A, D-18059 Rostock, Germany, Tel.: 0381 1281 246; E-mail: andreas.martin@catalysis.de

**S. Mayadevi**
Chemical Engineering & Process Development Division, CSIR-National Chemical Laboratory, Pune – 411 008, India, Tel.: +91 20 2590 2176; +91 20 2587 1356, E-mail: s.mayadevi@ncl.res.in ; maya0022@yahoo.co.in

**B. Pete Mcgrail**
Energy and Environment Directorate, Pacific Northwest National Laboratory, 902 Battelle Blvd, Richland, WA 99352, USA

**Radha Kishan Motkuri**
Energy and Environment Directorate, Pacific Northwest National Laboratory, 902 Battelle Blvd, Richland, WA 99352, USA, Tel.: +1 509-371-6484; E-mail: radhakishan.motkuri@pnnl.gov

**Banda Narsaiah**
Fluoroorganic Division, CSIR-Indian Institute of Chemical Technology, Hyderabad – 500 007, India

**Satish K. Nune**
Energy and Environment Directorate, Pacific Northwest National Laboratory, 902 Battelle Blvd, Richland, WA 99352, USA

**K. Praneeth**
Membrane Separations Group, Chemical Engineering Division, CSIR-Indian Institute of Chemical Technology (IICT), Hyderabad - 500 007, India

**P. S. Sai Prasad**
Inorganic & Physical Chemistry Division, CSIR-Indian Institute of Chemical Technology, Hyderabad – 500 007, India, Tel: +91 40 27193163; E-mail: saiprasad@iict.res.in

**R. B. N. Prasad**
Centre for Lipid Research, CSIR-Indian Institute of Chemical Technology, Uppal Road, Hyderabad – 500 007, India, Tel.: +91 40 27193370; E-mail: rbnprasad@iict.res.in

**K. V. Raghavan**
Reaction Engineering Laboratory, CSIR-Indian Institute of Chemical Technology, Uppal Road, Hyderabad – 500 007, India, Tel.: +91 40 27193119; E-mail: kondapuramiict@yahoo.com

**Yadla Rambabu**
Fluoroorganic Division, CSIR-Indian Institute of Chemical Technology, Hyderabad – 500 007, India

**K. Yamuna Rani**
Process Dynamics and Control Group, Chemical Engineering Division, CSIR-Indian Institute of Chemical Technology, Hyderabad – 500 007, India, Tel.: +91 40 2719 3121, E-mail: kyrani@iict.res.in

**Pamulaparthy Shanthan Rao**
Fluoroorganic Division, CSIR-Indian Institute of Chemical Technology, Hyderabad – 500 007, India. Tel.: +91 40 2719 3185; E-mail: psrao@iict.res.in

**Benjaram M. Reddy**
Inorganic and Physical Chemistry Division, CSIR-Indian Institute of Chemical Technology, Hyderabad – 500 007, India, Tel.: +91 40 2719 3510; E-mail: bmreddy@iict.res.in

**Lankela H. Reddy**
Inorganic and Physical Chemistry Division, CSIR-Indian Institute of Chemical Technology, Hyderabad – 500 007, India

**Post S. Reddy**
Organic and Biomolecular Chemistry Division, CSIR-Indian Institute of Chemical Technology, Hyderabad – 500 007, India

**Gangavaram V. M. Sharma**
Organic and Biomolecular Chemistry Division, CSIR-Indian Institute of Chemical Technology, Hyderabad – 500 007, India, Tel.: +91 40 2719 3154; E-mail: esmvee@iict.res.in

**I. Sreedhar**
Department of Chemical Engineering, BITS Pilani Hyderabad Campus, Hyderabad – 500 078, India

**A. Sri Hari Kumar**
Inorganic & Physical Chemistry Division, CSIR-Indian Institute of Chemical Technology, Hyderabad – 500 007, India

**Madabhushi Sridhar**
Fluoroorganic Division, CSIR-Indian Institute of Chemical Technology, Hyderabad – 500 007, India

**S. Sridhar**
Membrane Separations Group, Chemical Engineering Division, CSIR-Indian Institute of Chemical Technology (IICT), Hyderabad – 500 007, India, Tel.: +91 40 2719 3408; E-mail: sridhar11in@yahoo.com

**Darbha Srinivas**
Catalysis Division, CSIR–National Chemical Laboratory, Pune – 411008, India, Tel.: +91 20 2590 2018; E-mail: d.srinivas@ncl.res.in

**C. Sumana**
Process Dynamics and Control Group, Chemical Engineering Division, CSIR-Indian Institute of Chemical Technology, Hyderabad – 500 007, India

**Praveen Thallapally**
Fundamental and Computational Science, Directorate, Pacific Northwest National Laboratory, 902 Battelle Blvd, Richland, WA 99352, USA

**Pulikkeel Unnikrishnan**
Catalysis Division, CSIR–National Chemical Laboratory, Pune – 411 008, India

# LIST OF ABBREVIATIONS

| | |
|---|---|
| $(DHQ)_2PHAL$ | hydroquinine 1,4-phthalazinediyl diether |
| $(DHQD)_2PHAL$ | hydroquinidine 1,4-phthalazinediyl diether |
| A.P. | Andhra Pradesh |
| AHM | ammonium heptamolybdate |
| AMPA | molybdophosphoric acid |
| ANG | adsorbed natural gas |
| $B(Oi\text{-}Pr)_3$ | boron tri-isopropoxide |
| BET | Brunauer-Emmet-Teller |
| CaO | calcium oxide |
| CCS | capture and storage |
| CFC | chlorofluorocarbons |
| CFD | computational fluid dynamics |
| $CH_2Cl_2$ | dichloromethane |
| CNG | compressed natural gas |
| DEC | diethyl carbonate |
| DET | diethyl tartarate |
| DIPEA | N,N-Diisopropylethylamine |
| DMC | dimethyl carbonate |
| DME | dimethyl ether |
| DMF | dimethyl formamide |
| dppf | 1,1'- bis(diphenylphosphino) ferrocene |
| DTA | differential thermal analysis |
| EC | ethylene carbonate |
| ECTFE | ethylene chlorotrifluoro ethylene polymer |
| EDX | energy dispersive X-ray |
| EG | ethylene glycol |
| EOS | equation of state |
| $Et_3N$ | triethylamine |
| ETFE | ethylene tetrafluoroethylene polymer |
| FDA | Food and Drug Administration |
| FEP | fluorinated ethylene propylene polymer |
| FEP | fluorinated ethylene propylene polymer |
| FFA | free fatty acid |

| | |
|---|---|
| FP | frost points |
| GWP | global warming potential |
| $HBF_4$ | fluoroboric acid |
| HCFC | hydro chloro fluoro carbons |
| HCFC 133a | 1,1,1-trifluoro-2-chloroethane |
| HCl | hydrochloric acid |
| HFC | hydro fluoro carbon |
| HFC | hydro fluoro carbon |
| HFC-125 | 1,1,1,2,2-Pentafluoroethane |
| HFC-134a | 1,1,1,2-Tetrafluoroethane |
| HFC-143a | 1,1,1, -Trifluoroethane |
| HFC-236fa | 1,1,1,3,3,3-Hexafluoropropane |
| HFC-32 | difluoromethane |
| ICI | Imperial Chemical Industries |
| IICT | Indian Institute of Chemical Technology |
| IL | ionic liquids |
| $K_2CO_3$ | potassium carbonate |
| KF | potassium fluoride |
| LBL | layer-by-layer |
| LDMR | linear dimer |
| LICAT | Leibniz Institute for Catalysis at University Rostock |
| LTMR | linear tetramer |
| MAOS | microwave assisted organic synthesis |
| MeOH | methanol |
| MgO | magnesium oxide |
| MOFs | metal–organic frameworks |
| MOGs | metal-organic gels |
| MP11 | microperoxidase-11 |
| MPA | heteropoly molybdate |
| MPD | m-phenylenediamine |
| MTBE | methyl tert.-butyl ether |
| MTS | micelle template silicas |
| MWCO | molecular weight cut |
| *n*-BuLi | n-butyl lithium |
| NEERI | National Environmental Engineering Research Institute in India |
| NF | nanofiltration |
| NHC | heterocyclic carbine |
| NMR | nuclear magnetic resonance |

| | |
|---|---|
| OCT | Olefin Conversion Technology |
| ODH | oxidative dehydrogenation |
| ODP | ozone depleting potential |
| OH | octahedral |
| P(*o*-tolyl)$_3$ | Tris(o-tolyl)phosphine |
| P(*t*-Bu)$_3$ | Tri-tert-butylphosphine |
| PA | aromatic polyamide |
| PANI | polyaniline nanofibers |
| $P_c$ | critical pressure |
| PC | propylene carbonate |
| PCM | post-synthetic modification |
| PCPs | porous coordination polymers |
| PCTFE | polychlorotrifluoroethylene |
| Pd | palladium |
| $Pd(OAc)_2$ | palladium(II) acetate |
| $Pd(PPh_3)_4$ | tetrakis(triphenylphosphine)palladium(0) |
| Pd/C | palladium on carbon |
| $PdCl_2$ | palladium(II) chloride |
| $PdCl_2(PPh_3)_2$ | Bis(triphenylphosphine)palladium(II) dichloride |
| PES | polyether sulfone |
| PFA | perfluoroalkoxy polymers |
| PFC | perfluorocarbon |
| PG | propylene glycol |
| PI | process intensification |
| $PPh_3$ | triphenylphosphine |
| PTFE | polytetrafluoroethylene |
| PVA | polyvinyl alcohol |
| PVDF | polyvinylidene fluoride |
| QCMs | quartz crystal microbalances |
| RD | reactive distillation |
| ROMP | ring-opening metathesis polymerization |
| $RuCl_3$ | ruthenium(III) chloride |
| RWGS | reverse water-gas shift |
| SA | swelling agent |
| SAW | surface acoustic wave |
| SCFs | supercritical fluids |
| SMBS | sodium metabisulphite |
| SrO | strontium oxide |
| STY | space–time-yield |

| | |
|---|---|
| TA | tectonic acid |
| TAP | temporal-analysis-of-products |
| TB | tuberculosis |
| $T_c$ | temperature |
| TCE | trichloroethylene |
| TCTFE | 1,1,2-Trichloro-1,2,2-trifluoroethane |
| TDS | total dissolved solids |
| TEOS | tetraethyl orthosilicate |
| TFC | thin film composite |
| TFE | tetrafluoroethylene |
| TH | tetrahedral |
| TMG | tetramethylguanidine |
| TO | turnover |
| TOF | turnover frequency |
| TON | turnover number |
| UCLA | University of California Los Angeles |
| UF | ultrafiltration |
| VDF | vinylidene fluoride |
| WHO | World Health Organization |
| XPS | X-ray photoelectron spectroscopy |
| XRD | X-ray diffraction |
| XRF | X-ray fluorescence |
| $ZnBr_2$ | zinc bromide |
| $ZnCl_2$ | zinc chloride |

# LIST OF SYMBOLS

| | |
|---|---|
| $t_w$ | wall thickness |
| L | length |
| $T_{equi}$ | equilibrium thickness |
| V | coating plate velocity |
| $\rho$ | viscosity |
| g | gravity |
| $\eta$ | surface tension |
| $P_{max}$ | maximum load |
| $A_o$ | cross-sectional |
| dp | pore size diameter |
| $\gamma$ | surface tension of liquid |
| $\theta$ | contact angle of liquid |
| P | external pressure |
| $C_p$ | solute concentrations |
| $\Delta V^*$ | difference between the partial molar volume |
| $\varepsilon$ | dielectric constant |

# PREFACE

**Prof. M. M. Sharma, FRS, FNA**
Emeritus Professor of Eminence and
Former Director of Institute of Chemical Technology (ICT),
Nathalal Parekh Marg, Matunga, Mumbai – 400 019, India

This book, edited by the accomplished editors K. V. Raghavan and B. M. Reddy, is a multi-author endeavor and covers diverse topics. A majority of commercial processes in the chemical industry (CI) are catalytic and some noncatalytic processes, notably steam cracking of naphtha for olefins, which is practiced on a grand scale, are possible candidates for conversion to catalytic process. Although catalysis plays a pivotal role, even a magical role, major investment is in the separation process, including in biotechnology-based processes. Here the authors have emphasized process intensification (PI), along with that for the chemical part.

There is a general impression that emphasis on greener processes and process intensification has been of relatively recent origin, but in fact this is historically associated with the CI as illustrated by the replacement of Leblanc Process for soda ash by Solvay Process, Lead Chamber process for sulfuric acid by the contact process, ethylene oxide by the chlorohydrin process replaced by direct catalytic oxidation, ammoxidation of propylene to acrylonitrile replacing processes using HCN, etc. Even in the case of separation processes, notably in distillation, the bubble cap plates was replaced by sieve trays/valve trays and Raschig ring random packings by Intalox saddles, Paul rings and in later years by structured packings. Adsorptive processes provided yet another breakthrough including bulk removal of $CO_2$ in $H_2$/methanol/ammonia plants. Higee contactor provided yet another strategy. New design of impellers for better mixing, using less power, were ushered and even gas inducing impellers have been adopted on a large-scale; loop reactors have been introduced. The CI is highly science based, and innovations have been an integral part and some of these were serendipitous. A recent example is the oxidation of *n*-butane to maleic anhydride, replacing benzene as a feedstock where two carbon atoms go to $CO_2$. The most outstanding example of the discovery driven process is that of polyamide Nylon 6,6.

The editors have done a commendable job of bringing together a host of contributors with different expertise in a fairly consistent way. Membrane separations should have covered most recent developments connected with nanofiltration where even aggressive solvents can now be handled. There is a lot of talk about $CO_2$ utilization, and this book has given weight to this contemporary subject. However, we need to realize that we are talking of emissions of the order of 30 billion tones per annum! Urea continues to be a big consumer of $CO_2$. The use of supercritical $CO_2$ in organic synthesis has been duly covered but it is also relevant in PI in separation. It is refreshing to see fluorochemicals being covered in a useful way. Polyvinyl fluoride films are relevant in solar cells and should be mentioned.

Base catalyzed reaction has been covered with reference to transesterification to make biodiesel. A classical case of considerable practical importance is the side chain alkylation of toluene with propylene to give isobutyl benzene, apart from conversion of acetone to diacetone alcohol and isophorone and even here heterogenerous versions are in vogue or under consideration. Microwave assisted reactions are duly covered and it would have been useful to cover ultrasonically assisted reactions, particularly when solid reactant/catalyst is involved.

This book has made a sincere attempt and should appeal to persons connected with the CI, be it academics or research scientists. The subject is vast and for obvious reasons only limited topics can be covered. I do heartily recommend this book.

**— Prof. M. M. Sharma**

# INTRODUCTION

Although green process concepts were used in organic and inorganic chemical technologies since early part of 20th century, they were formally introduced in the early 1990s in the developed world as part of sustainable progress initiatives. Around 2000, the topic of green verses sustainable chemistry became a very prominent subject for the research community to explore new innovations. While green chemistry focused on the use of chemicals and processes that have little or no pollution potential, the sustainable chemistry laid stress on maintenance and continuity of ecological soundness of chemical processes. The Process Intensification (PI) is one of the most important and rapidly growing concepts impacting both green and sustainable chemistry and engineering. It has an excellent potential to improve existing chemical process technologies by creating new process options, which are more efficient and greener. Extensive open literature is currently available on new process options, improved process equipments and alternative feedstocks and solvents associated with PI.

This book is endowed with a unique collection of scientific articles contributed by bright and experienced researchers in green and sustainable chemistry. They demonstrate, in a meaningful way, the crucial role-played by catalysis in advancing the PI concepts. Though several books have been published recently on green catalytic options, very few of them deal with innovations pertaining to industrial catalysis, which lead to intensification of processes. This book focuses on this aspect to encourage new ways of thinking on catalytic reactions and their intensification. It will be an invaluable information source to postgraduate students, research scholars, scientists/engineers in industry as well as others interested in applied catalysis for minimizing the environment impact of chemical and allied processes. The other unique features of this book are its coverage of a range of catalytic PI options researched in recent years, providing new insights into process sustainability, covering the selected topics to a depth that allows the researchers to evolve appropriate strategies for commercial deployment. It is easily accessible to chemists and chemical engineers who are familiar with the homogeneous and heterogeneous catalysis and basic PI concepts.

For convenience of presentation, 12 chapters of the book are structured and divided into four parts. The Part I of the book deals with the engineering approaches to process intensification. Chapter 1 is contributed by Yamuna Rani and Sumana from India, two bright scientists with several creative ideas on process intensification. Apart from clarifying on definition, multifunctional operations, novel devices and modeling aspects, they have suggested a revised classification methodology for PI. Their detailed discussions on reactive and hybrid separations, micro and structured reactors, novel devices for mass, momentum and heat transfer, alternate PI configurations and modeling approaches to multifunctional operations are highly relevant to presentations made in other chapters. Chapter 2 by one of the editors and his research associates deals with an interesting subject of microenvironment management in heterogeneous catalysis to promote PI. It drew the inspiration from biocatalysis. This is for the first time such a subject is dealt with reference to PI. After defining the microenvironment of a heterogeneous catalyst and its positive effect on catalytic activity and process selectivity, an attempt is made by the authors to explain the role of catalyst microenvironment in controlling the course of oxidation, electrostatic interactions, aromatic nitration, asymmetric hydrogenation and selective synthesis of macrocycles. The latter studies by the authors led to the development of a novel technique for increasing the pore size of a mesoporous zeolite catalyst with the help of macrocycles of the suitable molecular size. The authors finally recommend a two-stage process restandardization strategy to achieve the microenvironment management goals. The above two chapters provide the much-needed background of PI strategies for application in catalytic organic synthesis.

Part II of the book focuses on the intensification of separation processes with high industrial relevance. Chapter 3 deals with the contemporary subject of metal organic frameworks (MOFs) with potential applications in molecular separations and sensing, catalysis, delivery systems for bioactive molecules and $CO_2$ capture for sequestration. The detailed analysis presented by Radha Kishan Motkuri et al., from USA, provides definitive indications that significant breakthroughs are most likely in the near future in its application in chemical processes and reactive separations. Their chemical versatility, pore tunability and functionalization of active sites are positive factors for employing them in reactive separations. There are good prospects for developing electro, mechano, sonochemical and microwave assisted synthesis for high throughput synthesis of MOFs for new material exploration.

Chapter 4 deals with an extremely important PI option viz., membrane separation. PI through this option is seen as a facilitator for the integrative design of separation phenomena on meso-scale. Sridhar et al., from India, have authored this chapter. Realizing the ability of polymeric membranes to create a favorable interface for mass transfer between two components of drinking water system by making use of their large exchange area endowed with independent fluid dynamics, they developed an industrially deployable defloridation technique for drinking water. It is particularly interesting that the developed technology was successfully scaled up and transferred to rural technicians in India with limited scientific and engineering knowledge base. This intensified process technology has immense socioeconomic relevance in developing countries.

Part III of the book highlights the important developments made in recent years in $CO_2$ facilitated chemical transformations with an eye on its large-scale utilization. The Chapter 5 highlights the need to pursue carbon capture and utilization (CCU) option much more vigorously than the carbon capture and storage (CCS) concept since large-scale storage of $CO_2$ has several practical problems. Unnikrishnan and Srinivas from India focussed their attention on catalytic conversion of $CO_2$ into fuels and downstream chemicals. Several options for $CO_2$ utilization have been brought out with potential industrial applications. With the help of an extensive literature review, they discused the $CO_2$ transformation into fuels viz., direct catalytic hydrogenation through Sabatier reaction into methane and long chain hydrocarbons and their subsequent conversion to industrial fuels through FT synthesis and $CO_2$ conversion to methanol, ethanol, higher alcohols and dimethyl ether and conventional and dry/tri reforming methods for syngas with $CO_2$ as one of the reactants. Novel biotechnologies for $CO_2$ transformation into liquid fuels also received attention. The authors also dealt in detail on the $CO_2$ conversion to bulk and specialty chemicals including organic carbamates, poly, cyclic and dialkyl carbonates, urea and its derivatives, formic and acetic acids and DMF. Finally, they stressed the need to enhance the market acceptability of $CO_2$ utilization technologies and developing innovative PI options to bring down material/energy consumptions and costs.

Chapter 6 authored by Mayadevi from India covers the important applications of supercritical (Sc) $CO_2$ in organic synthesis. The unique properties of $ScCO_2$ transform it into a chemically attractive and environmentally benign reaction medium for organic chemical synthesis. It is particularly useful for speeding up mass transfer limited heterogeneous catalytic processes in view of its density comparable to those of liquids and its contribution to higher diffusion coefficients. This chapter also highlights the $ScCO_2$ utilization in specific homogeneous and heterogeneous catalytic reactions, its role in influencing the mass transfer, reaction equilibrium and kinetics with reference to organic hydrogenation, hydroformylation, oxidation, C-C bond formation, alkylation, acetylation and esterification reactions. It finally deals with PI and scale up aspects.

Part IV of the book covers 6 chapters with focus on application of heterogeneous catalysis in a wide spectrum of organic chemical reactions including ammoxidation, catalytic/steam cracking, oxidative dehydrogenation, fluorination and transesterification. They are very relevant to fine, specialty and knowledge intensive segments of the chemical industry. Their current global production value exceeds UD 85 billion. A large toolbox of chemical reactions is available for their synthesis developed through traditional R&D approaches, which are more development than efficiency oriented. Their environmental burden is rather high compared to other segments of chemical industry. Significant investments are anticipated in the coming years in intensifying and greening these technologies. From this angle, the chapters covered in Part IV of this book assume importance.

Chapter 7, contributed by two eminent German researchers, reviews the current status of research on catalytic ammoxidation of hetero-aromatic compounds and industrial level successes. It also identifies the potential PI areas. The ammoxidation of hetero-aromatics is an important approach for the synthesis of various pharmaceuticals, agrochemicals, dyestuffs and other specialty products. Chapter 8 from Hari Kumar and Sai Prasad from India deals with the PI options currently available for catalytic steam cracking and oxidative dehydrogenation (ODH) of ethane to ethylene. Ever since the natural gas had become one of the preferred feedstocks for C1-C2 chemicals, the above processes have attracted large-scale industrial attention. The heterogeneous catalysis has played a major role in making ODH theoretically more attractive than catalytic/steam cracking options for ethane to ethylene conversion. However, PI through membrane assisted reactive separation, use of $CO_2$ as oxidant and more professional management of catalyst activity and specificity is expected to make ODH commercially also more attractive. Several scientific and stability barriers have to be overcome to achieve this objective.

Chaper 9 by Sharma et al., from India is concerned with basic research challenges and commercial successes achieved in a wide range of C-C and C-O bond forming reactions with transition metal catalysis. It also highlights the process intensification efforts of various kinds needed to achieve improved yields and selectivities. It is interesting to note that several innovative reactions developed by Nobel Laureates and well-known research groups in the world have received industrial acceptance due to their innovation content.

The PI options successfully tried are reaction integration, single pot synthesis, metal exchange in catalysis and others. The early involvement of knowledge leaders in process synthesis in most of the cases covered in this chapter resulted in the evolution of unique solutions for industrial deployment. Chapter 10 by Reddy et al. (one of the editors of this book) provides its undivided attention to microwave assisted organic synthesis, itself being a versatile PI tool. It has examined the mechanism of microwave heating, monomode verses multimode processing options, development and scale up trends in microwave reactors and important organic chemical processes successfully mediated by the microwave irradiation. There are good prospects for deriving the unique benefits offered by the microenvironment provided by the microwave cavities for achieving higher reaction specificity.

Chapter 11 by Shantan Rao and co-workers discusses the development of sustainable alternatives along with appropriate phase in and phase out strategies implemented from time to time by this sector for meeting the stringent environmental regulations specified by Montreal and Kyoto Protocols. The focus is on catalytic processes for the synthesis of hydrofluorocarbons, halons and fluoropolymers having high industrial relevance. It also highlights the successes achieved in the scale up of fluoro hydrocarbon technologies in India and the need to intensify future R&D efforts to develop a series of new molecules, which can meet the future environmental rigors. Chapter 12 discusses the innovative base catalytic options available for transesterification of lipid-based feedstocks for biodiesel production. The technological challenges in handling the multifeedstocks with high free fatty acid and unsaponifiables are highlighted.

To sum up, a systematic effort has been made in this book to review the recent advances made in evolving novel catalytic approaches to make chemical processes greener and environmentally sustainable. It has also highlighted the importance of novel separation processes to achieve sharper product cuts and the use of well-designed multiphase reactors for enhanced process performance. It has unequivocally stressed the need for chemists and chemical engineers to adopt innovative PI strategies to reduce the environmental burden of fine and specialty chemical processes.

# PART I

## SCIENTIFIC APPROACHES TO PROCESS INTENSIFICATION

CHAPTER 1

# CHEMICAL PROCESS INTENSIFICATION: AN ENGINEERING OVERVIEW

K. YAMUNA RANI and C. SUMANA

## CONTENTS

## 1.1 INTRODUCTION

Process intensification is one of the most promising development paths for chemical process industry. Stankiewicz and Moulijn [1] stated that the philosophy of process intensification is characterized by four words – smaller, cheaper, safer, and slicker. Smaller refers to miniaturized units for operation as well as reduction in the number of units due to integrated operations. Enhanced conversions due to equilibrium shift in integrated reactive separations, better separations owing to high surface to volume ratios in novel equipment and new modes of operations lead to significant improvements in economics and are therefore cheaper. Operations in miniaturized equipment need very low inventory and contribute to safer operations. All these factors together lead to smaller and more compact plants that are considered to be slicker and contributing to sustainable technological development.

The main motivating factors for process intensification (PI) can be attributed in the areas of sustainability, costs, safety, time to market, and company image. Costs are drastically reduced with reference to land, other investments, raw materials, utilities and waste processing. Safety is drastically improved due to 'smaller is safer' factor, better controlled process, elimination of hot spots and elimination of one or more components. Time taken to market or the time for scaling up is almost eliminated in certain intensified operations since higher scale of operations can be achieved not by increasing the size of equipment but by increasing their number. Since Process intensification leads to sustainable development, its application enhances a company's image.

Few industries have already ventured into PI and are reaping its benefits – Sulzer (KATAPAK), Nagaoka International Corporation, Koch-Glitsch, BHR group, Institute for Micro-technik, Mainz, Eastman Chemical, etc. DSM Technologies have estimated savings of 40 to 80% by application of PI principles to process and plant design consisting of six major chemicals covering polymers, life science products and performance materials. Sulzer has commercialized the concept of PI through static mixers, where mixing elements are made of heat transfer tubes, thereby simultaneously achieving mixing and heat removal/supply. These are found to be especially suitable in nitrations, neutralizations and polymerization reactions. Another innovative PI product of Sulzer is a support for catalytic material, KATAPAK-M®, which is also known for its mixing properties and good radial heat transfer characteristics. This has found applications in catalytic distillation in gas-phase exothermic oxidation processes. In nonreactive distillation processes, Nagaoka International Corporation has developed structured packings called Super X-Pack, which is claimed to reduce the height of a distillation column by a factor of 5 compared to a conventional column. BHR group has developed HEX reactors that provide favorable heat transfer conditions with large heat transfer areas suitable for highly exothermic reactions. They have used HEX reactors and have decreased by-product formation in one of the ICI acrylic processes by 75% and have reduced batch time from 18 hrs to 15 min in a fine chemical process. Micro-reactors, characterized by their extremely small dimensions, are being developed to perform various functions like mixing, catalytic reactions, heat transfer and separations and are especially attractive for reactions involving poisonous or explosive reagents, such as partial oxidation reactions. Institute for Mocrotechnik, Mainz is actively working in this area.

Ramshaw's group at New Castle University employed spinning disk reactors to achieve very fast liquid-liquid reactions such as nitrations, sulphonations and polymerizations. In one of Smith Kline Beecham's processes, the overwhelming outcomes achieved were 99.9% reduction in reaction time, 99% reduction in inventory and 93% reduction in impurity level. HiGee technology has successfully been applied for de-aeration of flood water in Chinese oil fields where the height of the unit was reduced from 30 m to 1 m. Dow chemical company has employed rotating packed beds in their hypochlorous acid technology. Another classic and well-cited PI application is Eastman Chemical Company's methyl acetate process, where almost 10–12 units have been reduced to a single unit consisting of distillation, extractive distillation, reactor and reactive distillation in a single unit with remarkable performance improvement.

These applications clearly illustrate the tremendous potential of PI in chemical process industry. An attempt is made in this chapter to present an engineering perspective of PI with special attention towards catalytic processes. The next section presents the commonly reported definitions and principles of PI, followed by some of the classifications for PI from various standpoints, while the subsequent three sections present details on multifunctional operations, novel devices and alternate PI configurations. The last section discusses modeling and other issues.

## 1.2 DEFINITIONS AND PRINCIPLES

### *1.2.1 VARIED DEFINITIONS*

Originally, PI was used synonymously with process improvement. Later, Ramshaw [2] described it as "devising an exceedingly compact plant, which reduces both the main plant as well as installation costs." Heggs [3] stated that "PI is concerned with order-of-magnitude reduction in process plant and equipment."

A more recent definition of PI [1] is given as "PI comprises novel equipment, processing techniques, and process development methods that, compared to conventional ones, offer substantial improvements in (bio) chemical manufacturing and processing." An alternate version of the same was reported by Stankiewicz and Moulijn [4]. Subsequently, Stankiewicz [5] defined PI as "consisting of development of innovative apparatuses and techniques that offer drastic improvements in chemical manufacturing and processing, substantially decreasing equipment volume, energy consumption, or waste formation and ultimately leading to cheaper, safer and sustainable technologies."

Grunewald and Agar [6] have defined PI and process integration from reaction engineering perspective as two widely used catch words encompassing a whole gamut of techniques for enhancing reactor performance. In this context, they quoted multifunctional coupling of reactions with unit operations, dynamic modulation of operating parameters, multistep synthesis in a single reactor and use of hybrid equipment, plants and processes as important examples. Charpentier [7] showed that PI is a path for the future of chemical and process engineering demands in the context of globalization and sustainability.

In the European road map, PI is considered to provide radically innovative principles ("paradigm shift") in process and equipment design, which can benefit the process (often with more than a factor two) and chain efficiency, capital and operating expenses, quality, wastes, process safety and more. Becht et al. [8] state that PI stands for an integrated approach for process and product innovation in chemical research and development and chemical engineering in order to sustain profitability even in the presence of increasing uncertainties.

### *1.2.2 PRINCIPLES*

Gerven and Stankiewicz [9] have gone a step further by providing a fundamental view on PI encompassing its generic principles and approaches, which are discussed below:

#### *1) MAXIMIZE THE EFFECTIVENESS OF INTRA AND INTERMOLECULAR EVENTS*

This implies that PI is not only about aiming at processes limited by their inherent kinetics, but is primarily about changing such kinetics. This is achieved by understanding the factors responsible for low conversions, selectivities, unwanted side products in a reactive event including number/frequency of collisions, mutual orientation of molecules, their energy levels, etc. This area generally falls within the scope of physicists and chemists.

### *2) GIVE EACH MOLECULE THE SAME PROCESSING EXPERIENCE*

This implies formation of uniform products with minimum waste or by-passing. Factors at macro, meso as well as micro scale play an important role here including residence time distribution, dead zones, by-passing, mixing and temperature gradients. Plug flow reactor is close to ideal behavior according to this principle compared to a stirred tank reactor.

### *3) OPTIMIZE THE DRIVING FORCES AT EVERY SCALE AND MAXIMIZE SPECIFIC SURFACE AREA TO WHICH THESE FORCES APPLY*

The driving forces with respect to mass and energy are to be enhanced to achieve better transport rates but driving force for momentum transfer should be minimized to avoid frictional losses and therefore process optimization is required for managing driving forces. PI is achieved by maximizing the specific surface area or surface area to volume ratio due to enhancement of transport rates across interfaces.

### *4) MAXIMIZE THE SYNERGISTIC EFFECTS FROM PARTIAL PROCESSES*

This refers at different scales – macro and molecular. At macro scale, this implies multi functionality such as reactive distillation, whereas at molecular scale, this refers to multi functionality of a catalyst such as reaction and adsorption.

Gerven and Stankiewicz [9] have illustrated how these principles are applied through different approaches and scales of PI. Four approaches have been defined by them viz., structure, energy, synergy and time representing spatial, thermodynamic, functional and temporal domains, respectively.

In the structural approach, application of first principle leads to molecular reactors, where catalysts themselves act as reactors to alter the kinetics and selectivity of chemical reactions. Application of second principle in this domain results in equipment such as static mixer where uniformity as well as intensive mixing is achieved simultaneously. By the third principle, enhanced mass transfer through enhanced specific surface area is achieved in structured reactors including monoliths, gauzes, foams, etc. The fourth principle regarding synergy of partial processes is achieved in combined operations such as reactive distillation in structured packings or in catalytic membrane reactors.

In the energy or thermodynamic domain, energy is to be transferred from source to recipient in the required form and amount, at the required moment and position. Application of first and second principles in this domain is seen in microwave-based reactions, photochemical reactions, etc., where the selectivity as well as uniformity are enhanced. Alternate forms of energies such as electric, magnetic and acoustic fields illustrate the applications of third principle by enhancing mass and heat transport between phases.

In the synergy or functional domain, the focus is on the fourth principle of maximizing the synergistic effects of partial processes. This is achieved at molecular scale through multifunctional catalysts, combination of different forms of energy at meso scale (e.g., microwave assisted photo-chemical processes), and multifunctional reactors at macro scale. Synergy is achieved in multifunctional reactors by shift in equilibrium by in-situ separation and removal, or through reactions with heat exchange in HEX reactors, etc.

In the time or temporal domain, periodicity in operation is incorporated to realize all four principles of PI. A combination of adsorption-reaction-desorption on catalyst surface by periodic forcing of temperatures and pressures demonstrates the application of first principle. Oscillatory baffled flow reactor enhances uniformity, and illustrates the second PI principle. The application examples for third and fourth PI principles are pulsation of feed in trickle bed reactors enhancing the mass transfer rates, and flow reversal in reversed flow reactors shifting the equilibrium beyond limitations respectively. Switching from batch to continuous processing also result in realization of second and third PI principles.

## 1.3 CLASSIFICATION

The most common and widely reported classification for PI was reported by Stankiewicz and Moulijn [1] in terms of PI *equipments* and *methods* or alternatively as PI *hardware* and *software*, by considering PI as a *Toolbox*. Dautzenberg and Mukherjee [10] proposed a classification based on multifunctionality at molecular scale (A), or at the reaction inter-phase (B), or at intra-reactor level (C), or at the inter-reactor (D) level. An alternative classification is proposed here cutting across the above two classifications (Fig. 1.1) by categorising into three classes viz., *multifunctional operations*, *novel devices* and *alternate PI configurations*.

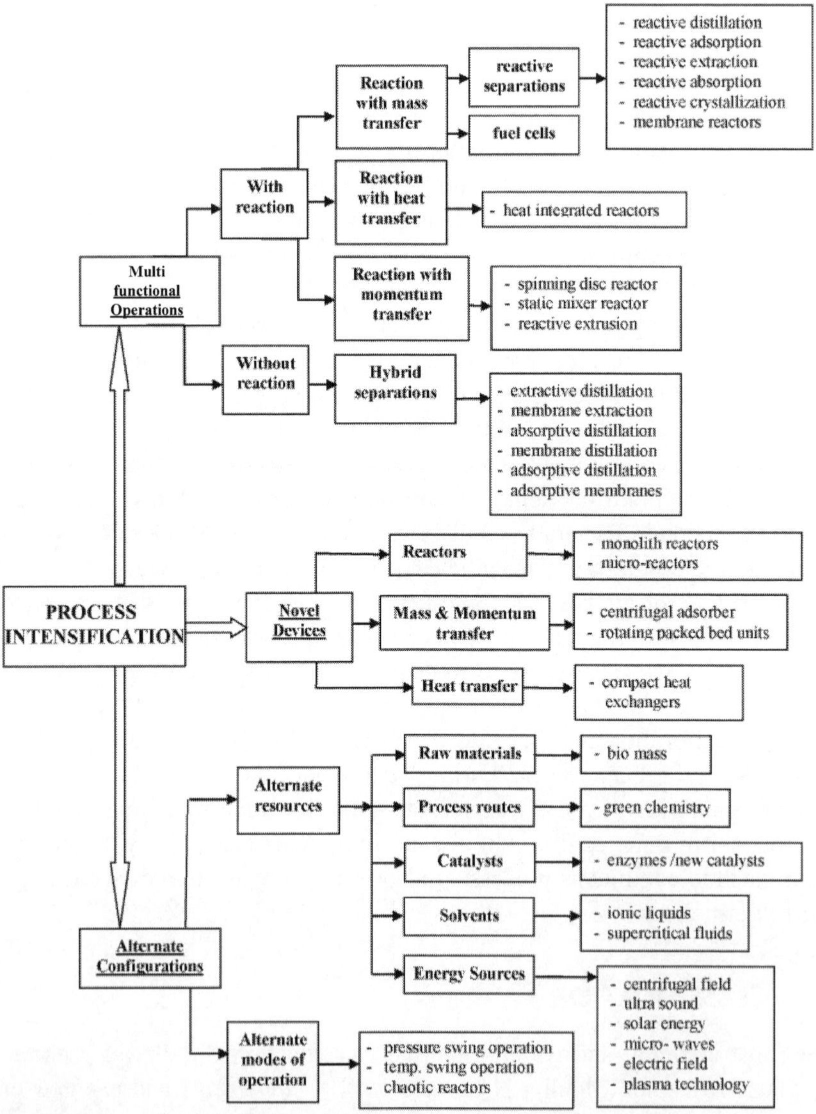


**FIGURE 1.1** Process intensification – an overview.

**Multi-functional operations** are classified as unit operations *with reaction* and *without reaction*. Further, *multifunctional reactive operations* are subdivided based on the transport phenomena relevant to the operation, namely *reaction with mass transfer*, *with heat transfer*, and *with momentum and mass* *transfer*. The reaction *with mass transfer* covers *reactive separations* consisting of reactive distillation, reactive adsorption, reactive extraction, reactive absorption, reactive crystallization, membrane reactors and fuel cells. The *reactions with heat transfer* include heat integrated reactors or HEX reactors, whereas *reactions with momentum and mass transfer* include configurations such as spinning disc reactors, static mixer reactors and rotating packed bed reactors. Most of these operations involve more than one phase and as such fall under the broad category of multiphase reactors. *Multi-functional nonreactive operations* include *hybrid separations* such as extractive distillation, membrane extraction, absorptive distillation, membrane distillation, adsorptive distillation, and adsorptive membranes.

**Novel devices** are further classified as *reactors and devices* for *heat, mass* and *momentum transfer*. Structured reactors and microreactors are typical categories of *novel reactors*. Compact heat exchangers are categorized as *novel heat transfer devices*, whereas centrifugal adsorbers and rotating packed bed units fall under the category of *novel mass and momentum transfer devices*.

**Alternate PI configurations** can further be classified as those using *alternate resources* and those based on *alternate modes of operation* in place of conventional resources and modes of operation. *Alternate resources* that can be used are different *raw materials* such as biomass, or *catalysts* such as enzymes, or *solvents* such as ionic liquids or supercritical fluids, or *process routes* based on green chemistry, or *energy source* such as centrifugal fields, ultrasound, solar energy, microwaves, electric field, magnetic field, etc. *Alternate modes of operation* include periodic operation or dynamic operation in symmetric or asymmetric manner, incorporation of advanced process synthesis and integration techniques to achieve intensified operations.

As seen from the above discussion, PI has a very wide scope and several books have been published on the subject. They include 'Re-engineering the chemical process plant – process intensification' by Stankiewicz and Moulijn [1], 'Modeling of process intensification' by Kiel et al. [11], 'Process intensification for green chemistry: engineering solutions for sustainable chemical processing' by Boodhoo and Harvey [12], 'Reactive distillation' by Sundmacher and Kienle [13], 'Micro-reactors' by Ehrfeld et al. [14], etc. The discussion in this chapter is limited to selected topics on multifunctional operations, novel devices and alternate PI configurations, which are dealt with more details in the next three sections.

## 1.4 MULTI-FUNCTIONAL OPERATIONS

Figure 1.1 shows different possible options of multifunctional operations as reactive and nonreactive operations that are further classified as per the associated transfer operation. The most popular among these options are reactive separations and hybrid separations. These multifunctional operations are discussed in detail in this section.

Industrial production processes generally consist of reaction section followed by downstream processing section comprising of various unit operations in series as shown in Fig. 1.2(a). In contrast to such conventional approach, it is also possible to combine and carry out more than one of the functions, that is, reactions, heat transfer and separation operations simultaneously in a single unit as shown in Fig. 1.2(b) and 1.2(c). Such integrated operations are termed as multifunctional operations, which forms a major class of PI approaches and hold promise of significant advantages over conventional processes. Their advantages include reduced equipment and operating costs, improved yields, faster reaction rates and enhanced separation efficiencies. The multifunctionality can be achieved through integration of reaction with separation operations or heat transfer or momentum transfer operations, or it can be through integration of more than one unit operations. Reaction processes performed simultaneously with separation operations in a single unit are called reactive separations, while integrating reaction with heat transfer results in heat integrated reactors, and combining more than one separation operations in a single unit results in hybrid separations. Out of all these multifunctional operations, as mentioned earlier, reactive and hybrid separations are dealt in detail further.

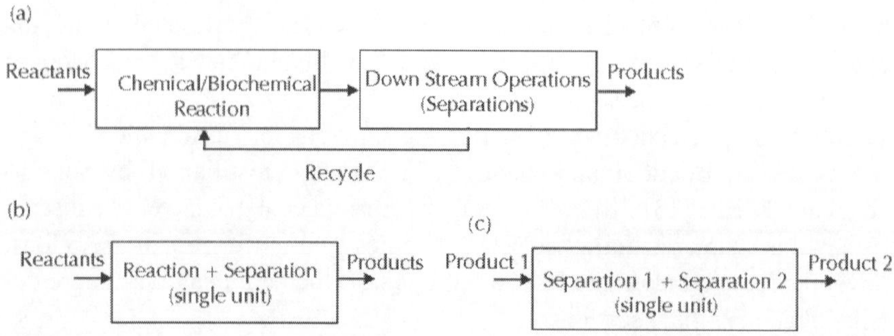


**FIGURE 1.2** (a) Conventional processes (b) Reactive separations, and (c) Hybrid separations.

### 1.4.1 REACTIVE SEPARATIONS

Reactive separations are multifunctional operations achieved through the integration of a chemical reaction with a selected separation operation within a single unit. Reactive separations fall under two categories. The first one is the simplest case, in, which there is no effect of reaction on separation or vice versa. In such cases, the potential benefits with the multifunctional reactors would be mainly reduced plant sizes (reduction in number of process units) leading to reduced capital and operating costs and improved heat management or energy utilization. The second case is the complex one, in, which there exists influence of reaction on separation operation and vice versa. In such reactive separations, the additional benefits include increased conversions, yields, selectivities, improved separation efficiencies, improved catalyst life, and also elimination of catalyst poisoning in some specific cases [1]. Some of the industrially relevant reactive separations include reactive distillation, reactive adsorption, reactive absorption, reactive extraction, reactive crystallization and reactive membrane separations or membrane reactors.

#### 1.4.1.1 REACTIVE DISTILLATION

Reactive distillation (RD) is one of the most important reactive separations with potential industrial applications. Here, both reaction and distillation take place within the same zone of a distillation column. It facilitates the instantaneous removal of products in pure form by using distillation principle. New vapor phase of products can be created by either using the heat of reaction in case of exothermic reactions, or by supplying external heat in case of endothermic reactions. Thus reactive distillation provides effective utilization of heat of reaction for product separation and thereby leads to significant reduction in utility consumption. In addition, in-situ removal of products results in improved conversions and yields in case of equilibrium limited reactions, thereby contributing significantly to the overall intensification of the existing process.

Methyl acetate production is a classical example of reactive distillation technology. A single reactive distillation column could replace the conventional process train consisting of 11 units. This significant reduction in number of units led to tremendous improvement in process economics via improved conversions and reduced equipment and operating costs. Various reactions that are suitable for reactive distillation are given in Table 1.1. These reactions include both catalytic (homogeneous/heterogeneous) and noncatalytic reactions [5, 15, 16].

Reactive distillation is generally carried out in a packed, tray or a bubble cap distillation column. However, the column internals and hardware that are required for reactive distillation may not be same as that of conventional distillation columns. The nature of the column internals mainly depends on whether the reaction is homogeneously or heterogeneously catalyzed or noncatalytic. The effective design of any reactive distillation column must fulfill the requirements of both reaction and distillation sections. If we consider the noncatalytic or homogeneously catalyzed processes, which are generally carried out in multitray type columns, the key design parameters include column diameter, liquid holdup and weir height. The reactive section requires larger column diameter compared to conventional distillation column in order to provide lower superficial vapor velocities and bubbly froth regime, which provide favorable conditions for the reaction. Also, the reactive tray should be designed such that it provides high liquid holdup for maximizing the conversion as demanded by the reaction and high interfacial area as required for good separation. Therefore, the weir heights used for reactive trays need to be higher than for nonreactive trays for providing higher liquid holdups [17].

**TABLE 1.1** Case studies for reactive distillation.

| S. No | Reaction (Catalyst) | Processes |
|---|---|---|
| 1. | Esterification (Amberlyst-15) | n-butyl acetate from acetic acid and n-butanol |
| | | amyl acetate from acetic acid and amyl alcohol |
| 2. | Transesterification (Amberlyst-15, KATAPAK-S) | butyl acetate and methanol from methyl acetate and n-butanol |
| 3. | Hydrolysis (Ion exchange resin) | formic acid and methanol from methyl formate |
| | | lactic acid and methanol from methyl lactate |
| 4. | Etherification (Beta-zeolite) | propylene glycol monoethyl ether from ethanol and propylene oxide |
| | | ethyl *tert* butyl ether from ethanol and *tert* butyl alcohol |
| 5. | Hydrogenation (Palladium based) | propylene from $C_3$ to $C_6$ |
| | | paraffins from olefins |

**TABLE 1.1** *(Continued)*

| S. No | Reaction (Catalyst) | Processes |
|---|---|---|
| 6. | Hydro-desulfurization (Group VIII metals on a support) | sulfur removal from a diesel boiling range petroleum fraction<br>hydrogen sulfide removal from naphtha |
| 7. | Dehydrogenation (Ru -Pd/C) | acetone and hydrogen from 2-propanol |
| 8. | Alkylation (Beta-zeolite) | cumene/isopropyltoluene from benzene or toluene and propylene<br>ethyl benzene from benzene and ethylene |
| 9. | Metathesis/Disproportionation (Cation exchange resin) | tetramethyl ethylene from isobutylene |
| 10. | Hydration/Dehydration (Amberlyst-15) | isobutene and water from *tert*-butyl alcohol<br>*tert* –butyl alcohol from isobutene and water |
| 11. | Carbonylation (Group VIII metal with a halide promoter) | acetic acid from methanol and carbon monoxide |
| 12. | Polymerization (hetero Brönsted acid catalyst) | polyamide and ammonia from amino nitrile and water |
| 13. | Amination | diethanol amine, monoethanol amine, tri ethanol amine, ethylene glycol from ethylene oxide and ammonia |
| 14. | Synthesis of carbonates (Homogeneous catalyst) | alkyl phenyl carbonate from dialkyl carbonate and phenol<br>2-hydroxyethyl methyl carbonate from dimethyl carbonate |
| 15. | Chiral separation | propylene glycol (R) and propylene oxide (S) from propylene oxide (R, S) |
| 16. | Acetalization (KATAMAX, Amberlyst- 15) | 2-methyl-1,3-ioxolane, water from ethylene glycol and acetaldehyde |

For heterogeneous catalytic processes conducted in RD columns, it is essential to provide catalyst particles with sufficient interfacial area for higher reaction rates and higher liquid hold up for better separation. The catalyst particles as such cannot be directly used as packing because they tend to form lumps and block the upward flow of the vapor and downward flow of liquid. To avoid such problems, the catalyst particles are generally placed on structured supports and within wire gauged envelopes. There are two types of structured packings that are commonly used in RD columns. The first type is called bale packing in which the catalyst particles are placed in pockets in a cloth belt enclosed in an open mesh knitted with stainless steel wire. The advantage with this type of packing is that its structure allows the swelling of catalyst particles, prevents breakage and loss of catalyst particles due to pressure drop, and avoids channeling and reduction in catalyst due to mechanical attrition. The second type of structured packing is KATAPAK, in, which the catalyst is present in solid particulate form and is sandwiched between corrugated sheets of wire gauze. The top and side views of these packings are shown in Fig. 1.3. Generally, the plates are made of wire gauze or metal screen material, but other types of material such as plastic gauze and ceramics are also used. Also, while designing RD columns for heterogeneously catalyzed reactions, care should be taken such that the requirements of reaction (small particle sizes and high catalyst loadings) and separation (large interfacial area between gas and liquid phases) are fulfilled [17].

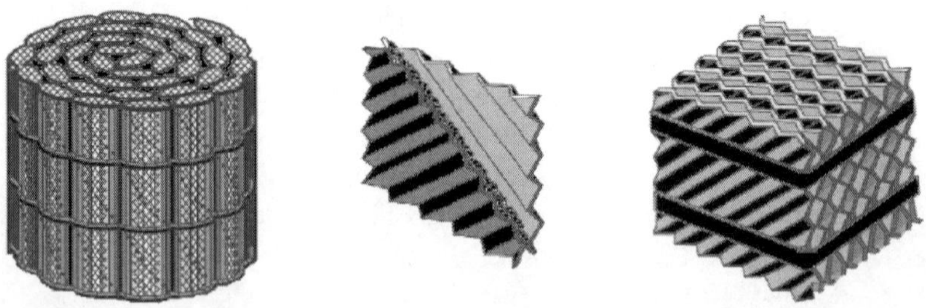

**FIGURE 1.3** Catalyst packing types for RD (a) bale (b) sandwich, and (c) collected sandwich cubes [17].

Though there are wide ranges of reactions that are suitable for RD, as shown in Table 1.1, not many reactions are commercially viable. The main factor that hinders the broader application of RD is the existence of a small feasible operation window. The overlap region in the pressure-temperature domain in which chemical reaction and distillation design are feasible is usually narrow [1]. Development of new types of catalysts would allow one to significantly broaden the feasible operation window for chemical reaction and distillation. In addition, the nonlinear dependencies among reaction kinetics and vapor liquid equilibria may lead to the formation of reactive azeotropes, and/or occurrence of multiple steady-states, which further may pose great challenges in terms of design, operation and control of RD columns.

### *1.4.1.2 MEMBRANE REACTORS*

Membrane reactors are multifunctional reactors in which chemical reactions are integrated with membrane separations resulting in intensified process operations compared to conventional processes. Membranes provide number of promising functionalities that aid in improving the overall performance of the individual processes when integrated with a reactor. They can be used as a means to introduce/separate/purify reactant or products, to provide a surface for reactions, to provide a structure for reaction medium, or to provide a support for specific catalysts. Integration of membranes with catalytic processes is driven by a number of advantages. The schematic of a typical membrane reactor is illustrated in Fig. 1.4. Membranes can be used for in-situ removal of one of the products, thus driving the reaction equilibrium shift in the desired direction in case of equilibrium reactions. In case of reactions with undesired products, the side reactions can be suppressed by the removal of products through membrane separation. Membranes can facilitate catalytic reactions by housing the catalyst particles on the surface or inside the membrane. A membrane, when integrated with reactor, also acts like a barrier to prevent two incompatible reactants from being on the same side of the reactor. In addition, incorporating a membrane into a reactor tends to reduce the number of operating units thereby enhancing the process economics.

Membrane sewage treatment system was one of the earliest commercial applications of membrane bioreactors for wastewater treatment in North America in late 1960's and later its application was extended to industrial wastewater treatment processes [18]. One of the commercial breakthroughs for membrane reactor technology is Russian vitamin K technology. Membranes are frequently employed in combination with a bioreactor, for instance an enzymatic pharmaceutical process [1]. Table 1.2 illustrates how specific properties of membranes are useful in membrane reactors in improving the overall process performance [19].

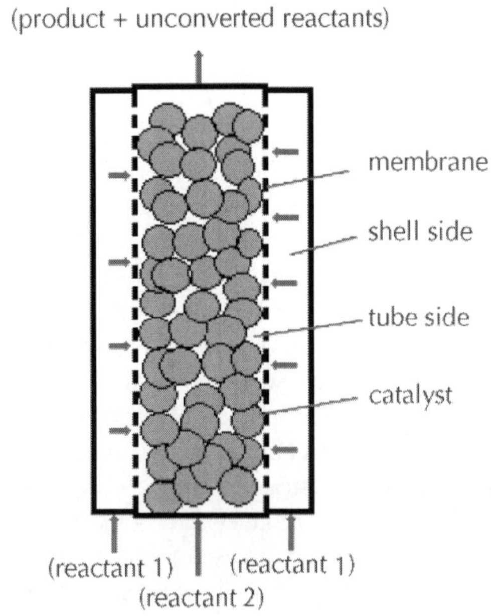


**FIGURE 1.4** Schematic of packed bed membrane reactors.

Though membrane reactors demonstrate promising results in intensifying most of the processes, their large-scale commercial implementation is very limited. The reasons could be primarily high price of membrane units and the problems associated with membranes such as low stability, low permeability, sealing, fouling, as well as mechanical and thermal fragileness, etc. Also the development of very thin, flawless membranes over large complex supports withstanding high pressure and high temperature conditions, as demanded by commercial scale processes, seems to be very crucial. Identification of key hurdles that affect the separation features of membrane reactors and thorough analysis of the associated heat and mass transport issues may definitely improve the performance of membrane reactors. Further, the recent developments in material science and engineering may surely improve the status of successful commercialization of large-scale catalytic membrane reactor technology on a wide basis in near future.

**TABLE 1.2** Applications of membrane reactors.

| S. No. | Function of membrane | Example (type of membrane) |
|---|---|---|
| 1. | Separation of products from reaction mixture | Catalytic decomposition of ammonia into nitrogen and hydrogen (palladium ceramic membrane) |
| | | Water gas shift reaction (palladium membrane reactor enclosing an iron chromium oxide catalyst) |
| | | Ethanol dehydrogenation to $CH_3CHO$ and hydrogen (palladium membrane (catalyst) Cu/ $SiO_2$ aq.) |
| | | Catalytic dehydrogenation of ethanol to acetaldehyde (palladium membrane) |
| | | Removal of alcohol from the fermentation broth in ethanol fermentation process (silicone capillary membranes) |
| 2. | Separation of a reactant from a mixed stream for introduction into reactor | Partial oxidation of methane to syngas (dense ceramic membranes) |
| | | Removal of organic pollutants such as aniline, 4-chloroaniline 3,4-dichloroaniline from wastewater for bio-degradation (silicone capillary membrane) |
| | | Recovery of diltiazem from an aqueous reaction mixture containing NaCl and $NaHCO_3$ (liquid membrane) |
| | | Removal of hydrogen from catalytic dehydrogenation of 1-butene to butadiene (surface catalyzed palladium membrane) |
| 3. | Controlled addition of a reactant or two reactants | Production of syngas from partial oxidation of methane (membrane tube packed with Rhodium based reforming catalyst) |
| | | Hydrogenation of nitrobenzene to aniline (tubular $\gamma$ $Al_2O_3$ membrane having a platinum catalyst) |

**TABLE 1.2** *(Continued)*

| S. No. | Function of membrane | Example (type of membrane) |
|---|---|---|
| 4. | Nondispersive phase contact | Extraction of products in fermentation process producing ethanol and acetone-butanol-ethanol (ABE) (microporous/porous hydrophobic membranes) |
| | | Enzymatic splitting of olive oil (lipase enzyme immobilized at the aqueous organic interface on hydrophobic microporous membrane) |
| 5. | Segregation of catalyst (Co-factor in a reactor) | Production of amino acid (ultra filtration) |
| | | Enantio selective addition of zinc to benzaldehyde (homogenously soluble catalyst retained by a solvent stable polyaramide ultrafiltration membrane) |
| 6. | Immobilization of catalyst on a membrane | Dehydrogenation reactions (immobilized Pt-Pd onto metal oxides such as Aluminum or Silica) |
| | | Hydrogenation of propene to propane (nanoporous membrane containing nano clusters of Palladium) |
| 7. | Membrane as the catalyst | Esterification of methanol to methyl acetate and n-butanol to butyl acetate (cation exchange membrane, Nafion membrane) |
| | | Dehydrogenation reaction (palladium membrane) |
| 8. | Membrane as the reaction medium | Oxifunctionalization of n-hexane to a mixture of hexanol and hexanone (silicone membrane acts as reaction medium and reactor) |
| 9. | Solid electrolyte membrane | Solid polymer electrolyte fuel cell (thin polymer membrane (Nafion)) |

#### *1.4.1.3 REACTIVE ADSORPTION*

Reactive adsorption is a multifunctional process that results by integrating any chemical reaction with adsorption operation in a single unit. Reactive adsorption is favorable for those reactions in which the product of interest is in fluid phase and needs to be separated instantaneously. It is also advantageous to those fluid-solid separations, where the incorporation of reaction improves the extent of adsorption. Similar to any other reactive separations, reactive adsorption is always better than the conventional scheme of reaction followed by adsorption in terms of the resulting reduction in capital costs and downstream processing steps as well as improved product purities. The basic idea of reactive adsorption is to remove the product in-situ from the reaction zone by using a selective adsorbent. Instantaneous product removal favors the forward reaction resulting in complete conversion of reactants. Further it may also yield benefit in energy and material savings. Adsorbent should be highly porous and should also possess some functional groups to enhance the retention of solute on the adsorbent.

Reactive adsorption has a promising role in resolving several environmental issues arising due to the release of contaminants to atmosphere from various industries. For example, removal of sulfur contaminants from petroleum products, which is a critical issue in petroleum industries, can be effectively solved by using reactive adsorption. In Table 1.3, several reactive adsorbents and their application for the removal of different sulfur based contaminants, organic and metal contaminants from industrial wastewater, and their application for treatment of air pollution is described [20]. The key design parameters of reactive adsorption processes include choice of reactive adsorbents, mode of regeneration and optimal operating conditions. The selection of a reactive adsorbent for removal of a particular adsorbate species mainly depends on its capacity for chemisorption and physiosorption. Chemisorption is the ability of adsorbent to capture the adsorbate by means of a chemical reaction (e.g., oxidation, hydroxylation ion-exchange, reduction, redox, electro-chemical reactions, etc.) whereas physiosorption is the ability of adsorbent to hold the adsorbate physically in its pores. The physical parameters such as mechanical strength, porosity and nature of support, etc. greatly influence the performance of reactive adsorbents. Desorption can be accomplished either by displacement with a compound of a higher affinity, or by variation of pressure or temperature. The choice of regeneration method for any particular system depends on economic as well as technical considerations.

**TABLE 1.3** Reactive adsorption for the removal of solid/liquid/gaseous impurities.

| S. No. | Adsorbate (contaminant) | Adsorbent | Major Findings |
|---|---|---|---|
| 1. | Dibenzothiophene (DBT) | Modified activated carbon (AC) | High adsorption capacity for large molecules |
| 2. | Thiophene | Zeolites | Adsorption is affected by zeolite structure |

**TABLE 1.3** *(Continued)*

| S. No. | Adsorbate (contaminant) | Adsorbent | Major Findings |
|---|---|---|---|
| 3. | Sulfur from diesel oil | Ni/ZnO | |
| 4. | Sulfur from digester gas | Wood-based activated carbons modified with $N_2$ | Prehumidified adsorbent has high capacities for $H_2S$ removal. |
| 5. | Sulfur from Gasoline | Ni/ZnO-$SiO_2$-$Al_2O_3$ | Adsorbent exhibits significant desulfurization |
| 6. | 4,6-dimethyl di benzo thiophene | Polymer-derived activated carbons | Major role of surface chemistry |
| 7. | Sulfur compounds from commercial diesel | Ni nanoparticles supported mesoporous silica | Adsorption capacity depends on Ni concentration and substrates pore structure |
| 8. | Pollutants from wastewater | Activated carbon modified with $H_2SO_4$ at 150–270°C | Chemical modification increases mesoporous volume for adsorption of large molecules |
| 9. | Phenol | Cobalt loaded activated carbon | 98% Phenol removal |
| 10. | Cadmium | Coconut shell derived activated carbon | Oxidation has effect on cadmium ion adsorption |
| 11. | Arsenate [As(V)] | Synthetic nanoscale zero-valent iron | For *in-situ* and *ex-situ* groundwater treatment |
| 12. | Silver | Microparticles of sulfodiphenylamine and diaminonaphthalene copolymers | Effectiveness of particle morphology in noble metallic ion recovery in wastewater |
| 13. | Ammonia | Cu-based MOF/Graphene composites | Moist conditions favor $NH_3$ adsorption |
| 14. | $H_2S$ | Catalytic carbonaceous adsorbents | Dependence of adsorbent capacity on surface chemistry, porosity and water content |
| | | Graphite oxide/ $Zr(OH)_4$ composites | Enhanced basicity promoting reactive adsorption |
| 15. | $NO_X$ | $Al_2O_3$ supported Rh nanoparticles | |
| 16. | Hydrogen cyanide | Modified activated carbon | Nitrogen in carbon matrix in basic environment enhances adsorbent efficiency |

Although reactive adsorption technologies are well proven at lab scale, they still need to be applied commercially for continuous large-scale applications. It is essential to develop new materials for reactive adsorbents with improved chemical and mechanical properties such that they fulfill necessary requirement of matching of operating conditions for both reaction and adsorption. The reusability of an adsorbent for a longer duration with sustainable adsorption capacity is another crucial parameter, which needs to be improved. Also for the processes where more than one contaminant is present in the effluent stream, it is essential to develop multicomponent reactive adsorption systems; however the interaction between adsorbates and adsorbent makes the selectivity in reactive adsorption process more complex.

## *1.4.2 HYBRID SEPARATIONS*

When two or more separation techniques are integrated in a single operation, such operations are termed as hybrid separations. Such integration may lead to intensified and improved separation processes. It may also be possible to achieve the extent of separations that cannot normally be achieved by any one of the single technique. Figure 1.5 shows some of the binary combinations of separation techniques leading to hybrid separations. It is also possible to have multiple combinations of separations. Their analysis is much more complex.

### *1.4.2.1 EXTRACTIVE DISTILLATION*

Extractive distillation can be generally used to separate close boiling liquids or azeotropes, which cannot be separated through conventional distillation process. A solvent is introduced into the distillation column to alter the relative volatility of the feed components, and to avoid the formation of azeotropes. The extracted less volatile components leave from the bottom, whereas more volatile components come out as top products in pure form. Extractive distillation can replace conventional distillation or extraction processes resulting in improved separations, reduced capital investment and energy consumption. Industrially, extractive distillation can be implemented for binary separations resolving the close boiling mixtures, namely m-xylene/ o-xylene, methylcyclohexane/toluene, propylene/propane, 1-butane/1,3-butadiene, and azeotropic mixtures such as iso propylether/acetone, ethyl acetate/ethanol/water, MTBE/ethanol, etc.

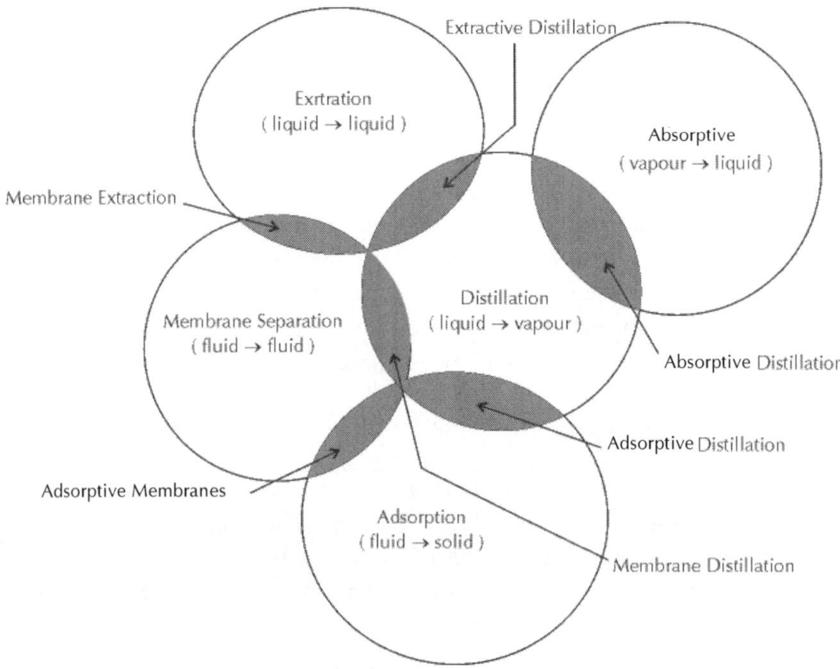


**FIGURE 1.5** Hybrid separations.

### *1.4.2.2 ADSORPTIVE DISTILLATION*

Adsorptive distillation is an integrated operation in, which adsorption is combined with distillation to separate the close boiling components or constant boiling liquid mixtures. It is a three-phase mass transfer operation in which the adsorbent, usually in the form of (fine) fluidized powder, is introduced into the column along with an inert carrier gas. The adsorbent selectively adsorbs one of the components and flows into the desorption column in, which the adsorbed component is desorbed. Thus adsorptive distillation is successful in separation and in avoiding the formation of azeotropes. Most commonly used adsorbents in the industry are silica gel, activated carbon, zeolite and alumina. Though adsorptive distillation has been reported long back, its industrial and commercial applications are very limited. However, potential application fields for adsorptive distillation include separation of toluene/methyl cyclohexane, naphtha reformates, p-xylene/ m-xylene, etc.

### 1.4.2.3 MEMBRANE DISTILLATION

Membrane distillation is an integrated mass-transfer operation by combining distillation with membrane separation. In this operation, the membrane is subjected to liquid/aqueous solution at different temperatures and compositions. The temperature difference across the membrane creates the difference in vapor pressures and thereby tends to increase the molecular transport through the pores of membrane. Membrane distillation possesses various advantages over conventional distillation or membrane separation. Membrane distillation technology has wide scope in water desalination, concentration of organic acids and bases ($H_2SO_4$, $H_3PO_4$, NaOH, $HNO_3$ and HCl), protein solutions, oil-water emulsions, sugar/sucrose solutions, etc. and for treatment of wastewater from various industries.

### 1.4.2.4 MEMBRANE ABSORPTION

Membrane absorption process is an integrated mass transfer operation in which one of the components in a multicomponent gaseous mixture can be selectively removed through a membrane and simultaneously dissolved in an absorbing liquid. It facilitates a bubble free gas-liquid mass transfer, which is most desirable for shear sensitive biological mixtures. Some of the important industrial applications of membrane absorption include removal of $CO_2$ from flue gas, removal of $H_2S$ from gas streams, recovery of cyanide from wastewater and absorption of $NH_3$ (in a polypropylene hollow-fiber column) with dilute $H_2SO_4$ as absorbent.

### 1.4.2.5 MEMBRANE ADSORPTION

Membrane adsorption is an integrated adsorption operation coupled with membrane separation. The separation efficiency of membranes is enhanced by combining it with the principle of adsorption. Specific adsorbent is placed in the inner pores of membranes, which selectively adsorbs and separates the required component from the mixture. When the mixture is subjected to the membrane with adsorbent, specific component gets attached to the adsorbent, which acts as a functional ligand and gets separated from the mixture. The dissolved molecules are carried directly to the adsorptive sites in the membranes by convective flow, which increases the throughput of the process. Potential applications of membrane adsorption reported in literature include separation of polynucleotides, separation of oligonucleotides and peptides, separation of small hydrophobic molecules, and separation of enantiomers.

#### 1.4.2.6 MEMBRANE EXTRACTION

Membrane extraction involves coupling of solid/liquid membrane separation with extraction operation. It is mainly used for separation of selective components in liquid streams. The molecules transported through the membrane get extracted into a solvent and are removed from the mixture. It is very useful for the removal of specific contaminants from wastewater and also for the removal of products from fermentation broths.

Although it is noted from the reported literature that hybrid separations hold considerable promise for application in various fields, their commercial implementation is not yet achieved due to inadequate validated thermodynamic and kinetic data, appropriate materials, scale-up capability, multidisciplinary design approach to process integration and lack of prototypes for demonstration. Reduction of degrees of freedom caused by the integration of these processes often leads to reduced operational flexibility. These issues need to be addressed for achieving higher success in commercial applications.

## 1.5 NOVEL DEVICES

They are further classified as novel reactors and devices for mass and momentum transfer. From industrial application perspective micro and monolithic reactors are more relevant and are accordingly given attention.

### *1.5.1 MICRO-REACTORS*

Micro-reaction technology fulfills four aspects of PI viz., shorter residence time, higher temperature, faster kinetics and smaller reaction volume, and is endowed with the continuous processing of reactants within defined parallel channels whose dimensions are less than 1000 μm [21]. At least one dimension of these channels is smaller than 1 mm, but is rarely smaller than 100 μm [22]. These narrow channels are generally formed from materials such as silicon, glass, stainless steel and ceramics or polymers. These channels are connected to a series of reservoirs containing reagents or solvents. Pumps are operated in laminar flow regime and mixing is achieved by diffusion. In a wider context, a combination of several microscale units such as microreactors, micromixers, microseparators, microheat exchangers and microanalyzers are used to facilitate accurate process control with respect to mixing, quenching and temperature management.

Some of the terms frequently associated with microreactor technology include microstructures, elements, units, stacks, housing devices and systems [14]. Micro-structures refer to channel structures, which is part of a flow channel configuration of a specific geometric form referred to as an element. A combination of an element, fluid lines connected to it along with the supporting base material is denoted as a unit. A combination of several units to enhance the throughput is called a stack, which is supported by top and bottom plates and the assembly is referred to as housing. These plates serve as connectors to the external units, and a unit embedded in housing is called a device. A network consisting of several devices making up a complete process is denoted as a system or a 'Micro-system'. A schematic representation of some of these components is shown in Fig. 1.6.

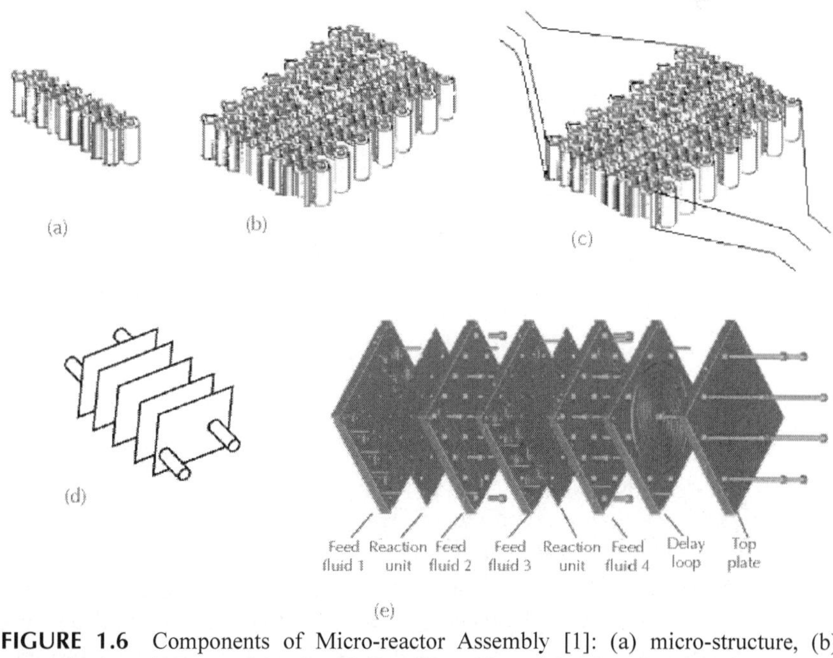


**FIGURE 1.6** Components of Micro-reactor Assembly [1]: (a) micro-structure, (b) rectangle shaped element, (c) unit with flow lines, (d) stack in a housing, (e) micro-reactor assembly.

Three basic process engineering effects are associated with microreactor technology viz., enhanced heat transfer in the direction of smallest dimension leading to very large specific heat transfer capacity in the case of highly exothermic, endothermic and explosive reactions, enhanced mass transfer due to high mixing efficiency leading to molecular mixing of reactants, and enhanced momentum transfer leading to large specific boundary layers at the wall surface as well as at the interface of two immiscible fluid phases [22].

Micro-reactors are suitable for fast reactions, which are not kinetically controlled, multiple reactions with undesirable products, and for highly exothermic/endothermic/multiphase processes, which play a critical role in the reaction progress. The classes of reactions falling under this category are liquid, gas and gas-liquid phase reactions, some of which are highlighted in Table 1.4. Most of the gas phase reactions are heterogeneously catalyzed with conversion occurring at gas/solid interface and transport limitations (both mass and heat) are quite prevalent. Some of the commonly reported methods for microfabrication of microreactors are LIGA technology, wet and dry etching processes, mechanical micromachining, microelectro discharge machining, and micromachining by means of laser radiation.

Effect of miniaturization in microreactors results in smaller devices requiring less space and with the facility for integration of several functional elements. The main impact is in the intensification of mass and heat transfer and in improving flow patterns. Decrease in linear dimensions gives rise to an increase in diffusional flux per unit volume or unit area. Heat transfer coefficients increase by an order of magnitude and mixing times are drastically reduced. Surface to volume ratios in microchannels range between 10,000 and 50,000 $m^2/m^3$ compared to typical laboratory and production vessels, which are of the order of 1000 $m^2/m^3$ and 100 $m^2/m^3$, respectively. Reactor volumes are significantly reduced by miniaturization as well as by change in operation from batch to continuous flow. For example, in case of an organo-metallic reaction, material hold up was reduced from a tank of 6,000 L to a volume of a few mL with 5 miniaturized mixers [14].

Increase in number of units rather than increasing the volume of a single unit offers several advantages. Micro-devices can be used in fast and cost-saving screening of materials and processes by integration of synthesis and analysis in one device. They offer production flexibility by changing the number of units in operation depending on the varying demand of the product. In the conventional scale up process, there is an intermediate stage of pilot scale experimentation before designing a commercial plant. On the other hand, with microdevices, it is possible to achieve the desired scale of operation by just enhancing their numbers by reducing both scale up time and efforts.

With reference to the application potential of microreactors and other microdevices, the numbering up facility leads to faster transfer of research results into production; smaller plant size for distributed production; lower costs for transportation, materials and energy; reduction of quantity of catalyst by several orders of magnitude; and more flexible response to market demands. Safety issues are efficiently handled owing to the low inventory of material in each unit thereby enabling reactions to be carried out in explosive regimes. In addition to increase in conversion and selectivity, it is also possible to achieve changes in product properties, especially molecular weight distribution in polymers and size distribution of droplets in semisolid pastes. Micro-reactors are promising tools for on-site and on-demand production.

**TABLE 1.4** Some liquid, gas and two-phase reactions in micro-reactors.

| Reaction | Intensified Equipment details | Improvements Achieved |
|---|---|---|
| Vitamin precursor synthesis* | Micro-reaction system with integrated static mixer heat exchangers, reaction channels and delay loop | Residence time of 30 s, yield of 95%, amount of byproducts reduced by > 50% |
| Poly methyl methacrylate (PMMA) formation* # | Micro-mixer with 36 micro-channels 25 μm width. | Fouling problems reduced, improved polymer uniformity, scale-up by numbering up concept |
| Organo-metallic reactions* | Micro-mixers with microchannels of 40 μm width | Yield increased to 95% and residence time reduced from 5hr to <10 s |
| $H_2/O_2$ reaction* | Micro-reactor with stainless steel platelets with Pd coatings | Safe operation of potentially explosive reaction |
| Partial oxidation of propene to acrolein* | Micro-reactor with copper platelets | Channel length of 4 μm sufficient to reach equilibrium |
| Selective partial hydrogenation of cyclic triene and benzene* | Reactor with microstructured Al foils | Yield increased by 45% for first reaction and improved Selectivity for second reaction |

**TABLE 1.4** *(Continued)*

| | | |
|---|---|---|
| Dehydration of derivatized alcohol* | Multi-tubular reactor and Micro-channel reactor | Selectivity of 85% & 96% |
| Hydrogenation of α methyl styrene to cumene* | Multi channel structure with T-configuration | Useful for gas/ liquid or liquid /liquid reactions |
| | Flat plate microreactor with porous wall and a catalytic wall opposite to it | high conversion |
| Absorption of ammonia into acidic aqueous solutions* | Micro-channel reactor | Residence times: 2 to 20 s |
| Absorption of $CO_2$ | Micro-bubble column with a static micromixer, reaction platelet and a microheat exchanger | Conversion increased by a factor of 2.5 with a residence time of 0.1 s |
| Direct fluorination of toluene dissolved in acetonitrile | Falling film microreactor | Safer process, Yield in microreactor 2.5 times higher than in a lab bubble column |

* [14], #[23].

### *1.5.2 STRUCTURED REACTORS*

Structured reactors have attracted lot of attention over the past decade. The problems associated with the conventional slurry and bubble column reactors are difficulty in separation of product and catalyst, and catalyst loss due to attrition. In trickle bed reactors the often-quoted disadvantages are maldistribution of reactants, hot spot and dry area formation, low contacting efficiency, possible run-away conditions, low gas and liquid velocities leading to low productivity and high operating costs. Several scale-up issues arise while designing them for commercial operations. These disadvantages can be effectively eliminated by employing structured reactors. They are endowed with uniform flow distribution, low pressure drop, enhanced mass transfer due to efficient gas-liquid-catalyst contact, small diffusion paths and easy scale-up.

Structured catalytic reactors are of two types viz., one in which a structured catalyst is employed and the other in, which catalyst is located in the structural framework. Monolith reactors represent the first category, whereas parallel passage, lateral flow, three level porosity and membrane encapsulated reactors are examples of the latter category. Types of structured packing include packing made of regular ceramic or metal support, referred to as 'monolith,' sandwich packing, and open cross flow structures. Other types of structuring at the level of catalyst and reactor include honey-comb type monoliths, gauzes, coated structured packings and foams.

Monolith reactors contain a single structure with many thin, vertical, parallel channels, separated from each other by walls. Channels can be of different geometries including rectangles, triangles, hexagons, etc., and are provided with internal fins to have a stabilizing effect on gas-liquid flow and to avoid flooding. A schematic representation of a monolith block is shown in Fig. 1.7, where $t_w$ denotes wall thickness and $L$ denotes the length from one channel wall center to a neighboring channel wall center. A monolith structure is characterized by the wall thickness and cell density (representing the number of channels per unit cross section). The cell density in monolith reactors varies from 100 to 1200 channels per square inch [24]. The void fraction ranges between 0.5 and 0.9 in monolith channels, whereas wall thickness varies between 0.006 and 0.05 cm. In order to achieve enhanced selectivity and productivity, the liquid must be distributed uniformly before it enters the monolith bed. Some of the distributor systems have been reported by Roy et al. [24] as showerheads, sieve plates, glass frit and liquid driver ejectors for down-flow and up-flow modes of operation. Monoliths are produced by extrusion of a paste containing catalyst particles or by extrusion of a support on, which the catalyst can be coated by wash coating, and are commercially available at reasonable cost. Fabrication and applications of monoliths are reported by [25].

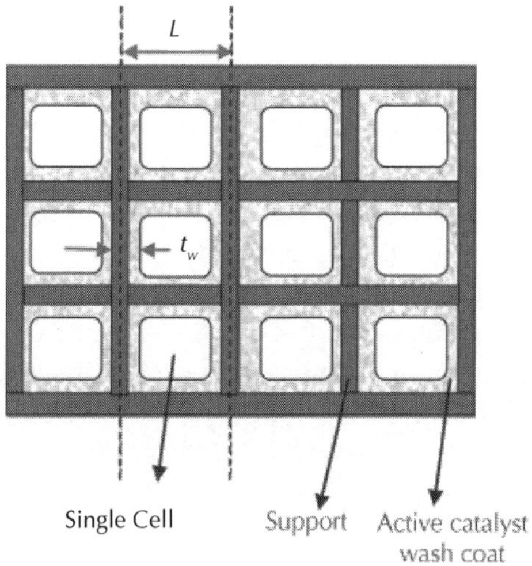


**FIGURE 1.7** Schematic representation of a monolith block.

Monolith reactors can be operated in batch and continuous modes as well as for cross-flow, countercurrent and cocurrent operations. The configurations of monolith reactors include froth, in-line, loop, and ejector facilitated monoliths [24]. They have the flexibility to facilitate both fast and slow reactions with different configurations. In slow reactions, the intrinsic reaction kinetics becomes the rate-determining step and efforts should be directed to enhance this speed by increasing the catalyst inventory through thicker walls. On the other hand, in fast reactions, which are controlled by mass transfer, intraparticle diffusion is more important, and this can be facilitated by thinner catalyst coatings and increasing cell densities while taking care not to increase pressure drop to unacceptable levels. Application of monolith reactors has been reported [26] in the hydrogenation step from anthraquinone to the corresponding hydroquinones for production of hydrogen peroxide. Reported commercial applications are for three-way catalysts in cars, selective catalytic reduction in power stations and ozone destruction in air planes [1]. Monolith reactors have also been designed as low $NO_X$ burners and for catalytic oxidation of volatile organic compounds because of low-pressure drop and resistance against dust. Roy et al. [24] illustrated better performance of monolith reactors over conventional counterparts for hydrogenation and oxidation reactions.

### 1.5.3 NOVEL DEVICES FOR HEAT TRANSFER

Compact heat exchangers are novel devices, which enhance heat transfer characteristics and are available in different configurations. They are mainly classified as tubular, plate, extended surface and regenerative types of heat exchangers. Plate heat exchangers and plate-fin heat exchangers have hydraulic diameters between 1 and 10 mm, whereas printed circuit heat exchangers, diffusion-bonded, super plastic-formed and microchannel heat exchangers have hydraulic diameters below 1 mm. The compactness factor or the surface-to-volume ratio in these heat exchangers is greater than 700 $m^2/m^3$, and could be greater than 1000 $m^2/m^3$ for heat exchangers with low hydraulic diameters. They can accommodate wide range of operating conditions between 4–1000 bar pressure and 70–1000°C temperature [1]. The application of different configurations of compact heat exchangers is found to be feed/effluent heat exchangers, process evaporators, etc. Single flow applications are found to be common in industry, whereas their utility is yet to be established in multiphase flow systems.

### 1.5.4 NOVEL DEVICES FOR MASS AND MOMENTUM TRANSFER

Novel devices enhancing mass and momentum transfer include centrifugal adsorbers and rotating packed beds. A centrifugal adsorber facilitates adsorption processes to occur under the influence of centrifugal force. A countercurrent flow is established between the liquid phase and the adsorbent by applying centrifugal force. In this process, it is possible to use adsorbent particles of size 10–50 μm to achieve short contact times and high processing capacities (10–50 $m^3/h$). Compared to conventional fixed bed and moving bed units, the size of centrifugal adsorber is very small and is accordingly compact. Rotating packed beds employ centrifugal or high gravity field in which the packed bed is placed in the center of a rotor and is suitable for gas/liquid or liquid/liquid separations. In gas/liquid separation, the liquid enters through the eye of the rotor, flows through the packing and leaves from the housing, whereas the gas enters at the other end of the packing and after coming in contact with the liquid in the packed bed, leaves from the eye of the rotor at the top. In liquid/liquid system, the heavy liquid takes the place of liquid in gas/liquid system, whereas the light liquid takes the place of gas. Rotating packed bed units can be useful in absorption, stripping, distillation, liquid-liquid extraction, crystallization, etc. They have found commercial applications for water de-aeration and hypochlorous acid synthesis processes. In case of the latter, very small units provided yields exceeding larger volume pilot scale units. These operations resulted in low-pressure drops and higher liquid side mass transfer coefficients. There is good scope for the use of rotating packed beds in other mass transfer as well as in multifunctional operations.

## 1.6 ALTERNATE PI CONFIGURATIONS

Another means of achieving PI is by employing alternate resources or through other modes of operation.

### 1.6.1 ALTERNATE RESOURCES

The conventional raw materials used in chemical process are derived predominantly from fossil fuels. Alternate raw materials such as biomass can be employed to promote greener processing. In this direction, the concept of biorefineries is quite relevant [27]. The future biorefineries can employ operations such as extractive distillation with ionic liquids and hyper-branched polymers, adsorption with molecular sieves and bio-based adsorbents, nano-filtration, extractive-fermentation, membrane pervaporation in bioreactors, and vacuum membrane distillation [28]. Similarly enzyme-catalyzed reactions are finding more applications in processes such as bio-transformations, biodiesel production, biomass hydrolysis as well as in biofuel cells [29].

Alternate reaction media such as ionic liquids or supercritical fluids have contributed to PI. Ionic liquids are being used as solvents to replace toxic solvents, and are also resulting in improved performance. They are also used as catalysts in some reactions like esterifications, and are contributing to PI by acting as support materials in membrane separations. Higher yields are achieved using supercritical fluids as solvents, and when they are used in crystallization process, it is found that they result in more desirable morphology for the crystals. Alternate process routes based on green chemistry also contribute to PI in terms of reduced number of steps in reaction or a combination of above criteria or by use of other means to improve atom efficiency.

Use of alternate energy source also contributes to PI, with examples in centrifugal fields, ultrasound, solar energy, microwaves, electric field, magnetic field, etc. Replacement of gravitational field by centrifugal force has resulted in applications such as spinning disc reactors, rotating packed beds, HiGee technology, etc. Ultrasound can be used to generate microbubbles in liquid reaction media acting as high-energy microreactors, which can find applications in catalyst regeneration. Solar reactors are being explored in water treatment plants and cycloaddition reactions for olefin preparation, whereas microwave energy is being used to facilitate extremely fast organic synthesis reactions. Electric fields are found to enhance liquid/liquid extraction rates and are also useful in bio-separations.

### 1.6.2 ALTERNATE MODES OF OPERATION

Non-conventional modes of operation also contribute to PI. They include periodic or dynamic operation in symmetric or asymmetric manner, incorporation of advanced process synthesis and integration techniques. Pressure/temperature swing adsorptions are the operations, which are conducted with a periodic swing in increase and decrease of pressure/temperature alternately. This not only enhances the efficiency but also uses the beds continuously through adsorption and desorption phases occurring simultaneously in adjacent beds. Another example is the well-known fluid catalytic cracking in the petroleum industry where catalytic reaction and catalyst regeneration are carried out simultaneously in adjacent units. Pulsating flows in trickle bed reactors are found to enhance mass transfer rates, and also result in improvement of product yields/selectivities. Chaotic reactors are being employed to achieve enhanced process performance in systems that exhibit multiple steady-states when it is necessary to operate at one of the unstable steady-state operating points.

Process integration and process synthesis are systematic methodologies using advanced mathematics and process optimization for PI. Further, utilization of dynamic and other modeling techniques using advanced process control strategies also leads to significant improvements in process performance. Product design is another way of achieving intensified processes where the end product itself is designed to exhibit certain desired features.

## 1.7 MODELING FOR PROCESS SIMULATION

Multi-functional operations as well as novel devices, discussed in previous sections, employ advanced chemical engineering concepts in heat transfer, mass transfer, reaction engineering, hydrodynamics, mixing, etc. Modeling of such operations provides immense chemical engineering challenges. An attempt is made in this section to discuss some modeling issues with respect to the intensified operations and devices.

### *1.7.1 MODELING OF MULTI-FUNCTIONAL OPERATIONS*

The existence of multifunctionality poses challenges in terms of modeling due to the reduction in degrees of freedom as well as nonlinear interactions between reaction and separation or between different separations. Further, the high computational requirement and numerical problems associated with solution of such complex models hinder the application of modeling approaches to industrial problems. Figure 1.8 highlights the major components of multifunctional operational modeling.

Homogeneous and heterogeneous kinetic models are important in multifunctional operations with reaction. Mass transfer models based either on equilibrium stage or rate-based concepts are employed in reactive and hybrid separations. Energy balances in adiabatic or nonadiabatic operations based on conductive, convective or radiative heat transfer play a key role in modeling heat integrated reactors as well as in reactive and hybrid separations. Momentum transfer based on macrolevel models such as ideal plug flow, complete mixing, dispersed flow, and tanks-in-series, and microlevel models that employ Naviers Stokes equations using computational fluid dynamics (CFD) are relevant in simulating the performance of reactive separations (especially with structured packing) and in reactions accompanied by momentum transfer such as static mixer reactor. Thermodynamic models based on ideal Raoult's law, simple relative volatility as well as nonideal equilibrium concepts employed in UNIQUAC, UNIFAC, Wilson, and NRTL methods are used in simulating multiphase systems that are encountered mostly in operations dominated by mass transfer.

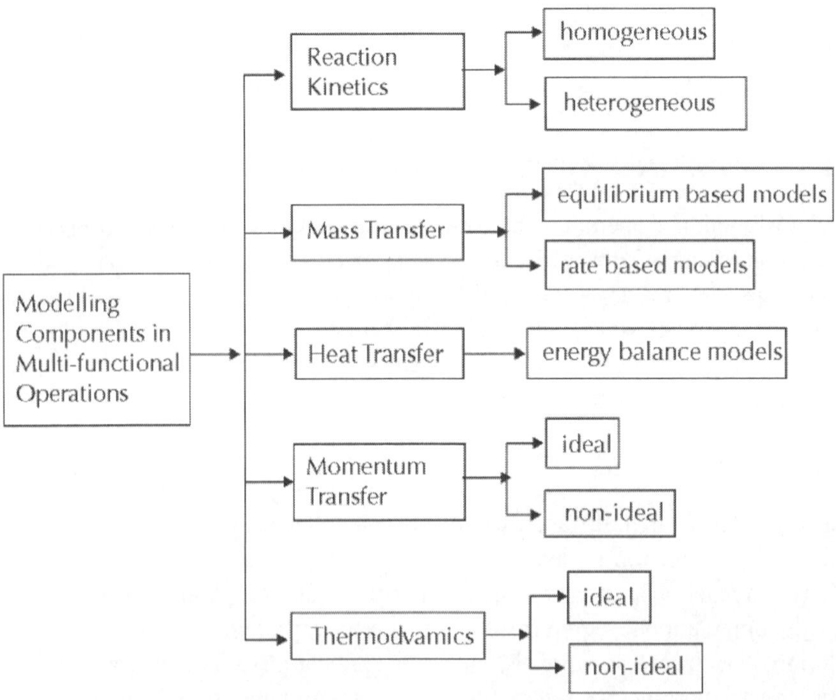


**FIGURE 1.8** Modeling aspects in multi-functional operations.

Modeling in multifunctional operations can also be categorized as steady-state, dynamic and data-driven modeling. The steady-state models based on equilibrium stage concept are employed by commercial software such as ASPEN PLUS for process design purposes. The CFD modeling is useful in evaluating the intrinsic mechanisms of mass, momentum, heat transfer and reaction kinetics with the help of rate based models. Very few studies have been reported on the development of dynamic models for reactive and hybrid separations, with the exception of reactive distillation [30] and reactive absorption [31]. Some studies are reported on data-driven modeling using nonlinear mathematic tools such as neural networks [32].

## 1.7.2 MODELING OF NOVEL DEVICES

### 1.7.2.1 MICRO-REACTORS

Modeling of microreactors needs to address special features involving laminar flow, small heat and mass transfer diffusion paths, high surface to volume ratio, and considerable contribution of heat transfer through solid wall [11]. Simulation of microreactors is generally carried out using CFD techniques. The continuum approach described by the Navier-Stokes equation and convection-diffusion equations for heat and mass transfer are best suited for their modeling. The gas flow regime is described by the Knudsen number, which is the ratio of the mean free path of the gas molecules and the characteristic length scale of the flow domain. Generally, Knudsen numbers in microreactors are less than 0.1. The model equations are similar to those employed for the macroscopic systems. However, the ranges of the parameters and the dimensionless groups are different, and one needs to pay attention to surface tension and viscous heating, which are ignored in macroscopic systems.

Sobienzuk et al. [33] discussed the flow regimes in closed and open channel microreactors under five different categories viz., bubbly, Taylor, Taylor annular, annular and churn flows. They have categorized these regimes with the help of superficial gas and liquid flow rates. Among the flow regimes in microchannels, slug flow, which is also called the Taylor flow, is the most common one. Nusselt number is an important dimensionless number representing the efficiency of heat transfer. The goal regarding heat transfer in microreactors is to achieve a temperature distribution as uniform as possible in order to suppress unwanted side reactions and to increase the selectivity of the process. The volume-averaging approach presented by Kiel [11] is suitable for comparison of thermal performance of different reactors and to evolve appropriate process designs. Mechanism of mixing as well as residence time distribution in microreactors can be established by Lagrangian particle tracking in combination with CFD computations of the flow field. Residence time distribution was shown to approach a delta like distribution of practically zero width [11]. Sobienzuk et al. [33] reported the mass transfer contributions in Taylor flow in terms of bubble to wall, bubble to slug, and slug to wall phenomena. The liquid and gas mass transfer coefficients in microdevices vary between $10^{-7}$ to $10^{-5}$ for liquids and $10^{-3}$ to $10^{-2}$ for gases, respectively.

Understanding the chemical kinetics is one of the key requirements for successful optimization and design of microreactors. Kinetics approximated using power law model are used in the simplified case. The complicated models based on Langmuir-Hinshelwood concept are employed in complex cases. When these models are combined with transport equations, the solution of the resulting set of equations is challenging due to nonlinearities and different time scales involved. Since heat and mass transfer limitations are eliminated in microreactors, they can be used to measure intrinsic rate constants of chemical reactions. An important dimensionless group characterising reactive flow is the Damkohler number representing the ratio of diffusive and reactive time scales. When Damkohler number is significantly smaller than 1, concentration profiles can be obtained using plug flow models incorporating Taylor-Aris dispersion [11].

### *1.7.2.2 STRUCTURED REACTORS*

Monoliths are the most common structured reactors. In monolith reactors, the factors influencing hydrodynamic parameters of flow regime are pressure drop, liquid hold-up, superficial gas and liquid velocities, physical properties of fluids, channel geometry and modes of operation such as cocurrent, countercurrent, up-flow and down-flow. High mass transfer rates are achieved in them due to high surface-to-volume ratio and a short diffusion path of gas through liquid. The flow pattern is generally found to be two phase slug flow (Taylor flow) or annular flow, as illustrated in Fig. 1.9. Depending on the mode of operation viz., upward or downward, alternate gas bubbles and liquid slugs move accordingly. The gas bubbles have diameter almost same as the channel diameter with the bubble length being several times more. In the region where gas bubbles exist, a very thin liquid film (30 to 70 μm) separates the gas bubbles and the channel wall. In annular flow, liquid flow rates are low and gas flow rates are high and liquid flows along the wall as a thin film, while the gas flows through the central part.

Compared to single-phase liquid flow, two phase gas-liquid Taylor flow leads to mass transfer of one order magnitude higher coupled with lower pressure drops. Maintaining proper flow rates (linear velocity > 0.1 m/s) can result in good distribution of liquid over the reactor cross section. Monolith catalytic packages, together with Sulzer type of contactors expand the region of operation with respect to flow rates, are more flexible, and result in more compact reactor systems [1]. The pressure drop in monolith reactors is influenced by wall friction, acceleration of gas phase, orifice effect at the entrance and between stacks, and by the gas-liquid distributor type. In most of the cases, wall friction is represented using Hagen-Poiselle law, and the contribution of gas phase to pressure drop is found to be small. In Taylor flow regime, the liquid hold up can be computed as the ratio of the average liquid slug length to the average total length of liquid slug and gas bubble neglecting the liquid in the thin film adjacent to gas bubbles. The average volume fraction of gas can be calculated as the ratio of gas flow rate to the total flow rate (gas and liquid) with a difference of about 15%. Roy et al. [24] reported several correlations for pressure drop valid in capillaries and monoliths, and for hold-up including those based on Lockhart-Martinelli correlations.

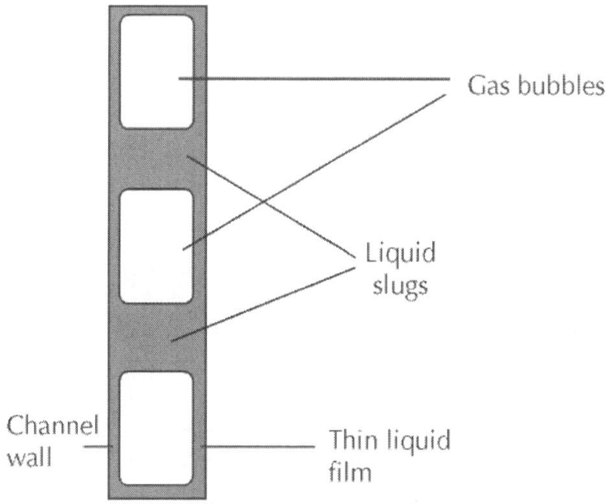


**FIGURE 1.9** Taylor flow in a monolith channel.

The mass transfer rate in monolith reactors is mainly controlled by four mechanisms – gas/solid, gas/liquid, liquid/solid, and pore diffusion. The reaction takes place on the surface and in the pores of the wash coat layer at the channel wall. The reactants are transferred from the bulk liquid and gas phases to the solid surface, and the reactants from the gas phase are also transferred from bulk gas phase to bulk liquid phase and from there to the solid surface. The three mass transfer coefficients involved are gas-liquid, liquid-solid and gas-solid. In addition to these factors, diffusion of the reactants into the pores and products out of the pores also influences the overall mass transfer rate. The correlations for liquid-solid and gas-liquid mass transfer have been reported by Roy et.al. [24], where Sherwood number is expressed as empirical functions of Reynolds number, Schmidt number, dimensionless channel length, liquid slug length, and unit cell length (bubble length + slug length). Thin film model was used to determine gas-solid mass transfer coefficient as the ratio of liquid diffusivity to film thickness mainly influenced by surface tension. Modeling in monolith reactors from reaction engineering viewpoint employs three distinct approaches. In the first approach, Taylor flow regime with three mass transfer rates discussed earlier (gas-liquid, liquid-solid, gas-solid) are considered, assuming plug flow in each phase and incorporating kinetic terms while defining gas-solid and liquid-solid fluxes. The second approach is based on annular flow regime, which is similar to trickle flow regime in packed beds, whereas the third approach is based on assumptions of plug flow in liquid phases in annular flow regime.

Modeling and simulation techniques for these novel devices such as microreactors and monolith reactors have not yet reached the maturity achieved in macro devices. Further development is necessary to formulate system level models that allow coupling of CFD simulations, which pave way for fast evaluation of process thereby providing impetus to widespread commercial applications of these new devices.

## 1.8 SUMMARY AND CONCLUSIONS

In this chapter, the process intensification has been presented from a chemical engineering perspective. It shows the tremendous scope for industrial applications in this area. After examining the reported definitions and classifications of PI, an attempt has been made to suggest a classification methodology based on multifunctionality, novelty in devices, and alternate-configurations.

Multi-functional operations are discussed under two sections of reactive separations and hybrid separation platforms. Reactive separations of reactive distillation, reactive adsorption, and membrane reactors are presented in more detail including their principles, advantages and applicability to different systems. Hybrid separations incorporating different unit operations are discussed briefly along with their application and scope.

Novel reactor configurations are discussed in terms of micro and monolith reactors. Examples of gas and liquid phase reactions and appropriate reactor configurations and process details are presented with a focus on PI. Structured reactors are discussed with their advantages compared to conventional reactors. Process aspects of monolith reactors including their applications in some industrially important reactions are discussed. A brief mention is made about novel devices for heat, mass and momentum transfer. The achievement of PI through alternate resources and modes of operation received adequate attention. The former include different raw materials, catalysts, solvents, process routes, and energy sources. The discussion on alternate modes of operation is predominantly based on periodic or dynamic operation and incorporation of advanced process synthesis and integration techniques to achieve PI. Modeling issues are presented in case of multifunctional operations, which are influenced by reaction kinetics, mass transfer, heat transfer, momentum transfer and thermodynamics. For novel devices including microreactors and structured reactors, modeling approaches based on hydrodynamic parameters like holdup and pressure drop, mass and heat transfer coefficients are discussed.

The review presented in this work has shown the potential of PI to expand fast both with respect to theoretical understanding and new applications. Reactive separations may be further intensified by use of rotating equipment providing enhanced mass transfer rates due to integration of high gravity fields. Another possibility could be integration of reactive and hybrid separations with microreactor technology. Such integrated intensification options may need further exploration to achieve the full potential of PI. The ever-growing number of research publications on PI in engineering as well as science journals provides the much needed comfort that these objectives will be fulfilled.

## KEYWORDS

- **Alternate PI configurations**
- **Hybrid separations**
- **Membrane reactors**
- **Micro-reactors**
- **Modeling of multifunctional operations**
- **Modeling of novel devices**
- **Multi-functional operations**
- **Novel devices**
- **Process intensification**
- **Reactive adsorption**
- **Reactive distillation**
- **Reactive separations**
- **Structured reactors**

## REFERENCES


1. Stankiewicz, A., and Moulijn, J.A. – Eds. *Re-engineering the Chemical Processing Plant: Process Intensification.* New York, Marcel Dekker Inc. (2004).
2. Ramshaw, C. HIGEE distillation – an example of process intensification. *Chem Eng (London)* **389**, 13–14 (1983).
3. Heggs P.J. Experimental techniques and correlations for heat exchanger surfaces: packed beds. *Chem Eng (London)* **394**, 13 (1983).
4. Stankiewicz, A. and Moulijn, J.A. Process intensification: transforming chemical engineering. *Chemical Engineering Progress* **96**, 22–33 (2000).
5. Stankiewicz, A. Reactive Separations for Process Intensification: An Industrial Perspective, *Chem. Eng. Process.* **42**, 137–144 (2003).
6. Grunewald, M. and Agar, D. W. Enhanced Catalyst Performance using Integrated Structured Facilities. *Chem. Eng. Sci.* **59**, 5519–5526 (2004).
7. Charpentier, J. C. In the frame of globalization and sustainability, process intensification, a path to the future of chemical and process engineering (molecules into money). *Chem. Eng. J.* **134**, 84–92 (2007).
8. Becht, S., Franke, R., Geisselman, A., and Hahn, H. An industrial view on process intensification. *Chem. Eng. Process.: Process Intens.* **48**, 329–332 (2009).
9. Gerven, T. G. and Stankiewicz, A. Structure, Energy, Synergy, Time – The Fundamentals of Process Intensification. *Ind. Eng. Chem. Res.* **48**, 2465–2474 (2009).
10. Dautzenberg, F. M. and Mukherjee, M. Process Intensification Using Multi-functional Reactors *Chem. Eng. Sci.* **56**, 251–267 (2001).
11. Keil, F. J., Ed. *Modeling of Process Intensification.* Weinheim, Wiley-VCH (2007).
12. Boodhoo, K. and Harvey, A. – Eds. *Process Intensification for Green Chemistry: Engineering Solutions for Sustainable Chemical Processing.* UK, Wiley (2013).
13. Sundmacher, A. and Kienle, A. – Eds. *Reactive distillation: Status and Future Directions.* Germany, Wiley-VCH Verlag GmbH & Co. (2002).
14. Ehrfeld W., Hessel V., and Löwe H. *Micro-reactors*. Germany, Weinheim:Wiley-VCH (2000).
15. Harmsen, G.J. Reactive distillation: the front-runner of industrial process intensification: a full review of commercial applications, research, scale-up, design and operation. *Chem. Eng. Process.* **46**, 774–780 (2007).
16. Hiwale, R.S., Bhate, N. V., Mahajani, Y.S., and Mahajani, S. M. Industrial Applications of Reactive Distillation: Recent Trends. *International Journal of Chemical Reactor Engineering*, **2** (1), R1, 1–52 (2004).
17. Krishna, R. Reactive separations: more ways to skin a cat. *Chem Eng Sci* **57**, 1491–1504 (2002).
18. Yang, W., Cicek N, and Ilg, J. State-of-the-art of membrane bioreactors: Worldwide research and commercial applications in North America. *Journal of Membrane Science* **270** (1–2), 201–211 (2006).
19. Sirkar, K. K., Shanbhag, P.V., and Kovvali, A.S. Membrane in a Reactor: A Functional Perspective. *Ind. Eng. Chem. Res.* **38**, 3715–3737 (1999).
20. Sharma, M., Vyas, R.K., and Singh, K. A review on reactive adsorption for potential environmental applications. *Adsorption* **19,** 161–188 (2013).

21. Frost, C.G. and Mutton, L. Heterogeneous Catalytic Synthesis Using Micro-reactor Technology. *Green Chemistry* **12**, 1687–1703 (2010).
22. Klemm, E., Doring, H., Giesselmann, A., and Schirrmeister, S. Micro-structured Reactors in Heterogeneous Catalysis. *Chem. Eng. Technol.* **30**, 1615–1621 (2007).
23. Ehrfeld, W., Golbig, K., Löwe, H., and Richter, T. Characterization of Mixing in Micro-mixers by a Test Reaction: Single Mixing Units and Mixer Arrays. *Ind. Eng. Chem. Res.* **38**, 1075–1082 (1999).
24. Roy, S., Bauer, T., Al-Dahhan, M., Lehner, P., and Turek, T. Monoliths as Multiphase Reactors: A Review. *AIChE J.* **50**, 2918–2938 (2004).
25. Williams, J. L. Monoliths structures, materials, properties and uses. *Catal Today* **69**, 3 (2001).
26. Albers, R.E., Nystrom, M., Siverstrom, M., Sellin, A., Dellve, A.C., Andersson, U., Herrmann, H., and Berglin, Th. Development of Monolith Based Process for $H_2O_2$ Production: Fom Idea to Large-scale Production. *Catal. Today* **69**, 247 (2001).
27. Clark, J. H. Perspective Green chemistry for the second generation biorefinery sustainable chemical manufacturing based on biomass. *J. Chem Technol Biotechnol.* **82**, 603–609 (2007).
28. Huang, H. J., Ramaswamy, S., Tschirner, U.W., and Rama Rao, B.V. A review of separation technologies in current and future biorefineries. *Sepn. Purif. Technol.* **62**, 1–21 (2008).
29. Ranganathan, S.V., Narasimhan, S.L., and Muthukumar, K. An overview of enzymatic production of biodiesel. *J. Bioresource Technology* **99**, 3975 (2008).
30. Sumana, C. and Venkateswarlu, Ch. Optimal selection of sensors for state estimation in a reactive distillation process *Journal of Process Control* **19** (6), 1024–1035 (2009).
31. Kenig, E.Y., Kuchka, L., and Gorak, A. Rigorous Modeling of Reactive Absorption Processes *Chem. Eng. Technol.* **26** (6), 631–646 (2003).
32. Teja, M.V., Ganesh, B., and Rani, K.Y. Modeling of a Reactive Batch Distillation Process using Time-Varying Neural Networks, in Proceedings of *INAE-DAE International Conference on advances in manufacturing technology (ICAMT 2008) for Young Engineers*, IIT Madras, Chennai, INDIA, Feb. 6–8, 2008.
33. Sobieszuk, P., Aubin, J., and Pohorecki, R. Hydrodynamics and Mass Transfer in Gas-Liquid Flows in Micro-reactors *Chem. Eng. Technol.* **35**, 1346–1358 (2012).

# CHAPTER 2

# MICROENVIRONMENT MANAGEMENT IN HETEROGENEOUS CATALYSIS

I. SREEDHAR, RADHA KISHAN MOTKURI, and K. V. RAGHAVAN

## CONTENTS

## 2.1 INTRODUCTION

The subject of microenvironment in catalytic reactions drew its inspiration from the water activity paradox encountered in biocatalysis in 1980s. The enzyme activity in biomedia with low water content was found to differ with that observed in dilute aqueous solutions [1]. The thermodynamic state of water was found to be an important factor influencing enzyme activity in aqueous media. Barzana et al. [2] demonstrated that an immobilized dehydrated enzyme oxidizes methanol and ethanol vapors at elevated temperature in the absence of water. Similar conclusions were drawn in another study on alcohol and aldehyde production in a solid-gas bio-reactor [3]. These investigations had shown that the presence or absence of aqueous environment around enzyme catalysts plays a key part in influencing their activity. The developments at the beginning of twenty-first century have firmly established the need to study the effect of microenvironment in solid-gas bioreactors by employing powerful analytical tools coupled with science driven industrial R&D [4]. Furthermore, artificial microenvironments in synthetic biology have shown great promise for tuning the properties of the industrially important enzymes to achieve the desired stability and activity even in chemical environments. Very recently Alamillo et al., [23] successfully tailored microenvironment for catalytic biomass conversion in inorganic-organic nanoreactors. Intercalation of polyvinyl pyrrolidone, a polar aprotic polymer, into the pores of SBA-15 silica catalyst significantly enhanced the selectivity to 5-hydroxy methyl furfural formation during the dehydration of fructose.

The importance of tailoring the microenvironment within and around the catalyst particles in heterogeneous catalysis was realized in organic nitrations, oxidations and hydrogenations in late 1990s. For convenience, the catalyst microenvironment can be defined as a specific micro area or volume of single or multiphase reaction medium within and around a catalyst particle distinguishable from its immediate surroundings by such phenomena as heat and mass transport and chemical/biotransformations. Though the positive effects of microenvironment on the catalytic activity and process selectivity have been acknowledged, the precise knowledge of its role and the mechanisms involved were not very well understood. An attempt is made in this chapter to analyze the recent developments in tailoring microenvironment of heterogeneous catalysts for controlling the course of specific cases of oxidation, hydrogenation, nitration, electrostatic interactions, and cyclization for macrocycle synthesis with the help of personal research contributions of the authors as well as other reported investigations.

## 2.2 CONTROLLING THE COURSE OF OXIDATIONS

The NANOCAT project was executed successfully by the industrial chemistry and reaction-engineering group of the Abo Akademi University in Finland in 2008. The investigators examined several organic reactions including oxidations over nanosized metal catalysts supported by organic matrices [5]. Studies on metal nanoparticles formed in various media and stabilized by different mechanisms have shown that the organization of nanoparticle crystal lattices, their surface characteristics and the chemical nature of microenvironment associated with them significantly influenced their catalytic activity and selectivity. In an interesting study, Aranjo et al., [6] encapsulated the ferric microperoxidase-11 (MP11) into a mesoporous silica MCM-41 to mimic enzymatic activity of hemoproteins. With the help of elemental and UV-visible spectrum analysis, they had shown that the heme group was in a hydrophobic microenvironment and Fe(III)-MPII-MCM41 exhibited specificity for hydrogen peroxide to be converted to a high valence oxidized intermediate. Phenol is able to act as the reducing agent for the encapsulated catalyst leading to the completion of a peroxidize cycle to produce 2,4-dihydroxy phenol.

## 2.3 PROMOTING ELECTROSTATIC INTERACTIONS

Walker [7] employed nanoparticle catalyst platforms to control the electrostatic interactions. He demonstrated that the self-assembled monolayers on the nanoparticles can be used to create an artificial microenvironment, which controls the chemical properties of organic molecules and their electrostatic interactions. This approach can have useful applications in catalytic dyads and triads.

## 2.4 ENHANCING PARA SELECTIVITY IN ORGANIC NITRATIONS

This is relatively the most intense area of research in exploring the microenvironments within and around catalyst particles. The authors examined [8] the prospects of tailoring the catalyst microenvironment while reviewing the scientific advances made in the sulfuric acid free toluene nitration employing solid acid catalysis. Haouas et al., [9] carried out toluene nitration with nitric acid and acetic anhydride employing H Beta Zeolite catalyst. They attributed the enhancement of para-selectivity in the mononitro toluenes to the transformation of lattice aluminum framework of the catalyst from a tetrahedral (TH) to an octahedral (OH) form and the effective coordination of the latter with acetyl nitrate formed on the zeolite surface (Fig. 2.1).

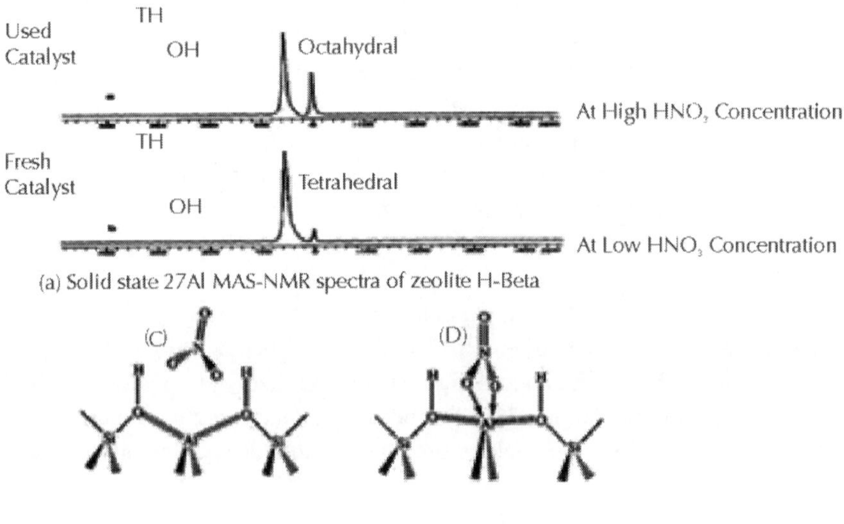


**FIGURE 2.1** Lattice aluminum transformations in zeolite beta during toluene nitration [9].

The authors [10] employed a two-stage procedure for process standardization to tailor the most conducive microenvironment for toluene nitration with higher para-selectivity to occur in the case of zeolite beta catalyst in a solid-vapor-liquid-liquid reaction medium. Figure 2.2 highlights the important features of the reaction system. The nitration was conducted in a semibatch reactor under reflux conditions. The first phase of the process standardization enabled the adjustment of $HNO_3$ – toluene phase dispersion to maximize toluene conversion and para selectivity of the nitrotoluenes formed. The liquid-liquid system offered interesting options for process optimization since the mass transfer characteristics of $HNO_3$ – toluene dispersion is greatly influenced by phase inversion phenomenon as reported by Zaldivar et al., [11]. It provided three types of reaction media viz., dispersed toluene in continuous $HNO_3$ phase, dispersed $HNO_3$ in continuous toluene phase and an ambivalent region in which either of these phases can exist depending upon phase volume ratio and energy inputs.

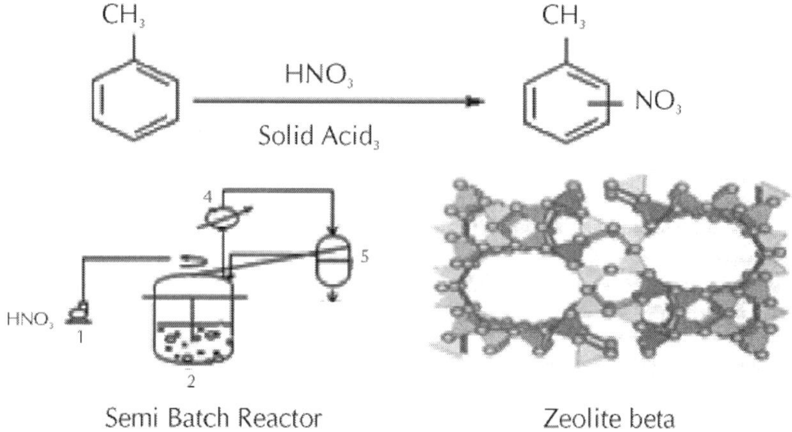


**FIGURE 2.2** Basic features of solid acid catalyzed toluene nitration.

The phase inversion in such systems can be predicted based on the hold up of dispersed phase and the changes in system properties [13]. The authors [12] estimated the phase inversion point, Sauter mean diameter of the droplets in the dispersed phase and mass transfer coefficients for the toluene-$HNO_3$ mixtures at various concentrations to characterize the toluene-$HNO_3$ dispersion. Their subsequent studies [14] have established the improved catalyst stability in the $HNO_3$ dispersed in toluene medium as compared to toluene dispersed in $HNO_3$ medium. Batch nitration experiments were made under reflux conditions covering a wide range of toluene volume fractions (0.1 to 0.95) to generate the conversion and para selectivity profiles for this volume fraction range (Fig. 2.3).

It shows that conversion of limiting reactant exceeds 80% in case of toluene dispersed in $HNO_3$ whereas maximum 60% conversion is achievable when $HNO_3$ is dispersed in toluene. Interestingly conversion of the limiting reactant dropped to 35% and increased again within the ambivalent region. The toluene-$HNO_3$ dispersion characteristics have relatively less effect on para-selectivity, which remained in the range of 0.7–0.8. The second phase process standardization studies were focussed on further enhancing the para selectivity in nitrotoluene mixture. A semibatch reactor is employed to facilitate continuous dosing of $HNO_3$ with its concentration in the range of 10 to 90%. The catalyst samples were subjected to solid state NMR spectroscopy to discover that the Al MAS-NMR spectrum of zeolite H Beta Catalyst had shown a strong signal at $HNO_3$ concentration exceeding 84% indicating the predominance of octahydrally coordinated aluminum similar to earlier findings of Haouas et al. [9].

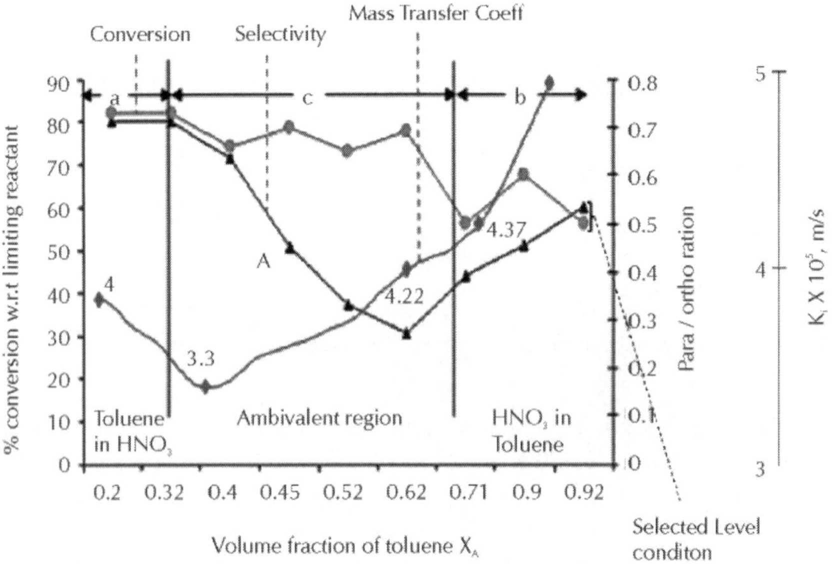


**FIGURE 2.3** Phase dispersion effects on toluene nitration [14].

The authors [13] also proposed a multiphase mass transport model for toluene nitration, the details of, which are given in Fig. 2.4. It is based on the formation of a thin aqueous film around the hydrophilic catalyst particles, which are dispersed in toluene medium. The model also accounted for the existence of vapor phase over the liquid-liquid-solid reaction medium. The major mass transfer resistances are offered by the liquid film around the catalyst particles and in the catalyst pores. The aqueous film and the liquid in the pores constitute the micro environment necessary to facilitate the desired level of lattice transformation in the catalyst particles. Figure 2.4 also shows the concept of the microenvironment within and around the catalyst particle. These studies have demonstrated that shape selectively effect of zeolite Beta catalyst is significantly enhanced by the specific microenvironment created within and around the catalyst particles. This has significantly enhanced the para-selectivity from 0.7 to 1.5. The microenvironment has also improved the accessibility of reactant molecules to the catalyst active sites.

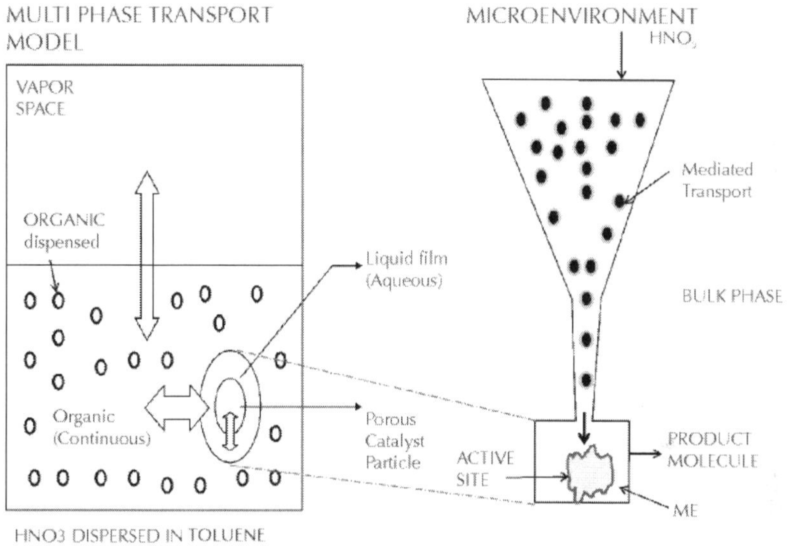


**FIGURE 2.4** Multiphase mass transport in vapor-liquid-liquid-solid system [13].

## 2.5 ENHANCING CATALYTIC PERFORMANCE OF ASYMMETRIC HYDROGENATIONS

Yang et al., [15] reported the successful development of a methodology for encapsulating transition metal complexes inside the nanocage of a catalyst. They exhibited high enantio-selectivity and catalytic activity in asymmetric synthesis of organic compounds. This development has brought in a new dimension to the creation of a microenvironment in catalytic processes. Bai et al., [16] reported that the catalytic activity of Ru-Ts DPEN confined in the nanocage of SBA-16 had greatly enhanced by modifying the nanocage through the creation of an amphiphilic microenvironment to achieve asymmetric transfer hydrogenation of acetophenone at $90^{+}$% conversion. It is found to be ten times more active than that with a hydrophobic reaction medium. Enhanced rate of diffusion of reactants in the nanocage has been identified as the major responsible factor.

## 2.6 SELECTIVE SYNTHESIS OF MACROCYCLES

Substituted calix[4]pyrrole and porphyrins represent an important class of macrocycles with several industrial end uses. Traditionally, homogeneous liquid phase acid catalysis based on HCl, PTSA, $BF_6$ etc., was employed to obtain high yields, however, suffered with inherent disadvantages of heavy dissolved salts and handling waste disposal burden. Until the year 2000, microporous zeolites were ineffective for larger molecules because of pore size limitations. The authors [17–22] for the first time employed mesoporous Al-MCM-41 as a heterogeneous catalyst in a liquid phase reaction between a pyrrole and a ketone to achieve macrocyclization, synthesizing caliz[4]pyrrole with >90% yield. On the other hand, by replacing the ketone with an aldehyde they achieved successful synthesis of porphyrins with yields 23.5, 16 and 40.1% with benzaldehyde, anisaldehyde and tolualdehyde respectively. The authors also studied the scientific challenges and mechanistic pathways for formation of macrocycles and effect of micro/mesopore management in zeolite catalysis.

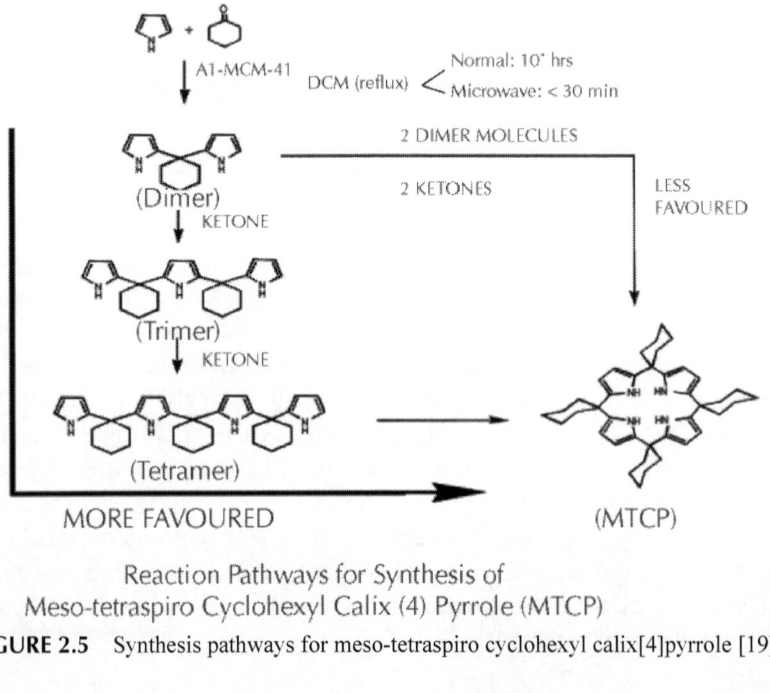


**FIGURE 2.5** Synthesis pathways for meso-tetraspiro cyclohexyl calix[4]pyrrole [19].

The influence of Al-MCH-41 catalyst microenvironment on the product distribution was investigated by varying the reaction conditions, nature of the solvent and Si/Al ratio of the MCM-41 structure. Also microwave irradiation under solvent and solve free conditions was applied to see the catalyst performance and finally concluded on the most favored macrocylization reaction path as shown in Fig. 2.5 for the synthesis of meso-tetraspirocyclohexyl calix[4]pyrrole (MTCP) with pyrrole and cyclohexanone as the reactants. Dichloromethane was employed as solvent. In order to understand the role of zeolitic structure and acidic character on the macrocylization reaction, various microporous zeolites including H-ZSM5 and HY were employed on pyrrole and cyclohexanone reaction. It is surprising to note that, the microporous zeolite HZSM-5 with a pore size of 5.6 Å showed the low yields of MTCP which has a size of ~14.20 Å while large pore HY zeolite (7.5 Å) yielded linear oligomers such as linear dimer (LDMR), trimer (LTRMR) and tetramer (LTMR) respectively. Further, to understand the reaction mechanism more precisely, both microporousHZSM-5 and Al-MCM-41 catalysts are surface poisoned with tetraethyl orthosilicate (TEOS) using chemical vapor deposition method and employed the surface poisoned catalysts for macrocyclization reactions. Similarly, in order to show that macrocyclization takes places inside the mesopores of Al-MCM-41 catalyst, its externals surface was made nonfunctional (poisoned) by treating with $Ph_2SiCl_2$. It is very interesting to note that the conversions and yield were not affected in case of Al-MCM-41 catalyst. Table 2.1 shows the effect of catalyst microenvironment on the product distribution. The best overall conversion is obtained with higher selectivity to MTCP in the case of normal reaction in the presence of solvent. In order to show that macrocyclization takes places inside the mesopores of AL-MCM-41 catalyst, its external surface was made nonfunctional. The conversion and yield of MTCP was found to be comparable though with slightly less dimer and more of trimer and tetramer formation. The microwave assisted solvent free and liquid phase synthesis resulted in low conversion and selectivity to MTCP and very high dimer formation in case of solvent free microwave irradiation. The process time was however drastically reduced (less than 10 min) in both cases of microwave irradiation.

In order to establish most favored course of reaction and to substantiate the need for tailoring the catalyst pore size and its microenvironment, the sorption of linear oligomers and MTCP in Al-MCM-41 was modeled by the authors [19] with the help of Monte Carlo simulation using MSI software on Octane Workstation supported by Cerius library data base. Drieding Buchart force field along with default atomic charges was employed to characterize the zeolite framework. For the sorbents, the force fields with AM1 derivated electrostatic potential charges were used. Vanderwaals and electrostatic interactions were considered. Figure 2.6 shows the Al-MCM-41 supercell sorbed with MTCP and intermediates, their average loading and stabilization energy and hexagonal cell structure of MTCP, which is a cyclic tetramer.

**TABLE 2.1** Influence of catalyst Al-MCM-41 and HZSM-5 microenvironment on the synthesis of meso-tetraspiro cyclohexyl calix[4]pyrrole (MTCP). Reproduced from Ref. [17] with permission from The Royal Society of Chemistry.

| S. No | Microenvironment | Conversion of Pyrrole, % | Yield (Wt. %) | | |
|---|---|---|---|---|---|
| | | | MTCP | Dimer | Others |
| 1. | Normal reaction in the presence of solvent (10 hrs) | 95.0 | 70.3 | 12.3 | 12.4 |
| 2. | Normal reaction in the presence of solvent (10 hrs) – HZSM-5 | 69.6 | 10.7 | 53.0 | 5.9 |
| 3. | Normal reaction with catalyst external surface passivated (10 hrs) – Al-MCM-41 | 93.0 | 64.7 | 7.5 | 20.8 |
| 4. | Normal reaction with catalyst external surface passivated (10 hrs) – HZSM-5 | 82.3 | – | 61.2 | 21.1 |
| 5. | Microwave irradiation in the presence of solvent (<10 min) – Al-MCM-41 | 64.3 | 38.3 | 1.2 | 24.7 |
| 6. | Microwave irradiation in the absence of solvent (<10 min) – Al-MCM-41 | 71.5 | 17.16 | 38.96 | 15.3 |

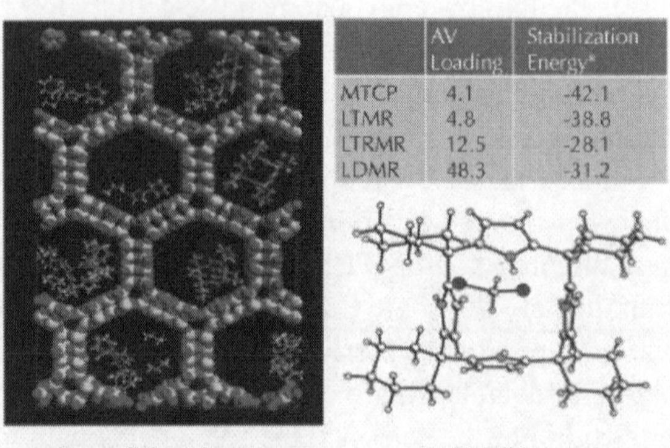


**FIGURE 2.6** Al-MCM-41 supercell sorbed with MTCP and intermediates. The inset shows average loading (number of molecules) and their stabilization energies in Kcal/mol. Reproduced from Ref. [17] with permission from The Royal Society of Chemistry.

It indicates that the average energy of MTCP is much lower than the oligomers viz., linear dimer (LDMR), trimer (LTRMR) and tetramer (LTMR). The LDMR was found to accumulate more than all others including MTCP during the initial phase of the reaction. However, on further reaction, LTRMR and LTMR are produced and the latter condenses readily with cyclohexanone to form more stabilized MTCP and very low yields of bigger macrocycles such as calix[$n$ = 5–7]pyrroles. These studies have also opened up the prospects for increasing the mesopore size of Al-MCM-41 to accommodate the formation of larger macrocycles. We attempted two different strategies for increasing the pore size of Al-MCM-41 by (i) varying the alkyl chain length from $C_{16}$ to $C_{22}$ of the surfactant molecule during the synthesis of Al-MCM-41 and (ii) also employing the auxiliary organic (expander) molecules as swelling agents in the synthesis process (Fig. 2.7). The first approach did not result a drastic increase in the pore size but our attempts to use MTCP as a swelling agent (SA) resulted the construction of larger pore Al-MCM-41 named as micelle template silica (MTS).

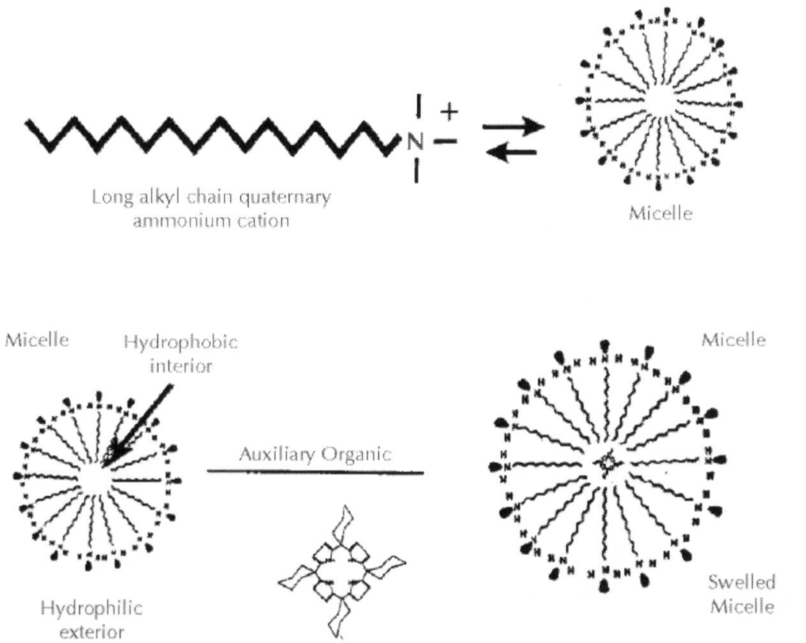


**FIGURE 2.7** Meso-tetraspirocyclohexyl calix[4]pyrrole (MTCP) as swelling agent for enhancing Al-MCM-41 pore size: Synthesis of micelle template silicas (MTS) [19].

The preparation methodology is similar to that of conventional Al-MCM-41 synthesis, but a specified amount of swelling agent is added during the synthesis with varying surfactant to SA ratio. Thus obtained MTS materials were characterized by powder XRD. It is confirmed the highly ordered hexagonal crystalline structures of the products and the total absence of any other crystalline phase in each of them with reference to four reflections d(100), d(110), d(200) and d(210) as shown in Fig. 2.8. From nitrogen adsorption spectra it can be seen that MTS-1 (CTAB/MTCP=0.5) and MTS-2 (CTAB/MTCP = 1) promote capillary condensation without any swelling agent as in the case of MCM-41. However, the condensation step is spread over a wider region, reflecting their broad pore size distribution. The broadening of the hysteresis loop at higher relative pressures above 0.4 indicates a more rigid mesoporous structure.

The pore size enhanced MTS materials were subjected to catalytic reactions and authors were able to synthesize cyclic oligomers of NYLON 6,6 with a molecular weight of 226, for the first time in the literature [19].

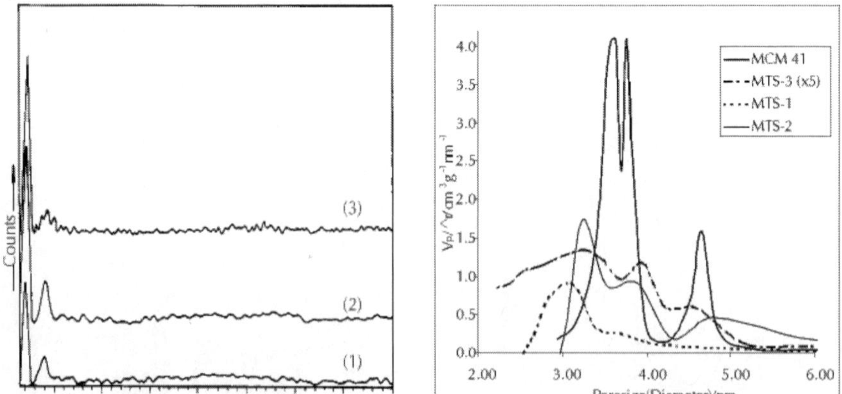


**FIGURE 2.8** (a) The XRD patterns of micelle template silica MTS-1(CTAB/MTCP=0.5). MTS-2 (CTAB/MTCP=1). MTS-3(CTAB/trimethylbenzene=1); (b) Poresize distribution studies compared with Al-MCM-41 material [19].

## 2.7 CONCLUSIONS

The examples and case studies cited in this chapter have shown that the physicochemical nature of microenvironment within and surrounding a catalyst particle can significantly influence its activity and regioselectivity. Though the micro and mesoporous structure of inorganic matrices can be tuned to promote shape selectivity, the specific microenvironment associated with the catalyst can further enhance its activity and selectivity with reference to a desired product. This aspect has been demonstrated in five specific catalytic reactions highlighted in this chapter. The enhanced performance of reaction system in the microenvironment can be attributed to one or more factors viz., influence on the electrostatic transformations on the catalyst matrix, maintaining the desired level of pH, acidity, hydroaffinity and other characteristics of reaction media, better accessibility of the reactants to the catalyst active sites and creation of micro or nanostructures within the catalyst pores. The microenvironment may be composed of liquid, vapor or solid phases or their combinations. The tailoring of microenvironment may require a two-stage process restandardization strategy with the first stage focussed towards the maximization of catalyst specific effects and the second stage to enhance them through the microenvironment restandardization. There is a very large potential for further development of the tailoring of microenvironment for process intensification with a good degree of predictability. The presented case studies show this promise.

## KEYWORDS

- **Microenvironment**
- **Heterogeneous catalysis**
- **Oxidation reactions**
- **Organic nitrations**
- **Para-selectivity in nitration**
- **Synthesis of Macrocycles**
- **Calix[4]pyrroles**
- **MCM-41**
- **Micelle template silica**

## ACKNOWLEDGMENT

The authors acknowledge the support provided by the respective Heads of Institutions to this work. They are thankful to Mr R R Yadu Krishna, Sr Steno, IICT, Hyderabad, India for his word processing and secretarial support for the manuscript preparation.

## REFERENCES

1. Drapron, R. Enzyme activity as a function of water activity: Properties of water in foods, Martinus Nijhoff Publishers, Dordrecht, The Netherlands (1985).
2. Barzana, E., Klibanov, A., and Karel, M. Ethanol oxidation in gaseous phase using Pichia pastorus. *Biotech Bio Eng* **34,** 1178–1185 (1989).
3. Pulvin, S., Parvaresh, F., and Thomas Ann, D. *NY Acad Sci* **545,** 434–439 (1988).
4. Lamare, S., Dominique, L.M., and Marianne, G. Solid/gas bioreactors: Powerful tools for fundamental research and efficient technology for industrial applications. *hal.archives-ouvertes.fr/docs/00/32/97/13/Doc/publi5.doc. (2002).*
5. Murzin, D.Y. Tailored nanosized metal catalyst for improving activity and selectivity via engineering of their structure and local environment. *NANOCATA Contract Program* No. 507721 (2008).
6. Arajno, J.C., Prieto, T., Prado, F.M., Trindade, F.J., and Nunes, G.L. Peroxidase catalytic cycle of MCM41 entrapped microperoxidase-11 as a mechanism for phenol oxidation, *J Nano Sci Nano Tech* **7**(10), 3643–52 (2007).
7. Walker, D.A., Designing nano scaled micro environments for molecular catalysis and synthetic biology, *Paper presented at the AICHE Annual Meeting, San Francisco, USA, November 3–8, 2013.*
8. Sreedhar, I., Singh, M., and Raghavan, K.V. Scientific advances in sulfuric acid free toluene nitration. *Catal Sci and Tech (Electronic Version)* **10.1.1039**/C3CY00337 (2013).
9. Haouas, M., Kogelbauer, A., and Prins, R. The effect of flexible lattice aluminum in zeolite beta during the nitration of toluene with nitric acid and acetic anhydrite. *Catal Today* **60,** 275–287 (2000).
10. Sreedhar, I., Reddy, K.S., Rama Krishna, M., Kulkrni, S.J., and Raghavan, K.V. Studies on para selectivity and yield enhancements in zeolite catalyzed toluene nitration. *Can J Chem Eng* **86,** 219–227 (2008).
11. Jaldivar, J.M., Alos, M.A., Molga, E., Hernandez, H., and Westerterp, H. R. The effect of phase inversion during semibatch aromatic nitrations. *Chem Eng & Processing* **34,** 529 (1995).
12. Leslie, Y.Y., Matar, O.K., Susana, E.P.O., and Geoffrey, F.H. A simple predictive tool for modeling phase inversion in liquid dispersion. *Chem Eng Sci* **57**, 1069–1072 (2002).
13. Sreedhar, I., Reddy, K.S.K., and Raghavan, K.V. Dispersion and kinetic studies of zeolite catalyzed toluene nitration. *Int J Chem Eng Res* **1(2),** 89–102 (2009).
14. Sreedhar, I., Reddy, K.S.K., and Raghavan, K.V. Studies on physical changes in zeolite beta in mononitration of toluene. *Kinetics and Catalysis* **50(1).** 131–137 (2009).
15. Yang, H.Q., Zhang, L., Su, W.G., Yag, Q.H., and Li, C. *J Catal* **248,** 204 (2007).
16. Bai, S., Yang, H.Q., Wang, P., Gao, J., Li, B., Yang, Q.H., and Li, C. Enhancement of catalytic performance in asymmetric transfer hydrogenation by microenvironment engineering of the nanocage. *Chem Commun* **46,** 8145–47 (2010).
17. Radha Kishan, M., Srinivas, N., Raghavan, K.V., Kulkarni, S.J., Sharma, J.A.R.P., and Vairamani, M. A novel shaper selective zeolite catalyzed synthesis of calix[4]pyrroles. *Chem Commun* 2226–2227 (2001).

18. Radha Kishan, M., Srinivas, N., Kukarni, S.J., Ramakrishna Prasad, M., Kamalakar, G., and Raghavan, K.V. Synthesis of macrocycles using molecular sieve catalysts. *Stud Surf Sci Catal* **135,** 174 (2001).
19. Radha Kishan, M. Synthesis of macrocycles over mesoporous moleculear sieves: Characterization, activity and mechanics. *PhD thesis submitted to School of Chemistry, University of Hyderabad, India, November 2002.*
20 (a) Radha Kishan, M., Radha Rani, V., Murthy, M.R.V.S., Sita Devi, P., Kulkarni, S.J., and Raghavan, K.V. Synthesis of calizpyrroles and porphyrins over molecular sieve catalysts. *J Mol Catal A: Chem* **223,** 263–267 (2004); (b) Radha Kishan, M., Radha Rani, V., Kulkarni, S.J., and Raghavan, K.V. A new environmentally friendly method for the synthesis of calix[4]pyrroles over molecular sieve catalysts. *J Mol Catal A: Chem* **237,** 155–160 (2005);
21. Prabhakar, S., Radha Kishan, M., Mirza, S.P., Raghavan, K.V., and Vairamani, M. Mass spectral study of meso-alkyl and meso-cycoalkyl calix[4]pyrroles under electron impact conditions: *Rapid Commun Mass Spectrometry* **18** (2004) 2077–2087.
22. Radha Kishan, M., Radha Rani, V., Sita Devi, P., Kulkarni, S.J., and Raghavan, K.V. A novel zeolite based stationary phases for synthesis and evaluation of porphyrins and calix[4]pyrroles. *J Mol Catal A: Chem* **269** (2007) 30–34.
23. Alamillo, R., Crisci, A.J., Marcel, J., Gallo, R., Scott, S.L., and Dumesic, J.A. A tailored microenvironment for catalytic biomass conversion in inorganic-organic nanoreactors. *Angew Chem Int Ed* **52,** 1–4(2013).

# PART II

# INTENSIFICATION OF SEPARATION PROCESSES

## CHAPTER 3

# METAL ORGANIC FRAMEWORKS–SYNTHESIS AND APPLICATIONS

RADHA KISHAN MOTKURI, JIAN LIU, CARLOS A FERNANDEZ, SATISH K NUNE, PRAVEEN THALLAPALLY, and B. PETE MCGRAIL

## CONTENTS

## 3.1 INTRODUCTION

Porous materials have received enduring interest in chemical science because of their suitability for many applications, including host materials for molecular separation and storage, catalysis, molecular sensing, magnetism, and drug delivery systems. These porous solids are broadly classified into two major categories: (i) amorphous and (ii) crystalline. Porous amorphous solids (e.g., plastics, gels) do not exhibit any ordered repeated units within their structures, but it is beneficial to work with them as they are usually less expensive and easy to prepare. Their primary disadvantages are the potentially wide range of molecular architectures with nonpredictable channels or topologies and the lack of long-range order that results in low mechanical stability. In contrast, the porous crystalline solids are more advantageous because of their ordered structures with reproducible pores/channels and topologies, which give them high thermal and mechanical stability. Nanoporous silica and zeolites are characteristic examples of such ordered porous solids that have predictable structural features with reproducible pores/channels, dimensions, and topologies. These porous solids are subdivided into three categories based on their pore size. According to the International Union of Pure and Applied Chemistry, the porosity of crystalline solids is usually given by the diameter of the pore size and is categorized as microporous (5–20 Å), mesoporous (20–500 Å), and macroporous (>500 Å) [1].

The inorganic microporous zeolites, both natural and synthetic, constructed from aluminosilicates have been extensively studied for decades for a wide variety of applications. Natural zeolites, which are predominantly ocean based, have various uses, but their application is limited due to the difficulty in eliminating impurities. However, synthetic zeolites have been used extensively for catalytic applications, adsorption, and separation because they can be systematically tailored under controlled conditions to meet user requirements. For example, zeolite 4Å ($Na_{12}Al_{12}Si_{12}O_{48}$) with pore size of 4Å was commercially used as a drying agent, while zeolite ZSM-5 served as a cracking agent. Similarly, Silicalite-1 was extensively used as a porous sorbent for removal of gasoline from drinking water [2]. Although zeolites have been widely used, they possess a number of drawbacks that limit their utility. The disadvantages include difficulty in controlling the synthesis conditions, the very limited number of structure and channel topologies (178 to date), and the fact that they are not easily changeable covalently bonded crystalline structures.

In the last decade, a new class of porous solids called metal-organic frameworks (MOFs) or porous coordination polymers (PCPs) was discovered. These porous solids exhibit unique and outstanding properties such as the largest pore volumes and highest surface areas known [3]. The tunable nanoporosity of MOFs places them far ahead of zeolites and other metal oxides. MOFs are considered as organic analogs of inorganic zeolites in, which oxygen atoms are replaced by rigid organic ligands/linkers that bridge the inorganic ions or ion clusters that self-assemble into one-, two-, and three-dimensional frameworks [4]. The advantage of this class of materials is that by carefully choosing the building blocks such as inorganic metal ions and organic ligands, it is possible to tailor the structures, pore sizes, and their topologies [5]. Because of the large selection of transition and lanthanide metal ions with diverse coordination geometries and the possibility of creating an even larger number of organic linkers via synthetic organic chemistry, an immensely diverse range of MOF framework architectures can be generated [6]. The schematic representations of the general classification of porous solids are shown in Fig. 3.1.

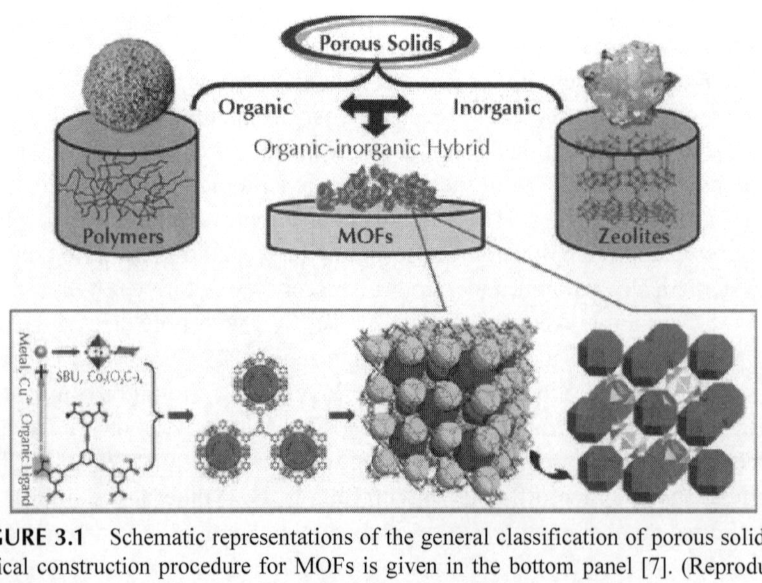


**FIGURE 3.1** Schematic representations of the general classification of porous solids. A typical construction procedure for MOFs is given in the bottom panel [7]. (Reproduced from Li, J. R. et al., *Chem Rev* **112,** 869–932 (2012), with permission).

The coordination polymers were discovered in the 1950s [8], and their discovery was followed by work reported by Jacobson [9], Bujoli [10], and Férey [11]. However, the major interest in these materials was boosted in the late 1990s by a pioneering material chemist Omar Yaghi and his co-workers. Yaghi and his team have been the most prolific group in synthesizing rigid coordination networks with metal cations and organic ligands. They started by replacing the 4,4′-bipyridine moieties with multidentate carboxylates ligands and named the resulting *neutral* frameworks as MOFs [12]. Amongst them, MOF-5 is the well studied in the literature and its structure aided the development of new concepts in design and tuning of MOFs. Later, because of the unique and outstanding properties of these architectures, an exponential increase in publications related to the development of new materials and their applications was observed. The historical developments in MOF synthesis are summarized in Fig. 3.2 [6].

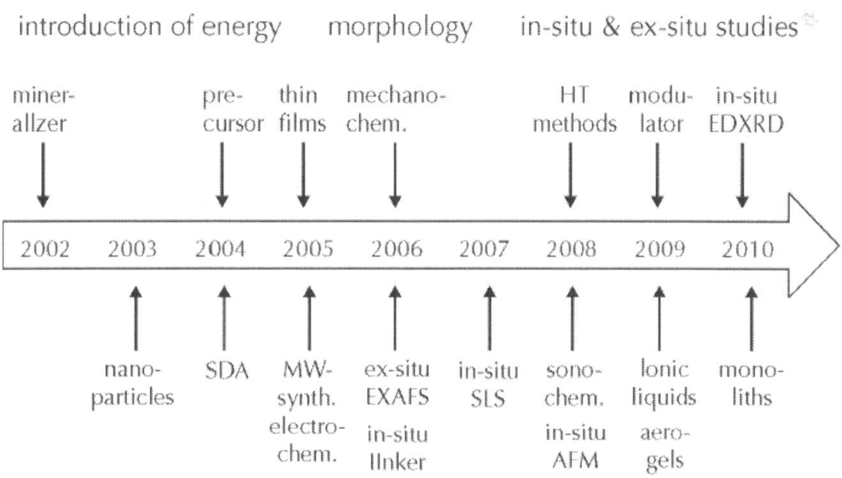


**FIGURE 3.2** Historical developments in MOF synthesis [6]. (Reproduced from Stock, N. et al., *Chem Rev* **112,** 933–969 (2012), with permission).

## 3.2 SYNTHESIS AND CHARACTERIZATION

The synthesis of MOFs is a moderately young field, and its evolution from well-developed solid-state/zeolite chemistry is noteworthy. Different types of synthesis methods were reported in the literature such as classical hydro (solvo) thermal synthesis, microwave heating [13], diffusion [14], ultrasonic [15], mechanochemical [16], electrochemical synthesis [17], and more recently spray-drying [18]. However, conventional hydrothermal or solvothermal heating of organic, inorganic precursors dissolved in a suitable solvent system in closed vessels under autogeneous conditions was well studied for major MOF synthesis. Both compositional parameters, such as solvent, pH, and molar concentration, and process parameters, such as temperature, pressure, and reaction time, will play a large role in the formation of framework structures. In a clear solution of dissolved reactants, the concentration gradient needs to be adjusted to exceed the critical nucleation concentration that enables the formation of crystal growth. The concentration gradient can be achieved by various methods, including changing the temperature of the system, solvent evaporation, layering, or slow diffusion of two or multiple solutions containing the reactants. Along with the powders, various morphologies such as thin films, membranes, and composites have been successfully synthesized for various applications (Fig. 3) [6].

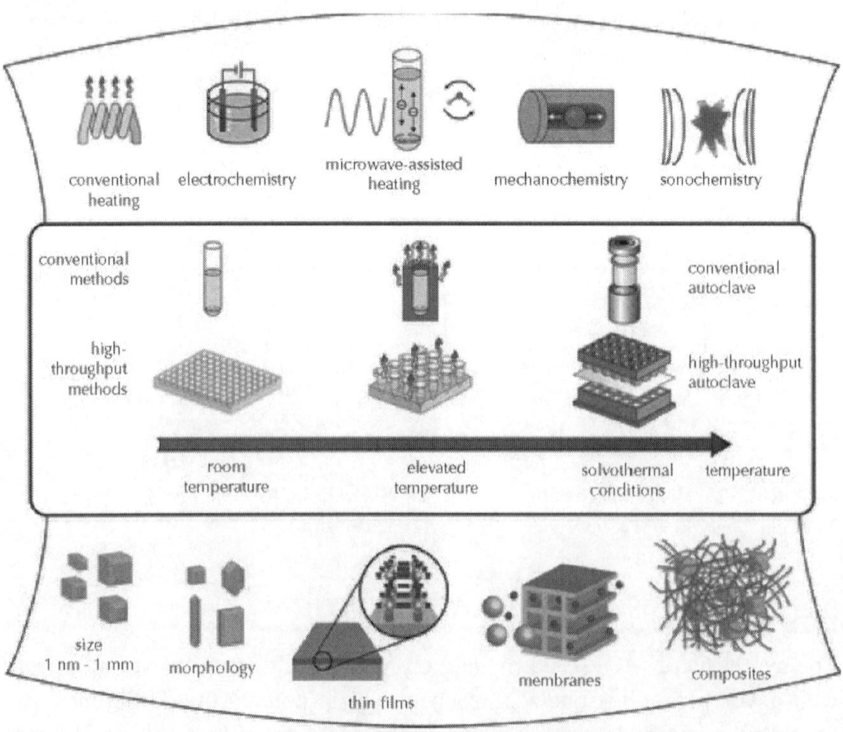


**FIGURE 3.3** Various synthesis methods and morphologies in MOFs [6]. (Reproduced from Stock, N. et al., *Chem Rev* **112,** 933–969 (2012), with permission).

In contrast to zeolites, MOFs have significantly weaker bonds (e.g., coordinative and hydrogen bonds) between the organic–inorganic connections. These weaker connections are responsible for MOF structural flexibility under external stimuli such as temperature [19], mechanical pressure [20], electric field, magnetic field, light, gas [21], and liquid/vapor [22] sorption. More specifically, sorption-induced structural deformation or "breathing" has been studied extensively to determine its effect on chemical, structural, and topological changes in MOFs in the presence or absence of the guest [23]. The different phases of a flexible MOF, MIL-53, resulting from increasing the amount adsorbed are shown in Fig. 3.4 [24]. Recently, it has been demonstrated that postsynthetic modification methods successfully showed the possibility to tune the flexibility in the material for tailored dynamic behavior in MOFs [25].

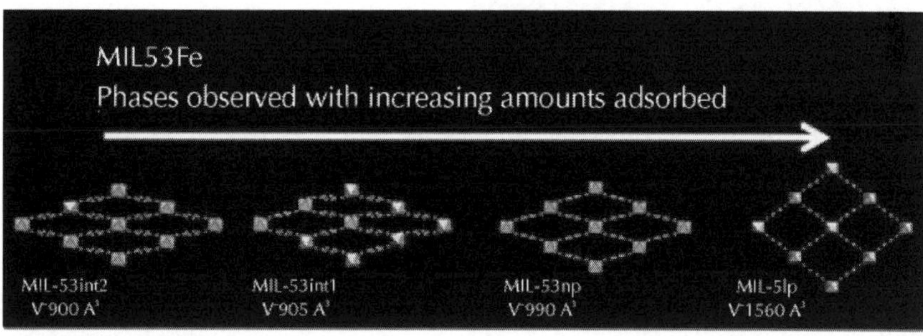


**FIGURE 3.4** Different phases of dynamic (flexible) metal-organic framework MIL-53 by increasing the amounts adsorbed. (Reproduced from Llewellyn, P.L. et al., *J Am Chem Soc* **131,** 13002–13008 (2009), with permission) [24].

Scalability is an important issue for any material for industrial applications. Large-scale synthesis of MOFs for industrial applications was successfully performed by BASF by considering the classical solvothermal synthetic approach, which is generally used in industrial production of zeolites [26]. In large-scale synthesis (kilotons/year), the availability and cost of reactants and the space–time-yield (STY) of synthesis play a crucial role. The higher the STY value (kg of MOF per $m^3$ of reaction mixture per day of synthesis), the greater the feasibility for MOF production, industrially. The inorganic salts (nitrates, oxides, sulfates) are readily available and inexpensive, while the organic linkers are usually expensive and become a significant component of the raw material costs. The STY values for Basolite M050 (Mg(II)/formic acid) and Basolite A520 (Al(III)/fumaricacid) are >3000 kg/($m^3$ D) while Basolite Z1200 (ZIF-8) has 160 kg/($m^3$ D). The typical zeolites that produced by BASF has a STY value of 50–150 kg/($m^3$ D) respectively [26]. Various characterization techniques have been successfully used to study the MOFs, most notably single crystal and powder X-ray diffraction (XRD) studies for getting the crystal morphology, arrangement of atoms and pore structure; BET and pore size distribution studies for getting the information of surface area and pore width and volume; Infrared spectroscopy studies for getting the information of functional groups in the MOFs structures; Thermogravimetric analysis for getting information of stability and optimal activation temperature ; Gas adsorption studies for getting the information of gas storage capacities, gas adsorption selectivities, and diffusivities of gas molecules in MOFs.

## 3.3 SORPTION APPLICATIONS

### *3.3.1 GAS STORAGE AND SEPARATION*

MOFs are widely considered as promising novel adsorbents for gas storage and separation due to their high surface areas, tunable pore size and structures, and versatile chemical compositions [28, 29].

### *3.3.2 HYDROGEN STORAGE AND SEPARATION*

Hydrogen is considered to be one of the best alternative fuels to fossil fuels because of its high energy density, nonpolluting combustion products, and natural abundance. However, $H_2$ is an extremely volatile gas under ambient conditions, resulting in a volumetric energy density that is much too low for practical applications. The goal therefore is to design lightweight materials that can reversibly and rapidly store $H_2$ near ambient conditions at a density equal to or greater than liquid hydrogen. MOFs have attracted considerable attention in the $H_2$ storage area in recent years because of their high working capacities [27].

One of the first MOFs investigated for $H_2$ storage was the cubic carboxylate-based framework MOF-5 or IRMOF-1, IRMOF-6, and IRMOF-8 (Fig. 3.5) [28]. With complete activation, MOF-5 can adsorbs 7.1 wt% $H_2$ at 77 K and 40 bar and 10.0 wt% at 100 bar. The latter value is corresponding to a record volumetric storage density of 66 g $L^{-1}$, which is near the density of liquid $H_2$ at 20.4 K and 1 bar [29]. This MOF material is the best cryogenic storage material currently known. In addition, it was demonstrated that $H_2$ can be loaded into MOF-5 within 2 min, and the sample can maintain a reversible capacity for at least 24 cycles [30]. Researchers have reported hydrogen storage data for more than 150 other microporous MOFs since MOF-5 was investigated as a novel adsorbent for $H_2$ [31]. Various strategies have been adopted to improve $H_2$ adsorption in MOFs. Open metal coordination sites (or unsaturated metal centers) in the structures of MOFs can increase the interaction between $H_2$ molecules and MOFs, resulting in an improved performance. This was first demonstrated in the MOF $Mn_3[(Mn_4Cl)_3(BTT)_8]_2$, which contains open $Mn^{2+}$ coordination sites and exhibits an isosteric heat of adsorption of 10.1 kJ $mol^{-1}$ at zero coverage [32].

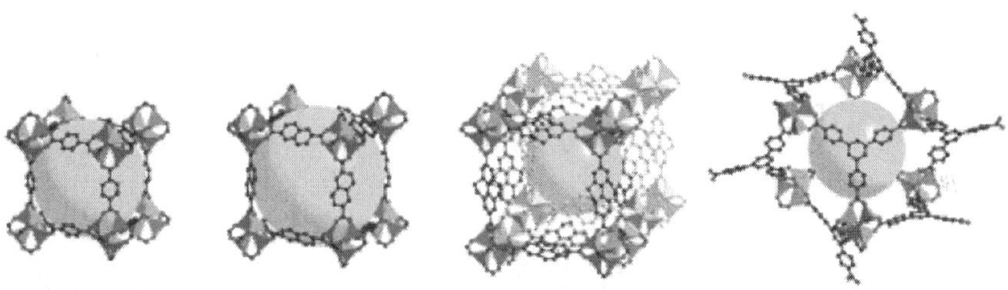


**FIGURE 3.5** MOFs, topologically the isoreticular MOFs (IRMOFs), are based upon an augmented simple cubic net (the boron net in $CaB_6$), while MOF-177 is based upon the augmented form of the (3,6)-coordinated net **qom**. The large pore of each structure is represented by a yellow sphere with diameter defined by the distance between the van der Waals surfaces of the framework atoms. Atom colors: C, black; O, red; Zn, blue tetrahedra; H, omitted [28a]. (Reproduced from Rowsell, J.L.C. et al., *Angew Chem Int Edit* **44,** 4670–4679 (2005), with permission).

In addition, surface area plays an important role in $H_2$ adsorption on solid materials. As a general rule, the maximum or saturated adsorbed amount of gas on a solid surface is dependent on its surface area [27a]. When the pressure of $H_2$ is not high enough to reach its saturated adsorption, the amount of the adsorbed $H_2$ is determined mainly by the interactions between molecules and MOF structures that can be positively related with the heats of adsorption [33]. Pore size can also affect the interaction between $H_2$ molecules and MOFs. It was found that MOFs having small pores with walls of high curvatures interact with $H_2$ molecules more strongly than large-pore MOFs [34]. The ideal pore size is slightly larger than the kinetic diameter of $H_2$ (2.8 Å) for low-pressure adsorption because, under that condition, $H_2$ molecules can interact with a larger portion of the framework, which increases the interaction energy between the frameworks and $H_2$ molecules [34b].

MOFs have displayed outstanding performance for cryogenic $H_2$ storage at 77 K and pressures up to 100 bar. Improvements in gravimetric capacity and in volumetric storage density are possible, provided advances in the synthetic chemistry. The challenge is to design ligands or surface functional groups that will lead to a high density of strong open metal sites, which are capable of binding more than one $H_2$ molecule with a binding enthalpy of about 20 kJ mol$^{-1}$. Thus, the creation of MOFs for practical applications in $H_2$ storage is still a difficult yet engaging challenge [27b].

On the other hand, MOFs are also promising adsorbents for the purification of $H_2$ due to their tunable porous structures. Molecular sieving using MOFs has been used to separate $H_2$ and $CH_4$ due to their different kinetic diameters ($H_2$: 2.8 Å; $CH_4$: 3.8 Å) [35]. In addition, Long's group has recently studied some representative MOFs, including MOF-177, $H_3[(Cu_4Cl)_{3-}(BTTri)_{8}]$, and Mg-MOF-74, for separating $CO_2$ and $H_2$ via pressure swing adsorption [36]. These measurements were performed under conditions (at 313 K and pressures up to 40 bar) close to practical application for precombustion $CO_2$ capture. These MOFs showed excellent separation performances, with evaluated $CO_2/H_2$ selectivities between 2 and 860 on an 20:80 $CO_2/H_2$ gas mixture. In addition to experimental observations, molecular simulations have been used to investigate the separation of $H_2$, $CO_2$, and $CH_4$ using MOFs [37].

### 3.3.3 METHANE STORAGE

Natural gas, which consists mainly of $CH_4$, is widely employed as feedstock for synthesis gas in many countries. It is currently stored as compressed natural gas (CNG) at 207 bar in pressure vessels, requiring an expensive multistage compression. An attractive alternative to CNG is adsorbed natural gas (ANG), in, which the gas is stored as an adsorbed phase in a porous solid at a lower pressure [38]. To promote the vehicular application of methane, the U.S. DOE has set the target for methane storage at 180 v(STP)/v under 35 bar, near ambient temperature, with the energy density of ANG being comparable to that of CNG currently used. An ideal material for $CH_4$ adsorption should have not only a large accessible surface area but also a large pore volume, a low framework density, and strong energetic interactions between the frame work and $CH_4$ molecules. GCMC simulation has been used as a screening tool to identify new candidates for methane storage and to guide the design of new materials.

A microporous PCN-14 MOF based on a predesigned anthracene derivative, 5, 5′-(9,10-anthracenediyl)-di-isophthalate, was synthesized for $CH_4$ uptake because it contains nanoporous cages as shown in Fig. 3.6. It has an estimated Langmuir surface area of 2176 $m^2/g$ and an estimated pore volume of 0.87 $cm^3/g$. More important, PCN-14 MOF exhibits an absolute methane-adsorption capacity of 230 v/v (28% higher than the DOE target of 180v/v at ambient temperature s) and heats of adsorption of methane of around 30 kJ/mol. These results indicate that PCN-14 is a promising adsorbent for $CH_4$ storage and can be used to store the fuel in the natural-gas-fueled vehicles.

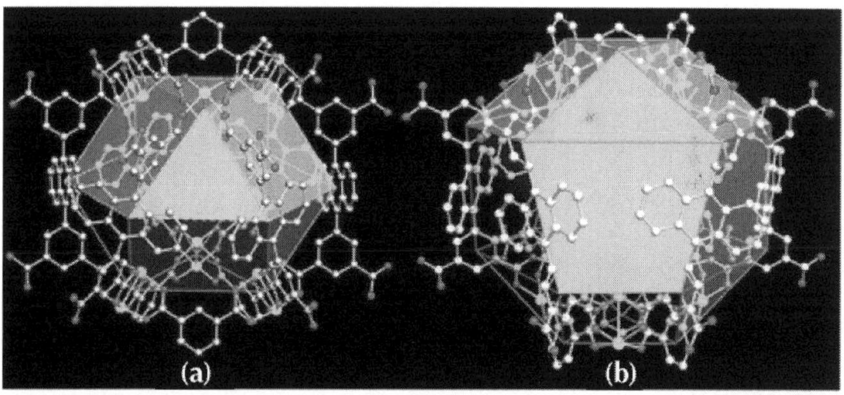


**FIGURE 3.6** The crystal structure of the PCN-14 MOF. (a) Squashed cuboctahedral cage and (b) nanoscopic cage with 18 vertices, 30 edges, and 20 faces. Color scheme: C, gray; Cu, turquoise; and O, red [39]. (Reproduced from Ma, S.Q. et al., *J Am Chem Soc* **130,** 1012–1016 (2008), with permission).

### *3.3.4 CARBON DIOXIDE CAPTURE FROM FLUE GAS*

The post combustion separation of $CO_2$ from power plant flue gas is of great interest to mitigate the global warming phenomenon [40]. The sorption of $CO_2$ using various MOFs and related porous materials is well documented in the literature [21, 41]. The $CO_2$ partial pressure in a typical flue gas is about 0.15 bar, which is much lower than atmospheric pressure. Therefore, it is important to study $CO_2$ adsorption in MOFs at low pressures. Yazaydin and colleagues used both experiments and simulation to screen MOFs for the highest $CO_2$ capacities at about 0.1 atm [42]. They found that Mg/DOBDC and Ni/DOBDC (also known as Mg-MOF-74 and Ni-MOF-74 or CPO-27-Mg and CPO-27-Ni) have the highest $CO_2$ capacities at 0.1 atm and 298 K, which are 5.95 mol $kg^{-1}$ and 4.07 mol $kg^{-1}$, as shown in Fig. 3.7.

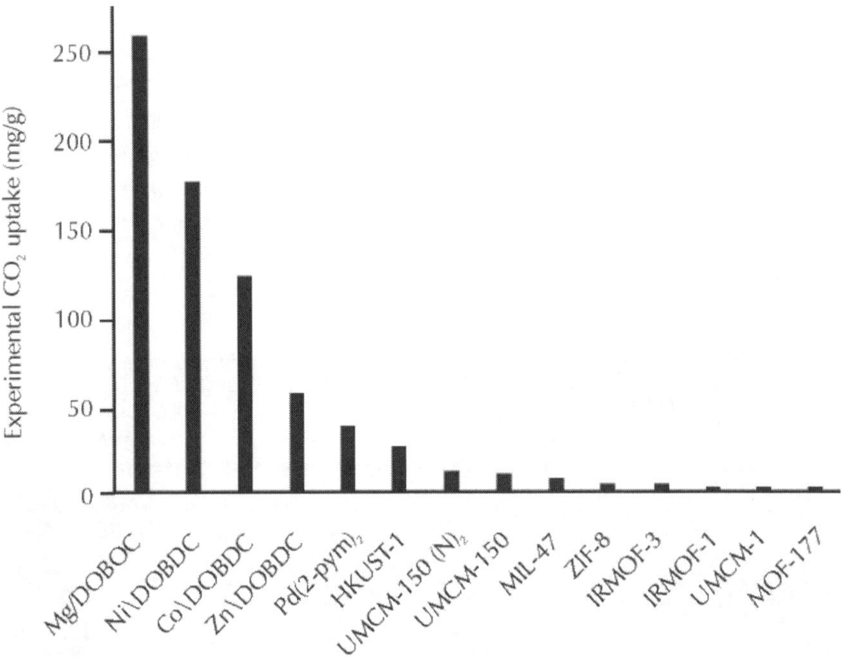


**FIGURE 3.7** Experimental $CO_2$ uptake in different MOFs at 0.1 bar. Data were obtained at 293–298 K [42]. (Reproduced from Yazaydin, A.O. et al., *J Am Chem Soc* **131,** 18198 (2009), with permission).

Instead of the surface area or the free volume, the authors found that MOFs such as Mg/DOBDC and Ni/DOBDC with a high density of open metal sites are promising candidates for $CO_2$ capture from flue gas in, which $CO_2$ partial pressure is about 0.1 atm. Caskey et al. [43] found that metal substitution in the DOBDC series can significantly impact their $CO_2$ capacities in the low-pressure region. The metal substitution effect may be caused by the differences in the ionic character of the metal–oxygen bonds in the DOBDC-series MOFs. Liu et al. found that Ni/DOBDC has a higher $CO_2$ capacity than NaX and 5A zeolites at 0.1 atm, and 25° C. In addition, water does not affect $CO_2$ adsorption in the Ni/DOBDC as much as in NaX and 5A zeolites, and it is much easier to remove water from Ni/DOBDC by heat regeneration [44]. Therefore, the Ni/DOBDC can adsorb more $CO_2$ than traditional zeolites under the same moist conditions.

Most flue gas is composed of $N_2$, so it is important to have a MOF that can selectively adsorb $CO_2$ over $N_2$. Motkuri et al. showed that Prussian blue derivatives represent a class of highly stable ordered crystalline structures that can selectively adsorb $CO_2$ over $N_2$, $CH_4$, and water present in flue and natural gas conditions [41]. Seven MOFs, including CuBTC, MIL-47 (V), IR-MOF-1, IRMOF-12, IRMOF-14, IRMOF-11, and IRMOF-13, were studied for the separation performance of $CO_2$ over $N_2$ by Liu and Smit using GCMC simulations [45]. In all the MOFs they considered, $CO_2$ is more preferentially adsorbed than $N_2$, with CuBTC showing the highest selectivity. Motkuri and team found that pore size plays an important role in the selective adsorption of $CO_2$ over $N_2$ [47]. The effect of the chemistry of the materials (i.e., effects of the electrostatic interaction) becomes less evident compared to the effects of pore size because both $CO_2$ and $N_2$ molecules have quadruple moments. As a result, the electrostatic interactions will help increase the adsorption for both of them. Wu et al. obtained a Li-modified IRMOF-1, chem-4Li MOF, by substituting all the hydrogen atoms with O–Li groups in the aromatic rings of IR-MOF-1.The chem-4Li MOF was found to have an extraordinarily high $CO_2$/$N_2$ selectivity of 395 ($CO_2$:$N_2$ = 15.6:84.4), which is two orders of magnitude larger than that of the original IRMOF-1 [46]. This selectivity is caused by the strong electrostatic interactions between the gas molecules and the lithium atoms in the framework. Thallapally and co-workers synthesized eight dissimilar metal-organic supramolecular isomers using one flexible tetrahedral organic linker, tectonic acid (TA), with different metals Zn, Cd, Co, Mg, and Cu [22, 47]. In the MOF constructed with Zn, addition of auxiliary ligands (pyrazine, 1, 4-diazabicyclo [2.2.2]octane or DABCO) yielded MOFs with no pores to mesopores. It is interesting to note that the auxiliary ligand bipyridineinvolved in the crystal formation resulted in a flexible MOF formation, TetZB (isomer A in Fig. 3.8) that showed a breathing phenomenon upon sorption of gasses/vapors. All other TA-based MOFs showed preferable sorption of $CO_2$ over $N_2$, $H_2$, and methane at room temperature.

Meanwhile, selectivity between $H_2O$ and $CO_2$ is important for using MOFs to separate $CO_2$ from flue gas because a typical flue gas is saturated with water vapor. Some research work shows that a small amount of water can help enhance $CO_2$ adsorption in HKUST-1 [48]. The enhanced $CO_2$ uptake is caused by interactions between the quadruple moment of $CO_2$ and the electric field created by the coordinated water molecules. However, further increasing the water loading on the HKUST-1 will result in considerable uncoordinated water molecules that block pore space and cause the HKUST-1 to adsorb less $CO_2$ than the dry sample. On the other hand, Liu et al. reported that neither the HKUST-1 nor the Ni/DOBDC MOF can adsorb any significant amount of $CO_2$ when water loadings are high, which means that the two

MOFs preferentially adsorb $H_2O$ over $CO_2$ although water does not affect $CO_2$ adsorption in HKUST-1 and Ni/DOBDC as much as it does in traditional zeolites [44].

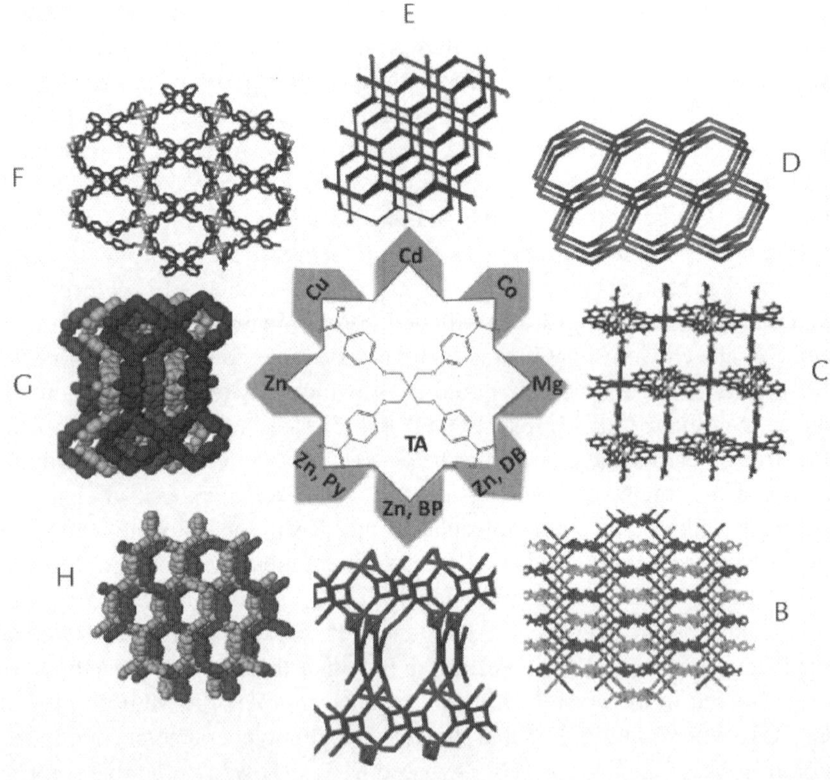


**FIGURE 3.8** Eight metal-organic supramolecular isomers formed with the flexible ligand TA with transition metals Zn, Cd, Co, Cu, and Mg. All the isomers showed the rigid framework while structure A (TetZB) showed flexible framework towards guest sorption [21, 22, 47]. Py = pyrazine, DB = 1,4-diazabicyclo[2.2.2]octane, and BP = bipyridine. (The structures reproduced from Motkuri et al. in referenced RSC and ACS journals, with permission).

A straightforward approach to reduce water adsorption in MOFs is to create more hydrophobic MOF pores. This can be done by either postmodification synthesis of attaching hydrophobic moieties or generating MOFs with hydrophobic ligands. Nune et al. reported the synthesis of thermally stable metal-organic gels (MOGs) using near-supercritical processing conditions; significant $CO_2$ uptake capabilities were observed for the first time [49]. Yang et al. synthesized some FMOFs in, which hydrogen atoms are replaced by fluorine atoms in all ligands [50]. FMOFs with fluoro-lined or fluoro-coated channels or cavities have enhanced thermal stability and hydrophobicity compared to their nonfluorous counterparts. Farha et al. replaced the coordinated solvent molecules in a three-dimensional MOF with various cavity modifiers, including pyridine and its derivatives [51]. The resulting tailored cavities show different adsorption properties, and this postsynthetic modification method can be adopted to cover hydrophilic surfaces in some MOFs with hydrophobic molecules, such as pyridine, to reduce $H_2O$ effects on $CO_2$ adsorption. An alternative engineering solution to mitigate water effects on $CO_2$ adsorption in MOFs is to install a guard bed loaded with desiccants in front of the main bed loaded with MOFs to remove most of the water and to take advantage of the MOFs' high $CO_2$ capacities and selectivities.

### *3.3.5 WATER SORPTION STUDIES FOR ADSORPTION COOLING APPLICATIONS*

Technologies based on the sorption of water are detailed in the literature for a number of applications, including adsorption-based heating and cooling, air-conditioning systems, desiccant dehumidifiers, and freshwater production. The main criterion is to find a best adsorbent–adsorbate (solid + vapor) pair that effectively works on sorption principles. Because of water has the highest mass-based evaporation enthalpy (2440 kJ/kg at 25° C), it would be a preferred adsorbate. Currently, inorganic zeolites and silicagel are widely used as adsorbents for water sorption, but these sorbents suffer with water adsorption at overly high relative pressures and also at high temperature for desorption, resulting in a large footprint and high capital cost. An ideal working pair with water would be a material that has high surface areas that can sorb huge water molecules at low partial pressures and moderate sorbate–sorbent interactions for low-temperature desorption in a cycle.

MOFs with extraordinary porosity in combination with tunable pore characteristics make a suitable choice as adsorbent material. Janiak et al. demonstrated a three-dimensional MOF, ISE-1, as an efficient adsorbent for heat transformation cycles for refrigeration, heat pumping, and storage [52]. The MOF ISE-1 has approximately 22 water molecules per formula unit with a potential solvent volume of 1621 Å$^3$ (52%) of the unit cell volume constructed from benzene-1,3,5-tricarboxylate and 1,2-bis(1,2,4-triazol-4-yl)ethane. ISE-

1 demonstrated good water stability and achieved a loading spread of about 210 g/kg, which is larger than that of five zeolites and silica gel that have been used in commercial heat pump applications. The stability and loading spread features may be due to the less hydrophilic nature of ISE-1 compared to silica and zeolites, which release water molecules (desorption) at lower temperature and lower partial pressures. The Janiak research group recently also successfully used a thermally stable porous MIL-100 (Fe, Al) for water adsorption/desorption of up to 0.75 g of water vapor per gram of MOF. Because of very good cycle stability and suitable hydrophilicity, MIL-100 proved itself a valuable addition to the pool of sorption materials used in heat pumps or sorption chillers [53]. Henninger et al. compared the water sorption capacities of HKUST-1 with that of zeolites and silica gel and showed that the highest water uptake for driving temperature of 95° C was mesoporous aluminophosphate AlPO-18 (0.253 g/g), while at 140° C, HKUST-1 showed a highest water uptake of 0.324 g/g [54]. They also showed the first results on integral heats of adsorption in the cycle. Ferey et al. have demonstrated that hierarchically porous MIL-100 and MIL-101 with mesoporous cages behave as advanced water adsorbents applicable to energy-efficient dehumidification systems that showed superior performance over commercial adsorbents such as zeolite NaX, SAPO-34, and silica gel [55]. On a large-scale evaluation, a honeycomb rotor coated with these MOFs showed huge sorption uptakes, even at 40° C, as well as high desorption rates below 80° C and almost two times higher efficiency in energy consumption for adsorptive dehumidification applications in commercial/residential buildings.

### 3.3.6 HARMFUL GAS REMOVAL

Harmful gasses in the environment pose a growing international security threat. Effective capture of these gasses is critical to protect lives and the environment [56]. Pioneering work has been published on the performance of six different MOFs such as MOF-5, IRMOF-3, MOF-74, MOF-177, MOF-199, and IRMOF-62 for dynamic adsorption capacities for eight harmful gasses: sulfur dioxide, ammonia, chlorine, tetrahydrothiophene, benzene, dichloromethane, ethylene oxide, and carbon monoxide and compared the same with BPL carbon respectively [56a]. The dynamic adsorption capacity of each MOF for each gas has been determined using a kinetic breakthrough method and compared with that of a Calgon BPL-activated carbon. Pore functionality was found to play a dominant role in determining the dynamic adsorption performance of the MOFs. For example, MOF-199, which featured a reactive functionality, outperformed BPL carbon in all cases except chlorine while MOF-74 and IRMOF-3 outperformed in sorption of sulfur dioxide and ammonia respectively [56a].

In addition, various MOFs and their composites have been used to remove $H_2S$, $SO_2$, $NH_3$, CO, NO, and benzene [57]. For efficient adsorptive removal, both the pore structures and open metal sites and functional groups of adsorbents are of great importance. The open metal sites of MOFs have been reported many times as the active sites for the removal of various toxic gasses. Petit and Bandosz reported composites of MOFs and a graphitic compound (graphite or graphite oxide, GO) for the adsorptive removal of $NH_3$, $H_2S$, and $NO_2$ under ambient conditions [58]. The open metal sites of porous MOFs coordinated with the oxygen groups of GO led to the formation of a new pore space in the interface, resulting in increases of more than 12% (for $NH_3$), 50% (for $H_2S$), and 4% (for $NO_2$) in the adsorption capacity for the GO/Cu-BTC composite.

A fluorinated MOF, FMOF-2 obtained from 2,2′-bis(4-carboxyphenyl) hexafluoropropane and zinc nitrate hexahydrate, was also reported for the adsorptive removal of toxic acidic gasses [59]. FMOF-2 was quite stable for the adsorption of $SO_2$ and $H_2S$ and has adsorption capacities of 14.0% and 8.3% for $SO_2$ and $H_2S$, respectively, at room temperature and 1 bar. Fernandez et al. showed that FMOF-2 also manifests "breathing" behavior during $SO_2$ and $CO_2$ adsorption, an uncommon feature in, which the framework structure expands and contracts upon adsorption and desorption of guest molecules [59, 60]. In addition to chemically toxic gas adsorption, the removal of radioactive gases such as Xe and Kr using MOFs has also been investigated. Thallapally et al. found that the Ni/DOBDC has a Xe adsorption capacity 10% higher than that of activated carbon at room temperature and 1 bar [61]. Meek et al. found that the uptakes of Xe and Kr can be affected by the linker polarizability from their results on a series of mono halogenated isoreticular MOFs [62].

Several MOFs and a benchmark-activated carbon sample were studied as adsorbents to separate low-concentration Xe and Kr from air [63]. Both the Ni/DOBDC and the HKUST-1 can selectively adsorb Xe and Kr from air, even at concentrations at the parts-per-million level. The Ni/DOBDC has a Xe capacity of 9.3 mmol/kg when the concentration of Xe is 1000 ppm in air. More important, the Ni/DOBDC is able to separate 400-ppm Xe from 40-ppm Kr mixture in air with a Xe/Kr overall selectivity of 7.3. The high Xe/Kr selectivity is due mainly to the strong unsaturated metal centers in the crystal structures of the Ni/DOBDC. In addition, the uniform cylindrical pores in the Ni/DOBDC are believed to be favorable to maximize the Xe/Kr selectivity [64]. Nevertheless, Fernandez et al. demonstrated for the first time that a

MOF material can selectively capture and separate Kr from Kr/Xe mixtures at moderate temperature via a temperature gating mechanism on a copper-based fluorinated MOF [65]. MOFs have shown great application potential in the gas capture and separations area, which will help in evaluating the MOFs for sorption, separation-related applications, and selective single-component sorption/separation applications, respectively.

## 3.4 HETEROGENEOUS CATALYSIS

Although MOF-based catalysis was proposed and experimentally proven in the last decade, substantial experimental exploration with considerable promise in catalysis has been reported very recently [66]. The use of MOFs as catalysts for industrial processes may still be far into the future, but it is likely that the growing opportunities demonstrated on the laboratory scale may trigger interest in scaling up for industrial use. The low thermal and chemical stability of MOFs, certainly limit their use in vapor-phase reactions (reactions carried out above 300° C) in industrial processes such as oil refining or petrochemical processes in, which zeolites are currently used. However, in liquid-phase reactions, MOFs can compete with or even outperform the existing zeolites, particularly the reactions carried out under mild conditions [67]. The main advantage of MOFs is their versatility in chemical composition, organic and inorganic building units, and the bifunctional metal/acid sites for insertion using isoreticular chemistry. MOFs can complement zeolites to perform the reactions at lower temperature s. As heterogeneous catalysts, the three known possibilities for building catalytic activity are (1) framework activity (2) encapsulation of active species, and (3) postsynthesis modification.

### *3.4.1 FRAMEWORK ACTIVITY*

One of the most widely explored strategies for configuring MOFs as heterogeneous catalysts is to take advantage of the exchangeable coordination positions around the metal ions. Another methodology would be to introduce the organic linker with acidic or basic groups as the active sites. Recently, the structural defects around the metal nodes also were explored for catalytic applications [68]. The framework activity is further divided into acidic and basic sites, as discussed below.

#### *3.4.1.1 LEWIS ACID CATALYSIS*

The porous material $[Cu_3(btc)_2(H_2O)_3]$, also known as HKUST-1, contains large cavities. The copper metal is coordinated to water molecules that can be removed easily under heat treatment, leaving open metal centers (Cu(II) sites act as Lewis acidic centers) [69]. Jacobs et al. investigated the behavior of HKUST-1 as a Lewis acidic catalyst by examining three sets of reactions, including isomerization of (i) alpha-pinene oxide (ii) rearrangement of alpha-bromoacetal, and (iii) cyclization of citronellal. Based on the experimental data and product selectivities, they concluded that HKUST-1 works primarily as a Lewis acid catalyst [70]. Kaskel et al. evaluated the Lewis acidity of both HKUST-1 and large-pore MIL-101 having the formula $[Cr_3F(H_2O)_2O(bdc)_3]$ on cyanosilyzation of benzaldehyde and concluded that the Cr(III) sites in MIL-101 showed greater activity than Cu(II) sites in HKUST-1. Moreover, the catalytic sites in MIL-101 are immune to the unwanted reduction of benzaldehyde during the reaction [71]. Long et al. used a manganese-based MOF with the formula $Mn_3[(Mn_4Cl)_3BTT_8(CH_3OH)_{10}]_2$ ($H_3BTT$ = 1,3,5-benzenetris(tetrazol-5-yl) having a three-dimensional pore structure with a pore diameter of 10 Å having two $Mn^{+2}$ sites. The MOF actively engaged in Lewis acid catalytic conversion of selected aldehydes and ketones with cyanotrimethylsilane to the corresponding cyanosilylated products with conversion yields above 90%, one of the highest yields reported in metal-organic frameworks (Fig. 3.9). The same MOF also was shown to catalyze the Mukiyama aldol reaction, which generally requires stronger Lewis acidic sites than cyanosilyzation reactions [72]. In another report, Gandara et al. documented highly Lewis acidic In (III) sites in a two-dimensional MOF that catalyzed acetalization of benzaldehyde with trimethyl orthoformate [73].

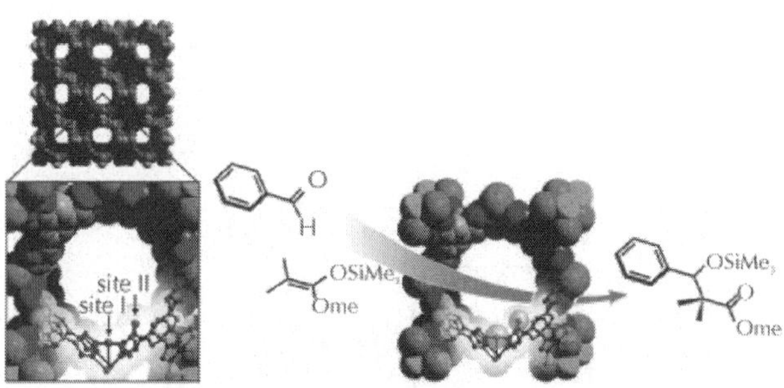


**FIGURE 3.9** (Left) A portion of the crystal structure of $Mn_3[(Mn_4Cl)_3BTT_8(CH_3OH)_{10}]_2$ ($H_3BTT$ = 1,3,5-benzenetris(tetrazol-5-yl) showing the two different types of $Mn^{+2}$ sites exposed in 10-Å-wide channels. Site I is five-coordinate, while site II is only two-coordinate; the separation between them is 3.4 Å. (Right) Lewis acid catalyzed conversion of selected aldehydes and ketones with cyanotrimethylsilane to the corresponding cyanosilylated products. Orange, green, gray, and blue spheres represent Mn, Cl, C, and N atoms, respectively [72]. Reproduced from Horike, S. et al., *J Am Chem Soc* **130**, 5854 (2008), with permission).

#### 3.4.1.2 BRØNSTED ACID CATALYSIS

Ferey et al. reported the Brønsted acid catalysis reaction on Friedal-Crafts benzylation using two different MIL-100 cations, Fe and Cr. Although the structure is identical $[M_3OF_{0.85}(OH)_{0.15}(H_2O)_2(btc)_2]$ in both cases, the Fe-based MIL-100 showed superior performance over that the Cr-based MIL-100 and even surpassed the reported HY and HBEA zeolites [74]. In MIL-100($Cr^{+3}$), the Cr-OH sites showed medium Brønsted acidic strength as confirmed by low-temperature CO chemisorption studies [75].

### 3.4.2 ORGANIC OR PSEUDO-ORGANIC LINKERS AS ACTIVE SITES

The metal complexes with functionalized ligands such as porphyrin building blocks as organic linkers reported the catalytic active in oxidation reactions. The porphyrinocarboxylate frameworks constructed from both Mn(III) and Zn (II) successfully showed for olefin epoxidation reactions [76]. Here, the metal coordinated to the porphyrin nitrogens performed as active sites when compared to inorganic nodes of the MOF. Amino-functionalized MOFs such as IRMOF-3 and [ZnF(Am2Taz)] synthesized from corresponding 2-aminoterepthalic acid and 3,5-diamino-1,2,4-triazole (Am2Taz) with zinc has been successfully used as base catalysts in the Knoevenagel reaction. Both MOFs showed the activity toward Aza-Michael condensation reactions at 25° C with a turnover number (TON) of 1.4 $h^{-1}$ and 0.15 $h^{-1}$ and fatty methyl ester transesterification with a TON of 3.3 $h^{-1}$ and 0.3 $h^{-1}$ at 130° C, respectively [77].

### 3.4.3 CATALYSIS WITH TAILORED MICROENVIRONMENT

Prepared post synthetic modifications of MOFs as described below have led to the creation of a new microenvironment inside their porous network for accelerating the rates of specific reactions and enhancing the selectivity to the desired product.

#### 3.4.3.1 ENCAPSULATION OF ACTIVE SPECIES

The encapsulation of active species such as metal complexes or metal nanoparticles (MNPs) inside the porous MOFs is of growing interest in the catalysis community because MOFs can provide a large pore volume that can accommodate guest molecules. Here the MOF porous framework is used as a support or host for the catalytic species, which was positioned in the cavity by noncovalent interactions using a ship-in-a-bottle or bottle-around-the-ship approach [68, 78]. The encapsulation of MNPs inside MOFs can be achieved by a stepwise process of particle infiltration followed by decomposition. Here, the pore size, shape, and channel structure govern the size and shape of the nanoparticle. It has also been reported that simple grinding of the organometallic complex and MOF could result in the encapsulation inside the MOF for aerobic oxidation of alcohols [79].

BASF is industrially producing MOF-5 mainly for encapsulating platinum, palladium, copper, and gold nanoparticles for catalysis applications. The Pd@MOF-5 is used as a heterogeneous catalyst in hydrogenation reactions, specifically hydrogenation of cyclooctene, while Cu@MOF-5 is used in methanol production from syngas [80].

Another example is encapsulation of metal porphyrins into the MOF cavities to be used for oxidation of cyclohexane to corresponding cyclohexanol and cyclohexanone. The reactivity was comparable to that of reactions carried out using a free metal porphyrin complex [81]. The mesoporous MOFs, chromium (III) terephthalate Cr-MIL-101 [82], and iron anologue Fe-MIL-101 [83] have a rigid zeotype crystal structure with quasi-spherical cages of 2.9 and 3.4 nm and have been demonstrated to have very good encapsulation capabilities. Because of their high thermal stabilities, huge surface areas, and large pore volumes, these materials played a greater role as host materials for active species. Kholdeeva et al. showed the encapsulation of polyoxometalates $[PW_{11}TiO_{39}]^{5-}$ and $[PW_{11}CoO_{39}]^{5-}$, which accelerates the oxidation of alpha-pinene to corresponding alcohol and ketones using hydrogen peroxide (Fig. 3.10) [84].

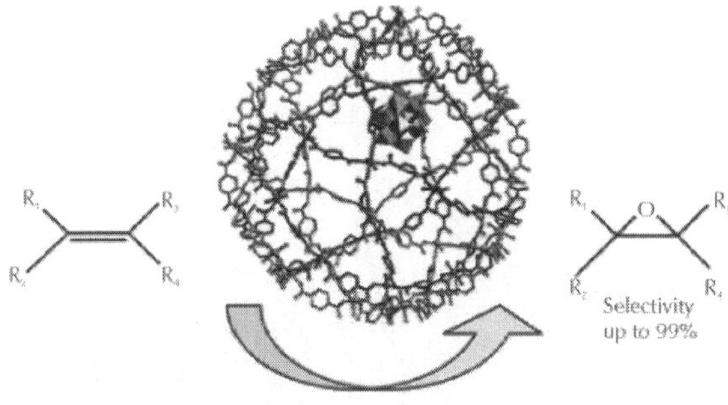


**FIGURE 3.10** The catalytic activities of polyoxotungstate-encapsulated MIL-101 ($PW_x$/MIL101) for selective oxidation of alkenes with aqueous hydrogen peroxide [85]. (Reproduced from N. V. Maksimchuk, N.V. et al., *Inorg Chem* **49**, 2920–2930 (2010), with permission).

### 3.4.3.2 POST SYNTHETIC MODIFICATION

A new synthetic strategy of incorporating active functional groups using a postsynthetic modification approach has been proposed recently as an alternative to presynthetic approaches discussed above. Cohen et al. used this approach to modify the amino functional groups in IRMOF-3 with alkyl anhydrides [86]. The amino groups within MOFs have been synthetically modified using carboxylic acid or isocyanates to introduce various functions. Similarly, Ferey et al. introduced thermally stable amino groups into MIL-101 framework to the unsaturated Cr (III) coordination centers that open new opportunities for introduction of organic functional groups [87]. He showed the successful introduction of ethylenediamnie (EDTA) (ED@MIL-101) and diethylenetriamine (APS@MIL-101) into the MIL-101 framework. The free terminal $NH_2$ groups hanging in the cavities showed Bronsted basicity in the Knoevenagel condensation reactions with 99.3% selectivity. Using a similar approach, Corma et al. introduced a Au(III) complex with a Schiff base, and the material showed superior activity and selectivity in domino-coupling reactions [88]. The catalyst is completely recyclable and showed higher catalytic activity when compared to gold-based homogeneous and supported catalysts. Eddaoudi et al. demonstrated a versatile platform by encapsulation of free-base porphyrin in a large pore of an In-HImDC-based *rho*-ZMOF, which post-synthetically modified by various transition metals to produce a wide array of encapsulated metalloporphyrins in MOFs and successfully tested for oxidation reactions (Fig. 3.11) [89].

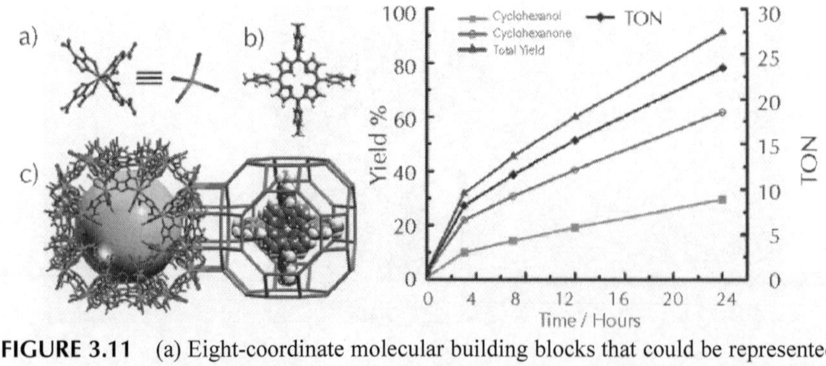


**FIGURE 3.11** (a) Eight-coordinate molecular building blocks that could be represented as tetrahedral building units (b) $[H_2TMPyP]^{4+}$ porphyrin (c) crystal structure of *rho*-ZMOF, schematic representation of $[H_2TMPyP]^{4+}$ porphyrin ring enclosed in *rho*-ZMOF α-cage (drawn to scale) (right) catalytic oxidation of cyclohexane at 65°C (right) [89]. (Reproduced with permission from Alkordi, M.H. et al., *J Am Chem Soc* **130,** 12639 (2008)).

### 3.4.4 MOFS FOR SENSING APPLICATIONS

A sensor is a device that responds to a physical or chemical stimulus (e.g., heat, light, sound, pressure, chemical vapor) and transmits a resulting impulse. As described earlier MOFs have high surface area and porosity, structural diversity, flexibility, and physical and chemical tailorability that render them excellent candidates for sensing applications. In MOF-based sensors, the MOF responds to external stimulation such as gravimetric, mechanical, optical or environmental changes by manifesting a change in structure and properties or by showing a guest-dependent response. Although the latter is the most common case, an example of structural change was observed in MOF-5. The framework volume of MOF-5 changes with external temperature, showing a linear negative thermal expansion ($-16 \times 10^{-6}\ K^{-1}$) between 4 K and 600 K. This expansion is due to the large-amplitude transverse vibration of the carboxylate groups. A large number of examples of MOF-based sensors show the guest-dependent response [90]. MOF samples have displayed optical, magnetic, or electronic responses based on the interactions between the framework structure and a guest molecule, which opens up a large number of applications of MOFs as sensing devices [91].

Depending on their response mechanism or route, MOF sensors can be classified into three groups: piezoresistive response, mass response, and optical response. The first relies on the stress induced by a guest molecule and takes advantage of the structural flexibility of MOF materials. Mass response takes advantage of the large surface area and available volume capacity of the pores to produce an important change in host mass when capturing a given analyte. Optical responses, including photoluminescence quenching, emission wavelength shifts, or simple color changes due to a guest-induced change in MOF optical absorption spectrum, are reported on extensively in the literature.

### 3.4.5 MOF SENSORS BASED ON PIEZORESISTIVITY

Sensors based on piezoresistive MOFs are promising sensing devices for detecting gas molecules and small organic molecules. One of the attractive features of MOF materials is the structural flexibility. A number of MOF materials exhibit expansion and contraction of the crystal lattice induced by guest–host interactions. This property can be evidenced by "gating effects" and "breathing effects" due to expansion and contraction of the framework upon incorporation of guest molecules. The unit cell dimensions of some MOFs can vary by as much as 10% when molecules are adsorbed within their pores [92]. Therefore, transduction mechanisms in, which distortions in a MOF thin film creates stress at the interface with a second material can be used to recognize and measure a number of analytes. By depositing a MOF film on the surface of a static microcantilever, stress at the cantilever surface results in bending that can be detected by means of a built-in piezoresistive sensor [93]. In this fashion, MOFs can show effective recognition chemistries for a variety of gases and vapors provided they interact with the MOF pore.

An example is HKUST-1, which, although small, shows adsorption-induced distortions when water, methanol, or ethanol are reversibly adsorbed. Allendorf et al. showed this property when a thin film of HKUST-1 was integrated with a microcantiveler surface [94]. The time-dependent responses to $H_2O$ are shown in Fig. 3.12. It was also demonstrated that the sensor responds to $CO_2$ when the MOF layer is dehydrated. The stress in this case was due to the coordination of $CO_2$ with unsaturated copper sites in HUKST-1. Because the framework distortions were quite small, the group concluded that higher sensitivities can be achieved using more flexible MOFs.

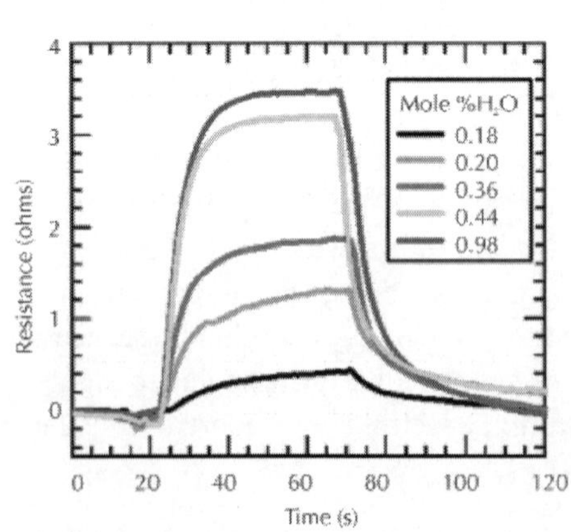


**FIGURE 3.12** Temporal response of the cantilever piezoresistive sensor to water vapor diluted in $N_2$ (room temperature, 1 atm) [94]. (Reproduced with permission Allendorf, M.D. et al., *J Am Chem Soc* **130,** 14404 (2008), with permission).

### 3.4.6 GRAVIMETRIC-BASED MOF SENSORS

Thin films of MOF materials can also be applied on quartz crystal microbalances (QCMs). In this fashion, it is possible to take advantage of the extremely high sensitivity to mass changes on the QCM surface when an analyte interacts with the cavities of the MOF. Detection limits on the order of the nanogram scale have been demonstrated [95]. MOF thin films grown on a QCM substrate can be widely adopted as a device to probe mass changes due to gas or vapor adsorption/desorption in the highly porous sorbent material. Liu et al. successfully used a stepwise liquid-phase epitaxial method to synthesize two homochiral thin films of $[Zn_2(\pm)cam_2dabco]_n$ (where $\pm$cam = (1R,3S)—($\pm$)-camphoric acid, and dabco = 1,4-diazabicyclo(2.2.2)octane) on self-assembled monolayers (SAMs) functionalized QCM substrates [96]. In another example, coating a QCM substrate with HKUST-1 showed to detect humidity variations via adsorption/desorption of water molecules on the sorbent film [97].

### 3.4.7 SURFACE ACOUSTIC WAVE-BASED MOF SENSORS:

Surface acoustic wave (SAW) sensors [93] are robust devices extensively employed for chemical sensing. Depositing MOF materials on the sensor surface can then be useful for detecting and measuring vapors and gases. A frequency shift of acoustic waves traveling parallel to the MOF surface, which are generated by an oscillator can be used to detect gas or vapor sorption. By growing a HKUST-1 film using a layer-by-layer (LBL) method on the quartz of 96.5-MHz devices, Robinson and collaborators reported humidity detection using SAWs (Fig. 3.13) [98]. Fast and reproducible detection on frost points (FP) as high as $10^\circ$ C and as low as $-70^\circ$ C (2.6 ppmv and 12,300 ppmv at an atmospheric pressure of 625 Torr, respectively) can be obtained. In addition, the group demonstrated three orders of magnitude better response to humidity using HKUST-1 as compared to the same coating on QCMs [97, 99]. The influence of film thickness on the SAW sensor response also was reported. Optimum coupling occurred between the MOF film and SAW surface; hence, the highest response of the HKUST-1 SAW sensor to water vapor was obtained for LBL coatings between 40 and 50 cycles (150-to 180-nm film thickness).

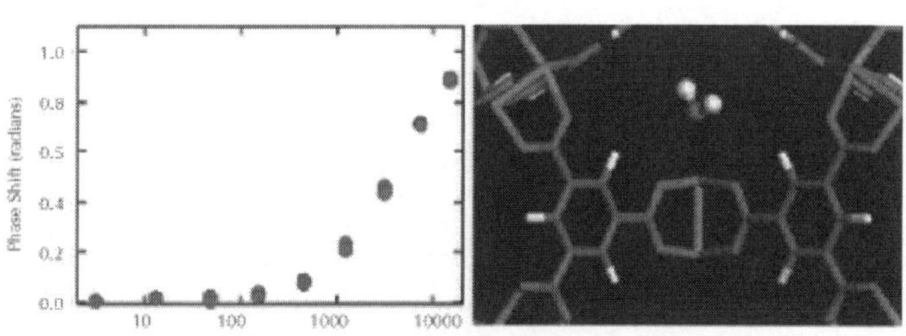


**FIGURE 3.13** Humidity detection (left) and a molecular model of $Cu_3(BTC)_2$ (HKUST-1) SAWs with water molecules (right) [98]. (Reproduced from Robinson, A.L. et al., *Anal Chem* **84,** 7043 (2012), with permission).

## 3.5 SENSORS BASED ON MOF OPTICAL PROPERTIES

### 3.5.1 PHOTOLUMINESCENCE (PL)-BASED MOF SENSORS

PL-MOFs are the most widely explored type of MOF sensing devices. The popularity of luminescence over other transduction mechanisms are a consequence of several key elements, including very high detection limits that can reach molecular level. In addition, there is no need for film fabrication or other processing, so conventional solvothermal synthesis is used to produce PL-MOFs. A very important advantage over other PL sensors is the porosity of MOFs. In addition, the possibility of adjusting MOF sorption properties offers a high degree of molecular specificity. As an example, Li et al. prepared a luminescent Ln-based MOF displaying efficient turn-on triggered by solvent vapors, showing good selectivity for DMF vapor [100].

In another example, Lanet al. reported the detection of trace nitroaromatic explosives (DNT, 2,4-dinitrotoluene and DMNB, 2,3-dimethyl-2,3-dinitrobutane) in the vapor phase using a highly luminescent MOF material $[Zn_2(bpdc)_2(bpee)]_n$ (bpdc = 4,40-biphenyldicarboxylate; bpee = 1,2-bipyridylethene) thin film where they described the sensing process as a redox quenching mechanism [101].

### 3.5.2 SOLVATOCHROMISM/VAPOCHROMISM

A visible change in a material's color is one of the simplest means of transducing a sensing signal. Solvatochromism and vapochromism refers to a large shift in the absorption spectrum of a material in response to a change in the identity of the solvent or vapor. As in any other optical active material, MOFs constructed from ligands that are chromosphors should behave in a similar fashion in the presence of vapors. The electronic transition responsible for the coloration entails charge transfer (i.e., a change in dipole moment upon excitation from the ground electronic state to the excited electronic state of the chromophoric component of the material) in a MOF material, the organic linker. If the ground state has the larger dipole moment, hypsochromic shifts (blue shifts) occur with increasing solvent polarity. On the other hand, if the excited state possesses a larger dipole moment than the ground state, it is preferentially stabilized by polar solvents, and bathochromic shifts (red shifts) are observed with increasing solvent polarity.

A square shaped nanotubular MOF with the stoichiometry $[[(WS_4Cu_4)I_2(dptz)_3]\cdot DMF]_n$ (124·DMF) in, which each $WS_4^{2-}$ anion chelates four Cu(I) cations was synthesized by Lu et al. and its optical properties studied. The sorbent material shows individual $WS_4Cu_4^{2+}$ units that are paired up by pyrazines that exhibit $\pi-\pi$ interactions at an interplanar distance of 3.578(2) Å. Also, each Cu(I) center is tetrahedrally coordinated. The 5.4-× 5.3-Å nanotubes along the c-axis contain DMF guests and, once DMF is removed, immersion of the MOF into various organic solvents manifested significant color changes. The group described a negative solvatochromic effect with a solvent-induced absorption band shift of 245 nm between $CH_3CN$ and $CHCl_3$ (Fig. 3.14) [102].

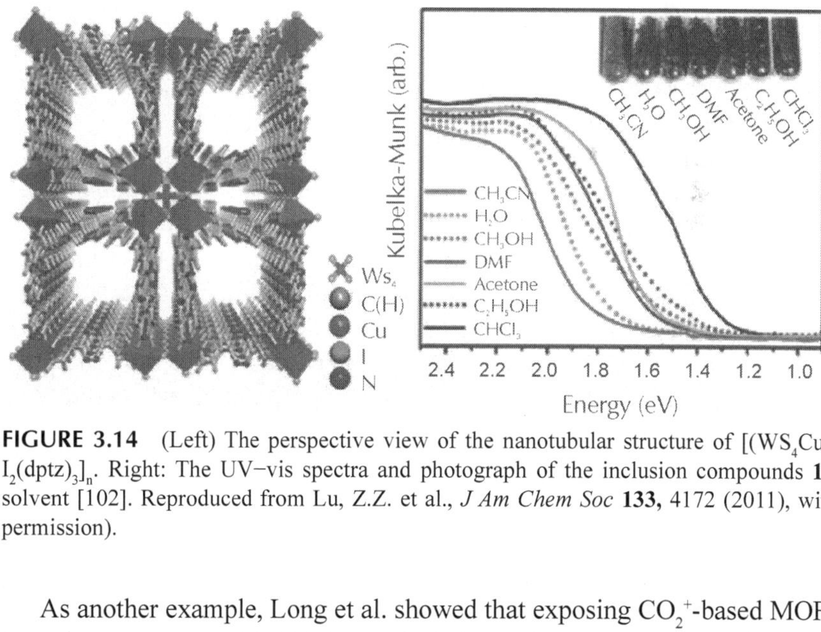


**FIGURE 3.14** (Left) The perspective view of the nanotubular structure of $[(WS_4Cu_4)I_2(dptz)_3]_n$. Right: The UV–vis spectra and photograph of the inclusion compounds 1⊃solvent [102]. Reproduced from Lu, Z.Z. et al., *J Am Chem Soc* **133,** 4172 (2011), with permission).

As another example, Long et al. showed that exposing $CO_2^+$-based MOFs to various vapors could shift the optical absorption across the visible region [103]. The explanation was a change in coordination environment from the as-synthesized octahedral to a tetrahedral geometry. Lee et al. proposed a similar color change mechanism on a MOF material that detects chloride ions derived from chlorine-containing vapors or gasses. The MOF contains $CO_2^+$ nodes coordinated to 1,2,4,5-tetra(2H-tetrazole-5-yl)-benzene (TTB) struts as well as Br-anions [104]. The as-synthesized material is characterized by a visible absorption peak at 475 nm, suggesting octahedral coordination of the $CO_2^+$ centers. Interestingly, exposure of the MOF material to chlorine-containing gasses, including HCl, $SOCl_2$ $(COCl)_2$, and $COCl_2$ (phosgene) show color changes from red to blue [104]. The blue color, which is due to a new absorption at 670 nm was attributed to tetrahedrally coordinated $CO_2^+$. To explain the change in metal coordination geometry, it was hypothesized that the Br-is replaced with Cl-originating from the reactive gasses.

## 3.6 MOFS FOR OIL SPILL CLEANUP

The world energy dependence on fossil energy continues to increase. Petroleum is the largest source of primary energy production in the world. The world petroleum consumption of 87,135,100 barrels/day (2010) poses a daunting task to use aggressive methods for producing and transporting the crude oil by various means including pipeline, ship, or barge to the rest of the world from production sites [105]. Production, transportation, and storage of oil present the risk of oil spill.

The first commercial oil spill dates back to 1967 in the United Kingdom; the most recent but not the worst (BP Deepwater Horizon Oil spill in the Gulf of Mexico) occurred in 2010, and its cleanup cost is estimated to be over $20 billion [106]. Oil spills not only cause financial burden but also pose an environmental damage risk. Oil is highly hydrophobic, does not mix with water, and forms a thick slick or droplets that float on water. With time the oil eventually becomes spreadout. Although no two oil spills are considered the same due to their location, weather conditions, and oil type, current oil spill cleanup methods include three main methods of response (i) booming (ii) skimming, and (iii) centrifuge in, which physical collection is routine (see Ref. [1], of Chapter 8). Use of dispersants that break the oil to speed up the natural biodegradation and burning of oil are considered as potential alternatives. However, the use of dispersants and burning often results in adverse effects on the environment. One another approach is to use the porous sorbent that can effectively absorb the oil [107, 108]. A porous sorbent for oil spill cleanup should have the following characteristic properties: (i) high rate of adsorption/absorption (ii) oil retention, and (iii) ease of application. Currently used sorbents include sand, organic clays, and cotton fibers. These adsorbents are cheap and easy to apply; however, they suffer from their affinity toward water, limiting their performance during cleanup.

In general, to be effective for oil cleanup, a porous material should be hydrophobic and olephilic. MOFs with excellent porosities and hydrophobicity are likely to be efficient hydrocarbon absorbers at low to high concentrations. Focusing on the potential industrial importance, Motkuri et al. recently reported for the first time breathing and adsorption of hydrocarbons using a novel flexible MOF (TetZB) [22]. They observed framework expansion and contraction upon uptake and release of hydrocarbons. Although framework expansion properties in flexible MOFs are not of interest for cleanup applications, they have potential uses in sensing and other applications. In general, typical hydrocarbon sorption experiments are conducted on gravimetric analyzers (Intelligent Gravimetric Analyzer (IGA), Hiden Instruments or duPont Model 990 TGA). Variable-temperature powder x-ray diffraction of TetZB has revealed that there is framework distortion (probably by contraction) by removing the coordinated solvent on metals. Figure 3.15 illustrates the adsorption and desorption of hydrocarbons on TetZBat room temperature. **TetZB** exhibits an open structure when the solvent is removed after activation. Upon exposure to the hydrocarbons, the narrow pores in the framework are saturated, while at high relative pressures, the framework undergoes expansion by opening access to the large pores [22].

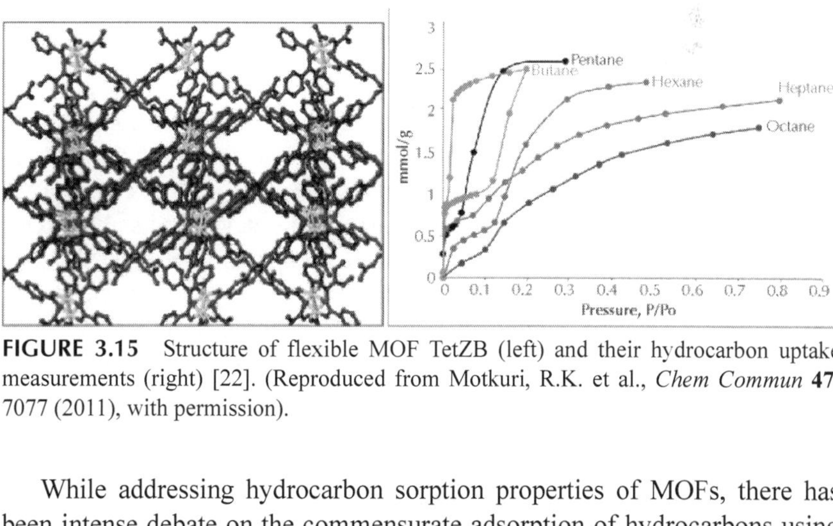


**FIGURE 3.15** Structure of flexible MOF TetZB (left) and their hydrocarbon uptake measurements (right) [22]. (Reproduced from Motkuri, R.K. et al., *Chem Commun* **47**, 7077 (2011), with permission).

While addressing hydrocarbon sorption properties of MOFs, there has been intense debate on the commensurate adsorption of hydrocarbons using MOFs; the mean number of adsorbed hydrocarbon molecules relates to the symmetry of the framework topology, which is nearly impossible in routinely studied/used zeolites [109]. Jing Li et al. reported that adsorption capacity and location of molecules are specifically controlled by cavity size, shape, symmetry, and channel features such as channel segment [109]. Three different adsorption mechanisms have been offered for the sorption and separation of hydrocarbons: (1) equilibrium (2) steric, and (3) kinetic. In some complex materials, it is possible that one or more mechanisms may operate at the same time. The equilibrium sorption and separation mechanism relies on the difference in the relative quantities of various hydrocarbons; the size and shape of the pores in the adsorbate contribute to the steric mechanism, and the difference in the rate of adsorption of different hydrocarbons contribute to the kinetic mechanism [109].

Generally, MOFs have high affinity toward water, limiting their applications; however, the recent Deepwater Horizon oil spill devastation has increased awareness of the need for new adsorbents. Significant efforts have been devoted to develop new hydrophobic MOFs with high thermal stability, high selectivity, and fast regenerability. A prominent approach to circumvent this problem is the utilization of postsynthetic modification (PCM) of MOFs to produce moisture resistant and superhydrophobic MOFs [110]. This approach garnered significant attention because it is an important tool to chemically modify or introduce various functional groups to produce new MOFs with completely new chemical and physical properties. It means chemical modification is performed on the final MOF materials rather than in the precursor used for MOF construction. This approach is particularly attractive because (1) the use of solvothermal conditions for MOF synthesis limits the presence of functional groups on the ligand and (2) unlike any inorganic porous materials, MOFs have organic component that can be functionally modified, rendering change in physical and chemical properties [110]. To circumvent the posed by water-unstable MOFs, Yang et al. recently developed stable MOFs using novel ligands that contain perfluorinated groups (Fig. 3.16) [111]. These new MOFs (FMOFs) are hydrogen free and fluorine rich, not only bestowing greater stability but also offering favorable hydrocarbon sorption properties while inducing hydrophobic characteristics. FMOF-1 is the first known MOF with an exceptional hydrophobic nature while offering affinity for typical aromatic hydrocarbons and aliphatic oil components. To the best of our knowledge, this finding is definitely a significant scientific achievement, considering the fact that many known MOFs are not stable in the presence of water or they uptake very large quantities of water that compromise the uptake of hydrocarbons during cleanup. Water adsorption isotherms clearly illustrate that the FMOF-1 is significantly hydrophobic, while routinely studied/used inorganic porous materials such as activated carbon and zeolite-5A are hydrophilic, absorbing water even at very low relative humidity [111]. FMOF-1 is not only hydrophobic but also very stable under aggressive conditions. FMOF-1 has been soaked in distiled water for several days; powder X-ray diffraction images of soaked materials resemble the untreated material, indicating the exceptional stability.

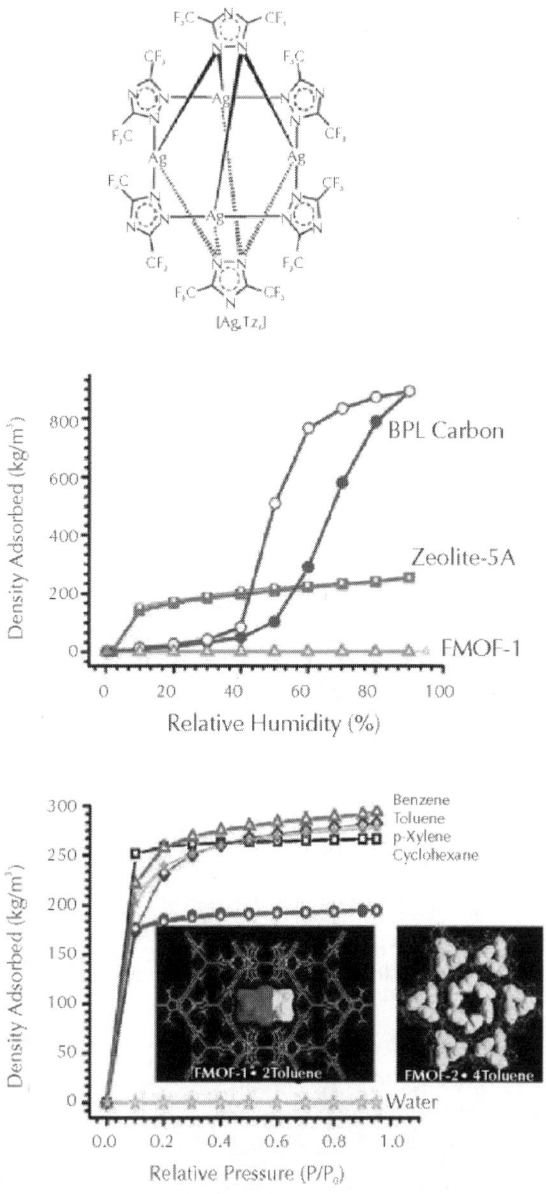


**FIGURE 3.16** (Left) Building blocks of FMOF-1 (middle) water sorption studies in FMOF-1, BPC-Carbon and Zeolite-5A; (right) water and hydrocarbon sorption studies in FMOF-1 [111]. (Reproduced from Yang, C. et al., *J Am Chem Soc* **133,** 18094 (2011), with permission).

In addition to oil cleanup, MOFs have been extensively investigated for hydrocarbon separation applications due to the fact that currently available separation strategies are energy-intensive processes and are contributing to the largest segment of industrial production costs [112]. A recent report using the flexible feature that we reported earlier unveils the exceptional selectivity of flexible MOFs based on the differences in their gate opening pressure [22, 113]. The selectivity depends on the chain length of the hydrocarbon and its specific interaction with framework [113].

## 3.7 CONCLUSIONS

Although the synthesis of MOFs was initially established for making new compounds with interesting properties, the field has matured well and is slowly widening its scope. The surface areas and pore properties of MOFs demonstrate that these materials have large applicability in both fundamental and industrial arenas. The chemical versatility, pore tunability and tailoring of active sites will lead to a broader range of applications. Although the conventional hydro/solvothermal synthesis was well used in MOF synthesis, alternative and new methods such as electro, mechano, sonochemical, and microwave-assisted synthesis are just emerging in this area. However, greater care must be taken while considering the crucial points in the synthesis, specifically the reaction conditions and input energy requirements. Recent developments of high-throughput methods may provide the tools needed in establishing the standard synthetic protocol for each MOF, but establishing these protocols takes time. With the developments in myriads of MOFs and their applications, it is time to take them from the laboratory to the industrial world and focus on applying this knowledge to industrial scenarios for real-world applications. Although there is some progress in scale-up of MOFs for industrial production and use, much still needs to be learned from zeolite chemistry for commercial success.

In gas capture and separation-related topics, the current status of MOF research is focused primarily on exploring new materials for single sorption properties that cannot be viewed as new sorbent materials for practical separation applications. The breakthrough experiments with mixed gases (including vapors) will be a valuable means for better understanding the materials for long-term potential and applicability. In the heterogeneous catalysis area, MOFs still are unable to compete with zeolites in high-temperature reactions, but these materials are capable of replacing zeolites in catalysis reactions that use mild reaction conditions in fine chemical and enantio-selective synthesis.

MOF materials can respond to external stimulation in a number of ways including changes in structure, as well as in their optical or mechanical properties, which can be employed for the fabrication of a number of sensing devices. Although some examples have been shown, their application in this area is still in the early stage. Because MOFs offer very attractive properties, there is a great potential for developing novel MOF-based sensors whose detection limits will depend on the type of stimulation the material is responsive to. Another attractive feature of MOFs is that their physical and chemical properties can be altered by postsynthesis modification; a technique not possible in any other class of porous inorganic materials thus will help in developing suitable materials for required applications.

Most of the current technological options for the recovery and separation of industrially important gases including $CO_2$ and $H_2$ entail the formation of molecular complexes, through physical and/or chemical interactions, that must then be reversed through significant energy input, which is wasteful both thermodynamically and dynamically. While continued improvement in the above technologies can be expected with further R & D, the new options with novel materials like MOFs and their analogs could provide significant breakthroughs in intensification of both chemical processes and separations by tuning to vary cavity size and functionalizations to create a more effective microenvironment inside their porous structure and to enhance accessibility and interactions with molecules confined within the tailored environment. As there is a great deal of flexibility in tailoring their structural framework and functionalizations, it is likely that MOFs will play much more important role in the future to achieve higher level of process intensification leading to greener process.

## KEYWORDS

- **Adsorption chiller**
- **$CO_2$ capture**
- **Gas capture and separation**
- **Heterogeneous catalysis**
- **Metal-organic frameworks**
- **Oil spill cleanup**
- **Sensing applications, sensors**
- **Water sorption**

## ACKNOWLEDGMENTS

R. K. Motkuri would like to thank Dr. K.V. Raghavan for the support over the years. This review was not possible without support from U.S. Department of Energy's office of Advanced Research Projects Agency for Energy (ARPA-e) under Building Energy Efficiency through Innovative Thermodevices (BEET-IT) program. Pacific North-west National Laboratory is a multiprogramming laboratory operated by Battelle Memorial Institute for the U.S. Department of Energy under contract DE-AC05–76RL01830.

## REFERENCES

1. Sing, K. S. W., Everett, D. H., Haul, R. A. W., Moscou, L., Pierotti, R. A., Rouquerol, J., and Siemieniewska, T. Reporting physisorption data for gas solid systems with special reference to the determination of surface-area and porosity (Recommendations 1984). *Pure Appl Chem* **57,** 603–619 (1985).
2. (a) Corma, A. From microporous to mesoporous molecular sieve materials and their use in catalysis. *Chem Rev* **97,** 2373–2419 (1997); (b) Erdem-Senatalar, A., Bergendahl, J. A., Giaya, A., and Thompson, R. W. Adsorption of methyl tertiary butyl ether on hydrophobic molecular sieves. *Environ Eng Sci* **21,** 722–729 (2004).
3. (a) Kitagawa, S., Kitaura, R., and Noro, S. Functional porous coordination polymers. *Angew Chem Int Edit* **43,** 2334–2375 (2004); (b) Rowsell, J. L. C. and Yaghi, O. M. Metal-organic frameworks: a new class of porous materials. *Micropor Mesopor Mat* **73,** 3–14 (2004); (c) Ferey, G. Hybrid porous solids: past, present, future. *Chem Soc Rev* **37,** 191– 214 (2008).
4. (a) Tranchemontagne, D. J., Mendoza-Cortes, J. L., O'Keeffe, M., and Yaghi, O. M., Secondary building units, nets and bonding in the chemistry of metal-organic frameworks. *Chem Soc Rev* **38,** 1257–1283 (2009); (b) Natarajan, S. and Mahata, P. Metal-organic framework structures-how closely are they related to classical inorganic structures? *Chem Soc Rev* **38,** 2304–2318 (2009).
5. (a) Furukawa, H., Ko, N., Go, Y. B., Aratani, N., Choi, S. B., Choi, E., Yazaydin, A. O., Snurr, R. Q., O'Keeffe, M., Kim, J., and Yaghi, O. M. Ultrahigh porosity in metal-organic frameworks. *Science* **329,** 424–428 (2010); (b) Deng, H. X., Grunder, S., Cordova, K. E., Valente, C., Furukawa, H., Hmadeh, M., Gandara, F., Whalley, A. C., Liu, Z., Asahina, S., Kazumori, H., O'Keeffe, M., Terasaki, O., Stoddart, J. F., and Yaghi, O. M. Large-pore apertures in a series of metal-organic frameworks. *Science* **336,** 1018–1023 (2012).
6. Stock, N. and Biswas, S. Synthesis of metal-organic frameworks (MOFs): Routes to various MOF topologies, morphologies, and composites. *Chem Rev* **112,** 933–969 (2012).
7. Li, J. R., Sculley, J., and Zhou, H. C. Metal-organic frameworks for separations. *Chem Rev* **112,** 869–932 (2012).
8. Knobloch, F. W. and Rauscher, W. H. Coordination polymers of Copper(II) prepared at liquid-liquid interfaces. *J Polym Sci* **38,** 261–262 (1959).

9. Johnson, J. W., Jacobson, A. J., Butler, W. M., Rosenthal, S. E., Brody, J. F., and Lewandowski, J. T. Molecular recognition of alcohols by layered compounds with alternating organic and inorganic layers. *J Am Chem Soc* **111,** 381–383 (1989).
10. Drumel, S., Janvier, P., Barboux, P., Bujolidoeuff, M., and Bujoli, B. Synthesis, structure, and reactivity of some functionalized zinc and Copper(II) phosphonates. *Inorg Chem* **34,** 148–156 (1995).
11. Riou, D., Roubeau, O., and Ferey, G. Composite microporous compounds. Part I: Synthesis and structure determination of two new vanadium alkyldiphosphonates (MIL-2 and MIL-3) with three-dimensional open frameworks. *Micropor Mesopor Mat* **23,** 23–31(1998).
12. Li, H., Eddaoudi, M., O'Keeffe, M., and Yaghi, O. M. Design and synthesis of an exceptionally stable and highly porous metal-organic framework. *Nature* **402,** 276–279 (1999).
13. Klinowski, J., Paz, F. A. A., Silva, P., and Rocha, J. Microwave-assisted synthesis of metal-organic frameworks. *Dalton T* **40,** 321–330 (2011).
14. Li, C. P., Tian, Y. L., and Guo, Y. M. Solvent-regulated assembly of 1-D and 2-D Zn(II) coordination polymers with tetrabromoterephthalate. *Inorg Chem Commun* **11,** 1405–1408 (2008).
15. Li, Z. Q., Qiu, L. G., Xu, T., Wu, Y., Wang, W., Wu, Z. Y., and Jiang, X. Ultrasonic synthesis of the microporous metal-organic framework Cu(3)(BTC)(2) at ambient temperature and pressure: An efficient and environmentally friendly method. *Mater Lett* **63,** 78–80 (2009).
16. Klimakow, M., Klobes, P., Thunemann, A. F., Rademann, K., and Emmerling, F. Mechanochemical synthesis of metal-organic frameworks: A fast and facile approach toward quantitative yields and high specific surface areas. *Chem Mater* **22,** 5216–5221 (2010).
17. Joaristi, A. M., Juan-Alcaniz, J., Serra-Crespo, P., Kapteijn, F., and Gascon, J. Electrochemical synthesis of some archetypical $Zn^{2+}$, $Cu^{2+}$, and $Al^{3+}$ metal organic frameworks. *Cryst Growth Des* **12,** 3489–3498 (2012).
18. Carne-Sanchez, A., Imaz, I., Cano-Sarabia, M., and Maspoch, D. A spray-drying strategy for synthesis of nanoscale metal-organic frameworks and their assembly into hollow superstructures. *Nat Chem* **5,** 203–211 (2013).
19. Liu, Y., Her, J. H., Dailly, A., Ramirez-Cuesta, A. J., Neumann, D. A., and Brown, C. M. Reversible structural transition in MIL-53 with large temperature hysteresis. *J Am Chem Soc* **130,** 11813–11818 (2008).
20. Beurroies, I., Boulhout, M., Llewellyn, P. L., Kuchta, B., Ferey, G., Serre, C., and Denoyel, R. Using pressure to provoke the structural transition of metal-organic frameworks. *Angew Chem Int Edit* **49,** 7526–7529 (2010).
21. Thallapally, P. K., Tian, J., Kishan, M. R., Fernandez, C. A., Dalgarno, S. J., McGrail, P. B., Warren, J. E., and Atwood, J. L. Flexible (breathing) interpenetrated metal-organic frameworks for $CO_2$ separation applications. *J Am Chem Soc* **130,** 16842–16843 (2008).
22. Motkuri, R. K., Thallapally, P. K., Nune, S. K., Fernandez, C. A., McGrail, B. P., and Atwood, J. L. Role of hydrocarbons in pore expansion and contraction of a flexible metal-organic framework. *Chem Commun* **47,** 7077–7079 (2011).

23. (a) Kitagawa, S. and Uemura, K. Dynamic porous properties of coordination polymers inspired by hydrogen bonds. *Chem Soc Rev* **34,** 109–119 (2005); (b) Ferey, G. and Serre, C. Large breathing effects in three-dimensional porous hybrid matter: facts, analyzes, rules and consequences. *Chem Soc Rev* **38,** 1380–1399 (2009).
24. (a) Llewellyn, P. L., Horcajada, P., Maurin, G., Devic, T., Rosenbach, N., Bourrelly, S., Serre, C., Vincent, D., Loera-Serna, S., Filinchuk, Y., and Ferey, G. Complex adsorption of short linear alkanes in the flexible metal-organic-framework MIL-53(Fe). *J Am Chem Soc* **131**, 13002–13008 (2009); (b) Coudert, F. X., Boutin, A., Jeffroy, M., Mellot-Draznieks, C., and Fuchs, A. H. Thermodynamic methods and models to study flexible metal-organic frameworks. *Chemphyschem* **12,** 247–258 (2011).
25. Wang, Z. Q. and Cohen, S. M. Postsynthetic modification of metal-organic frameworks. *Chem Soc Rev* **38,** 1315–1329 (2009).
26. Yilmaz, B., Trukhan, N., and Muller, U. Industrial outlook on zeolites and metal organic frameworks. *Chinese J Catal* **33,** 3–10 (2012).
27. (a) Zhao, D., Yuan, D. Q., and Zhou, H. C. The current status of hydrogen storage in metal-organic frameworks. *Energ Environ Sci* **1,** 222–235 (2008); (b) Murray, L. J., Dinca, M., and Long, J. R. Hydrogen storage in metal-organic frameworks. *Chem Soc Rev* **38,** 1294–1314 (2009); (c) Rowsell, J. L. C. and Yaghi, O. M. Strategies for hydrogen storage in metal-organic frameworks. *Angew Chem Int Edit* **44,** 4670–4679 (2005).
28. (a) Rowsell, J. L. C., Eckert, J., and Yaghi, O. M. Characterization of H-2 binding sites in prototypical metal-organic frameworks by inelastic neutron scattering. *J Am Chem Soc* **127,** 14904–14910 (2005); (b) Ye, Q., Song, Y. M., Wang, G. X., Chen, K., Fu, D. W., Chan, P. W. H., Zhu, J. S., Huang, S. D., and Xiong, R. G. Ferroelectric metal-organic framework with a high dielectric constant. *J Am Chem Soc* **128,** 6554–6555 (2006); (c) Rosi, N. L., Eckert, J., Eddaoudi, M., Vodak, D. T., Kim, J., O'Keeffe, M., and Yaghi, O. M. Hydrogen storage in microporous metal-organic frameworks. *Science* **300,** 1127–1129 (2003).
29. Alvaro, M., Carbonell, E., Ferrer, B., Xamena, F. X. L. I., and Garcia, H. Semiconductor behavior of a metal-organic framework (MOF). *Chem-Eur J* **13,** 5106–5112 (2007).
30. Klontzas, E., Mavrandonakis, A., Froudakis, G. E., Carissan, Y., and Klopper, W. Molecular hydrogen interaction with IRMOF-1: A multiscale theoretical study. *J Phys Chem C* **111,** 13635–13640 (2007).
31. (a) Younglove, B. A. Thermo-physical properties of fluids.1. Argon, Ethylene, Para-Hydrogen, Nitrogen, Nitrogen Trifluoride, and Oxygen. *J Phys Chem Ref Data* **11,** 1–353 (1982); (b) Kaye, S. S., Dailly, A., Yaghi, O. M., and Long, J. R. Impact of preparation and handling on the hydrogen storage properties of Zn4O(1,4-benzenedicarboxylate)(3) (MOF-5). *J Am Chem Soc* **129,** 14176–14177 (2007).
32. van den Berg, A. W. C. and Arean, C. O. Materials for hydrogen storage: current research trends and perspectives. *Chem Commun* 668–681 (2008).
33. (a) Dinca, M., Dailly, A., Liu, Y., Brown, C. M., Neumann, D. A., and Long, J. R. Hydrogen storage in a microporous metal-organic framework with exposed $Mn^{2+}$ coordination sites. *J Am Chem Soc* **128,** 16876–16883 (2006); (b) Hu, Y. H. and Zhang, L. Hydrogen storage in metal-organic frameworks. *Adv Mater* **22,** E117–E130 (2010).

34. (a) Morris, R. E. and Wheatley, P. S. Gas storage in nanoporous materials. *Angew Chem Int Edit* **47,** 4966–4981 (2008); (b) Collins, D. J. and Zhou, H. C. Hydrogen storage in metal-organic frameworks. *J Mater Chem* **17,** 3154–3160 (2007).
35. Cheon, Y. E. and Suh, M. P. Multifunctional fourfold interpenetrating diamondoid network: Gas separation and fabrication of palladium nanoparticles. *Chem-Eur J* **14,** 3961– 3967 (2008).
36. Herm, Z. R., Swisher, J. A., Smit, B., Krishna, R., and Long, J. R. Metal-organic frameworks as adsorbents for hydrogen purification and precombustion carbon dioxide capture. *J Am Chem Soc* **133,** 5664–5667 (2011).
37. (a) Yang, Q. Y. and Zhong, C. L. Molecular simulation of carbon dioxide/methane/hydrogen mixture adsorption in metal-organic frameworks. *J Phys Chem B* **110,** 17776–17783 (**2006**); (b) Babarao, R. and Jiang, J. W. Unprecedentedly high selective adsorption of gas mixtures in rho zeolite-like metal-organic framework: A molecular simulation study. *J Am Chem Soc* **131,** 11417–11425 (2009); (c) Liu, Y. H., Liu, D. H., Yang, Q. Y., Zhong, C. L., and Mi, J. G. Comparative study of separation performance of COFs and MOFs for $CH_4$/$CO_2$/H-2 mixtures. *Ind Eng Chem Res* **49,** 2902– 2906 (2010).
38. Duren, T., Sarkisov, L., Yaghi, O. M., and Snurr, R. Q. Design of new materials for methane storage. *Langmuir* **20,** 2683–2689 (2004).
39. Ma, S. Q., Sun, D. F., Simmons, J. M., Collier, C. D., Yuan, D. Q., and Zhou, H. C. Metal-organic framework from an anthracene derivative containing nanoscopic cages exhibiting high methane uptake. *J Am Chem Soc* **130,** 1012–1016 (2008).
40. Zou, R. Q., Abdel-Fattah, A. I., Xu, H. W., Zhao, Y. S., and Hickmott, D. D. Storage and separation applications of nanoporous metal-organic frameworks. *Crystengcomm* **12,** 1337–1353 (2010).
41. Motkuri, R. K., Thallapally, P. K., McGrail, B. P., and Ghorishi, S. B. Dehydrated Prussian blues for $CO_2$ storage and separation applications. *Crystengcomm* **12,** 4003–4006 (2010).
42. Yazaydin, A. O., Snurr, R. Q., Park, T. H., Koh, K., Liu, J., LeVan, M. D., Benin, A. I., Jakubczak, P., Lanuza, M., Galloway, D. B., Low, J. J., and Willis, R. R. Screening of metal-organic frameworks for carbon dioxide capture from flue gas using a combined experimental and modeling approach. *J Am Chem Soc* **131,** 18198–18199 (2009).
43. Caskey, S. R., Wong-Foy, A. G., and Matzger, A. J. Dramatic tuning of carbon dioxide uptake via metal substitution in a coordination polymer with cylindrical pores. *J Am Chem Soc* **130,** 10870–10871 (2008).
44. Liu, J., Wang, Y., Benin, A. I., Jakubczak, P., Willis, R. R., and LeVan, M. D. $CO_2$/$H_2O$ adsorption equilibrium and rates on metal-organic frameworks: HKUST-1 and Ni/DOBDC. *Langmuir* **26,** 14301–14307 (2010).
45. Liu, B. and Smit, B. Comparative molecular simulation study of $CO_2$/$N_2$ and $CH_4$/$N_2$ separation in zeolites and metal-organic frameworks. *Langmuir* **25,** 5918–5926 (2009).
46. Wu, D., Xu, Q., Liu, D. H., and Zhong, C. L. Exceptional $CO_2$ capture capability and molecular-level segregation in a Li-modified metal-organic framework. *J Phys Chem C* **114,** 16611–16617 (2010).
47. (a) Kishan, M. R., Tian, J., Thallapally, P. K., Fernandez, C. A., Dalgarno, S. J., Warren, J. E., McGrail, B. P., and Atwood, J. L. Flexible metal-organic supramolecular

isomers for gas separation. *Chem Commun* **46,** 538–540 (2010); (b) Tian, J. A., Motkuri, R. K., Thallapally, P. K., and McGrail, B. P. Metal-organic framework isomers with diamondoid networks constructed of a semirigid tetrahedral linker. *Cryst Growth Des* **10,** 5327–5333 (2010); (c) Tian, J. A., Motkuri, R. K., and Thallapally, P. K. Generation of 2D and 3D (PtS, adamantanoid) nets with a flexible tetrahedral building block. *Cryst Growth Des* **10,** 3843–3846 (2010).

48. Yazaydin, A. O., Benin, A. I., Faheem, S. A., Jakubczak, P., Low, J. J., Willis, R. R., and Snurr, R. Q. Enhanced $CO_2$ adsorption in metal-organic frameworks via occupation of open-metal sites by coordinated water molecules. *Chem Mater* **21,** 1425–1430 (2009).
49. Nune, S. K., Thallapally, P. K., and McGrail, B. P. Metal organic gels (MOGs): a new class of sorbents for $CO_2$ separation applications. *J Mater Chem* **20,** 7623–7625 (2010).
50. Yang, C., Wang, X. P., and Omary, M. A. Fluorous metal-organic frameworks for high-density gas adsorption. *J Am Chem Soc* **129,** 15454–15455 (2007).
51. Farha, O. K., Mulfort, K. L., and Hupp, J. T. An example of node-based postassembly elaboration of a hydrogen-sorbing, metal-organic framework material. *Inorg Chem* **47,** 10223–10225 (2008).
52. Henninger, S. K., Habib, H. A., and Janiak, C. MOFs as adsorbents for low temperature heating and cooling applications. *J Am Chem Soc* **131,** 2776–2777 (2009).
53. Jeremias, F., Khutia, A., Henninger, S. K., and Janiak, C. MIL-100 (Al, Fe) as water adsorbents for heat transformation purposes-a promising application. *J Mater Chem* **22,** 10148–10151 (2012).
54. Henninger, S. K., Schmidt, F. P., and Henning, H. M. Water adsorption characteristics of novel materials for heat transformation applications. *Appl Therm Eng* **30,** 1692–1702 (2010).
55. Seo, Y. K., Yoon, J. W., Lee, J. S., Hwang, Y. K., Jun, C. H., Chang, J. S., Wuttke, S., Bazin, P., Vimont, A., Daturi, M., Bourrelly, S., Llewellyn, P. L., Horcajada, P., Serre, C., and Ferey, G. Energy-efficient dehumidification over hierachically porous metal-organic frameworks as advanced water adsorbents. *Adv Mater* **24,** 806–810 (2012).
56. (a) Britt, D., Tranchemontagne, D., and Yaghi, O. M. Metal-organic frameworks with high capacity and selectivity for harmful gases. *P Natl Acad Sci USA* **105,** 11623–11627 (2008); (b) Khan, N. A., Hasan, Z., and Jhung, S. H. Adsorptive removal of hazardous materials using metal-organic frameworks (MOFs): A review. *J Hazard Mater* **244,** 444–456 (2013); (c) Thallapally, P. K., Motkuri, R. K., Fernandez, C. A., McGrail, B. P., and Behrooz, G. S. Prussian blue analogs for $CO_2$ and $SO_2$ capture and separation applications. *Inorg Chem* **49,** 4909–4915 (2010).
57. (a) Petit, C. and Bandosz, T. J. Enhanced adsorption of ammonia on metal-organic framework/graphite oxide composites: analysis of surface interactions. *Adv Funct Mater* **20,** 111–118 (2010); (b) Peterson, G. W., Wagner, G. W., Balboa, A., Mahle, J., Sewell, T., and Karwacki, C. J. Ammonia vapor removal by Cu-3(BTC)(2) and its characterization by MAS NMR. *J Phys Chem C* **113,** 13906–13917 (2009); (c) Sumida, K., Rogow, D. L., Mason, J. A., McDonald, T. M., Bloch, E. D., Herm, Z. R., Bae, T. H., and Long, J. R. Carbon dioxide capture in metal-organic frameworks. *Chem Rev* **112,** 724–781 (2012); (d) Zhao, Z. X., Li, X. M., Huang, S. S., Xia, Q. B., and Li, Z. Adsorption and diffusion of benzene on chromium-based metal organic

framework MIL-101 synthesized by microwave irradiation. *Ind Eng Chem Res* **50,** 2254–2261 (2011); (e) Glover, T. G., Peterson, G. W., Schindler, B. J., Britt, D., and Yaghi, O. MOF-74 building unit has a direct impact on toxic gas adsorption. *Chem Eng Sci* **66,** 163–170 (2011); (f) Hamon, L., Serre, C., Devic, T., Loiseau, T., Millange, F., Ferey, G., and De Weireld, G. Comparative study of hydrogen sulfide adsorption in the MIL-53(Al, Cr, Fe), MIL-47(V), MIL-100(Cr), and MIL-101(Cr) metal-organic frameworks at room temperature. *J Am Chem Soc* **131,** 8775–8777 (2009); (g) Karra, J. R. and Walton, K. S. Effect of open metal sites on adsorption of polar and nonpolar molecules in metal-organic framework Cu-BTC. *Langmuir* **24,** 8620–8626 (2008); (h) Hamon, L., Leclerc, H., Ghoufi, A., Oliviero, L., Travert, A., Lavalley, J. C., Devic, T., Serre, C., Ferey, G., De Weireld, G., Vimont, A., and Maurin, G. Molecular insight into the adsorption of $H_2S$ in the flexible MIL-53(Cr) and rigid MIL-47(V) MOFs: Infrared spectroscopy combined to molecular simulations. *J Phys Chem C* **115,** 2047–2056 (2011).

58. Petit, C. and Bandosz, T. J. Exploring the coordination chemistry of MOF-graphite oxide composites and their applications as adsorbents. *Dalton T* **41,** 4027–4035 (2012).
59. Fernandez, C. A., Thallapally, P. K., Motkuri, R. K., Nune, S. K., Sumrak, J. C., Tian, J., and Liu, J. Gas-induced expansion and contraction of a fluorinated metal-organic framework. *Cryst Growth Des* **10,** 1037–1039 (2010).
60. Fernandez, C. A., Thallapally, P. K., and McGrail, B. P. Insights into the temperature dependent "breathing" of a flexible fluorinated metal-organic framework. *Chemphyschem* **13,** 3275–3281 (2012).
61. Thallapally, P. K., Grate, J. W., and Motkuri, R. K. Facile xenon capture and release at room temperature using a metal-organic framework: a comparison with activated charcoal. *Chem Commun* **48,** 347–349 (2012).
62. Meek, S. T., Teich-McGoldrick, S. L., Perry, J. J., Greathouse, J. A., and Allendorf, M. D. Effects of polarizability on the adsorption of noble gases at low pressures in monohalogenated isoreticular metal-organic frameworks. *J Phys Chem C* **116,** 19765–19772 (2012).
63. Liu, J., Thallapally, P. K., and Strachan, D. Metal-organic frameworks for removal of Xe and Kr from nuclear fuel reprocessing plants. *Langmuir* **28,** 11584–11589 (2012).
64. Sikora, B. J., Wilmer, C. E., Greenfield, M. L., and Snurr, R. Q. Thermodynamic analysis of Xe/Kr selectivity in over 137000 hypothetical metal-organic frameworks. *Chem Sci* **3,** 2217–2223 (2012).
65. Fernandez, C. A., Liu, J., Thallapally, P. K., and Strachan, D. M. Switching Kr/Xe selectivity with temperature in a metal-organic framework. *J Am Chem Soc* **134,** 9046–9049 (2012).
66. (a) Lee, J., Farha, O. K., Roberts, J., Scheidt, K. A., Nguyen, S. T., and Hupp, J. T. Metal-organic framework materials as catalysts. *Chem Soc Rev* **38,** 1450–1459 (2009); (b) Thallapally, P. K., Fernandez, C. A., Motkuri, R. K., Nune, S. K., Liu, J., and Peden, C. H. F. Micro and mesoporous metal-organic frameworks for catalysis applications. *Dalton T* **39,** 1692–1694 (2010).
67. Dhakshinamoorthy, A., Alvaro, M., Corma, A., and Garcia, H. Delineating similarities and dissimilarities in the use of metal organic frameworks and zeolites as heterogeneous catalysts for organic reactions. *Dalton T* **40,** 6344–6360 (2011).

68. Vermoortele, F., Ameloot, R., Alaerts, L., Matthessen, R., Carlier, B., Fernandez, E. V. R., Gascon, J., Kapteijn, F., and De Vos, D. E. Tuning the catalytic performance of metal-organic frameworks in fine chemistry by active site engineering. *J Mater Chem* **22,** 10313– 10321 (2012).
69. Chui, S. S. Y., Lo, S. M. F., Charmant, J. P. H., Orpen, A. G., and Williams, I. D. A chemically functionalizable nanoporous material [Cu-3(TMA)(2)($H_2O$)(3)](n). *Science* **283,** 1148–1150 (1999).
70. Alaerts, L., Seguin, E., Poelman, H., Thibault-Starzyk, F., Jacobs, P. A., and De Vos, D. E. Probing the Lewis acidity and catalytic activity of the metal-organic framework [Cu-3(btc)(2)] (BTC = benzene-1,3,5-tricarboxylate). *Chem-Eur J* **12,** 7353–7363 (2006).
71. Schlichte, K., Kratzke, T., and Kaskel, S. Improved synthesis, thermal stability and catalytic properties of the metal-organic framework compound CU3(BTC)(2). *Micropor Mesopor Mat* **73,** 81–88 (2004).
72. Horike, S., Dinca, M., Tamaki, K., and Long, J. R. Size-selective lewis acid catalysis in a microporous metal-organic framework with exposed Mn(2+) coordination sites. *J Am Chem Soc* **130,** 5854–5855 (2008).
73. Gandara, F., Gornez-Lor, B., Gutierrez-Puebla, E., Iglesias, M., Monge, M. A., Proserpio, D. M., and Snejko, N. An indium layered MOF as recyclable lewis acid catalyst. *Chem Mater* **20,** 72–76 (2008).
74. Horcajada, P., Surble, S., Serre, C., Hong, D. Y., Seo, Y. K., Chang, J. S., Greneche, J. M., Margiolaki, I., and Ferey, G. Synthesis and catalytic properties of MIL-100(Fe), an iron(III) carboxylate with large pores. *Chem Commun* 2820–2822 (2007).
75. (a) Vimont, A., Goupil, J. M., Lavalley, J. C., Daturi, M., Surble, S., Serre, C., Millange, F., Ferey, G., and Audebrand, N. Investigation of acid sites in a zeotypic giant pores chromium(III) carboxylate. *J Am Chem Soc* **128,** 3218–3227 (2006); (b) Vimont, A., Leclerc, H., Mauge, F., Daturi, M., Lavalley, J. C., Surble, S., Serre, C., and Ferey, G. Creation of controlled Bronsted acidity on a zeotypic mesoporous chromium(III) carboxylate by grafting water and alcohol molecules. *J Phys Chem C* **111,** 383–388 (2007).
76. Suslick, K. S., Bhyrappa, P., Chou, J. H., Kosal, M. E., Nakagaki, S., Smithenry, D. W., and Wilson, S. R. Microporous porphyrin solids. *Accounts Chem Res* **38,** 283–291 (2005).
77. Savonnet, M., Aguado, S., Ravon, U., Bazer-Bachi, D., Lecocq, V., Bats, N., Pinel, C., and Farrusseng, D. Solvent free base catalysis and transesterifiation over basic functionalised Metal-organic frameworks. *Green Chem* **11,** 1729–1732 (2009).
78. Juan-Alcaniz, J., Gascon, J., and Kapteijn, F. Metal-organic frameworks as scaffolds for the encapsulation of active species: state-of-the-art and future perspectives. *J Mater Chem* **22,** 10102–10118 (2012).
79. Ishida, T., Nagaoka, M., Akita, T., and Haruta, M. Deposition of gold clusters on porous coordination polymers by solid grinding and their catalytic activity in aerobic oxidation of alcohols. *Chem-Eur J* **14,** 8456–8460 (2008).
80. Hermes, S., Schroter, M. K., Schmid, R., Khodeir, L., Muhler, M., Tissler, A., Fischer, R. W., and Fischer, R. A. Metal@MOF: Loading of highly porous coordination polymers host lattices by metal organic chemical vapor deposition. *Angew Chem Int Edit* **44,** 6237–6241 (2005).

81. Rahiman, A. K., Rajesh, K., Bharathi, K. S., Sreedaran, S., and Narayanan, V. Catalytic oxidation of alkenes by manganese(III) porphyrin-encapsulated Al, V, Si-mesoporous molecular sieves. *Inorg Chim Acta* **362,** 1491–1500 (2009).
82. Ferey, G. A chromium terephthalate-based solid with unusually large pore volumes and surface area (vol 309, pg 2040, 2005). *Science* **310,** 1119–1119 (2005).
83. Taylor-Pashow, K. M. L., Della Rocca, J., Xie, Z. G., Tran, S., and Lin, W. B. Postsynthetic modifications of iron-carboxylate nanoscale metal-organic frameworks for imaging and drug delivery. *J Am Chem Soc* **131,** 14261–14263 (2009).
84. Maksimchuk, N. V., Timofeeva, M. N., Melgunov, M. S., Shmakov, A. N., Chesalov, Y. A., Dybtsev, D. N., Fedin, V. P., and Kholdeeva, O. A. Heterogeneous selective oxidation catalysts based on coordination polymer MIL-101 and transition metal-substituted polyoxometalates. *J Catal* **257,** 315–323 (2008).
85. Maksimchuk, N. V., Kovalenko, K. A., Arzumanov, S. S., Chesalov, Y. A., Melgunov, M. S., Stepanov, A. G., Fedin, V. P., and Kholdeeva, O. A. Hybrid polyoxotungstate/MIL-101 materials: synthesis, characterization, and catalysis of $H_2O_2$ based alkene epoxidation. *Inorg Chem* **49,** 2920–2930 (2010).
86. (a) Wang, Z. Q. and Cohen, S. M. Postsynthetic covalent modification of a neutral metal-organic framework. *J Am Chem Soc* **129,** 12368–12369 (2007); (b) Tanabe, K. K., Wang, Z. Q., and Cohen, S. M. Systematic functionalization of a metal-organic framework via a postsynthetic modification approach. *J Am Chem Soc* **130,** 8508–8517 (2008).
87. Hong, D. Y., Hwang, Y. K., Serre, C., Ferey, G., and Chang, J. S. Porous chromium terephthalate MIL-101 with coordinatively unsaturated sites: Surface functionalization, encapsulation, sorption and catalysis. *Adv Funct Mater* **19,** 1537–1552 (2009).
88. Zhang, X., Xamena, F. X. L. I., and Corma, A. Gold(III) – metal organic framework bridges the gap between homogeneous and heterogeneous gold catalysts. *J Catal* **265,**155– 160 (2009).
89. Alkordi, M. H., Liu, Y. L., Larsen, R. W., Eubank, J. F., and Eddaoudi, M. Zeolite-like metal-organic frameworks as platforms for applications: On metalloporphyrin-based catalysts. *J Am Chem Soc* **130**, 12639–12641 (2008).
90. Zhou, W., Wu, H., Yildirim, T., Simpson, J. R., and Walker, A. R. H. Origin of the exceptional negative thermal expansion in metal-organic framework-5 Zn(4)O(1,4-benzenedicarboxylate)(3). *Phys Rev B* **78,** 054114 (2008).
91. (a) Takashima, Y., Martinez, V. M., Furukawa, S., Kondo, M., Shimomura, S., Uehara, H., Nakahama, M., Sugimoto, K., and Kitagawa, S. Molecular decoding using luminescence from an entangled porous framework. *Nat Commun* **2** (2011); (b) Harbuzaru, B. V., Corma, A., Rey, F., Atienzar, P., Jorda, J. L., Garcia, H., Ananias, D., Carlos, L. D., and Rocha, J. Metal-organic nanoporous structures with anisotropic photoluminescence and magnetic properties and their use as sensors. *Angew Chem Int Edit* **47,** 1080–1083 (2008); (c) Achmann, S., Hagen, G., Kita, J., Malkowsky, I. M., Kiener, C., and Moos, R. Metal-organic frameworks for sensing applications in the gas phase. *Sensors-Basel* **9,** 1574–1589 (2009); (d) Southon, P. D., Liu, L., Fellows, E. A., Price, D. J., Halder, G. J., Chapman, K. W., Moubaraki, B., Murray, K. S., Letard, J. F., and Kepert, C. J. Dynamic interplay between spin-crossover and host-guest function in a nanoporous metal-organic framework material. *J Am Chem Soc* **131,** 10998–11009 (2009); (e) Maspoch, D., Ruiz-Molina, D., Wurst, K., Domingo, N., Cavallini, M.,

Biscarini, F., Tejada, J., Rovira, C., and Veciana, J. A nanoporous molecular magnet with reversible solvent-induced mechanical and magnetic properties. *Nat Mater* **2,** 190–195 (2003).

92. Serre, C., Mellot-Draznieks, C., Surble, S., Audebrand, N., Filinchuk, Y., and Ferey, G. Role of solvent-host interactions that lead to very large swelling of hybrid frameworks. *Science* **315,** 1828–1831 (2007).
93. Goeders, K. M., Colton, J. S., and Bottomley, L. A. Microcantilevers: Sensing chemical interactions via mechanical motion. *Chem Rev* **108,** 522–542 (2008).
94. Allendorf, M. D., Houk, R. J. T., Andruszkiewicz, L., Talin, A. A., Pikarsky, J., Choudhury, A., Gall, K. A., and Hesketh, P. J. Stress-induced chemical detection using flexible metal-organic frameworks. *J Am Chem Soc* **130,** 14404–14405 (2008).
95. Liu, B., Ma, M. Y., Zacher, D., Betard, A., Yusenko, K., Metzler-Nolte, N., Woll, C., and Fischer, R. A. Chemistry of SURMOFs: Layer-selective installation of functional groups and postsynthetic covalent modification probed by fluorescence microscopy. *J Am Chem Soc* **133,** 1734–1737 (2011).
96. Liu, B., Shekhah, O., Arslan, H. K., Liu, J. X., Woll, C., and Fischer, R. A. Enantiopure metal-organic framework thin films: Oriented SURMOF growth and enantioselective adsorption. *Angew Chem Int Edit* **51,** 807–810 (2012).
97. Biemmi, E., Darga, A., Stock, N., and Bein, T. Direct growth of Cu-3(BTC)(2)($H_2O$)(3)center dot xH(2)O thin films on modified QCM-gold electrodes-Water sorption isotherms. *Micropor Mesopor Mat* **114,** 380–386 (2008).
98. Robinson, A. L., Stavila, V. L., Zeitler, T. R., White, M. I., Thornberg, S. M., Greathouse, J. A., and Allendorf, M. D. Ultrasensitive humidity detection using metal-organic framework-coated microsensors. *Anal Chem* **84,** 7043–7051 (2012).
99. Ameloot, R., Stappers, L., Fransaer, J., Alaerts, L., Sels, B. F., and De Vos, D. E. Patterned growth of metal-organic framework coatings by electrochemical synthesis. *Chem Mater* **21,** 2580–2582 (2009).
100. Li, Y., Zhang, S. S., and Song, D. T. A luminescent metal-organic framework as a turn-on sensor for DMF vapor. *Angew Chem Int Edit* **52,** 710–713 (2013).
101. Lan, A. J., Li, K. H., Wu, H. H., Olson, D. H., Emge, T. J., Ki, W., Hong, M. C., and Li, J. A luminescent microporous metal-organic framework for the fast and reversible detection of high explosives. *Angew Chem Int Edit* **48,** 2334–2338 (2009).
102. Lu, Z. Z., Zhang, R., Li, Y. Z., Guo, Z. J., and Zheng, H. G. Solvatochromic behavior of a nanotubular metal-organic framework for sensing small molecules. *J Am Chem Soc* **133,** 4172–4174 (2011).
103. Beauvais, L. G., Shores, M. P., and Long, J. R. Cyano-bridged Re(6)Q(8) (Q = S, Se) cluster-Cobalt(II) framework materials: Versatile solid chemical sensors. *J Am Chem Soc* **122,** 2763–2772 (2000).
104. Lee, H., Jung, S. H., Han, W. S., Moon, J. H., Kang, S., Lee, J. Y., Jung, J. H., and Shinkai, S. A Chromo-fluorogenic tetrazole-based $CoBr_2$ coordination polymer gel as a highly sensitive and selective chemosensor for volatile gases containing chloride. *Chem-Eur J* **17,** 2823–2827 (2011).
105. Orcajo, B., Muruzabal, F., Isasmendi, M. C., Gutierrez, N., Sanchez, M., Orive, G., and Anitua, E. The use of plasma rich in growth factors (PRGF-Endoret) in the treatment of a severe mal perforant ulcer in the foot of a person with diabetes. *Diabetes Res Clin Pr 93*, E65–E67 (2011).

106. Dave, D. and Ghaly, A. E. Remediation technologies for marine oil spills: A critical review and comparative analysis. *American Journal of Environmental Sciences* **7,** 423–440 (2011).
107. (a) Adebajo, M. O., Frost, R. L., Kloprogge, J. T., Carmody, O., and Kokot, S. Porous materials for oil spill cleanup: A review of synthesis and absorbing properties. *J Porous Mat* **10,** 159–170 (2003); (b) Tao, S. Y., Wa'ng, Y. C., and An, Y. L. Superwetting monolithic $SiO_2$ with hierarchical structure for oil removal. *J Mater Chem* **21,** 11901– 11907 (2011); (c) Venkatanarasimhan, S. and Raghavachari, D. Epoxidized natural rubber-magnetite nanocomposites for oil spill recovery. *J Mater Chem A* **1,** 868–876 (2013); (d) Wang, X. S., Liu, J., Bonefont, J. M., Yuan, D. Q., Thallapally, P. K., and Ma, S. Q. A porous covalent porphyrin framework with exceptional uptake capacity of saturated hydrocarbons for oil spill cleanup. *Chem Commun* **49,** 1533–1535 (2013); (e) Zhu, Q., Pan, Q. M., and Liu, F. T. Facile removal and collection of oils from water surfaces through superhydrophobic and superoleophilic sponges. *J Phys Chem C* **115,** 17464–17470 (2011).
108. USE OF SORBENT MATERIALS IN OIL SPILL RESPONSE *Technical Report from The International Tanker Owners Pollution Federation Limited (ITOPF)* (2013).
109. Wu, H. H., Gong, Q. H., Olson, D. H., and Li, J. Commensurate adsorption of hydrocarbons and alcohols in microporous metal organic frameworks. *Chem Rev* **112,** 836– 868 (2012).
110. Cohen, S. M. Postsynthetic methods for the functionalization of metal-organic frameworks. *Chem Rev* **112,** 970–1000 (2012).
111. Yang, C., Kaipa, U., Mather, Q. Z., Wang, X. P., Nesterov, V., Venero, A. F., and Omary, M. A. Fluorous metal-organic frameworks with superior adsorption and hydrophobic properties toward oil spill cleanup and hydrocarbon storage. *J Am Chem Soc* **133,** 18094– 18097 (2011).
112. Maes, M., Alaerts, L., Vermoortele, F., Ameloot, R., Couck, S., Finsy, V., Denayer, J. F. M., and De Vos, D. E. Separation of C-5-hydrocarbons on microporous materials: Complementary performance of MOFs and Zeolites. *J Am Chem Soc* **132,** 2284–2292 (**2010**).
113. Nijem, N., Wu, H. H., Canepa, P., Marti, A., Balkus, K. J., Thonhauser, T., Li, J., and Chabal, Y. J. Tuning the gate opening pressure of metal-organic frameworks (MOFs) for the selective separation of hydrocarbons. *J Am Chem Soc* **134,** 15201–15204 (2012).

## CHAPTER 4

# MEMBRANE FACILITATED DEFLUORIDATION OF WATER: PROCESS INTENSIFICATION AND SCALE UP

S. SRIDHAR, K. PRANEETH, D. MANJUNATH, and SURESH K. BHARGAVA

## CONTENTS

## 4.1 INTRODUCTION

Polymeric membranes became a viable means of brackish water treatment in the late 1960s with the development of high performance systems. An increasing scarcity for portable fresh water fueled a further push to new membrane options like UF, NF, ED and RO. Compare to UF and NF based membrane systems, RO gained further prominence service. They are effectively nonporous and therefore exclude particles and even low molar mass species such as salt ions and organics. The defluoridation of drinking water has become a major scientific challenge world over in 1990s. The conventional approaches based on flocculation and adsorptions have multiple disadvantages in their application in all equipped rural areas particularly in developing countries. India has been facing the problem for quite some time.

This research team at the Indian Institute of Chemical Technology at Hyderabad in India undertook the challenging task of applying RO concept for defluoridation of drinking water in rural areas. Scientific investigations were undertaken to select the most efficient membrane system, which could intensity the defluoridation operation controlled by solution-diffusion mechanism of mass transfer wherein sorption of $H_2O$ molecules had risen due to their affinity for the hydrophilic membrane system resulting in greater flux. After extensive process standardization studies on laboratory scale, a pilot defluoridation plant was designed for deployment in villages. This paper highlights the experiences of the research team in the intensification and scarcity of a defluoridation facility based on membrane separation.

### 4.1.1 *BACKGROUND ON FLUORIDE CONTAMINATION IN WATER*

Fluoride exists fairly abundantly in the earth's crust as fluorite mineral ore and can enter groundwater by natural processes such as leaching. The soil at the foot of mountains is particularly likely to be high in fluoride due to weathering and leaching of bedrock with high fluoride content. While innocuous and even beneficial in small amounts, fluoride can become toxic when consumed in excess by humans and animals resulting in varying degrees of fluorosis diseases. The known fluoride belts on land include: one that stretches from Syria through Jordan, Egypt, Libya, Algeria, Sudan and Kenya, and another that stretches from Turkey through Iraq, Iran, Afghanistan, India, northern Thailand and China. Figure 4.1 shows the countries having excess fluoride concentration in drinking water resources. There are similar belts in the North America and Japan [1].

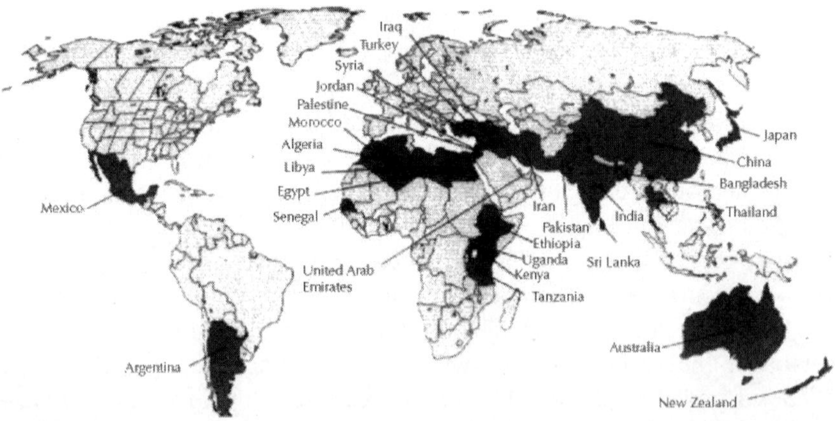


**FIGURE 4.1** Countries affected by endemic fluorosis due to the presence of excess fluoride in drinking water resources [2].

Fluorosis is endemic in large areas of India, mostly rural. This is because more than 90% of the rural population in India predominantly uses ground water for drinking, cooking and other domestic purposes. Nearly 95 million people in 160 districts and 17 states across India drink water contaminated by excess fluoride (over 1.5 mg/L). Figure 4.2 shows the different states of India affected with high concentrations of fluoride in water. Many states in India have ground waters between 1 and 5 mg F/L and in some areas, fluoride concentration exceeds 21 mg F/L. High fluoride levels of more than 4 mg/L are found in Punjab, Haryana, Rajasthan, Gujarat, Madhya Pradesh, Andhra Pradesh and Tamil Nadu. The Ministry of Health, Govt. of India has prescribed 1 mg/L and 2 mg/L as the permissive and excessive limits, respectively. The Indian Standard Specification for drinking water gives a desirable limit of 0.6–1.2 mg/L (Indian Standard Institute, 1983).

Defluoridation refers to methods of water treatment that reduce the concentration of fluoride in the water, normally, in order to make it safe for human consumption and if possible for animal intake as well. In an effort to combat this public health problem, various technologies have been developed and explored in both the developing and developed world [4].

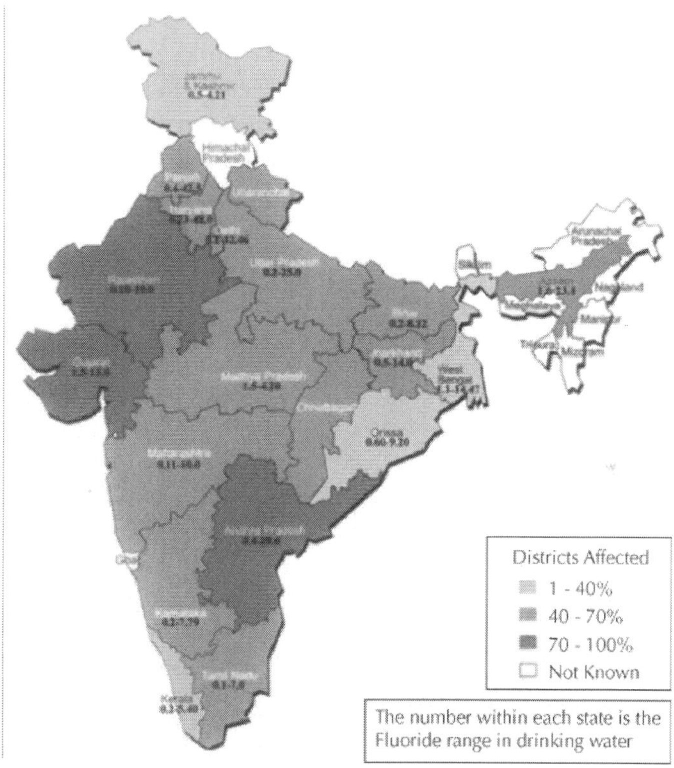


**FIGURE 4.2** Indian states with excess fluoride concentration in drinking water [3].

## *4.1.2 METHODS FOR DEFLUORIDATION OF WATER*

Similar to treatment protocols of other chemical contaminants including arsenic, fluoride cannot be removed by typical water treatment means. Boiling, UV treatment, most methods of filtration, and most chemical treatment options are ineffective to remove fluoride from water. Synthetic ion exchange and precipitation processes, activated alumina filters, and reverse osmosis are typically used to remove fluoride from water in the developed world [5, 6]. There are no universally accepted used defluoridation techniques in the developing world with long felt need for development of appropriate technologies.

### *4.1.2.1 ACTIVATED ALUMINA*

Activated alumina is an effective adsorbent. It is activated through the dehydration of aluminum hydroxides at high temperature (300–600° C). Activated alumina has been used in defluoridation [6] since 1936. Though it is often used for large-scale defluoridation in the developed world, it has not been considered as an appropriate technology for use in the developing world because of its cost, non availability, regeneration requirements and hazards related to its leaching into the water. Activated alumina (Fig. 4.3) is used in an adsorption process with very high fluoride removal efficiency as it is capable of treating water containing fluoride in the range 4–20 mg/L) [6]. Though configurations may vary, activated alumina is typically used in a gravity fed column filter.

**FIGURE 4.3** Activated alumina.

### 4.1.2.2 BONE CHAR

Bone char is the oldest known agent for water defluoridation used since 1940s [7]. It has also been employed successfully for the removal of arsenic from ground water. It is produced from animal bones (Fig. 4.4) that have undergone calcination or pyrolysis processes. Though raw bones have some defluoridation value, it is low and limited by the various organics obstructing the interfaces where chemical reactions with the fluoride can take place [8].

The uptake of fluoride onto the bone char surface is one of the major defluoridation methods. The process involves ion exchange in which carbonate radical of the apatite comprising bone char, and $Ca(PO_4)_6.CaCO_3$, gets replaced by fluoride to form an insoluble fluorapatite. Bone char produced by carbonizing zone at temperature of 1100–1600° C had superior qualities than those of unprocessed bone as a defluoridation agent [8].

**FIGURE 4.4** Bone char.

#### 4.1.2.3 CONTACT PRECIPITATION

It is a technique by which fluoride is removed from water through the addition of calcium and phosphate compounds followed by bringing the water in contact with a bone charcoal medium.

#### 4.1.2.4 DEGREASED AND ALKALI TREATED BONES

Degreased and alkali treated bones are effective in the removal of fluoride from 3.5 to 10 mg/L to less than 0.2 mg/L. Bones containing calcium phosphate have a greater affinity for fluoride. The bone is degreased, dried and powdered for use as a contact bed for removal of fluoride from water. The exhausted bed is regenerated with aqueous sodium hydroxide.

#### 4.1.2.5 SYNTHETIC TRI-CALCIUM PHOSPHATE

It is prepared by reacting phosphoric acid with lime. The medium is regenerated with 1% NaOH solution followed by mild acid rinse.

#### 4.1.2.6 FLOREX

A mixture of tri-calcium phosphate and hydroxy-apatite (Florex), exhibited a fluoride removal capacity of 600 mg/L and is regenerated with 1.5% sodium hydroxide solution. Owing to high attritional losses, Florex was not successful in the long run and pilot plants trials with this material were abandoned [9].

#### 4.1.2.7 ACTIVATED CARBON

The activated carbons prepared from a variety of carbonaceous sources showed different extents of fluoride removal capacities after alum impregnation. Alkali digested alum that was impregnated with paddy husk carbon was found to be an efficient defluoridation agent. Investigations have shown that carbonized saw dust when quenched in 2% alum solution forms excellent adsorbent media. Separation is stoichiometric and equilibrium gets established between carbon and fluoride. On exhaustion (after continued use) the carbon can be regenerated by passing 0.2 to 0.5% alum solutions. Activated carbon prepared from cotton, coffee and coconut wastes was tried for defluoridation but all these materials proved to be of academic interest only.

#### 4.1.2.8 LIME

Natural occurrence of magnesium in raw water enables fluoride removal due to formation of $Mg(OH)_2$ flocs in the presence of lime. In this case, the water must be treated to a caustic alkalinity of 30 mg F-/L, a pH of $\geq$10.5 with option of recarbonation. Magnesia and calcined magnesite have also been tried for fluoride removal from water and their separation capacity was reported to improve with temperature.

#### 4.1.2.9 ION EXCHANGE RESINS

Strong base-exchange resins remove fluoride either through hydroxyl or chloride exchange cycle along with other anions. Since the proportional concentration of fluoride relative to other anions present in water such as $SO_4^{-2}$ or $CO_3^-$ is very small, the efficiency of such resins works out to be quite poor. Some inorganic ion exchangers such as complex metal chloride silicates, prepared by reacting barium or ferric chloride with silicic acid, also exchange fluoride for chloride.

Cation exchange resins impregnated with alum solution have also been found to act as defluoridation agents. Alum treated cation exchange resins like 'Avaram Bark' were reported to work effectively in removing fluoride. Polystyrene-based anion exchange resins and strongly basic quaternary ammonium resins are known to remove fluoride from water along with other anions [9].

#### *4.1.2.10 MAGNESIA*

Investigations were conducted to study the usefulness of magnesia for fluoride removal. Crystalline magnesium hydroxide was obtained by reacting a magnesium salt with milk of lime. The precipitate was filtered, washed, dried and subsequently calcined at 1000° C for 3 h to obtain magnesia. Varying quantities of magnesia were added to 1 L aliquots of water samples and stirred for 30 min using a jar test facility. A typical groundwater containing 10 mg/L fluoride, 60 mg/L hardness, 500 mg/L alkalinity and 7.6 pH was treated with magnesia (MgO) concentrations of 10–1,500 mg/L. The treated water showed a pH value above 9. The average fluoride concentration in the filtrate was 5.8 mg/L for a dosage of 1000 mg/L of MgO. A dose of 1500 mg/L magnesia and a contact period of 3 h was required to reduce the fluoride content in the water to 1 mg/L.The high initial cost, large dosing concentrations, alkaline pH of the treated water and complexity of the preparation of magnesia are the inhibiting factors for field application.

#### *4.1.2.11 ELECTROCOAGULATION/ELECTROCHEMICAL METHODS*

Electrocoagulation process with bipolar aluminum electrodes has been used for defluoridation. The influence of parameters such as interelectrode distance, fluoride concentration, temperature and pH of the solution were investigated and optimized with synthetic water in batch mode. A technology for defluoridation through electrochemical route was developed. The process involves adsorption of fluoride with freshly precipitated aluminum hydroxide, which is generated by the anodic dissolution of aluminum or its alloys in an electrochemical cell. The above technology is power intensive inhibiting its application in remote rural environment use of solar power is an option.

#### *4.1.2.12 NALGONDA TECHNIQUE*

The Nalgonda technique is a means of fluoride removal that depends on the flocculation, sedimentation, and filtration of fluoride with the addition of alum (aluminum sulfate) and lime. This technique was developed by the National

Environmental Engineering Research Institute in India (NEERI) in 1975 in response to concerns over fluorosis disease in India.

While alum ($Al_2(SO_4)_3 18H_2O$) is commonly used as a flocculant in general water treatment, the amounts used in defluoridation are much higher, atleast 150 mg/mg$F^-$. As is typical with flocculation processes, the water must be thoroughly stirred to ensure dispersal of the flocculating agent. Since the reaction results in an excess of $H^+$ ions, Lime ($Ca(OH)_2$) is added to the water during the process to help maintain a neutral pH and hasten the settling of the sediment. The amount of lime added is typically 5% of mass of the aluminum sulfate added [10].

After waiting for the sediment to settle, filtration is then carried out to ensure that none of the created sediments is consumed. The actual dosing amounts of alum and lime to be added can vary depending on the initial alkalinity, fluoride concentration, and the quality of lime used. The amount of alum added must be carefully monitored as residual aluminum can cause neurological, cardiovascular, and respiratory problems its level in drinking water must be maintained at 0.2 mg/L. Although less serious, residual sulfates must also be monitored as they can cause poor taste and gastrointestinal discomfort when in excess of 800 mg/L of alum.

### 4.1.3 OVERVIEW OF MEMBRANE SEPARATION TECHNOLOGIES

The industrial application of synthetic membranes started in the 1960s, although the earliest recorded study of membrane application can be traced back to the middle of the eighteenth century. With nearly 50 years of development, membrane processes have found utility in diverse areas like drinking water purification, brackish water desalination, wastewater reclamation, food and beverage production, gas and vapor separation, energy conversion and storage, air pollution control, hazardous industrial waste treatment, hemodialysis and protein concentration. Membrane technology has greatly enhanced its application in restructuring production processes, protecting the environment and maintaining public health [11].

### 4.1.4 MEMBRANE BASED DEFLUORIDATION TECHNIQUES

Groundwater containing dissolved salts beyond the permissible limits is harmful and unsuitable for human consumption. Excessive fluoride concentration in groundwater is a major problem in many parts of the world. Fluorineis, a natural constituent of the environment, belongs to the halogen group of compounds, with high electronegativity. In its salt form, it enters the human body as fluoride through food, tobacco, industrial exposure, drugs, and cosmetics. World Health Organization (WHO) and IS: 10500 recommend that the fluoride content in drinking water should be less than 1.5 mg/L.

The main source of fluoride in ground water is fluoride-bearing rocks such as fluorospar, fluorite, cryolite, fluorapatite and hydroxylapatite [12]. The flouride content in ground water is a function of several factors including hydrogeological environment, solubility of fluoride minerals in water, velocity of flowing water, pH, temperature as well as concentrations of other species such as calcium and bicarbonate ions in water [13, 14].

### *4.1.4.1 REVERSE OSMOSIS (RO) PROCESS*

The RO is the reversal of water flow through a semipermeable membrane by the application of a pressure greater than the natural osmotic pressure that gets exerted on the raw water from the opposite direction. Figure 4.5 depicts the principle of RO process.

Semipermeable membrane selectively allows certain species to pass through it while retaining others. In actual practice, a mixture of solutes can pass through the membrane, but at significantly different rates. In RO, the solvent, which is mostly water passes through the membrane at a much faster rate than the dissolved solids. The net effect is that a solute-solvent separation occurs, with pure water being the product.

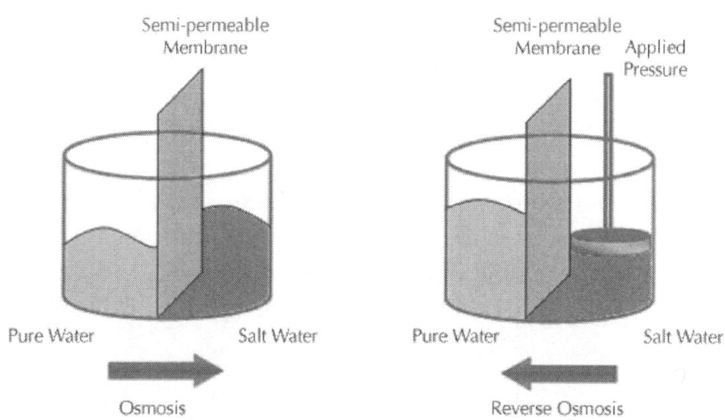


**FIGURE 4.5** Principle of RO.

$H_2O$ molecules move from pure water-side to the salty water-side exerting osmotic pressure on the membrane until equilibrium is achieved. Application of an external pressure on the salt solution side equivalent to its osmotic pressure will also cause equilibrium to be established (Fig. 4.5). Additional pressure will raise the chemical potential of the water in the salt solution and cause the solvent to flow to the pure water-side, which now has a lower chemical potential.

The driving force for the reverse osmosis process is the applied pressure. The amount of energy required for osmotic separation is directly related to the salinity of the solution. Thus, more energy is required to produce the same amount of water from solutions containing higher salt concentrations.

#### *4.1.4.2 NANOFILTRATION (NF)*

NF is a pressure driven cross flow filtration technology with the nominal pore size of the membrane in the range of 1–10 nanometers. Its range of operation has between UF and RO [15]. NF membranes are related by molecular weight cut it (MWCO) rather than nominal pore size. Its trans membrane pressure drop (upto 3 Mpa) is lower than that of RO reducing the operating cost significantly. It is mainly used for the separation of ions or fluid mixtures. It is commonly used for desalination of brackish water. There are reports of using polymeric NF in conjunction with RO to treat brackish ground water.

In aqueous solutions, charged NF membranes allow separation of specific ionic species of opposite charge and repel like charged ions [16]. It is believed that molecular sieving or steric hindrance is the dominant rejection mechanism in NF for colloids and large molecules, while the physicochemical interactions of solute and membrane become increasingly important for ions and low molecular weight organics [17]. Figure 4.6 shows a hypothetical polymeric membrane with –CONH groups attached to the surface of the membrane, which is brought in contact with an aqueous solution of salts. Here, the separation occurs by molecular sieving only wherein monovalents salts such as NaCl pass through whereas bivalent ones like $MgSO_4$ undergo maximum retention. There are no reports of NF application in defluoridation of water.

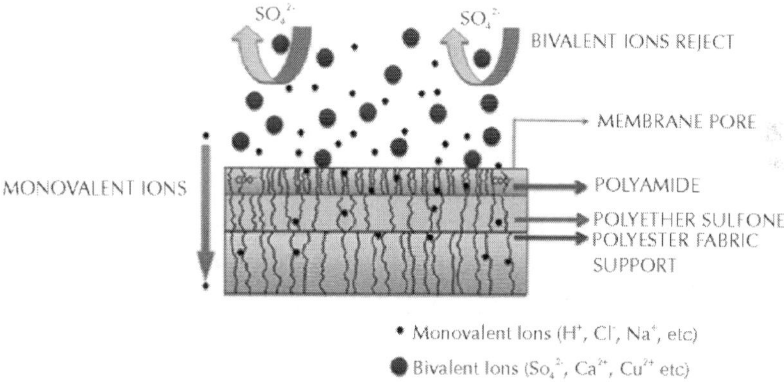


**FIGURE 4.6** Principle of nanofiltration.

### *4.1.4.3 ULTRAFILTRATION (UF)*

UF is a pressure driven molecular sieving process. Due to the size of the pores, the particles that are too large to pass through are physically blocked out, while the dissolved salts and essential minerals pass through the membrane (Fig. 4.7). This process is useful for clarification and disinfection of water but cannot remove dissolved contaminants like fluoride. UF can however be employed for pretreatment to RO process to reduce turbidity and microbial load on the membrane.

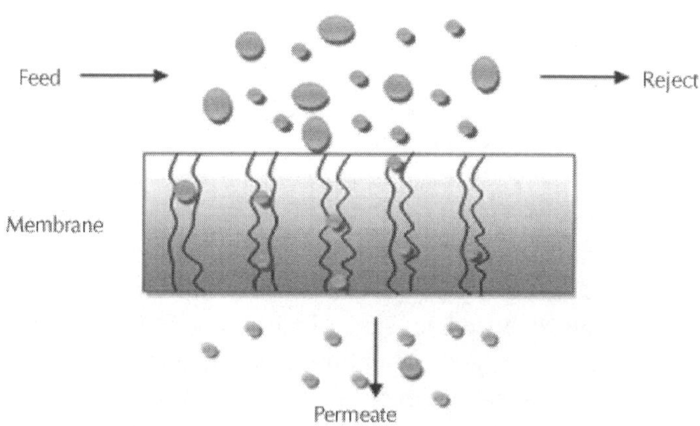


**FIGURE 4.7** Principle of ultrafiltration.

### 4.1.5 SCOPE OF PRESENT WORK

Ground water forms a major source of drinking water in urban as well as rural areas. Many the ground water in some rural areas of Andhra Pradesh state in India are unfit for human consumption because of high fluoride (>1.5 mg/L) and high salinity (>1500 mg/L). More than 80% of the rural population uses ground water for drinking. Since quality of public health depends on the quality of water, an attempt has been made to provide in-depth information on fluoride content in ground water from a cross-section of localities in different districts of Andhra Pradesh, which are severely affected by fluorosis, especially villages of the worst affected Nalgonda district as well as adjoining Warangal and Mahboobnagar besides Prakasam district.

The main aim of the paper is to present the application of RO based membrane technology for defluoridation of contaminated groundwater in remote villages to mitigate the sufferings of the rural poor. Operating parameters such as feed concentration, pressure, temperature and cross-flow velocity were varied to study their effect on membrane performance. Efforts were made in disposing or recycling the reject stream through the design of a zero wastewater treatment system. Running cost of RO process for 10 L operation duration per day was calculated and presented in this study.

## 4.2 EXPERIMENTAL

### 4.2.1 MATERIALS

Spiral wound module of 1 $m^2$ effective separation area of thin film composite (TFC) Polyamide RO membrane was fabricated with the help of Permionics Membranes Pvt. Ltd., Vadodara, India. *m*-phenylenediamine (MPD) and trimesoyl chloride (TMC) obtained from AVRA Synthesis Pvt. Ltd (Hyderabad, India) were used without further purification. Piperazine was purchased from Sigma-Aldrich, USA. Ground water samples collected from two different places of Andhra Pradesh (A.P.), India was used for experimental trials. Potassium dichromate, ferrous ammonium sulfate, mercuric sulfate, sulfuric acid, ferroin indicator for COD analysis, sodium thiosulfate, Wrinkler's reagent, $MnSO_4$, potassium iodide, starch indicator for BOD analysis, citric acid, HCl, EDTA, NaOH, and sodium metabisulphite (SMBS) for washing the membranes were purchased from SD Fine Chemicals Ltd., Mumbai, India. Deionized water for cleaning and feed preparation was generated from the same RO system. BOD incubator (RCI-S.NO-313 India), COD incubator

(DRB 200 COD Reactor, Germany), Colorimeter (Hach-DR-890) for detailed water analysis were procured from M\s Hach, Bangalore, India. Conductivity meter (Global-DCM-900) and pH meter (Global-DPH-504) were purchased from Global electronics, Hyderabad, India.

### 4.2.2 MEMBRANE FORMATION

RO membranes are formed as an ultrathin aromatic polyamide (PA) layer through interfacial polymerization, as shown in Fig. 4.8 (a), on an ultra porous polyether sulfone (PES), made by phase inversion technique [18, 19]. The PES substrate is soaked in 1% aqueous solution of MPD for 1 min. After draining off excess water, the substrate was immersed in hexane bath containing 0.1% TMC for 30 sec. The membrane is then heated in an oven at 70° C for 5 min to obtain RO membrane of 1 Å pore diameter, which corresponds to a MWCO of 50 Da. Similar procedure is followed for preparing NF membrane except that MPD was replaced with piperazine reagent to obtain a pore size of 15–20 Å with MWCO 400 Da, as depicted in Fig. 4.8 (b).

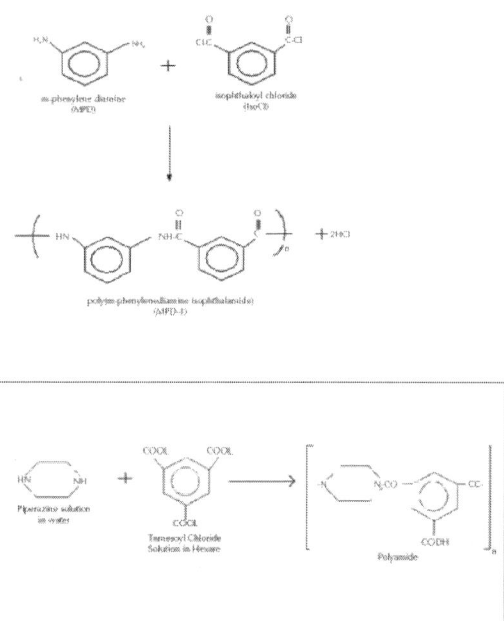

**FIGURE 4.8** Interfacial polymerization reaction schemes for (a) RO and (b) NF membranes.

The NF membrane surface was further functionalized with 0.75% w/v Polyvinyl alcohol (PVA) solution to reduce the MWCO to 250 Da for better rejection of fluoride. Figures 4.9a and 4.9b represent the physical structures of RO polyamide and functionalized (FPA) 250 NF membranes.

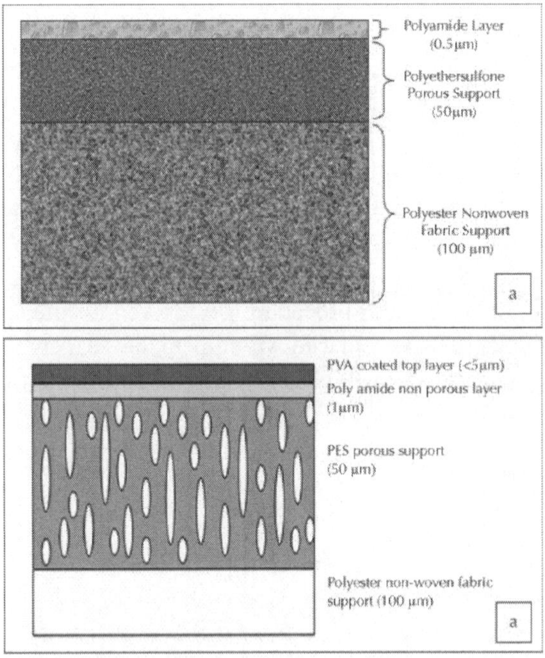


**FIGURE 4.9** Physical structure of (a) RO and (b) NF composite membrane.

Polyamide composite membrane is selected for RO due to its high mechanical strength and robust nature imported by its aromatic backbone. It exhibits high rejection for TDS and other contaminants and is primarily known for its selective permeability to water and relative impermeability to various dissolved impurities including fluoride and other salt ions. Moreover, polyamide is not susceptible to biological fouling and can operate over a wide range of pH conditions (2–11) and temperature up to 45° C.

#### *4.2.2.1 SUBSTRATE PREPARATION BY PHASE INVERSION*

The ultraporous flat sheet substrate for membrane casting are prepared by dissolution of polysulfone in DMF and is cast on a suitable nonwoven polyester fabric followed by phase inversion in a non solvent bath for 5–10 min. The resultant in the polymer is transformed from liquid to solid state (Fig. 4.10) by employing 15% w/v PES solution in DMF solvent containing 3% propionic acid. It is then cast to 45–50 μm thickness film followed by gelation in ice-cold water bath for duration of 10 min to obtain ultraporous substrate. Water washing for 24 h is carried out to remove residual DMF solvent, which could affect permeation characteristics. The pore diameter and pore size distribution can be controlled by varying the viscosity and additive concentrations in the dope. Porous substrates are also prepared by sintering, stretching, leaching, and trace etching techniques employing polysulfone, polyacrylonitrile and cellulose acetate.

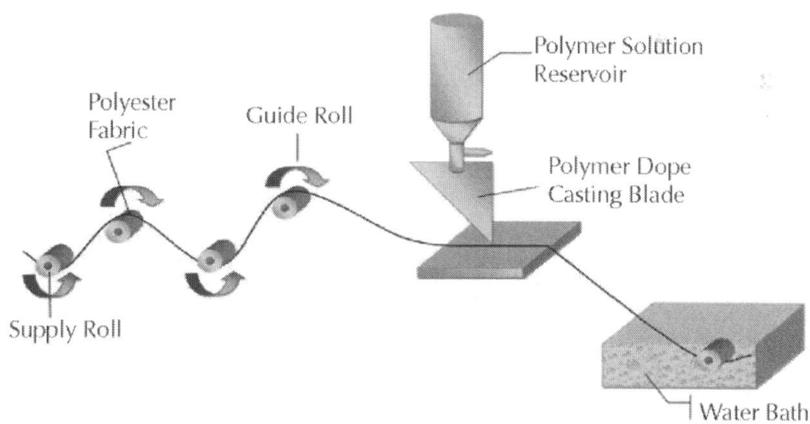


**FIGURE 4.10** Schematic of phase inversion process for preparation of ultraporous PES substrate.

### *4.2.2.2 SOLUTION CASTING AND SOLVENT EVAPORATION*

The membrane-casting unit shown in Fig. 4.11 is used for membrane preparation through solution casting and solvent evaporation for preparing dense membranes. Figure 4.11(a) represents the schematic drawing of hand casting unit whereas Fig. 4.11(b) represents the laboratory automated unit wherein thickness of the membrane formed is more uniform and can be more accurately adjusted to the desired value using microguages. Homogeneous bubble-free solutions of polymers in appropriate low-boiling solvents are cast to the desired thicknesses on a clean glass plate followed by complete evaporation of the solvent. Thickness of the membrane is fixed using a doctor's blade to vary the gap between the movable metallic bar and glass plate. The glass plate is made to slide through the gap to spread the solution uniformly. The volatile solvent evaporates, leaving behind the defect-free dense polymer membrane for further applications. Thin dense film can be cast on composite substrates to prepare TFC RO membranes. Another variation of this process involves partial vaporization of the volatile solvent in which a polymer is dissolved, followed by precipitation in a nonsolvent water bath. Partial evaporation produces a thin skin layer capable of rejecting salts and allowing water to pass through by solution-diffusion mechanism. In fact the first RO membrane was asymmetric cellulose developed by Loeb and Sourirajan [20].

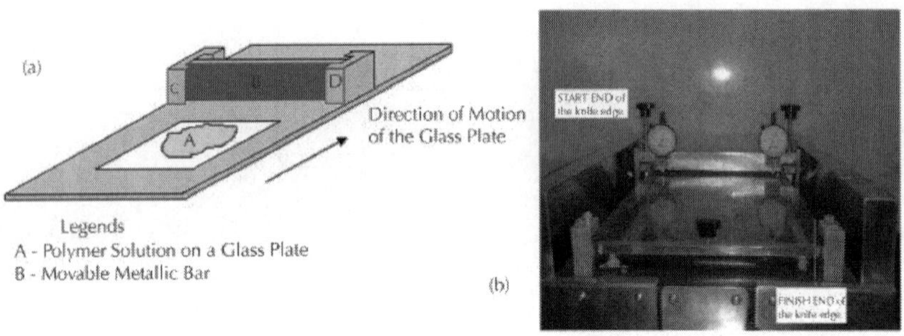


**FIGURE 4.11** (a) Schematic of a membrane casting apparatus; (b) Photograph of casting unit.

#### 4.2.2.3 DIP COATING

Composite membranes are also prepared by the dip coating technique to obtain a dense surface layer. An ultraporous flat sheet substrate is immersed in a lean coating solution containing the desired polymer for a specific duration and then removed from the bath, which results in a thin layer of polymer solution adhering to the substrate. The sheet is dried at room temperature or in an oven. The modeling of the coating of metal plates with a viscous liquid was attempted by employing Navier Stokes equation (the Eq. (1)) to describe the thickness of the coating layer:

$$T_{\text{equi}} = \frac{2}{3}\sqrt{\frac{\eta V}{\rho g}} \tag{1}$$

Where $T_{\text{equi}}$ is the equilibrium thickness, $V$ is the coating plate velocity, $\rho$ is the viscosity, $g$ represents gravity and $\eta$ is the surface tension.

#### 4.2.2.4 IMMERSION PRECIPITATION

In this case, the polymer is dissolved in a suitable solvent and the bubble free solution is cast on a nonwoven fabric substrate and immediately immersed in a nonsolvent like water. The porous membrane is formed by exchange of solvent with nonsolvent.

#### 4.2.2.5 INTERFACIAL POLYMERIZATION

Interfacial polymerization can be used for creating a nonporous ultrathin skin layer on a porous substrate such as PES through a polymerization reaction occurring between two reactive monomers at the interface of two immiscible solvents. The porous substrate is initially saturated in an aqueous solution containing a reactant such as a diamine, followed by immersion in an immiscible organic solvent such as hexane containing a reagent like an acid chloride. A dense top layer forms on the porous substrate at the interface between water and the organic solvent when the two monomers react with each other, as shown in Fig. 4.12 after which the composite is heated to crosslink the selective layer. The active layer is produced by saturating the ultraporous PES matrix with a 1% solution of aqueous MPD for 1 min, immersing it in a 0.1% solution of isophthaloyl chloride (ISPC) or TMC in hexane for 30 seconds followed by heating for a period of 30 min at 65–70°C. Flux can improved through addition of 1–5% of DMSO or NMP or DMA as a solvent in the amine bath. The rejection is expected to be minimal twice MPD is insoluble in hexane and ISPC is insoluble in water, and polymerization takes place exclusively at the interface. The skin formed provides a diffusion barrier for both monomers, hence the reaction rate decreases rapidly. Table 4.1 depicts the types of composite membranes produced by the interfacial polymerization method in this work as well as those reported in literature.

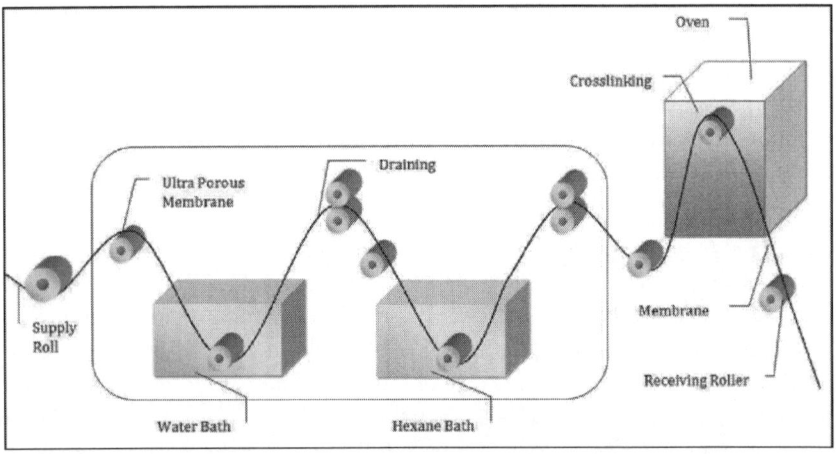


**FIGURE 4.12** Schematic illustration interfacial polymerization to fabricate RO membrane.

**TABLE 4.1** Composite membranes produced by interfacial polymerization.

| Membrane Type | Aqueous Reactant | Crosslinking agent | Flux (L/ m²h) | Rejection (%) | Reference |
|---|---|---|---|---|---|
| Polyamide | *m*-phenylenediamine (1%) | Isophthaloyl chloride (0.2%) in Hexane | 65.0* | 91.0 | Present work |
| Polyamide | Piperazine (0.5%) | Terephathaloyl chloride (0.1%) in Hexane | 65.5* | 92.7 | -do- |
| Polyamide | *p*-phenylenediamine (2%) | Terephathaloyl chloride (0.5%) in Hexane | 60.6* | 25.0 | -do- |
| Polyamide | *p*-phenylenediamine (1%) | Terephathaloyl chloride (0.1%) in Hexane | 38.2* | 32.0 | -do- |
| Polyester | Resorcinol (1%) | Terephathaloyl chloride (0.1%) in Hexane | 30.0* | 25.0 | 20 |
| Polyester | Sorbitol (1%) | Terephathaloyl chloride (0.1%) in Hexane | 32.7* | 24.3 | 20 |
| Polyamide | *m*-phenylenediamine (1%) | Trimesoyl chloride (0.1%) in Hexane | 38.0* | 60.0 | 20 |
| Polyamide | Piperazine (1%) | Toluylenediisocyanate (0.5%) in Hexane | 33.0* | 25.0 | 20 |
| Polyester | Resorcinol | Isophthaloyl chloride in Hexane | 75.0† | 40.0 | 20 |
| Polyamide | Piperazine | Isophthaloyl chloride in Hexane | 45.8† | 98.0 | 21 |

**TABLE 4.1** *(Continued)*

| Membrane Type | Aqueous Reactant | Crosslinking agent | Flux (L/ $m^2h$) | Rejection (%) | Reference |
|---|---|---|---|---|---|
| Polyester | Sorbitol | Terephthaloyal chloride in Hexane | 37.5† | 97.0 | 21 |
| Polyamide | Polyethyleneimine | Terephthaloyal chloride in Hexane | 50.0† | 96.0 | 21 |
| Polyurea (NS 100) | Polyethyleneimine | Toluylenediisocyanate in Hexane | 37.5† | 99.4 | 21 |
| Polyamide (FT 30) | *m*-phenylenediamine | Trimesoyl chloride in Hexane | 75.0• | 98.0 | 21 |
| Polyamide (PA 300) | Polyethyleneimine | Isophthaloyl chloride in Hexane | 50.0† | 99.4 | 21 |
| Polyamine | 1,3-diaminobenzene | Formaldehyde vapor | 75.0† | 97.0 | 21 |

Conditions: * Δp = 10 bar, 25° C, † Δp = 100 bar, 25° C, • Δp = 40 bar, 25° C.

Both the ground water samples collected from two different locations of Andhra Pradesh state contained fluoride owing to hydrogeological conditions but there was no evidence of any contamination by industrial effluents or domestic sewage as proven by detailed water analysis given in Table 4.2.

**TABLE 4.2** Characteristics of feed and permeate samples.

| Parameters | Water samples | | | |
|---|---|---|---|---|
| | Prakasam Dist. | | Nalgonda Dist. | |
| | Characteristics | | | |
| | Feed | Permeate | Feed | Permeate |
| TDS (ppm*) | 821.1 | 29.6 | 502 | 18.5 |
| Turbidity (FAU) | 5 | Nil | 83 | Nil |
| Color (Pt-Co) | 15 | Nil | 129 | Nil |
| Total hardness (mg/L) | 700 | 40 | 480 | 35 |
| Total alkalinity (mg/L) | 340 | 20 | 280 | 10 |
| Calcium (mg/L) | 126 | 10 | – | – |
| Magnesium (mg/L) | 90 | 3 | – | – |

**TABLE 4.2** *(Continued)*

| Parameters | Water samples | | | |
|---|---|---|---|---|
| | Prakasam Dist. | | Nalgonda Dist. | |
| | Characteristics | | | |
| | Feed | Permeate | Feed | Permeate |
| Sodium (mg/L) | 120 | 3 | – | – |
| Potassium (mg/L) | 2 | <1 | – | – |
| Chloride (mg/L) | 250 | 25 | 280 | 23 |
| Sulphate (mg/L) | 156 | 6 | 150 | 3 |
| Silica (mg/L) | 18 | 1 | – | – |
| Iron (mg/L) | 0.09 | 0.02 | – | – |
| Nitrate (mg/L) | 30 | 2 | – | – |
| Fluoride (mg/L) | 3.2 | 0.2 | 2.8 | 0.5 |
| COD (mg/L) | – | – | – | – |
| BOD (mg/L) | – | – | – | – |
| Total coliforms (MPN/100 mL) | 1100 | Nil | 12000 | Nil |

*1ppm=1 mg/L.

### 4.2.3 CHARACTERIZATION OF MEMBRANES

#### 4.2.3.1 SCANNING ELECTRON MICROSCOPY

The surface and cross-sectional morphologies of polymeric membranes were determined by using scanning electron microscopy (SEM). The membranes were cut in liquid nitrogen to obtain smooth cross-sections, followed by gold sputtering before scanning on a SEM instrument with potential of 10 kV (Hitachi S2150 microscope, Ibaraki, Japan).

#### 4.2.3.2 TENSILE STRENGTH

Tensile strength and percentage elongation at break are intensive properties that reveal the mechanical stability of a TFC membrane. The tensile strength of a membrane is the maximum amount of stress that it can be subjected to before failure. The maximum load applied in breaking a tensile test piece divided by its original cross-sectional area is measured in Newton/mm$^2$ or tons/in$^2$ by employing (Eq. (2)).

$$R_m = \frac{P_{max}}{A_0} \tag{2}$$

where $P_{max}$ is the maximum load and $A_o$ is the cross-sectional area of the sample. RO membranes should possess adequate tensile strength and percentage elongation in order to have potential for commercialization.

#### 4.2.3.3 PORE SIZE MEASUREMENT

The membrane was fully wetted with isopropanol, with a surface tension 21.7 dynes/cm, and mounted on the sample chamber that was then sealed. Pure nitrogen was then allowed to flow into the chamber gradually. When the increased nitrogen pressure reached a point that overcame the capillary flow of the fluid within the largest pore, the bubble point was found (Eq. (3)).

$$dp = \frac{4\gamma COS\theta}{P} \tag{3}$$

where $dp$ is the pore size diameter, $\gamma$ is the surface tension of liquid, $\theta$ is the contact angle of liquid and $P$ is the external pressure.

#### 4.2.3.4 SOLUTE TRANSPORT

The flat sheet membranes were tested using aqueous solutions containing nonionic macromolecules, that is, PEG or Dextran of various molecular weights. First, pure water permeation flux was measured. Subsequently, the solute retention test was performed with PEG or Dextran of increasing molecular weights. The initial feed solute concentration was noted and the solute separation was calculated using Eq. (4):

$$a = (1-(C_p/C_f)) \times 100 \tag{4}$$

where $C_p$ and $C_f$ are the solute concentrations in permeate and feed, respectively.

### 4.2.4 EXPERIMENTAL SETUP

A pilot RO system of 250 L/h capacity is skid mounted with a feed tank of maximum 100 L capacity. A low pressure centrifugal pump is provided for pretreatment of the feed through a polypropylene micron rope cartridge of 5 μm pore size to prevent entry of suspended particles capable of damaging the membrane. A high pressure triplex plunger pump, capable of maintaining a pressure up to 25 bar is installed at the upstream side of the prefilter cartridge to transport the feed to the RO module. This pump is run by a 1.5 HP single-phase motor. A restricting needle valve is provided on the concentrate (reject) outlet of the membrane, to pressurize the feed liquid to a desired value as indicated by the pressure gauge installed upstream of the needle valve. A coil type heat exchanger is provided in the reject line to bring down the temperature of the reject, which gets heated rapidly by continuous pressurization and recirculation. The reject stream exiting the membrane gets distributed into three concentric coils of different radii, which provide large heat transfer area for its sufficient cooling. Ice cold water is circulated through the shell side of the heat exchanger using a water circulation pump to ensure a relatively uniform feed temperature (26–28° C). The flow diagram of RO/NF process is provided in Fig. 4.13. Permeate and reject flow rates are measured by two glass rotameters containing metal floats. Permeate is collected in marked buckets with the help of a polyethylene tube. All the other connections are made with 1/2 inches O.D. stainless steel 316 piping in the high pressure region whereas the low pressure lines are constituted of PVC braided tubing for transporting permeate and enable reject recycle.

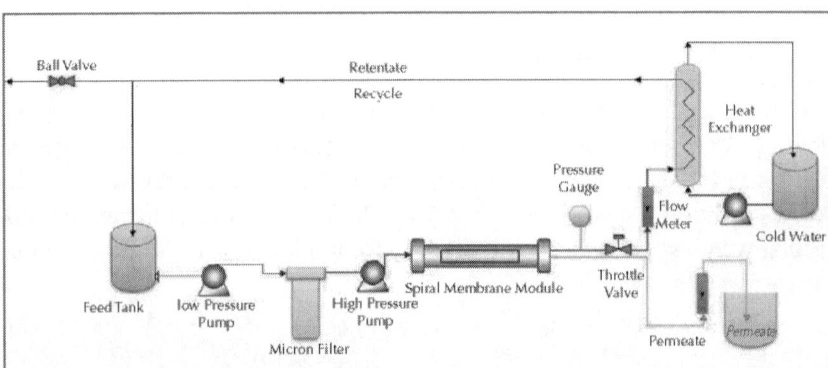


**FIGURE 4.13** Flow diagram of RO/NF pilot scale setup.

### 4.2.4.1 *OPERATING PROCEDURE*

Before introducing the feed into the RO system, the membrane was cleaned and wetted with deionized water. Initially feed water was passed through the micron prefilter cartridge using low-pressure pump to remove coarse solids.

The pretreated water was then pressurized through the membrane using the high-pressure pump.

The needle valve in the reject line is throttled to obtain a suitable pressure in the range of 7 to 25 bar. The initial permeate was recycled until a steady permeate flux was obtained. The reject was partially circulated back to the feed tank through the heat exchanger. The flow rates of the two streams such as permeate and reject streams were measured at regular time intervals using the rotameters. The conductivity, total dissolved solids (TDS) and flux values of feed and permeate were recorded as a function of time until the desired water recovery was achieved.

### 4.2.5 MEMBRANE CLEANING AND MAINTENANCE

At the end of each operation, the membrane was washed for 10 min with tap water to displace most of the impurities present in the pressure housing as well as feed, permeate and reject lines. This system was operated at a pressure of about 3 bar with total disposal of permeate and reject for 5 min to flush out the impurities present in the pipelines. For the remaining period the exiting streams were recycled to the feed tank with repeated pressurising and depressurising to ensure efficient cleaning.

Suspended solids, dissolved salts microbes and organic materials present in the feed water accumulate either on the membrane surface as scales or clog the pores resulting in membrane fouling. Soluble heavy metals like iron can get oxidized within the modules to form deposits and similar problems can occur with silica or even colloidal sulfur [22]. In order to prevent biological fouling by fungi, algae or microbes, the membrane is placed in sodium metabisulphite (SMBS) aqueous medium (0.5% w/v) whereas the removal of mineral scales or metal salt precipitates is done using an aqueous solution of citric acid/HCl (1%w/v), which is circulated through the system for 10 min. An aqueous solution containing tetrasodium EDTA (0.5% w/v) + NaOH (1%w/v) and sodium lauryl sulfate (0.1%) is used for the removal of organic foulants and polishing of membrane surface. Biological fouling causes permanent damage, which on most occasions requires membrane replacement. Therefore, the membrane was always stored in a preservative and maintained in wet condition.

### 4.2.6 ANALYTICAL PROCEDURES

The color and turbidity present in the feed and permeate samples were analyzed using a digital colorimeter (Hach-DR-890, Bangalore, India). The conductivity and pH of the samples were determined by a digital conductivity meter (Model DCM-900, Global Electronics, Hyderabad, India) and pH meter (Model DPH-504, Global Electronics). The extent of E. coli bacteria present in the samples were estimated by coliform test [21]. The TDS was measured using digital conductivity/TDS meter (Global-DCM-900).

The COD values of feed and permeate samples were measured by gravimetric analysis. COD was measured by open reflux method, following standard methodologies [24]. The BOD was measured by noting the difference of the oxygen concentration between the sample before and after incubating it for 5 days [25].

#### 4.2.6.1 PERMEATE FLUX (J)

The permeate volume was measured during the separation process. The permeate flux was calculated by dividing the permeate volume by the effective membrane area and sampling time (Eq (5)).

$$\text{Permeate flux} \quad (J) = \frac{\textit{Permeate volume collected}}{\textit{Membrane area X Time}} \tag{5}$$

#### 4.2.6.2 REJECTION EFFICIENCY

It indicates the separation performance of a membrane and is denoted in terms of percentage rejection of TDS. The percentage of rejection is calculated by Eq. (6).

$$\text{Rejection \%} = [1-(\frac{C_P}{C_F})] \text{ x } 100 \tag{6}$$

where $C_P$=concentration of the solute in permeate; $C_F$=Concentration of the solute in feed.

#### 4.2.6.3 WATER RECOVERY (%)

Percentage of Recovery was calculated as shown in Eq. (7)

$$\%Recovery = \frac{Q_P}{Q_f} \times 100 \tag{7}$$

where, $Q_p$ and $Q_f$ are flow rates of permeate and feed, respectively.

## 4.3 RESULTS AND DISCUSSIONS

### 4.3.1 MEMBRANE CHARACTERIZATION

#### 4.3.1.1 DISTILLED WATER FLUX

Distilled water was used as the feed at ambient temperature to study the permeation characteristics of TFC RO polyamide membrane. The relationship between hydrostatic pressure applied in the range 5–24 bar and resultant pure water flux is shown in Fig. 4.14. At a pressure of 5 bar, membrane exhibited a flux of 32 $L/m^2$ h, which further enhanced to 130 $L/m^2h$ at 24 bar. Usually, pure water flux is linearly related to TMP over the pressure range explored.

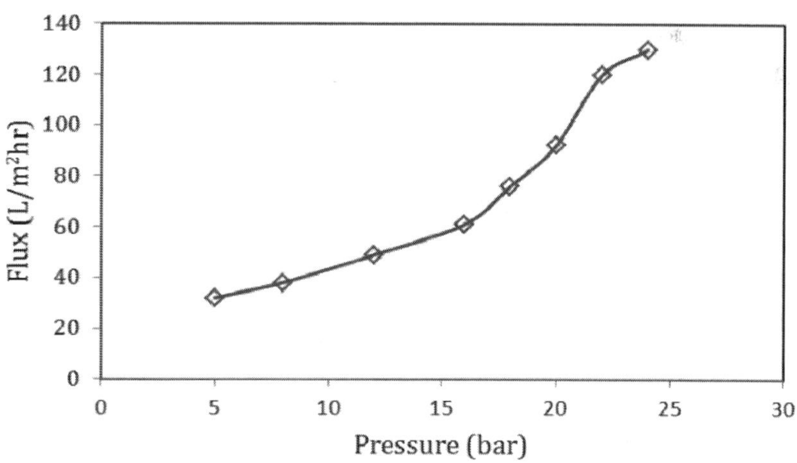


**FIGURE 4.14** Pure water flux through RO Polyamide membrane at different feed pressures.

#### 4.3.1.2 FLUX AND REJECTION OF $CaF_2$ AND NaF FEED SOLUTIONS

Initial experimental studies on flux and fluoride rejection were conducted using synthesized TFC RO membrane. Feed solutions prepared from $CaF_2$ and NaF were pressurized at 10 bar across the membrane. The TDS of feed sample was 820 ppm. The fluoride concentration in feed and permeate, flux and reject streams are listed in Table 4.3 for constant water recovery of 50% in permeate.

**TABLE 4.3** Results of laboratory defluoridation studies.

| Feed | Fluoride Concentration (mg/L) | | Flux ($L/m^2h$) | % Fluoride Rejection |
|---|---|---|---|---|
| | Feed | Permeate | | |
| $CaF_2$ | 5.1 | 0.34 | 48 | 93.3 |
| NaF | 30.6 | 2.94 | 45 | 90.4 |

The effect of pressure on flux and permeate fluoride concentration is shown in Fig. 4.15 for a feed fluoride concentration of 5.1 mg/L prepared by addition of $CaF_2$ salt in tap water having 820 ppm TDS. Increase in flux from 4 to 45 $L/m^2h$ was observed when the pressure was raised from 5 to 10 bar with simultaneous decrease in fluoride ion concentration in permeate from 3.5 to 1.1 mg/L was reported. The observations can be attributed to solution-diffusion mechanism of mass transfer wherein sorption of $H_2O$ molecules rises due to affinity for the hydrophilic membrane resulting in greater flux [27]. Comparatively, the flux of solute molecules remains more or less the same due to lack of any interaction with –CONH groups of the barrier. Water molecules are also much smaller than fluoride and other ions present in the feed and are expected to have greater diffusivity resulting in higher water flux and improved fluoride as well as TDS rejection at higher pressures.

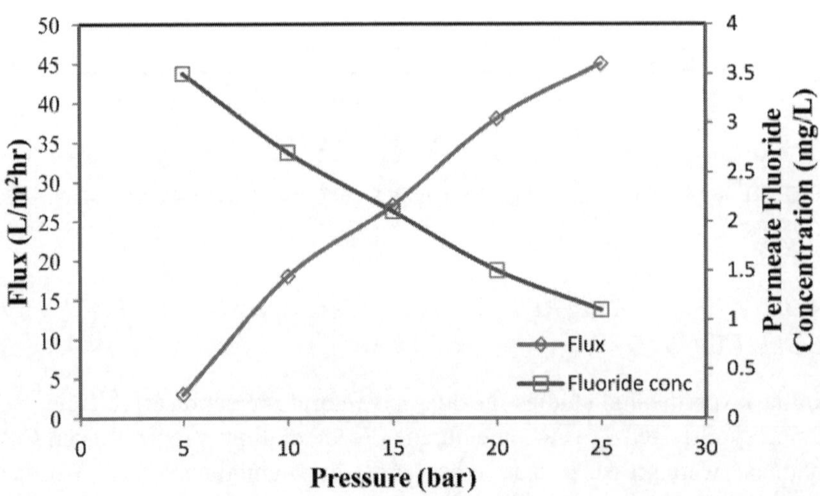


**FIGURE 4.15** Effect of pressure on flux and permeate fluoride concentration using synthetic feed solution.

#### 4.3.1.3 SEM STUDIES

Figure 16(a) and (b) show the SEM surface images of TFC RO membrane before and after fouling, respectively. From Fig. 16(a) it can be seen that the surface morphology appears to be clear with visible pores, while mineral salts and metal precipitates appear to form scales on the surface of the membrane used for defluoridation for a period of three years. This results in blockage of pores of 1Å diameter with inorganic and organic foulants. The difference of the surface morphology between the TFC membranes confirms that fouled membrane surface layer becomes denser. After three years of utilization, a dense cake layer formation over the surface of membrane is observed.

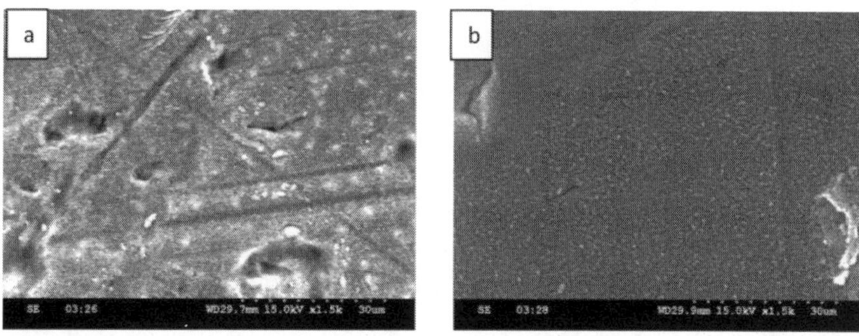


**FIGURE 4.16** SEM surface images of TFC polyamide membranes.

### 4.3.2 EFFECT OF OPERATING PARAMETERS ON MEMBRANE PERFORMANCE

Property of the membranes was observed for parameters such as turbidity, color and pH. Permeate obtained from RO membrane was found to undergo 6 Log reduction in E. Coli bacteria and further processed by UV treatment to remove residual bacteria and produce disinfected water for drinking.

#### 4.3.2.1 EFFECT OF FEED CONCENTRATION

The effect of feed concentration on flux and rejection is depicted in Fig. 4.17. As expected, a rise in the feed concentration resulted in a decrease in flux and percentage of impurity rejection. This is due to increase in concentration polarization and osmotic pressure at the membrane surface. Higher solute concentrations at the feed side induce an osmotic pressure gradient, which opposes the externally applied pressure gradient. Therefore, the driving force for the process ($\Delta p$–$\Delta\pi$) gets reduced, which causes a gradual decrease in flux.

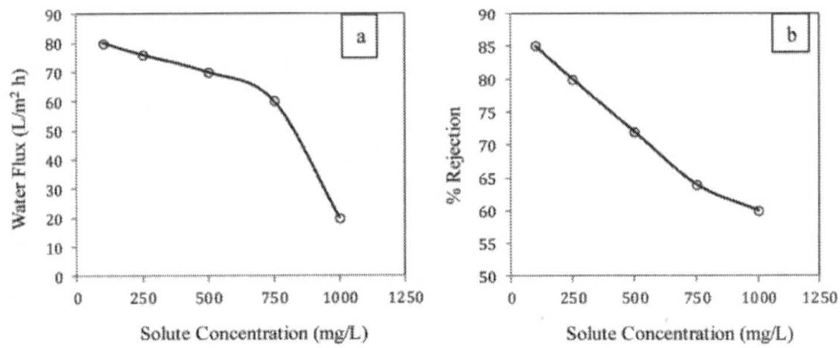


**FIGURE 4.17** Effect of feed concentration on (a) water flux and (b) percentage of rejection.

### 4.3.2.2 EFFECT OF FEED FLOW RATE

The effect of flow rate on flux and percentage of rejection is presented in Fig. 4.18, which depicts an increase in flux and percentage of rejection.

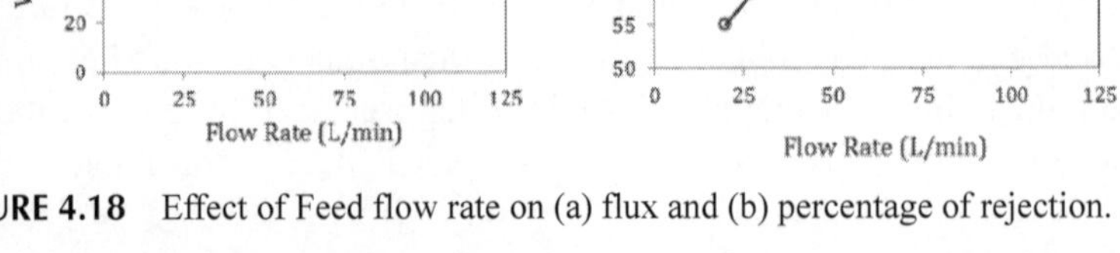


**FIGURE 4.18** Effect of Feed flow rate on (a) flux and (b) percentage of rejection.

### 4.3.2.3 EFFECT OF PRESSURE

Feed water pressure affects both the water flux and salt rejection of RO membranes. RO technology involves application of pressure to the feed water stream to overcome the natural osmotic pressure. Pressure in excess of the osmotic pressure is applied to the concentrated solution and the flow of water is reversed. As shown in Fig. 4.15 and explained in Section 4.3.1.2, water flux and fluoride rejection across the membrane increases with pressure. However, unlike in Fig. 4.15, the studies pertaining to Fig. 4.19 were carried out with actual ground water samples collected from villages.

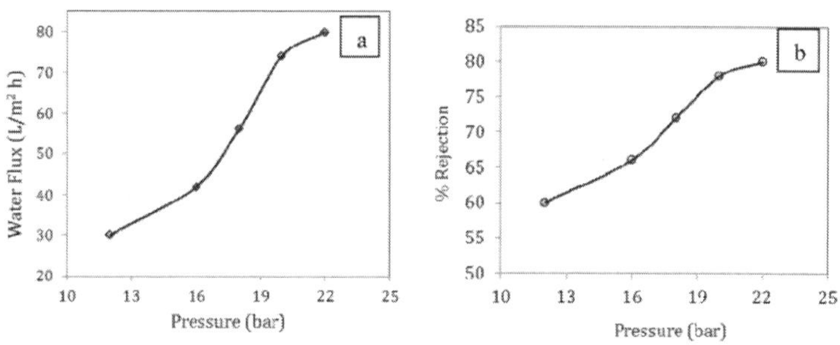


**FIGURE 4.19** Effect of feed pressure on (a) flux and (b) percentage of rejection.

## 4.3.2.4 EFFECT OF TEMPERATURE

As Fig. 4.20 demonstrates, membrane productivity is very sensitive to changes in feed water temperature. As feed temperature increases, overall flux increases almost linearly, primarily due to the higher diffusion rates of both water and solute through the membrane. Increased feed temperature also results in lower rejection or higher salt passage, which occurs since more energy gets dissipated to the solute molecules. This is due to a higher diffusion rate for salt through the membrane. The ability of a membrane to tolerate elevated temperature increases its operating latitude improves cleaning efficiency. However, in case of TFC polyamide RO membrane, the operating temperature must be restricted below 45°C to ensure long life.

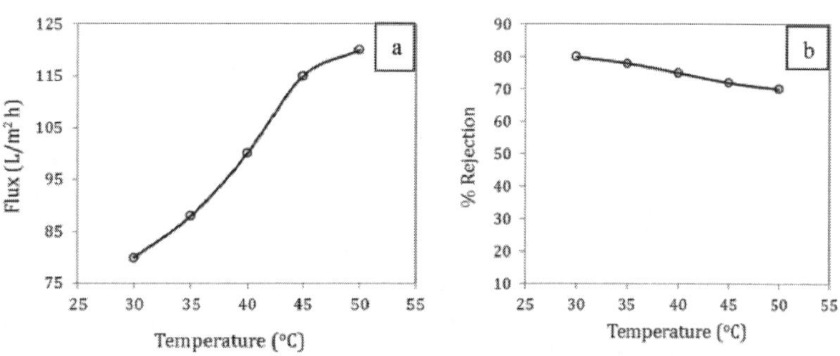


**FIGURE 4.20** Effect of feed temperature on (a) flux and (b) percentage of rejection.

### 4.3.3 COMPARATIVE PERFORMANCE OF RO AND NF

A batch of 80 L of feed volume (sample 1) containing 18 mg/L of fluoride and TDS of 820 ppm was introduced to a feed tank. Pressure on the feed side was adjusted to 15 bar and maintained constant during the run. The feed solution is processed through RO polyamide (PA) membrane and Functionalized Polyamide (HPA-250) Nanofiltration (NF) membrane. The reject or pass stream flow rate was maintained constant at 17 L/min. A recovery of 80% of volume in permeate with respect to the feed was attained and collected in a permeate tank. The flux of RO membrane was found to be 27 L/m$^2$h whereas it was 46 L/m$^2$h for NF membrane. The comparision of between RO and NF membranes are revealed in Table 4.4. Similarly, Sample 2 of fluoride concentration 1.05 mg/L and TDS of 610 ppm, tested at same operating conditions and summarized in Table 4.4.

**TABLE 4.4** Comparison of RO and NF processes for fluoride separation and water flux.

| S. No. | Reverse osmosis | | | | Nanofiltration | | |
|---|---|---|---|---|---|---|---|
| | Feed (mg/L) | Permeate (mg/L) | Flux (L/m$^2$hr) | Rejection (%) | Permeate (mg/L) | Flux (L/m$^2$hr) | Rejection (%) |
| Sample 1 | 18 | 0.36 | 27 | 98 | 1.2 | 38 | 93 |
| Sample 2 | 1.05 | 0.021 | 42 | 98 | 0.049 | 56 | 95.3 |

Though the flux in NF process is higher, its rejection is much lower than RO. However, NF can be applied wherever the ground water contains low fluoride concentrations ($\leq$ 2 mg/L). In specific rural areas where ground water is characterized by low TDS ($\leq$ 700 mg/L) and moderate fluoride, NF can be employed.

### 4.3.4 MODULAR CONFIGURATIONS

Membranes are never scaled-up as a single flat sheet, because its low surface area per unit volume ratio often results in high investment costs. Modular configurations are fabricated to enable large membrane surfaces to be housed in the smallest possible volumes. Their choice depends on cost, module packing density, risk of fouling and feasibility of cleaning. The most common commercially available membrane modules include plate-and-frame, tubular, spiral-wound and hollow fiber elements.

Plate-and-frame modules (Fig. 4.21(a)) consist of stacks of flat sheet membrane placed on supports, each membrane and support are separated by spacers, which direct the feed across each membrane and channel the permeate out of the module [26–29]. While this module is resistant to fouling, it has a low membrane surface area per element or packing density, which makes it expensive and can limit its use in areas with space restrictions. Tubular membrane elements (Fig. 4.21(b)) consist of membrane tubes (typically 1.3 cm ID) supported within perforated stainless steel tubes with feed flowing through the tubes and permeate from the shell or vice versa [26, 29]. While these elements are also fouling resistant and are easy to clean, the modules have a low packing density and can be expensive to operate because of the high feed flow rates necessary. Because of the plate-and-frame and tubular element disadvantages, these modules are used primarily for highly fouling feeds.

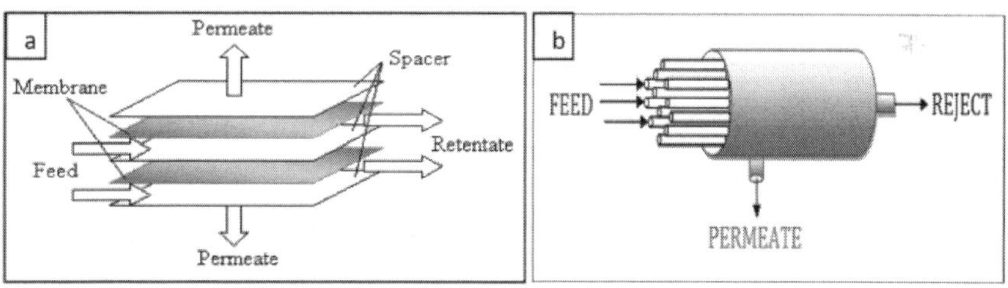


**FIGURE 4.21** (a) Plate and frame module; (b) Tubular membrane module.

The most widely used membrane modules are the spiral-wound and hollow-fiber elements. A spiral wound element (Fig. 4.22) consists of flat sheets of membrane separated by spacers that are rolled around a perforated central permeate collection tube. The feed is channeled across these rolled membrane sheets, permeates through the membrane, and gets collected in the central tube [26, 29]. This type of element has a high packing density, moderate fouling resistance with lower capital and operating costs compared to plate-and-frame or tubular modules. Moreover, spiral wound elements can withstand high pressures and are the preferred configuration for both RO and NF processes. Hollow fiber membrane modules consist of a large numbers of fine hollow fiber membranes with an outer diameter up to 0.5 mm placed in a housing or pressure vessel made of metal or fiber reinforced plastic (FRP) [26–29]. These elements have an extremely high packing density and so can have high permeate production rates per module. However, these modules are highly prone to fouling and need backwashing, flushing and chemical cleaning. Hollow fiber modules are being increasingly preferred for UF and MF.

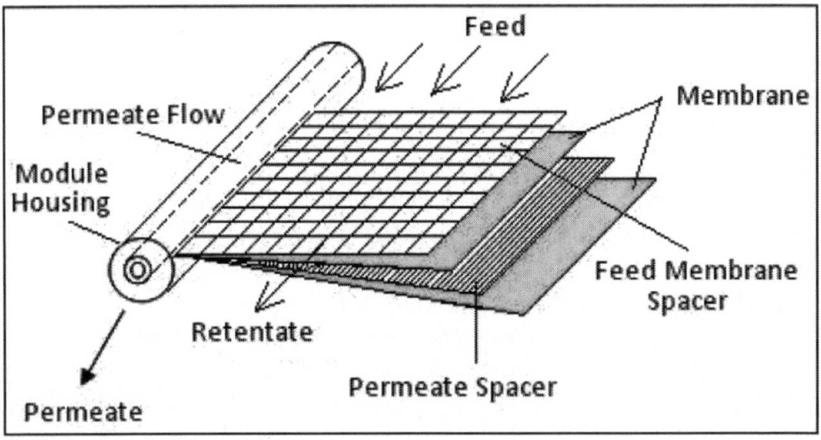


**FIGURE 4.22** Spiral wound membrane module.

In the present work, a spiral wound membrane module has been selected with TFC polyamide membranes inside a FRP cylindrical housing.

## 4.4 PILOT SCALE RO STUDIES IN RURAL ENVIRONMENT

Seven villages in the Andhra Pradesh state, in India were selected for the installation of pilot plants of 1000 L/h water treatment capacity each. The fluoride content in ground water in these villages ranged from 1.8 to 5.0 mg/L with TDS varying between 512 to 825 ppm.

### *4.4.1 DESIGN OF MEMBRANE SYSTEM*

The membrane system was designed based on the available feed water quality, product flow or recovery and required product quality. The fluoride in the raw water, which varied between 1.8 to 5.0 mg/L is brought down to a concentration of less than 1.5 mg/L as per WHO standards. The pilot plant is equipped with an effective pretreatment system consisting of sand filtration, activated carbon treatment, antiscalant dosing and micron cartridge filtration besides post treatment by UV radiation and ozonation for long-term storage. Moreover, the option of blending the permeate with a small fraction of raw water is also made available to maintain a total dissolved solids (TDS) level of at least 50 ppm in the product water to ensure sufficient supply of minerals for human consumption. It may be noted that the blending water goes through all the pretreatment stages except for antiscalant dosing besides all post treatment steps but does not pass through the membrane. In regions where fluoride presence in ground water is very high, blending is restricted to ensure that $F^-$ in permeate does not exceed 1.5 mg/L. A simulation program was designed to maintain the rate of blending and recovery of RO membrane. The view of Microsoft excel simulation program is shown in Fig. 4.23. Mode of safe reject disposal and reuse is dealt with in Section 4.5.

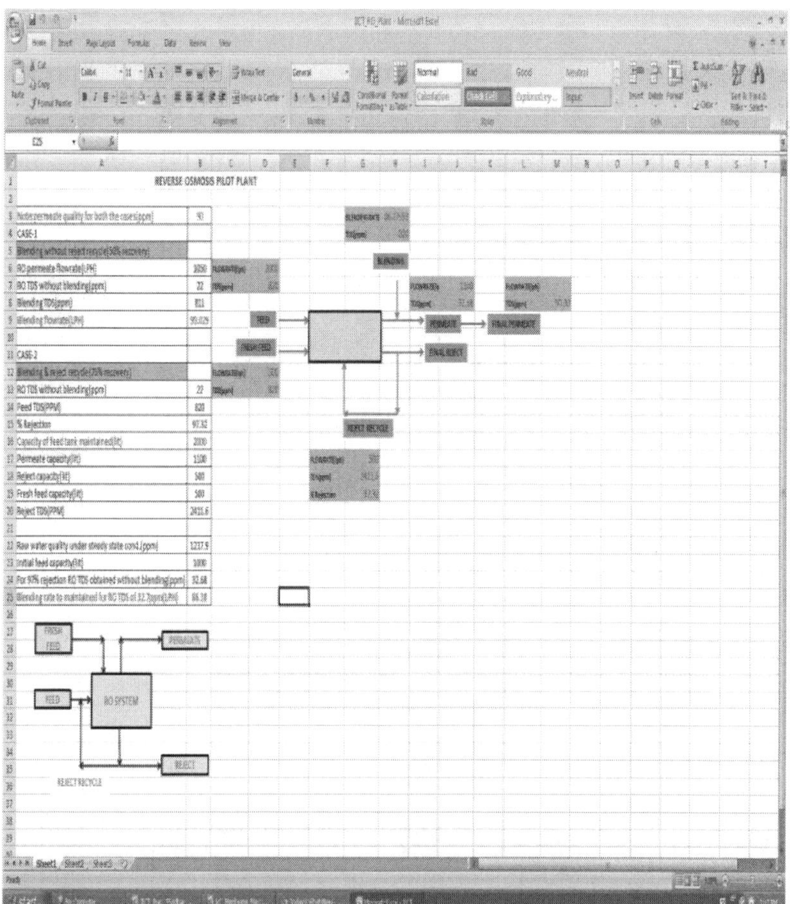


**FIGURE 4.23** Simulation for operation of RO plant.

### 4.4.1.1 PROCESS FLOW DIAGRAM FOR 1000 L/H CAPACITY

The feed tank is filled with raw water and then pumped by low pressure feed pump through a prefilter assembly consisting of sand filter, activated carbon column and micron filter to remove suspended solids as shown in Fig. 4.24. The preparing and instrumentation diagram of the pilot plant is shown in Fig. 4.25. The high pressure pump is then employed to transport the feed through the spiral wound membrane module and the system pressure is maintained at a constant value by means of a needle valve. Characteristics of TFC RO polyamide membrane are listed in Table 4.5. Permeate from the membrane system is further post treated by UV treatment and ozonation before being pumped out for drinking.

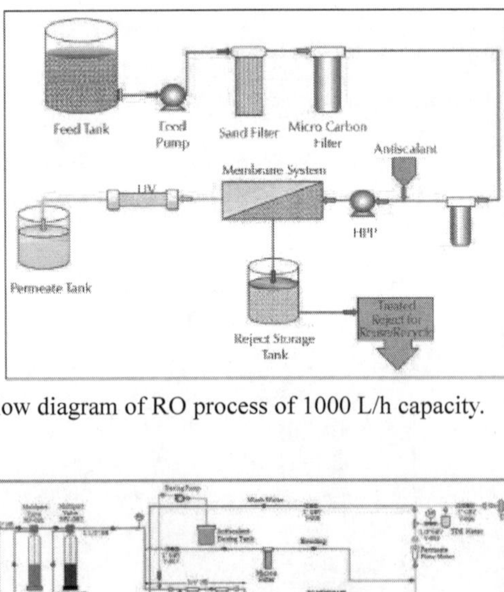


**FIGURE 4.24** Flow diagram of RO process of 1000 L/h capacity.

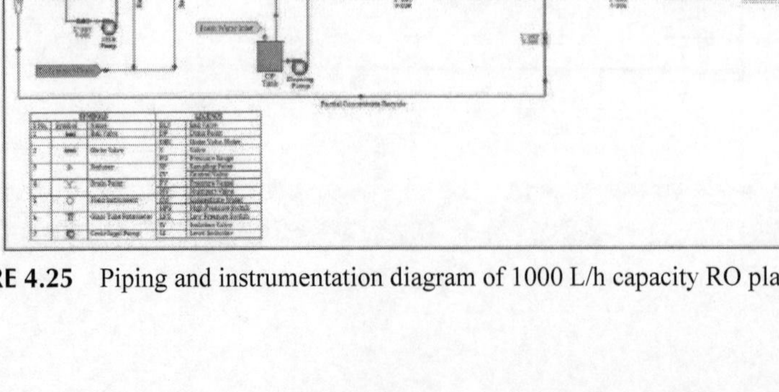

**FIGURE 4.25** Piping and instrumentation diagram of 1000 L/h capacity RO plant.

A skid-mounted system was built incorporating five commercial TFC polyamide RO membranes inside two FRP cylindrical housings as shown in Fig. 4.25. A feed tank of 3000 L capacity was provided for raw water storage and supplies the feed water to the pretreatment assembly includes sand filter column and activated carbon filter column by 1 HP raw water pump. The main components of the membrane system are prefiltration assembly, high pressure pump, membrane assembly and post treatment assembly. Pretreatment constitutes sand filtration, activated carbon filter and microfiltration. A polypropylene prefilter rope cartridge of 5 μm pore size was mounted at the upstream side of the membrane system to prevent entry of any suspended solid particles. The system is suitably instrumented to measure the pressure, total dissolved solids (TDS), flow rate and pH for permeate and reject. The pressure is measured with high precision and accurate pressure gauges while the TDS and pH of permeate are continuously measured by digital meters. Blending and partial reject recycle have been provided to maintain adequate product TDS levels and achieve high water recovery. Post treatment assembly constitutes UV and Ozonation connected online for disinfection of product water.

**TABLE 4.5** List of typical characteristics of RO membrane [Permeate flow and salt rejection based on the following test conditions: 2000 ppm solute concentration 100 psi (0.7 Mpa) feed pressure, temperature 25°C, 15% recovery, sucrose rejection is at 5% feed concentration)].

| Product Name | TFC Polyamide RO Membrane |
|---|---|
| Size | 8″dia * 40″ Long |
| Area | 36 $m^2$ |
| Permeate water flow rate | 27 $L/m^2h$ |
| $MgSO_4$ Rejection | 99% |
| $Na_2SO_4$ Rejection | 99% |
| NaCl Rejection | 93% |
| Sucrose Rejection | 99.9% |

### *4.4.1.2 PIPING AND INSTRUMENTATION DIAGRAM WITH PROCESS DESCRIPTION*

The pretreated water is pumped through the membrane module assembly arranged in series by a 3.5HP single phase high pressure pump of maximum capacity 30 bar. A restricting needle valve was provided on the concentrate outlet to pressurize the feed liquid to a desired value indicated by a pressure gauge installed at a position prior to that of the valve. Permeate is collected in the intermediate storage tank in which the RO treated water is remineralized and adequate product TDS levels prescribed by BIS standards are maintained by blending with pretreated feed water entering at a fixed calculated flow rate. Finally, after UV treatment and ozonation the water is pumped to a final storage tank. The reject from the membrane system is partially recycled to feed tank and partially disposed through a two valve arrangement provided on the reject line to recycle part of the reject back to feed tank and pump the other part for the reject post treatment. Fresh raw water is added continuously to make up for the decrease in total volume in the feed tank. A high cross-flow velocity of 2000 L/h of feed flow was maintained to minimize concentration polarization and fouling. Electrochemical dosing pump mounted on the dosing skid is provided to inject the dosing chemical continuously into feed water line. The dosing chemical is prepared by adding 150 mL of the concentrate to a PP tank containing 100 L purified water. Dosing pump is started from the system panel after adjusting the dosing rate to 5 mg/L or as required to inhibit scale formation. Permeate and concentrate flow rates were measured using accurately calibrated rotameters containing metal floats. The photograph of a model typical RO plant (1000 L/h capacity) designed and installed at CSIR-IICT at Hyderabad, India premises of 1000 L/h product capacity is shown in Fig. 4.26.

**FIGURE 4.26** Digital image of 1000 L/h capacity RO plant.

## 4.4.2 TREATMENT OF CONTAMINATED GROUNDWATER IN VILLAGES

Groundwater containing moderate TDS and high fluoride in different villages of Andhra Pradesh, India were treated by RO based defluoridation technology. Analytical report of feed and permeate samples is presented in Table 4.6. The observations pertain to an operating pressure of 15 bar and water recovery of 50-70%. The RO plants installed by CSIR-IICT at different villages are listed in the table and have been successfully for the past 3-4 years. Regular membrane cleaning and storage accompanied by system maintenance was found to extend the life of the plant.

The conductivity values were ranging between 0.8 to 1.2 mS/cm for feed samples at different villages listed above in Table 4.6. The treated water samples were exhibiting conductivity in the range 0.05–0.08 mS/cm.

**TABLE 4.6** Plants installed by CSIR-IICT in different villages of A.P., India.

| S. No. | Village | Feed | | | | Permeate | | | |
|---|---|---|---|---|---|---|---|---|---|
| | | Fluoride (mg/L) | TDS (ppm) | Turbidity (FAU) | pH | Fluoride (mg/L) | TDS (ppm) | Turbidity (FAU) | pH |
| 1. | Allahpur | 2.46 | 615 | 1 | 8.2 | 0.23 | 53 | 0 | 7.1 |
| 2. | Marlapadu | 3.20 | 512 | 2 | 7.9 | 0.31 | 46 | 0 | 7.3 |
| 3. | Gaggalapally | 1.80 | 825 | 5 | 8.3 | 0.15 | 33 | 0 | 6.8 |
| 4. | Nerimetta | 2.03 | 622 | 1 | 8.4 | 0.27 | 45 | 0 | 7.1 |
| 5. | Rachakonda | 3.60 | 637 | 6 | 8.9 | 0.22 | 38 | 0 | 7.2 |
| 6. | Mylaram | 5.00 | 820 | 3 | 8.5 | 0.50 | 42 | 0 | 7.1 |
| 7. | Karamchedu | 3.20 | 650 | 5 | 8.2 | 0.20 | 35 | 0 | 7.3 |

## 4.4.3 PROCESS ECONOMIC EVALUATION

### 4.4.3.1 CAPITAL COST FOR 1000 L/H CAPACITY RO PLANT

The complete list of equipment and corresponding costs for 1000 L/h capacity RO plant is presented in Table 4.7, in, which the total price for the major accessories such as high-pressure pump, TFC Polyamide membrane module with housing are given as 1200 USD and 2190 USD, respectively. The total capital investment is approximately 5500 USD for the RO system without considering the cost of storage tanks.

### 4.4.3.2 OPERATION AND MAINTENANCE COST FOR RO SYSTEM

Table 4.8 provides the complete operation and maintenance cost of RO system, which includes power, membrane module and micron filter cartridge replacement costs, chemical consumption for cleaning and storage, etc. Feed capacity and recovery were assumed to be 2 $m^3/h$ and 60%, respectively. The operational time for system was assumed to be 10 h per day in view of the fact that electric power supply in remote villages is restricted to a few hours per day. Depreciation costs were taken as 10% of the total capital investment while life of hardware is expected to be 10 years. Membrane life was found to be 3 years as per data collected at village sites where plants have been operated.

**TABLE 4.7** List of equipment and corresponding costs.

| S. No. | Item | Specifications | Quantity | Total Cost (USD) |
|---|---|---|---|---|
| 1. | Raw water storage tank | 3000 L capacity, HDPE | 1 No | 200 |
| 2. | Raw water Pump | 1HP (Make: Nanfang, Model: CHL 2–50), 3.5 $kg/cm^2$ discharge pressure | 1 No | 116 |
| 3. | Pressure Sand Filter | 4 ft dia×5.6 ft long | 1 No | 150 |
| 4. | Activated Carbon Filter | 4 ft dia×5.6 ft long | 1 No | 200 |
| 5. | Antiscalent Dosing Pump | ½ HP (Make: Injecta, Model: Olimpia 05–05) | 1 No | 90 |
| 6. | Antiscalent Dosing Tank | 100 L, MOC:PP | 1 No | 20 |
| 7. | Micron Cartridge Filter | Dimensions: 4″x20″, 5μm | 1 No | 25 |
| 8. | Micron Cartridge Filter for Blending line | 5μm pore size | 1 No | 15 |
| 9. | High Pressure Pump | 3HP Single phase (Make: WILO/CRI) Max Pressure: 30 $kg/cm^2$ | 1 No | 1200 |
| 10. | Reverse Osmosis Membranes | 4″ dia *x* 40″ length | 5 Nos | 2000 |
| 11. | Membrane Housing | 4″ dia *x* 120″ length | 1 No | 100 |
| 12. | Membrane Housing | 4″ dia *x* 80″ length | 1 No | 90 |

**TABLE 4.7** *(Continued)*

| S. No. | Item | Specifications | Quantity | Total Cost (USD) |
|---|---|---|---|---|
| 13. | Cleaning-In-Place Tank | 200 L | 1 No | 40 |
| 14. | Intermediate Storage Tank | 1000 L (MOC: HDPE) | 1 No | 100 |
| 15. | UV Feed Pump | 1 HP (Make: Nanfang, Model: CHL 2–30) | 1 No | 100 |
| 16. | UV System | 1 mt | 1 No | 200 |
| 17. | Pipes and Fittings* | MOC: UPVC<br>1 ½" dia for pretreatment section<br>1″ dia for high pressure line | 1 No | 200 |
| 18. | Permeate water storage Tank | 2000 L capacity | 1 No | 200 |
| 19. | Ozonator | 1 Gram Capacity | 1 No | 100 |
| 20. | 3-Phase Control Panel | SS 350/500 | 1 No | 170 |
| 21. | Skid | Stainless steel 316 | 1 No | 200 |

**TABLE 4.8** Operation and maintenance cost of RO based defluoridation system.

| | Value |
|---|---|
| Feed capacity ($m^3$/h) | 2 |
| Permeate capacity ($m^3$/h) | 1.2 |
| Recovery offered | 60% |
| Operating Cost Estimation | |
| Module Replacement Cost | |
| Number of modules (4″ dia, 40″ long) | 5 |
| Price per module (USD) | 400 |
| Total module replacement cost (USD) | 2000 |
| Duration of replacement (Years) | 3 |
| No. of working h/day | 10 |
| Cost/hr (USD) | 0.083 |
| Cartridge Replacement Cost | |
| No of cartridges | 2 |
| Price per cartridge (USD) | 12 |

**TABLE 4.8** *(Continued)*

| | Value |
|---|---|
| Total cartridge replacement cost (USD) | 24 |
| Duration of replacement (days) | 180 |
| No. of working hours per day | 22 |
| Cost/hr (USD) | 0.006 |
| **Power Cost** | |
| Feed pump (KW) | 0.745 |
| Dosing systems (KW) | 0.015 |
| High pressure pump (KW) | 2.237 |
| UV lamp (KW) | 0.072 |
| Total power consumption-KW | 3.06 |
| Hourly cost (USD) (4.5 Rs/unit) | 0.27 |
| Chemical Consumption | |
| Anti-scalant dosing (mg/L) | 5 |
| Dosage (L/h) | 0.01 |
| Cost/lit (USD) | 6.4 |
| Hourly cost (USD) | 0.064 |
| CIP Chemicals (EDTA, NaOH, Citric acid) | |
| Frequency (days) | 15 |
| Total cost of CIP per hour (USD) | 0.045 |
| Total operating cost per hour (USD) | 0.42 |
| Total operating cost per year assuming 22 h of operation/day (USD) | 1533 |
| Depreciation cost (Assuming 10% of capital cost) (USD) | 505.6 |
| Labor cost per year + Raw water (USD) | 2660 |
| Total cost per year (USD) | 4698.6 |
| Permeate | |
| Quantity (L/h) | 1200 |
| Operation time (h) | 10 |
| Quantity of permeate generated in 1 year (L/yr) | 4,380,000 |
| Cost/L of permeate (USD) {Indian Rupees} | 0.00107 {0.05} |
| If sold at $4\times10^{-3}$ USD/L | |
| Annual Profit (USD) | 12822 |
| Pay Back Period (Yr) | 0.8 |

The price of treated water is 0.00107 USD/L of purified water produced (Table 4.8). RO price is competitive as compared to adsorption and chemical precipitation technique and more economical than ion-exchange process, which requires cumbersome regeneration procedures by acid and alkali treatment.

### 4.4.4 METHOD FOR REJECT TREATMENT AND REUSE

The reuse of untreated RO reject water has raised serious environmental concerns. Direct disposal of this water into a water-body may invigorate the growth of aquatic masses causing environmental problems. Disposing on land has potential of concentrating salts and nutrient compounds in soil and contaminating groundwater with excess fluoride. Reject water produced from these processes requires further treatment for removing concentrated fluoride and other salts before its safe disposal. A study was carried out to investigate a process that has potential of removing the salts from the reject wastewater.

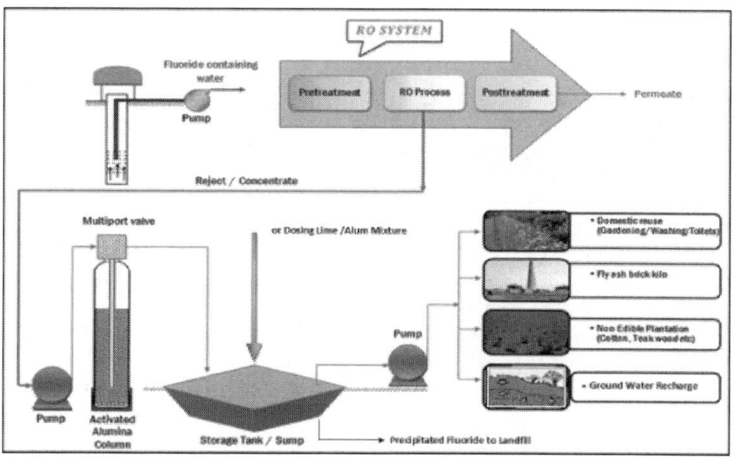


**FIGURE 4.27** Method of RO reject treatment and reuse.

Figure 4.27 exhibits a comprehensive method for treatment and reuse of reject stream coming from a defluoridation plant. The reject can be pumped through an activated alumina column particularly to remove the fluoride present in the water. Alternately, the water is stored in a tank/sump where the other salts present are precipitated by adding lime + alum mixture. The clear water is decanted at regular intervals and solid precipitate is safely disposed to landfill.

The treated water may be reused for domestic purposes like gardening, washing, etc. or for fly ash kilns or for nonedible plantations (like cotton, teak plantation, etc.) or for ground water recharge.

### 4.4.5 MAINTENANCE OF DEFLUORIDATION PLANTS

Figure 4.28 shows the images of CSIR-IICT team's visit to villages. IICT designed and commissioned the first plant of 600 L/h capacity at Mylaram village of Nalgonda District in July, 2005, which proved to be a model for replication by other organizations, and widely appreciated by the Governor of Andhra Pradesh, press and the masses due to the painstaking efforts put in by the authors.

**FIGURE 4.28** Rural folk collecting RO treated water.

Later the capacities were increased to 1000 L/h as well as 4000 L/h. In some regions the team has made more than 30 visits to overcome hurdles during installation such as poor logistics, low voltage, lack of power supply and shed to house the plant. Other challenges have been spreading awareness about fluorosis disease, plant maintenance in remote areas besides reject disposal or reuse.

## 4.5 CONCLUSIONS

1. The thin film composite membrane acts like a third phase and splits water molecules and dissolved fluoride into two different phases, permeate and reject. Due to preferential affinity and smaller molecular size, $H_2O$ penetrates the membrane whereas the non-interacting and bigger $F^-$ cannot pass through easily.
2. Amongst the various methods for fluoride removal from water, membrane based RO process is ideal for community-based application whereas adsorption through activated alumina as well as natural adsorbents is more suitable for domestic use. However, RO can remove all impurities in a single step including excessive hardness and harmful microbes if any, whereas adsorbents can target only one of the species like $F^-$.
3. RO process needs remineralization either through blending or addition of mixture of salts to permeate since the RO permeate is devoid of important minerals and contains low TDS content, which needs to be boosted to levels sufficiently healthy for the human heart and body. pH of RO permeate generally falls due to dissociation of carbonates and bicarbonates into $CO_2$ and sometimes alkali dosing or aeration may be required.
4. The reject from a RO process needs to be effectively treated either through chemical precipitation or adsorption such that the water can be recycled for domestic washing, nonedible plantation, fly ash brick kiln or ground water recharging. Without an effective reject treatment scheme, the RO based defluoridation process would be an incomplete temporary solution since the concentrated fluoride present in the waste stream would go back into the ground water and contaminate it.
5. Nanofiltration gives higher flux in comparision to RO but lower $F^-$ rejection and hence suitable only for low fluoride concentrations while RO is more suitable for high fluoride concentrations. For ground water containing low TDS with high fluoride, chemical precipitation followed by ultrafiltration to remove turbidity and microbial content would be a feasible process in future.
6. With increasing F-concentration in raw water, the recovery would reduce due to rising osmotic pressure. Applied pressure has a positive effect on both flux and $F^-$ rejection, which increase with increasing $\Delta p$. In contrast, high TDS concentration has a negative impact on both parameters whereas rising temperature induces greater $F^-$ permeation.

High cross flow velocity brought about by greater feed water flow rate also results in positive performance by the membrane in terms of $F^-$ rejection due to minimization of concentration polarization.

7. Conventional defluoridation systems employed in India such as adsorption using activated alumina or Nalgonda technique, which involves chemical precipitation with lime-alum mixture cannot remove all the impurities in a single step like RO does. RO can be scale-up to higher community-based capacities, which is not feasible in case of the aforementioned methods. Moreover, RO can also remove microbial content, which the conventional methods cannot. The price of water treated by RO is competent with other methods and more economical than ion-exchange process, which requires cumbersome regeneration procedures.
8. Membrane can be maintained well through cleaning with a mixture of 1% NaOH + 0.5% EDTA in tap water, which removes stubborn foulants. Alternate washing with 1% citric acid can remove metal salts and mineral precipitates from the membrane surface. The membrane must be stored in 0.5% SMBS ($NaS_2O_5$) during shutdown periods as well as over every weekend to prevent biological fouling, which is otherwise irreversible.
9. Maintenance of membrane plants in villages is a big challenge due to remoteness of fluorosis affected regions, and lack of logistics such as regular power supply. However, the major hurdle is the lack of awareness among rural folk regarding the causes of fluorosis and the technology leads available in purifying fluoride contained water. Through this chapter, the authors aim to spread awareness and information, which is sorely lacking especially causes of fluorosis and measure for its prevention and control. A population of 1 lakh in rural areas of Andhra Pradesh State of India have benefited from application of RO based defluoridation of ground water implemented by CSIR-IICT.

## ACKNOWLEDGMENTS

The authors are grateful to CSIR, New Delhi, India for grant of funding under XII Five Year Plan MATES project. The second author KP is thankful to RMIT, Melbourne, Australia for research fellowship. The support of Department of Science and Technology, New Delhi is acknowledged for sponsoring installation of five of the plants.

## REFERENCES

1. Fawell, J., Chilton, J., Dahi, E., and Magara, Y. Fluoride in Drinking-water, IWA Publishing: London (2006).
2. Ghada, F. and Suzanne, E. Physicochemical and geochemical characteristics of raw marine sediment used in fluoride removal. *J Environ Sci Health, Part A Toxic/Hazardous Subst Environ Engg* **45,** 1601–1615 (2010).
3. Bhattacharyya, D. and Williams, M. Introduction and Definitions-Reverse Osmosis Membrane Handbook; W. Ho.; Sirkar, K., Eds.; VanNostrand Reinhold: New York, pp. 265 (1992).
4. Naohito, K., Ogata, F., Hisato, T., Isao, Y. Removal of fluoride ion by bone char produced from animal biomass. *J Oleo Sci* **58,** 529–535 (2009).
5. Christoffesen, J., Christoffersen, M. R., Larsen R., and Møller, I. J. Regeneration by surface-coating of bone char used for defluoridation of water. *Water Res* **25,** 227–229 (1991).
6. Kaseva, M. E. Optimization of regenerated bone char for fluoride removal in drinking water: a case study in tanzania. *J Water Health* **4,** 139–147 (2006).
7. Amir, H., Sadia, N., and Qaisar, M. Strategies for low-cost water defluoridation of drinking water-a review of progress. *J Chem Soc Pak* **32,** 550–555 (2010).
8. Mjengera, H. and Mkongo, G. Appropriate deflouridation technology for use in flourotic areas in Tanzania. *Phy Chem Earth* **28,** 1097–1104 (2003).
9. Baker, R. W. Overview of membrane science and technology. In: Membrane technology and applications, 2nd ed.; Wiley: England (2004).
10. Meenakshi, V. K., Garg, K., and Renuka, A. Ground water quality in some villages in Haryana, India: focus on fluoride and fluorosis. *J Hazardous Mat* **106B,** 85–97 (2004).
11. Agarwal, V., Vaish, A.K., and Vaish, P. Ground water quality: focus on fluoride and fluorosis in Rajasthan. *Curr Sci* **739,** 743–746 (1977).
12. Chandra, S. J., Thergaonkar, V. P, and Sharma, R. Water quality and dental fluorosis. *Ind J Pub* **H115,** 47–51 (1981).
13. Hillis, P. Membrane technology in water and wastewater treatment, Royal Society of Chemistry: London (2000).
14. Hilal, N., Al-Zoubi, H., Darwish, N. A., Mohamma, A. W., and Abu Arabi, M. A comprehensive review of nanofiltration membranes: treatment, pretreatment, modeling, and atomic force microscopy. *Desalination* **170,** 281–308 (2004).
15. Noble, R. D., and Stern, S. A. Membrane separation and technology-principles and applications, Elsevier, Amsterdam, The Netherlands (1995).
16. Sridhar, S. and Rao, A. G. Treating effluents in bulk drug industry using reverse osmosis. *Pharma Bioworld.* 91–97 (2004).
17. Sridhar, S. and Smitha, B. Performance and prospects of nanofiltration for separations in pharmaceutical manufacture. *Management/Environment Pharma Bioworld* 78–83 (2004).
18. Cadotte, J. E., King, R. S., Majerle, R. J., and Petersen, R. J. Interfacial synthesis in the preparation of reverse osmosis membranes. *J Macromol Sci-Chem* **15,** 727–755 (1981).
19. Kesting, R. E. Phase inversion membranes; Lloyd, D. R., Ed.; ACS Symposium Series: Washington, DC (1985).

20. Kucera, J. Properly apply reverse osmosis. *Chem Eng Prog* **93,** 54–61 (1997).
21. Rautenbach, R., Albrecht, R. *Membrane Processes*, John Wiley & Sons Inc., New York (1989).
22. Clesceri, L.S., Greenberg, A. E., and Eaton, A. D., Standard methods for the examination of water and waste-water. American Public Health Association. Washington (2005).
23. Trivedi, R. K. and Goel, P. K. Chemical and biological methods for water pollution studies. environmental publications, Karad (1986).
24. Pangarkar, B. L., Sane, M. G., and Guddad, M. Reverse osmosis and membrane distillation for desalination of groundwater: a review. *ISRN Mater Sci* 1–9 (2011).
25. Allegrezza, A. Commercial reverse osmosis membranes and modules in reverse osmosis technology. Parekh, B. ed.; Marcel Dekker, Inc.: New York (1988).
26. Baker, R. Membrane and module preparation in membrane separation systems, US DOE Report, DOE/ER/30133-H1, Vol. **2** (1990).
27. Strathmann, H. Synthetic membranes and their preparation in handbook of industrial membrane technology, Porter, M., ed.; Noyes Publications: Park Ridge, NJ (1990).

# PART III

## CARBON DIOXIDE FACILITATED CHEMICAL TRANSFORMATIONS

## CHAPTER 5

# CATALYTIC CONVERSION OF $CO_2$ INTO FUELS AND CHEMICALS: A GREEN CCU OPTION

PULIKKEEL UNNIKRISHNAN and DARBHA SRINIVAS

## CONTENTS

## 5.1 INTRODUCTION

At present, the usable quantities of $CO_2$ are considerably smaller than its emissions from power generation and other sources. On the most optimistic front, it is estimated that chemical industries can convert at most around 6–7% of global emissions for bulk and fine chemical manufacture and 10% into synthetic fuels. The demand is likely to increase on long-term basis. Notwithstanding this, the utilization of $CO_2$ is still a welcome measure on a medium term basis as part of global strategy to create economic value for it and minimize its environmental impact as a greenhouse gas. Also significant drawbacks are associated with CCS option to capture $CO_2$ from industrial scale power plants and store it in geological formations. They include unforeseen leakages, long-term liability, problems with public acceptance and high cost of $CO_2$ capture.

The CCU concept is based on carbon capture and utilization through, which $CO_2$ molecules end up as industrially useful chemical molecules. This option can build on current post combustion technologies to offset the cost of additional $CO_2$ capture plant investment or even make the process profitable. There are thus good arguments for investing in CCU alongside CCS. This chapter indentifies the potential CCU options.

Carbon dioxide is the end product of chemical combustion processes. Thermodynamic stability and low reactivity (due to kinetic origin) of $CO_2$ are the causes for the initial general lack of interest in the chemistry and application of this molecule. However, the kinetic barrier can be overcome by activating $CO_2$ over catalyst surfaces and by photochemical, electrochemical, thermochemical and biochemical means. Currently, about 6.8 Gt of $CO_2$ is produced per year from combustion and other industrial and human related activities. Its concentration in the atmosphere has been growing continually and in the past decade, its average annual increase was reported to be 2.1 ppm per year [1]. At present the concentration of $CO_2$ in the atmosphere is 400 ppm, which is far above its upper safety limit of 350 ppm. $CO_2$ is a greenhouse gas and has adverse effects on environment and human health. Hence, $CO_2$ recycling is a priority to control global emission of this dominant greenhouse gas and also to reduce our dependence on nonrenewable petroleum feedstock in producing fuels and chemicals [2].

Photosynthetic process in green plants forms organic compounds and molecular oxygen from low-energy carbon dioxide by using water and sunlight. A rough estimate indicates that *ca.* 200 billion tons of biomass is produced in this way per annum. In terms of chemical throughput, photosynthesis is by far the most important synthetic process. Currently, $CO_2$ is used in beverage industry, fire extinguisher technology, refrigeration, enhanced oil recovery and supercritical $CO_2$ extraction. In inorganic chemical industry, it is used to manufacture $Na_2CO_3$/$NaHCO_3$ (the Solvay process), $CaCO_3$ and other carbonates. $CO_2$ is an attractive chemical feedstock. There are several motivations for producing fuels and chemicals from $CO_2$ whenever possible. $CO_2$ is a renewable feedstock of practically inexhaustible supply. It is an inexpensive, low-toxic and nonflammable feedstock that can frequently replace toxic chemicals such as phosgene and isocyanates in chemicals synthesis. The production of chemicals from $CO_2$ can lead to totally new materials (such as polymers) with novel properties. The new routes using $CO_2$ to existing chemical intermediates and products could be more efficient and economical than the conventional methods. There are only a handful of industrial chemical practices that use $CO_2$ as a chemical building block. They are: urea synthesis ($CO_2$ consumption: 107 million tons/year), manufacture of salicylic and para-hydroxy benzoic acids ($CO_2$ consumption: 30,000 tons/year), cyclic and polycarbonates synthesis (ca. 1.5 million tons/year) and production of methanol ($CO_2$ consumption: 2 million tons/year) [3]. Presently, only a small fraction of $CO_2$ produced (about 0.5–1%) is used in global chemical industry. As stated earlier, synthesis of fuels and chemicals from $CO_2$ could certainly have some positive impact on the global carbon balance and environment. Thus discovery of many more processes, which use $CO_2$ as a raw material, is highly desirable. Presently, a number of reactions and processes are under investigation (Fig. 5.1). There are excellent reviews on $CO_2$ activation and utilization [4–11]. This chapter is an overview on the subject of $CO_2$ as a raw material for production of fuels and commodity chemicals. The application of $CO_2$ as a solvent and as a catalyst precursor is also presented briefly. This chapter is confined mainly to catalytic conversion of $CO_2$ to some valuable products. No attempt is made to cover the literature on photo, electro and biocatalytic means of $CO_2$ transformation. A few technologies announced that are under pilot or exploratory stage combining all the above means of converting $CO_2$ are presented at the end of this chapter.

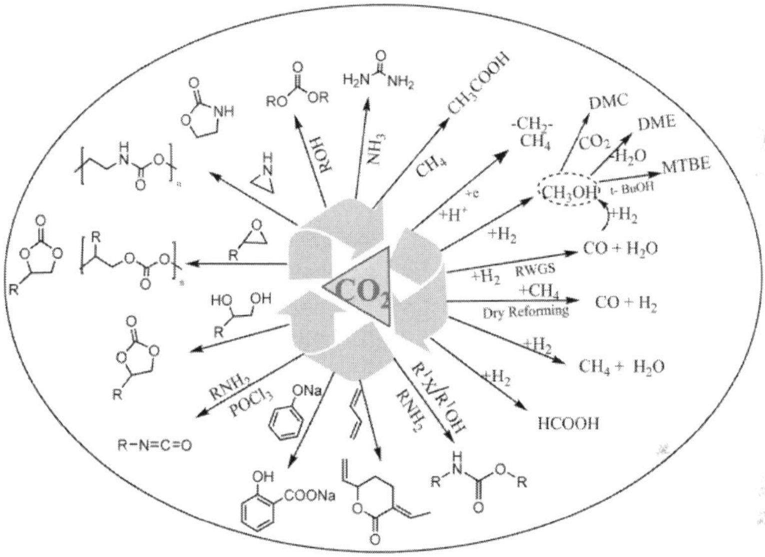

**FIGURE 5.1** Potential reactions for $CO_2$ based downstream products.

## 5.2 CARBON DIOXIDE CONVERSION INTO FUELS

The worldwide requirement of fuels is increasing day by day. Utilization of $CO_2$ for energy production endures an interesting research area because of its sustainability and economic profit. Catalytic $CO_2$ hydrogenation produces a variety of chemicals and fuels. Direct hydrogenation of $CO_2$ and hydrogenation of CO produced from $CO_2$ are the two different ways to manufacture fuels from $CO_2$. Biocatalytic routes are also available for fuel production from $CO_2$, but they are expensive compared to other catalytic routes. This section explains details of $CO_2$ conversion into fuels such as methanol, higher alcohols, hydrocarbons, formic acid, CO and hydrogen.

### 5.2.1 METHANOL

Methanol ($CH_3OH$) is one of the most flexible commodity chemicals. The worldwide consumption of methanol is growing constantly and projected to increase by 2% from 2008 to 2013 [12]. It is used as an alternative fuel, fuel additive and platform chemical for various downstream products. It burns at lower temperature and has high octane rating and high heat of vaporization. $CO_2$ conversion to methanol (Eq. (1)) is challenging due to the formation of by-products like CO, hydrocarbon and higher alcohols along with methanol. Most important challenge is designing of selective catalyst for methanol. This reaction is exothermic in nature; decrease in reaction temperature and increase in $CO_2$ pressure favor methanol formation.

$$CO_2 + 3H_2 \rightarrow CH_3OH + H_2O \quad -49 KJ/mol\ (\Delta H) \quad (1)$$

Cu/ZnO-based catalysts used in the production of methanol from syn-gas are effective also for this reaction. On the basis of various studies, it has been concluded that addition of ZnO to a Cu-based catalyst produces active sites, which have control over the specific activities for methanol synthesis [14, 15]. $Cu^+$ is more active than elemental copper for methanol synthesis. But the yield of methanol is comparatively lower than that obtained in syn gas conversion. The purity of the crude methanol produced from $CO_2$ is 99.9 wt%, which is higher than that of commercial methanol production from syngas [13]. Promoter, precursor and nature of support oxide have crucial role on catalyst performance. A large number of research publications appeared in the last few decades on the effects of preparation methods, support, promoters, active sites and surface morphology of such catalysts in methanol synthesis [13, 16–20]. The Cu/ZnO multicomponent catalysts (Cu/ZnO/$ZrO_2$/$Al_2O_3$ and Cu/ZnO/$ZrO_2$/$Al_2O_3$/$Ga_2O_3$) created a positive impact on $CO_2$ conversion and methanol selectivity [21, 22]. But these catalysts are not attractive from commercial point of view due to their instability. Their long-term stability has been significantly improved by addition of a small amount of colloidal silica to it [23–25]. The addition of $SiO_2$ to the catalyst retards crystallization of the components of the catalyst and improves the stability. In addition to this, calcination temperature also plays a major role in catalyst stability. Copper surface area, decreasing crystallinity and ZnO addition are important factors for developing Cu-based catalysts [26, 27]. Use of $Al_2O_3$ or $ZrO_2$ resulted in a better dispersion of copper particles in the catalyst, eventually improving the surface area of copper. $Ga_2O_3$ or $Cr_2O_3$ increased the specific activity per unit Cu surface area of the catalyst in the Cu-based multicomponent catalysts for methanol production [18]. Zirconia is a strong and thermally resistant material used as efficient catalyst support due to its high stability under reducing or oxidizing conditions. Catalytic activity of a $ZrO_2$ supported catalyst has been compared to other supporting material [28–34]. In zirconia-supported Cu catalysts, $CO_2$ is adsorbed on zirconia to form bicarbonate and finally, gets hydrogenated to form methanol along with small amount of by-products [35]. Monoclinic-$ZrO_2$-supported copper catalysts are more active than tetragonal-$ZrO_2$-supported copper due to variation in the concentration of active intermediate adsorption sites [36]. Water formed during $CO_2$ hydrogenation hinders the methanol formation rate [24]. Addition of other components like Ga, B and Al decreases the adsorption rate of water and shows better Cu dispersion and $ZrO_2$ concentration on the surface, which results in an increase in catalyst activity [25, 37–39]. In the case of $Ga_2O_3$-supported catalysts, the particle size of $Ga_2O_3$ plays a role in their promoting effect. Smaller the particle size of $Ga_2O_3$, better the activity in methanol synthesis from $CO_2$. Small $Ga_2O_3$ particles predominantly form $Cu^+$. Higher dispersion of copper and smaller particle size of gallium favor the formation of $Cu^+$, which determines the high selectivity for methanol and low selectivity of CO [25, 40].

Copper surface, such as Cu(111) and Cu(100) plays a dominant part in methanol synthesis. Fujita et al. [41] reported high activity and selectivity (67.2%) to methanol formation using Cu/ZnO catalyst with high dispersion of Cu and dominant Cu surface [Cu (111) and Cu (100)] formation. Ponce et al. [42] developed nano crystalline (NC) particles of copper by solvated metal atom dispersion technique and used for converting $CO_2$ into methanol. They found that Cu/pentane/NC–ZnO sample exhibits maximum $CO_2$ conversion. Gotti and Prins [43] explained the close contact between Cu and basic metal additives such as Zn, La and Ca for describing the strong promoting effect of basic metal oxides on Cu/$SiO_2$ catalyst. Low temperature synthesis of methanol in liquid medium has also undergone thorough investigation [44]. The objective of such research was to lower thermodynamic hindrance of the reaction, improve heat transfer efficiency and $CO_2$ conversion and to lower the operation cost. Ultrafine copper boride doped with Zr, Cr and Th were reported as effective catalysts for liquid phase hydrogenation of $CO_2$ [19].

Besides Cu catalyst, other metals like Au, Ag and Pd are also active for this reaction [45, 46]. Pd/ZnO supported on multiwalled carbon nano tubes acts as a good catalyst for hydrogenation of $CO_2$ to methanol. Active Pd supported on $Al_2O_3$, $SiO_2$, $TiO_2$ and MgO for $CO_2$ hydrogenation were reported by Solymosi et al. [47] but methanol selectivity was insufficient. Tsubaki and Fujimoto [48] studied the high temperature reduction on some reducible oxides like Pd/$CeO_2$, Pd/$TiO_2$ and Pd/$La_2O_3$ and found that Pd/$CeO_2$ reduced with hydrogen at 500°C exhibits high activity and long life time in methanol synthesis from $CO_2$ and $H_2$. The reason for such high activity and stability is strong metal-support interaction. Previous reports showed that $CO_2$ cann't be adsorbed on Pd at room temperature. Oxygen defects formed at the surface as well as in the bulk phase of the support in the high temperature reduced catalyst helps to adsorb $CO_2$ on it. Pd/$Ga_2O_3$ catalyst also showed high methanol selectivity due to formation of new Pd-Ga alloy active site. Dissociated hydrogen from the Pd metal makes a bond with adsorbed intermediate on $Ga_2O_3$ surface and finally forms methanol with high selectivity. The doping effect of Pd on Cu-ZnO based catalyst is highly profitable for the synthesis of methanol via hydrogenation of $CO_2$. Addition of Pd increases the activity of the catalyst and it acts like reservoir for spillover hydrogen, which results in an increase the reduction of copper oxide. Ultimately, it acts against oxidizing effects of $CO_2$ or $H_2O$ on the copper surface [59]. General possible mechanism for methanol synthesis by the low temperature method using oxide supported metal catalysts is shown in Fig. 5.2.

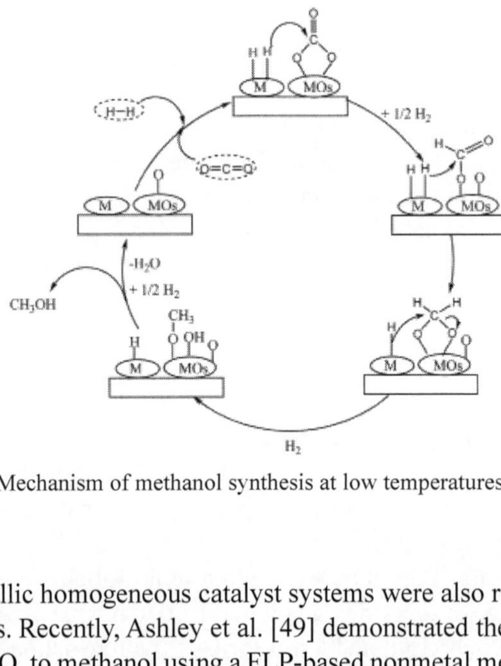


**FIGURE 5.2** Mechanism of methanol synthesis at low temperatures by oxide-supported metal catalysts.

Non-metallic homogeneous catalyst systems were also reported for methanol synthesis. Recently, Ashley et al. [49] demonstrated the selective hydrogenation of $CO_2$ to methanol using a FLP-based nonmetal mediated procedure at low pressures (1–2 atm). N-Heterocyclic carbine (NHC) was found to be an effective organic catalyst for methanol synthesis from $CO_2$ reduction with silane. Compared to transition metal catalyst, NHC is more efficient at ambient reaction conditions [50, 51]. Table 5.1 lists catalytic activities of different heterogeneous catalysts employed for methanol synthesis from $CO_2$. It shows that maximum $CO_2$ conversion of 25.9%, methanol selectivity of 99.5% and methanol yield of 378 mg/g-cat h could be achieved. The space velocities were tried between 1800 and 18,000 $h^{-1}$ and the temperature from 170 to 270°C.

**TABLE 5.1** Activity of some solid catalysts for $CO_2$ hydrogenation to methanol.

| Catalyst | Catalyst reduction temperature (°C) | Space velocity ($h^{-1}$) | Pressure (MPa) | Time on stream (h) | Reaction temperature (°C) | $CO_2$ conversion (%) | Methanol selectivity (%) | Methanol yield (mg/g-cat h) | Ref. |
|---|---|---|---|---|---|---|---|---|---|
| $Cu/ZrO_2$ | - | 5400 | 2 | - | 240 | 9.6 | 50 | 360 | 29 |
| Cu/ZnO | 240 | - | 6 | 20 | 240 | 10 | 60 | 288 | 52 |
| Cu/ZnO–NC | 350 | 4800 | 2 | - | 240 | 14 | - | 140 | 53 |
| $Cu/ZnO/ZrO_2$ | 200 | 3300 | 8 | - | 220 | 21.0 | 68.0 | 179 | 46 |
| $Cu/ZnO/Al_2O_3$[a] | - | - | 5 | 20 | 170 | 25.9 | 72.9 | - | 44 |
| $Cu/ZnO/Al_2O_3$[b] | 270 | 3600 | 2.6 | - | 260 | 15.8 | 23.3 | - | 54 |
| $SiO_2/Cu/ZnO/Al_2O_3$ | 270 | 3600 | 2.6 | - | 260 | 20.2 | 27.2 | - | 54 |
| $TiO_2/Cu/ZnO/Al_2O_3$ | 270 | 3600 | 2.6 | - | 260 | 16.1 | 25.3 | - | 54 |
| $Cu/ZnO/Al_2O_3$[c] | 280 | - | 4 | 20 | 200 | 4.0 | 90.0 | 256 | 55 |
| $Cu/ZnO/Al_2O_3/ZrO_2$ | 350 | 4000 | 5 | 24 | 250 | 23.5 | 55.5 | 180 | 56 |
| $Cu/Ga_2O_3/ZnO$ | 300 | 18000 | 2 | 20 | 270 | 6 | 88 | 378 | 40 |
| $Cu/Ga_2O_3/ZrO_2$ | 250 | 2500 | 2 | - | 250 | 13.7 | 75.6 | 61 | 37 |
| $Cu/Zn/Ga_2O_3/ZrO_2$ | 200 | 3300 | 8 | 2500 | 250 | - | 75.0 | 324 | 38 |
| $Cu/Zn/Ga_2O_3/SiO_2$ | 300 | 18000 | 2 | 20 | 270 | 5.6 | 99.5 | 350 | 40 |
| $Cu/B_2O_3/ZrO_2$ | 250 | 2500 | 2 | - | 250 | 15.8 | 67.3 | 58 | 37 |
| $Au/Fe_2O_3$ | - | 3000 | 5 | 12 | 250 | 18.4 | 29.3 | - | 57 |
| $Au/TiO_2$ | - | 3000 | 5 | 12 | 250 | 18.6 | 6.5 | - | 57 |
| Au/ZnO | - | 3000 | 5 | 12 | 250 | 8.2 | 48.8 | - | 57 |
| $Au/ZnO/ZrO_2$ | 200 | 3300 | 8 | - | 220 | 1.5 | 100 | 13 | 46 |
| $Ag/ZnO/ZrO_2$ | 200 | 3300 | 8 | - | 220 | 2.0 | 97.0 | 15 | 46 |
| Pd/ZnO/MWCNT | 500 | 1800 | 3 | - | 250 | 6.3 | 99.6 | 37 | 53 |
| $Ga_2O_3$- Pd/ $SiO_2$ | 450 | 7800 | 3 | - | 250 | - | 70 | 253 | 58 |

[a]Catalyst prepared by oxalate gel co-precipitation , [b]conventional co-precipitation, [c]co-precipitation at constant pH (ca. 7) and constant temperature (70 °C). NC = Nano catalyst

### 5.2.2 ETHANOL AND HIGHER ALCOHOLS

Ethanol is a high-octane fuel, which is mainly used as a gasoline additive. It is a more desirable product than methanol in $CO_2$ hydrogenation because of its ease to handle and transport. It is one of the largest producing alternative fuels from renewable source by fermentation process. Environmental concerns about methyl tert-butyl ether (MTBE) additive and price hiking of petroleum based fuels lead to an immense demand for ethanol. It can be produced either directly from syn-gas or from the homologation of methanol. But the latter method is not cost effective. Ethanol generation from both $CO_2$ and CO is represented in Eqs. (2) and (3), respectively. Direct synthesis of ethanol from $CO_2$ received much attention in recent years. The reaction can be considered as the combination of reverse water-gas shift (RWGS) reaction followed by direct formation of ethanol from syngas [60].

$$CO_2 + 6H_2 \leftrightarrow C_2H_5OH + 3H_2O \quad (2)$$

$$CH_3OH + CO + 2H_2 \leftrightarrow C_2H_5OH + H_2O \quad (3)$$

Hence, developing catalysts, which contain the active centers for both types of reactions at the same time giving overall activity are highly important. The catalyst should contain the active sites, which assist the partial reduction of $CO_2$ to CO and C-C bond formation. Fe-based Cu/ZnO is one of the catalysts, in, which Cu/ZnO helps the formation of methanol [61, 62]. As discussed in the above section, Pd and Ga addition as promoter helps to maintain the active oxidation state of Cu, which finally reflects in ethanol yield. The effect of potassium in the support on activity of Fe/Cu/ZnO was studied by Takagawa et al. [63]. In their work, introduction of $K_2CO_3$ in Fe/Cu/ZnO showed an increase in selectivity for ethanol from 6 to 20% at 300° C at 7 MPa $CO_2$ pressure [63].

Rhodium based catalysts were also active for these reaction. Rh/$SiO_2$ catalysts for $CO_2$ hydrogenation was examined by Kusama et al. [64] Several metal additives were used for improving the activity of Rh/$SiO_2$. Lithium was found to be the most effective promoter, reaching $CO_2$ conversion of 7% with 15.5% ethanol selectivity. With the help of in-situ FTIR measurements, the authors proved that $CO_2$ was hydrogenated to ethanol via CO intermediate. Rh-Fe/$SiO_2$ catalyst increases the catalyst activity enormously because $Fe^{3+}$ changes the electronic state of Rh that was confirmed by in-situ FTIR and XPS techniques [65]. Kurakata et al. [66,67] studied the activity [$Rh_{10}$Se] supported on metal oxides for ethanol from $CO_2$. Among several, [$Rh_{10}$Se]/$TiO_2$

showed good activity towards ethanol. Promoting effects of anionic Se on such cluster complex catalyst was probed by XAFS. Ethanol synthesis doesn't occur via CO over the $Rh_{10}Se/TiO_2$ catalyst because the reaction between CO and $H_2$ occurs very slowly and the products distribution is entirely different from that of $CO_2$ hydrogenation. Carbon dioxide was converted into ethanol by using composite catalysts involving three different kinds of elementary catalytic functions (partial reduction of $CO_2$ to CO, C-C bond formation, and OH group insertion). For this purpose, supported Rh catalyst, Fe based Fisher-Tropsch catalyst and Cu-based methanol synthesis catalyst was combined by different ways, but ethanol selectivity was found low.

Hydroformylation of olefins is one way to develop higher alcohols from $CO_2$. Actually, this reaction occurs through two steps. First step involves the formation of CO form $CO_2$ and $H_2$ by RWGS reaction followed by hydroformylation of CO in the second step results in the formation of higher alcohols. Rhuthenium carbonyl complex and Ru-based clusters were reported to be highly active for this reaction [68–71].

### *5.2.3 DIMETHYL ETHER*

Dimethyl ether (DME) is the derivative of methanol and used as a potential alternative for diesel oil due to its high cetane number and better combustion properties [72]. Conventional way of synthesizing DME is simply by dehydration of methanol under controlled reaction conditions using acid catalysts [73]. Besides conventional methods, $CO_2$ or CO hydrogenation reaction was also employed for DME production. The possible ways of DME formation is represented in Scheme 5.1. DME can be produce from $CO_2$ either through two-step process or through one-step process. The two-step process involves $CO_2$ hydrogenation to methanol using metal oxide catalyst followed by dehydration with acid catalyst. Bifunctional catalyst was used for the one-step synthesis, which lowers the thermodynamic limitation than the formerly mentioned method [74, 75]. Recent reports show that copper based shell membrane is also an active catalyst for the one-step DME synthesis from $CO_2$ [83]. Pd shows good promoting effect in the Cu based system. The activity of Cu on different supports is listed in the Table 5.2. Development of an efficient bifunctional catalyst for direct DME synthesis from $CO_2$ at moderate reaction conditions is an important task for this reaction.

**TABLE 5.2** Catalytic activities of Cu-based catalysts for direct synthesis of DME from $CO_2$.

| Catalyst | GHSV ($h^{-1}$) | Pressure (MPa) | Time on stream (h) | Reaction temperature (°C) | $CO_2$ conversion (%) | DME selectivity (%) | DME yield (mg/g-cat h) | Ref. |
|---|---|---|---|---|---|---|---|---|
| $Cu\text{-}ZrO_2$/HZSM-5 | 25000 | 5 | 12 | 250 | 14 | 50 | 413 | 76 |
| $Cu\text{-}ZrO_2$-CNT/HZSM-5 | 25000 | 5 | 12 | 250 | 16.6 | 50.3 | 493 | 76 |
| $Cu\text{-}ZrO_2$-PdCNT/HZSM-5 | 25000 | 5 | 12 | 250 | 18.9 | 51.8 | 579 | 76 |
| Cu–ZnO–$ZrO_2$/H-ZSM5 | 10000 | 3 | - | 240 | 16.1 | 33.9 | 430 | 77, 78 |
| CuZnAl oxide /Amorph. $SiO_2\text{-}Al_2O_3$ | 1800 | 3 | - | 266 | 47.1 | 42.7 | - | 83 |
| CuO–$TiO_2$–$ZrO_2$/HZSM-5 | 1500 | 3 | 4 | 250 | 15.6 | 47.5 | - | 79 |
| 0.50 Pd/CuO–ZnO–$Al_2O_3$–$ZrO_2$/HZSM-5 | 1800 | 3 | - | 200 | 18.7 | 73.6 | - | 80 |
| Cu–Mo/HZSM-5 | 1500 | 2 | - | 240 | 12.4 | 77.2 | - | 81 |
| CuO-ZnO-$Al_2O_3$/NaHZSM-5 | - | 4 | 6 | 275 | 35 | - | - | 82 |

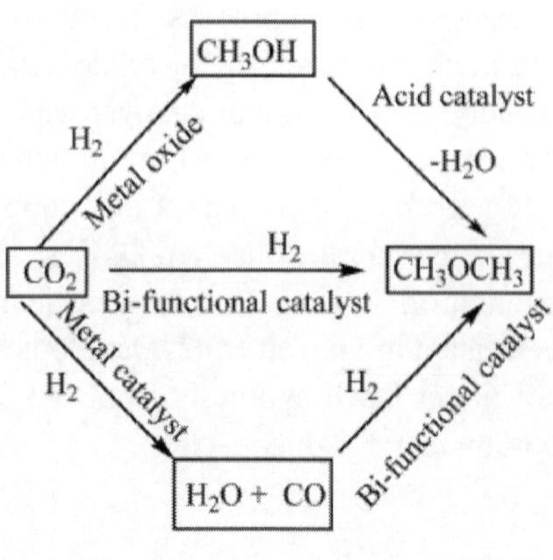


**SCHEME 5.1** Alternative ways of DME formation.

## 5.2.4 HYDROCARBONS

### 5.2.4.1 METHANE

Methane is produced by direct catalytic hydrogenation of $CO_2$ through Sabatier reaction (the Eq. (4)).

$$CO_2 + 4H_2 \rightarrow CH_4 + 2H_2O \quad (4)$$

Methane is important for electricity generation by burning it as a fuel in a gas turbine or steam boiler and also used in industrial chemical processes. $CO_2$ methanation reaction is thermodynamically favorable and kinetically limited due the highly oxidized state of carbon dioxide. The use of efficient metallic catalysts helps to remove such restriction. Ni, Ru, Pt and Pd on various supporting carriers are the main active catalysts for this reaction. Most of the catalytic methanation reactions of $CO_2$ have been carried out in fixed-bed reactors. Ni catalysts are most extensively investigated for this reaction not only due to their activity but low cost [84–86].

Various oxides such as $TiO_2$, $SiO_2$, $Al_2O_3$, $CeO_2$, $ZrO_2$ and $CeO_2$-$ZrO_2$ were used as supports for the metallic nickel. Dispersion of metal on oxide support plays a major role in determining the rate of a chemical reaction [86]. Amorphous silica obtained from RHA was found to be a very active support for Ni catalyst because of its high specific surface area, melting point and porosity [87]. Further, amorphous silica was used as a raw material for synthesizing a set of silica-alumina composite supports for nickel-based catalysts [88]. In Ni/RHA–$Al_2O_3$ synthesized via ion exchange method, alumina trapped in the NiO particles increases the activation energy of reduction. It causes the restriction in the reduction of NiO into the highly dispersed and active $Ni^{\circ}$ form. Mesoporosity and high surface area give an advantage of better metal dispersion on the support. Ni/RHA–$Al_2O_3$ catalyst, prepared by the incipient wetness impregnation method shows good catalytic activity as a result of mesoporous structure and high surface area [89]. Du et al. [90] used Ni/MCM-41 catalysts with different loading of Ni to achieve $CO_2$ methanation. Among different Ni concentrations, 3 wt% Ni/MCM-41 gave high selectivity (96%) and space time yield (91.4 $g.Kg^{-1}.h^{-1}$). Space time yield increases with temperature and attains 633 $g.Kg^{-1}.h^{-1}$ at 350° C. Ni-Al alloy with high nickel content showed 100% methane selectivity. Ni-Fe alloy also gave good activity to $CO_2$ hydrogenation [91, 92]. $ZrO_2$ is the one of the effective supports due to the acid-base nature facilitating $CO_2$ adsorption ability. Tetragonal zirconia is a more favorable phase for $CO_2$ adsorption and activation [93]. Ceria-zirconia-supported Ni ($Ni/Ce_{0.72}Zr_{0.28}O_2$) has excellent activity with 75.9% $CO_2$ conversion and

99.1% methane selectivity. In addition to transition metals, activity of rare earth modified Ni-based catalyst was also investigated [94, 95]. Mechanistic investigation of $CO_2$ methanation reaction with noble metal catalysts indicates either dissociation of $CO_2$ into CO or direct carbonate or formate intermediate formation in the reaction pathway [96–98]. In the former, CO obtained from $CO_2$ dissociation again dissociated into carbon and oxygen on metal surface. Further hydrogenation results the formation of methane, but this route is less possible. In recent studies, Aldana et al. [99] investigated the $CO_2$ methanation reaction mechanism on Ni/$CeO_2$-$ZrO_2$ catalysts with the help of operando IR technique and proved that the main mechanism does not require any CO as an intermediate (Fig. 5.3).

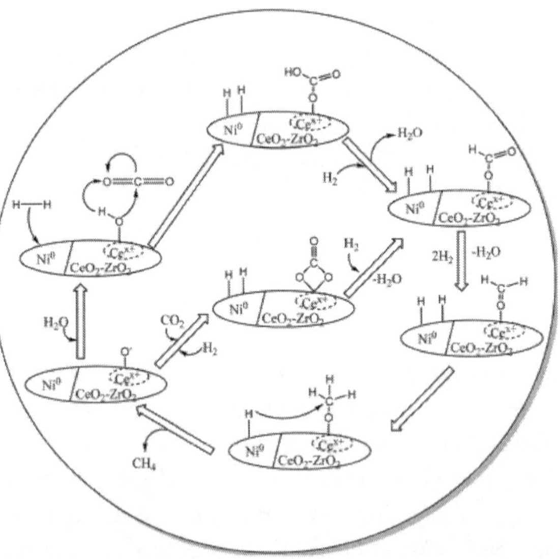


**FIGURE 5.3** Reaction mechanism of $CO_2$ methanation with Ni/$CeO_2$-$ZrO_2$ catalyst.

The major drawback of the nickel catalyst is poisoning of Ni surface by the interaction with CO to form nickel carbonyl at low temperature. Noble metal based catalysts are more active and stable catalyst in comparison with Ni based catalysts [100]. Among Ru nanoparticle dispersed on various supports ($Al_2O_3$, $MgAl_2O_4$, MgO, C, etc.), Ru/$Al_2O_3$ showed highest catalytic activity (turnover frequency, TOF = 16.5 x $10^3$ $s^{-1}$) [101, 102]. Yttrium addition to the Ru-based catalyst enhances the activity and stability for methanation reaction [103]. Pd/Mg-$SiO_2$ and platinum titanate nanotubes were also found active for this reaction [104, 105]. Methanation of $CO_2$ with high methane selectivity was obtained even at room temperature and atmospheric pressure, when Rh/γ-$Al_2O_3$ was used as a catalyst [106–108]. Overall particle size, operating temperature and support play a crucial role in methanation reaction. The apparent activation energy for $CO_2$ methanation reaction using different transition metals on oxide support is presented in Table 5.3. The lowest activation energy of 9.3 k cal/mol was obtained with Ni-Fe-MgO catalyst and the highest value was reported for 15% Co/$SiO_2$. Due to its large hydrogen consumption, the $CO_2$ methantion reactions are still economically challenging.

**TABLE 5.3** Activation energy of oxide-supported transition metal catalysts for $CO_2$ methanation reaction.

| Catalyst | Reactant composition ($CO_2$ : $H_2$) | GHSV ($h^{-1}$) | Total pressure (KPa) | Temperature (°C) | TOF ($s^{-1}$) *$10^2$ | Activation energy- $E_a$ ($kcal.mol^{-1}$) | Ref. |
|---|---|---|---|---|---|---|---|
| 5% Rh/ $TiO_2$ | 1:4 | - | - | 100-165 | 1.58 | 16.3 | 108 |
| 3% Rh/ $TiO_2$ | 1:4 | - | - | 100-165 | 2.27 | 17.0 | 108 |
| 3% Rh/ $Al_2O_3$ | 1:4 | - | - | 150-200 | 1.85 | 14.6 | 109 |
| 2.3% Rh/ $ZrO_2$ | 1:10 | 3600 | - | 160-240 | - | 14.9 | 110 |
| 2.3% Rh/ $ZrO_2$ | 1:10 | 3600 | - | 170-240 | - | 15.9 | 110 |
| 2.3% Rh/ MgO | 1:10 | 3600 | - | 260-320 | - | 23.1 | 110 |
| Rh foil | 1:3 | - | 101 | 175-300 | - | 17.0 | 111 |
| Rh foil+$TiO_2$ | 1:3 | - | 101 | 175-300 | - | 17.0 | 111 |
| 0.5%Ru/ $Al_2O_3$ | 1:2-1:4 | - | 101 | 200-300 | - | 17.0 | 112 |
| 0.5%Ru/ $SiO_2$ | 1:4 | 1720-5965 | 101 | 229-293 | 72 | 17.2 | 113 |
| Ru-MgO | 1:5 | - | 80 | 80-180 | - | 9.8 | 85 |
| 3%Ni/ $Al_2O_3$ | - | - | - | 252 | - | 17.0 | 114 |
| 3%Ni/$SiO_2$ | 1:4 | 16000-32000 | 140 | 227-302 | 24 | 19.1 | 115 |
| Ni-Fe-MgO | 1:5 | - | 80 | 90-120 | - | 9.3 | 85 |
| 15%Fe/ $SiO_2$ | 1:4 | 1470-6370 | 101 | 253-299 | 0.83 | 32.0 | 113 |
| 15% Co/ $SiO_2$ | 1:4 | 450-9620 | 101 | 180-222 | 690 | 41.3 | 113 |

#### 5.2.4.2 LONG-CHAIN HYDROCARBONS

The synthesis of hydrocarbons from $H_2$ and $CO_2$ is a key technology to convert $CO_2$ into useful fuels. $CO_2$ conversion to long chain hydrocarbons is more challenging in thermodynamic and kinetic perspectives. Alkane products, $C_nH_{(2n+2)}$ are applied in diesel and jet fuel. FT synthesis is generally used for hydrocarbon production, where its formation occurs through CO hydrogenation. In modified methods, $CO_2$-CO mixture is used instead of CO to produce hydrocarbons. There are mainly two ways of synthesizing hydrocarbons from $CO_2$, that is, (1) RWGS reaction followed by FT synthesis and (2) $CO_2$ hydrogenation to methanol and its further conversion to gasoline. RWGS produces CO and water. Splitting of water to energy rich hydrogen can be done electrochemicaly or photochemicaly. It can used for the further hydrogenation of $CO_2$. This is one of the possible ways to maintain the carbon cycle.

Transition metal catalysts are commonly employed for both categories. Fe and Co based FT catalysts are applicable for first type, where Co-based catalytic reaction occurs at lower temperature and Fe catalyst reaction takes place at high temperature [116–118]. Use of supports and promoters induce a significant impact on $CO_2$ hydrogenation. Addition of $CeO_2$ to Fe-based catalyst increases the hydrocarbon ($C_2$–$C_5$) yield because of the facile redox properties of $CeO_2$ [119, 120]. But increase of $CeO_2$ concentration adversely effects the $CO_2$ conversion as a consequence of blocking active sites in Fe by $CeO_2$. Co-catalyst prefers the formation of methane than long-chain hydrocarbon. Different oxide supports, especially $Al_2O_3$ act as stabilizer and avoid sintering of active metal particles during hydrogenation reaction. Zeolite was also reported as an active support in hydrocarbon synthesis. The characteristic properties of zeolite (pore structure, acidity, etc.) help to produce different hydrocarbon products, for example Fe-Zn-Zr/Zeolites catalyst is selective for iso-alkane formation [121]. In the second approach, $CO_2$ and $H_2$ react in presence of Cu-Zn-based catalysts to produce methanol, which is transformed into gasoline.

### 5.2.5 CONVENTIONAL, DRY AND TRI REFORMING METHODS FOR SYNGAS

Hydrogen is a clean source of energy for transportation. However, it is bound in water or hydrocarbons (petroleum and natural gas). A traditional $H_2$ separation process causes the emission of greenhouse gas viz., $CO_2$. At present, worldwide hydrogen production generates about 5% of the global $CO_2$ emissions [122]. The utilization of $CO_2$ for hydrogen production using methane can significantly contribute to the reduction of atmospheric concentration of green house gases. $CO_2$ reforming of $CH_4$ produces syngas (mixture of CO and $H_2$), which can be converted to variety of chemicals and fuels. This reaction is highly endothermic in nature, and thus requires a larger amount of energy. Another drawback of this reaction is that both the reactants are carbonaceous and may poison the active catalyst surface quickly by carbon deposition [123]. Designing of an efficient catalyst for $CO_2$ reforming of methane to produce syngas is the major challenge in this area of research. As discussed in the previous sections, highly dispersed metal on support can eliminate sintering and reduce coke deposition. In addition to metal dispersion, structural and geometrical confinement of support is also a key factor for catalyst activity.

#### 5.2.5.1 DRY AND TRI REFORMING CONCEPTS

The concept of dry reforming was first introduced by Fischer and Tropsch using different base metal catalysts. It is a unique green process for reforming of hydrocarbons powered by two reagents that contain carbon viz., $CO_2$ and the hydrocarbon to obtain syngas as a product. It is an ideal process to use the syngas produced as a raw material for the production of important liquid fuels through FT synthesis. Further studies showed that noble metals (Ru, Rh, Pd, Pt and Ir) and some group VIII non-noble metals like nickel and cobalt are catalytically active towards dry reforming reaction [124]. As discussed in earlier sections, noble metal catalysts have higher resistance to carbon deposition than the traditional nickel catalysts, but they are not feasible catalyst due their less abundance and high cost [125, 126]. Generally, $Al_2O_3$, MgO, $TiO_2$, $ZrO_2$ and $SiO_2$ are used as support and $CeO_2$, $La_2O_3$ and MnO as promoters for the Ni catalysts [127, 128]. Several Ni-based systems such as Ni/$CeAlO_3$-$Al_2O_3$ [129], Ni/$LaAlO_3$–$Al_2O_3$ [130], Ni-substituted barium hexaaluminate [131], nanocrystalline NiMnO/$SiO_2$ [132] and NiCe/$SiO_2$ [133] have been developed recently with improved stability and activity in dry reforming reactions. The mechanism of formation of syngas from $CO_2$ reforming is still under controversy. The main disadvantage of dry reforming of hydrocarbons is the significant formation of coke on the catalyst surface, which reduces its useful life. The presence of $CO_2$ is found to increase coke production.

Zhang et al. [134] introduced a bimodal catalyst support for $CO_2$ reforming reactions. This catalyst contains both macropores and mesopores, where macropores positively influence the mass transfer of reactant and products and mesopores improve the dispersion of the metal. $SiO_2$-$SiO_2$ biomodal support used for this reaction was prepared by impregnation of silica nanoparticles into the macropores [135]. Development of mesopores occurred through polycondensation of nanoparticles during calcination. $Pt/SiO_2$-$SiO_2$, $Pt/ZrO_2$-$SiO_2$ and $Pt/Al_2O_3$-$SiO_2$ bimodal support catalysts were used for this reaction. Among these, $Pt/Al_2O_3$-$SiO_2$ is the best catalyst [136, 137]. Solid solution catalysts (NiO–MgO or CoO–MgO) were found to be a stable system for $CO_2$ reformation reactions [138]. Activity of such catalyst depends on the composition, preparation conditions and properties. The reduced solid-solution catalyst exhibited excellent conversion of $CO_2$ with high selectivity to CO and $H_2$ at atmospheric pressure. The conversion and selectivity remained unchanged for longer reaction time. This indicates the stability of Ni-MgO solid solution catalyst. At the same time Ni-$Al_2O_3$ showed fast deactivation, although it exhibited high initial conversion and selectivity. Surface basicity and small particles of metal on the surface are the main factors affecting stability, through inhibition of the carbon deposition on the surface. Recently, Kumar et al. [139] reported the light assisted cogeneration of syngas from $CO_2$ and $H_2O$ using homogeneous CO evolving electro catalyst and heterogeneous hydrogen evolving photo-electrode.

Tri-reforming of methane, which is a synergetic combination of endothermic dry and steam reforming and exothermic partial oxidation, was attempted for the synthesis of industrially useful synthesis gas with desired $H_2/CO$ ratios using $CO_2$ in flue gas of power plants without $CO_2$ seperation [332]. The tri reforming represents a new way of thinking both for conversion and utilization of $CO_2$. It was conducted in a fixed bed reactor using nickel catalyst impregnated on MgO, $ZrO_2$, $CeO_2$-$ZrO_2$ and Mg-$CeO_2$-$ZrO_2$ supports to obtain methane and $CO_2$ conversions exceeding 97% and 80%, respectively, for producing syngas with $H_2/CO$ ratios of 1.5–2.0 at 800–850°C under atmosphereic pressure. Among all the catalysts tested for tri-reforming, their ability to enhance the conversion of $CO_2$ follows the order of Ni/MgO ▪ Ni/MgO/$CeZrO_2$ ▪ Ni/$CeO_2$ $\approx$ Ni/$ZrO_2$ $\approx$ Ni/$Al_2O_3$ $\approx$ Ni/$CeZrO_2$. Tri-reforming was found to consume less amount of energy as compared to steam reforming of methane. It is interesting to note that tri-reforming contributes to practically no carbon formation and no appreciable catalyst deactivation. More studies are required for understanding $CO_2$-$CH_4$ interactions in the presence of steam and oxygen during tri-reforming.

### 5.2.5.2 SOLAR LIGHT SPLITTING OF $CO_2$ AND WATER

Photosynthesis in plants creates so many organic compounds and biomass. The reaction between $CO_2$ and water in the presence of solar light is responsible for such chemical formation. Solar light splits water as well as $CO_2$ to form $H_2$ and CO. FT reaction yields methanol and also forms hydrocarbons, formic acid, etc. $CO_2$ could be converted into a fuel substitute by catalytic hydrogenation, and combined with solar energy; it would be able to control the greenhouse effect through a carbon neutral cycle (Fig. 5.4).

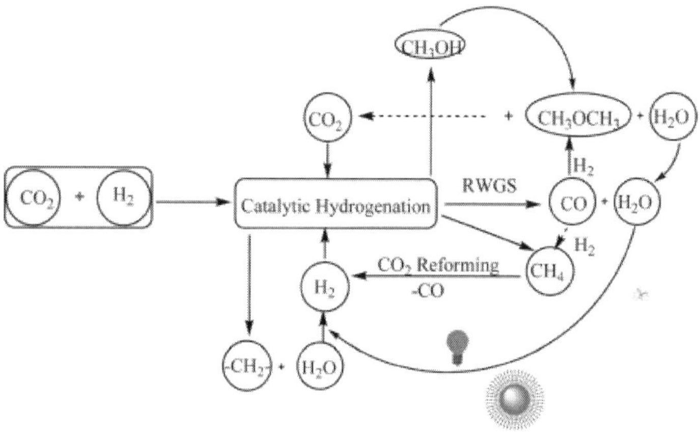


FIGURE 5.4 A carbon neutral cycle for catalytic hydrogenation of $CO_2$.

## 5.3 CARBON DIOXIDE CONVERSION INTO BULK AND SPECIALITY CHEMICALS

$CO_2$ conversion into chemicals has received immense research interest as an effective way to replace phosgene-based routes for the functionalization of organic substrates to value added chemicals as well as to reduce $CO_2$ concentration in the atmosphere. Several routes were investigated for the direct utilization of $CO_2$ into chemicals including commercial processes such as urea, salicylic acid, cyclic carbonates, and polycarbonates production [140]. Generally metal/metal oxide catalysts are employed for $CO_2$ activation and useful conversions. Activation of $CO_2$ occurred either through the metal coordination or through weak interaction between the active species and $CO_2$.

### 5.3.1 ORGANIC CARBONATES

#### 5.3.1.1 DIALKYL CARBONATE

Dialkyl carbonates such as, dimethyl carbonate (DMC) and diethyl carbonate (DEC) have a variety of industrial and synthetic applications as a chemical intermediate. Worldwide requirement of dimethyl carbonate is increasing steadily because of versatile application in the field of fuel technology, pharmaceutical, electrochemical and catalytic reactions [141–144]. It has been widely used as a solvent and a green reagent. It has received considerable attention in sustainable energy development in recent years [145–147]. DMC is a fuel oil additive owing to the high octane number and reducing particulate emission from diesel engines [146–148]. Its synthesis by traditional methods, such as phosgenation of methanol, oxidative carbonylation of methanol are highly unsecured because of toxic-corrosive nature of reagents. DMC production technologies from renewable feedstock are one of the important challenges faced in the current century. Recycling of $CO_2$ is possible in DMC synthesis (Fig. 5.5). Two routes for the synthesis of DMC using $CO_2$ as raw material have been proposed. They are (1) direct synthesis from methanol and $CO_2$ (Scheme 5.2) and (2) one-pot synthesis from oxiranes, methanol and $CO_2$.

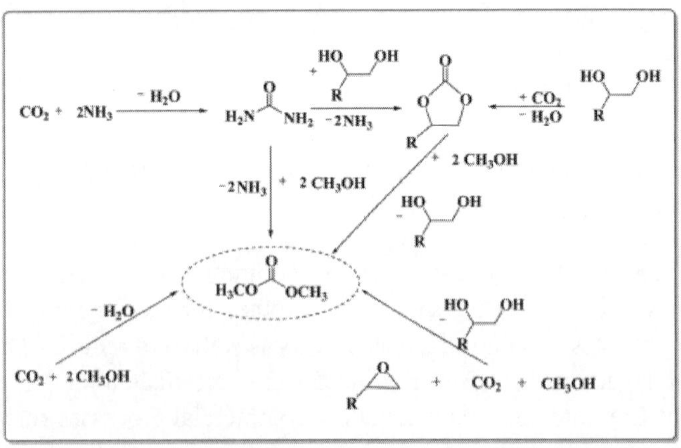


**FIGURE 5.5** Recycling of $CO_2$ in DMC synthesis.

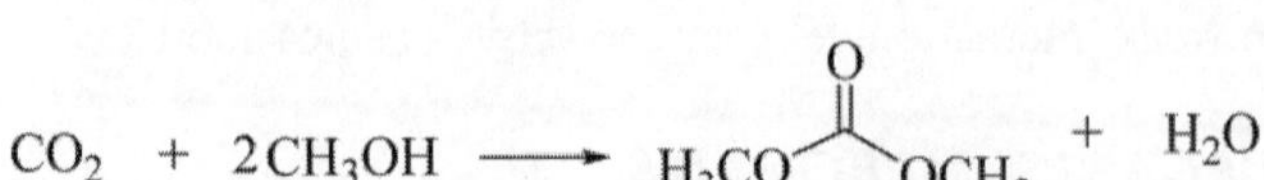

**SCHEME 5.2** Dimethyl carbonate synthesis from methanol and $CO_2$.

DMC synthesis from methanol and $CO_2$ was intensively studied over a wide range of catalysts with or with out use of water trapping agent in batch as well as in fixed bed reactors. Catalyst deactivation and unfavorable thermodynamics lead to low DMC productivity in this route. Both acidic and basic sites are of inevitable requirement for the direct synthesis of DMC. While the basic sites activate methanol and carbon dioxide, acidic sites generate methyl group. Fang and Fujimoto [149] reported the use of base catalysts with $CH_3I$ as a source of methyl group. Generally, $ZrO_2$-based catalysts showed good activity towards this reaction. $ZrO_2$ catalyst was prepared by calcination of zirconium hydroxide at different temperatures. The catalyst prepared by calcination at 400°C showed highest DMC formation rate [150]. In general, the reaction scheme of direct synthesis of DMC from methanol and $CO_2$ using acid-base bifunctional catalysts is represented below in Eqs. (5)–(10).

$$CH_3OH \rightarrow CH_3O^- + H^+ \text{ (on basic site)} \quad (5)$$

$$CO_2 \rightarrow CO_2 \text{ (on basic site)} \quad (6)$$

$$CH_3O^- + CO_2 \rightarrow CH_3OCO_2^- \text{ (on basic site)} \quad (7)$$

$$CH_3OH \rightarrow CH_3^+ + OH^- \text{(on acidic site)} \quad (8)$$

$$CH_3OCO_2^- + CH_3^+ \rightarrow (CH_3O)_2CO \quad (9)$$

$$H^+ + OH^- \rightarrow H_2O. \quad (10)$$

Xei and Bell [151] investigated the reaction mechanism of DMC synthesis from methanol and $CO_2$ over zirconia catalyst with the help of an in-situ Raman spectroscopy. On further detailed study using in-situ IR technique over monoclinic zirconia they identified the nature of the reaction sites required for each step and involvement of weakly and strongly adsorbed $CO_2$ in methyl carbonate formation [152]. The addition of phosphoric acid to $ZrO_2$ catalysts enhanced its catalytic activity [153]. $H_3PO_4/ZrO_2$ (P/Zr = 0.05) calcined at 400° C showed better activity at 130° C and 5 MPa initial $CO_2$ pressure [154] Jiang et al. [155] prepared $H_3PW_{12}O_{40}/ZrO_2$ catalysts via a sol-gel method and tested their activity in these reactions. They found that DMC yield obtained by using $H_3PW_{12}O_{40}/ZrO_2$ was higher than that with $ZrO_2$ under the same experimental conditions.

$CeO_2$–$ZrO_2$ [Ce/(Ce + Zr) = 0.2 and 0.33] binary catalysts were also investigated for DMC synthesis from methanol and $CO_2$ [156]. In all of these cases, methanol conversion was lower because the coproduced water shifts the equilibrium toward the reactants side. In order to overcome these problems, water trapping agents including molecular sieves [157] and 2, 2-dimethoxy propane

[158] were used along with methanol and $CO_2$. But at high temperature and pressure, their effect was negligible. Butylene oxide was used as an effective and recyclable water trap for direct synthesis of DMC from methanol and $CO_2$ over various catalysts [159]. In addition to $CeO_2$–$ZrO_2$, $CeO_2$ [160], $CeO_xTi_{1-x}O_2$ [161] and $Ga_2O_3/Ce_{0.6}Zr_{0.4}O_2$ [162] catalysts were also employed for DMC synthesis. Catalytic activity data of different catalysts is compiled in Table 5.4.

**TABLE 5.4** Activities of different solid catalysts for direct synthesis of DMC from $CO_2$ and methanol in batch reactors.

| Catalyst | Reaction temperature (°C) | Reactant (mmol) | | Reaction time (h) | Dehydration agent | DMC yield (mmol) | Ref. |
|---|---|---|---|---|---|---|---|
| | | $CO_2$ | $CH_3OH$ | | | | |
| $CeO_2$-$ZrO_2$((Ce/(Ce + Zr) =0.2) | 110 | 200 | 192 | 16 | - | 0.78 | 183 |
| $CeO_2$-$ZrO_2$((Ce/(Ce + Zr) =0.2) | 110 | 200 | 192 | 140 | 2,2-Dimethoxy propane | 6.9 | 183 |
| Sn-SBA-15 | 150 | 910 | 500 | 15 | - | 0.33 | 185 |
| Sn-SBA-$CH_3$ | 150 | 910 | 500 | 15 | - | 0.62 | 185 |
| $H_3PW_{12}O_{40}/Ce_{0.1}Ti_{0.9}O_2$ | 170 | 200 | 200 | 12 | - | 5.0 | 186 |
| $H_3PW_{12}O_{40}/CeO_2$ | 170 | 200 | 200 | 12 | - | 1.2 | 186 |
| $H_3PW_{12}O_{40}/TiO_2$ | 170 | 200 | 200 | 12 | - | 0.8 | 186 |
| $ZrO_2$ | 170 | 250 | 82 | 16 | - | 0.42 | 150 |
| $ZrO_2$–MgO | 120 | 1000 | 247 | 9 | [$C_4$-mim][MeO][b] | 108.9 | 173 |
| $H_3PW_{12}O_{40}/ZrO_2$ | 100 | 175 | 140 | 3.5 | - | 2.83 | 155 |
| n-$Bu_2Sn(OCH_3)_2$ | 150 | 820 | 500 | 15 | - | 3.8 | 187 |
| $K_2CO_3$[a] | 70 | 140 | 198 | 4 | - | 4.04 | 188 |
| $KHCO_3$[a] | 70 | 140 | 198 | 4 | - | 2.0 | 188 |
| $Na_2CO_3$[a] | 70 | 140 | 198 | 4 | - | 0.97 | 188 |

[a]$CH_3I$ was used as a promoter. [b][$C_4$-mim][MeO] = 1-butyl-3-methyl imidazolium methoxide.

Wu et al. [163] investigated this reaction in fixed-bed microgaseous reactor using Cu–Ni/$V_2O_5$–$SiO_2$ as catalyst. Several catalysts such as Cu–KF/MgSiO [164], $H_3PO_4/V_2O_5$ [165], organo tin-grafted SBA-15 [166], n-$Bu_2Sn(OMe)_2$ supported on polystyrene [167], and Cu-Ni bi-metallic catalyst [168] were also studied for the direct synthesis of DMC in a continuous flow reactor. A new photo-assistant synthetic process was developed by Wang et al. [169]

using supported copper-(Ni, V and O) semiconductor oxides as catalyst. This study showed that UV irradiation reduces reaction time and increases the yield of DMC compared to the (nonphoto-assisted) normal reaction. In spite of all these efforts, direct conversion of $CO_2$ into DMC is still far from commercialization. Improving the productivity by designing non-deactivating and highly efficient catalysts and development of effective dehydration methodologies are the major challenges in the DMC synthesis research.

Synthesis of DMC by the reaction of ethylene or propylene oxide (EO or PO), $CO_2$ and methanol in a two-step process was reported with a variety of catalysts [170, 171]. To avoid multi process steps, a one-pot synthesis route was also developed (Scheme 5.3). Bhanage et al. [171] reported the activity and selectivity of basic metal oxide catalyst for this reaction. MgO was found to be more efficient than other metal oxides. They also specified the role of strong and moderate basic sites on the reaction activity with the help of temperature-programmed desorption of carbon dioxide ($CO_2$-TPD). Wang and his group explored the one-pot synthesis of DMC under supercritical $CO_2$ condition using a catalyst mixture of KI and $K_2CO_3$ [172]. Calcined samples of KI/ZnO and $K_2CO_3$-KI/ZnO were found to be effective catalysts for oxirane conversion to DMC [174]. $Re(CO)_5Cl/K_2CO_3$ [175], KOH/4A-molecular sieve [176], n-$Bu_4NBr$/n-$Bu_3N$ [177], [bmim]/$BF_4$ [178], Mg containing smectite [179] and inorganic base/phosphonium halide functionalized polyethylene glycol [180] were also reported as catalysts for the one-pot synthesis of DMC from EO or PO. In recent years, ionic liquids have also considerable attention for this reaction. A basic ionic liquid choline hydroxide supported on MgO catalyst is active, giving DMC yield of 65.4% at 2.5 MPa and 120° C in 6 h reaction [181]. Bifunctional quaternary ammonium ionic liquids exhibited good catalytic efficiency with EO conversion as high as 99% and DMC selectivity of 74% [182].

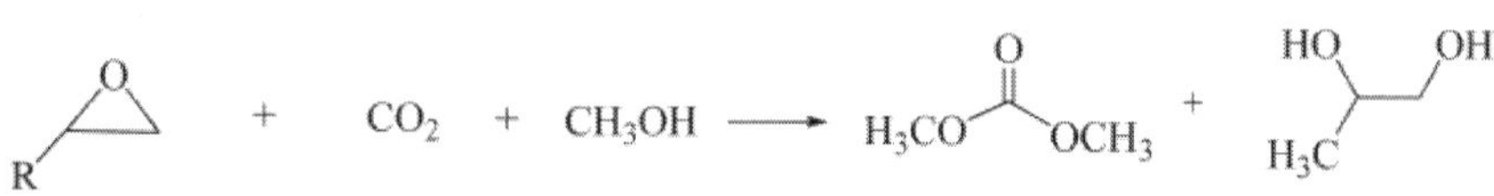


**SCHEME 5.3** Dimethyl carbonate synthesis from EO/PO, $CO_2$ and methanol.

Direct synthesis of diethyl carbonate (DEC) from $CO_2$ has not received much attention due to low carbon dioxide conversion and diethyl carbonate selectivity. Higher reaction temperature doesn't facilitate DEC synthesis because of the exothermic nature of the reaction. $CeO_2$–$ZrO_2$ [Ce/(Ce + Zr) = 0.2] solid solution is the first reported heterogeneous catalyst for the selective DEC synthesis from ethanol and $CO_2$ [183]. Further, Wang et al. [184] conducted a detailed examination of activity of $CeZrO_2$ catalyst for DEC synthesis and suggested that Ce/Zr ratio is a key factor for tuning crystal structure and surface acidty-basicity of the catalyst. The mechanisms involved in the formation of DEC are similar to DMC formation mechanism.

### *5.3.1.2 DIPHENYL CARBONATE*

Diphenyl carbonate (DPC) is an essential raw material for the production of aromatic polycarbonates. It is also used as a viscosity enhancer in polyamides and polyesters. Conventional way of preparing DPC is by phosgenation of phenol, which is a highly toxic route. Currently employed nonphosgene routes e.g., oxidative carbonylation of phenol has high investment costs and low DPC yield. Transesterification of dimethyl oxalate or dimethyl carbonate with phenol is an alternative way to prepare DPC [189–194]. However, its synthesis by the direct utilization of $CO_2$ and phenol is an attractive greener approach. Lewis acid (zinc halides) in the presence of carbon tetrachloride ($CCl_4$) and $K_2CO_3$ is a widely used catalyst system for the direct synthesis of DPC [195–197]. Good yield of DPC (with $ZnCl_2$ catalyst) at 120° C and 4 MPa pressure have been reported [197]. The use of harmful organic solvent like $CCl_4$ in the synthesis is nonecofriendly. Moreover, hydrogen chloride is formed during the reaction between $CO_2$ and phenol by zinc chloride catalyst, which leads to pollution and corrosion. Fan et al. [198] suggested a new route for DPC formation from phenol by Lewis acid catalysts and methanol (Scheme 5.4) using $CO_2$ as a reactant and solvent. Co(salen)(OAc) complex containing a quaternary phosphonium salt unit anchored on the ligand showed excellent catalytic activity compared to normal Lewis acid catalysts under mild conditions. DPC yield of 19% was achieved at 8 MPa and 90° C in a reaction of 10 h. A tentative mechanism with Lewis acid catalyst in the presence of methanol is depicted in the Fig. 5.6.

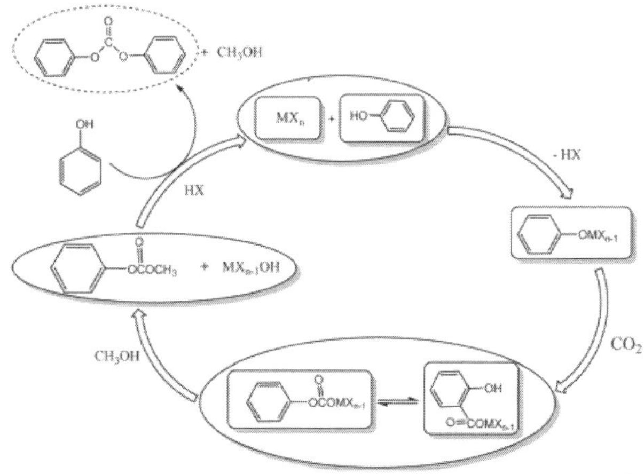


**FIGURE 5.6** Mechanism of DPC formation employing lewis acid catalyst.

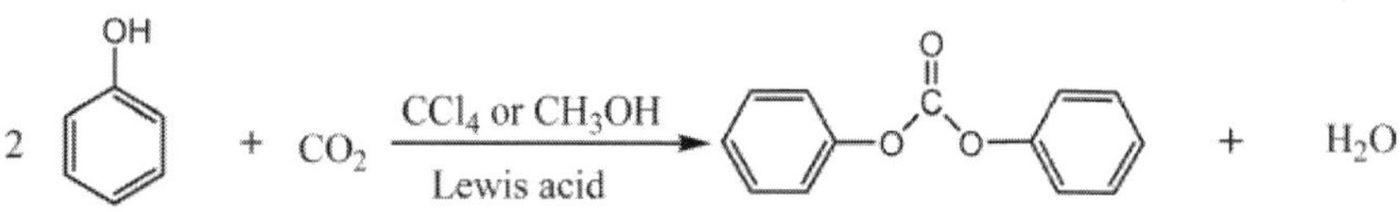


**SCHEME 5.4** Diphenyl carbonate synthesis from phenol and $CO_2$.

## 5.3.2 CYCLIC CARBONATES

### 5.3.2.1 FROM EPOXIDES

Cyclic carbonates are important raw materials for polycarbonates. They are used as electrolytes in lithium ion batteries, ecofriendly polar aprotic solvents and chemical precursor in organic synthesis. The properties such as low toxicity, biodegradability and high boiling point extends their applications in dye industry, textiles, and polyacrylonitrile fibers manufacturing. They can be used also as additives to fuels, lubes and hydraulic fluids [3]. Traditional phosgene route for cyclic carbonates is unsafe. In place of phosgene, carbon dioxide can be used by inserting it in to C-O bond of strained heterocyclic ring of an epoxide. Cyclic carbonates synthesis by the coupling reaction of $CO_2$ with epoxide or oxetane (Fig. 5.7 and Scheme 5.5), where epoxide creates five membered and oxetane giving six membered derivatives of cyclic carbonate, is an atom efficient environmental process. Cyclic carbonate research is getting a remarkable attention. Halide-containing organic or inorganic salts, including tetraethyl ammonium bromide or KI is generally used for commercial cyclic carbonate synthesis [2]. The current commercial catalysts are ambiguous to satisfy environmental concerns. Catalytic activity of various homogeneous catalysts including quaternary ammonium salts, and alkali metal salts, ionic liquids, metal complexes, silica based materials, metal oxides and onium salt were investigated for this reaction.

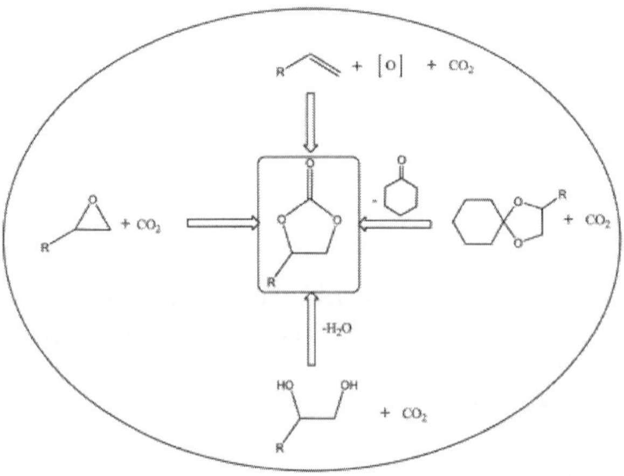


**FIGURE 5.7** Cyclic carbonate synthesis by the coupled reactions of $CO_2$.

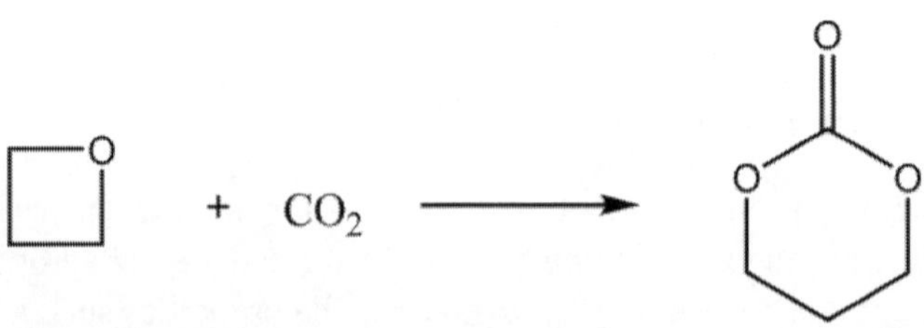


**SCHEME 5.5** Cyclic Carbonate Synthesis from $CO_2$.

This reaction usually requires high temperature and pressures. Homogeneous catalysts, such as $CH_3SnBr_3$, $Ph_4SbBr$, and n-$Bu_3SnI$ are effective at low temperature s, but high concentrations of catalyst is required. Catalyst separation and recycle are some issues [199–201]. Organo metallic complexes of zinc, Ni (0) and Cu (I), exhibit higher catalytic activity, they are excessively affected by air and moisture [202, 203]. Our group reported several modified silica-based heterogeneous systems for this reaction (Table 5.5) [240–242]. Among them, organo-functionalized metallosilicate catalyst, Ti(Al)-SBA-15-grafted adenine is more attractive in terms of high activity in the production of a variety of cyclic carbonates in high yields (80–90%) at moderate reaction conditions (120° C, 0.69 MPa) [242]. Metal oxide and mixed oxide catalysts were also investigated for this reaction, but the use of organic solvents is necessary to get good carbonate yield. The activity comparison of different heterogeneous catalyst system used for cycloaddition of $CO_2$ to epoxides is presented in the Table 5.5.

**TABLE 5.5** Cyclic carbonate synthesis from epoxide and $CO_2$.

| Catalyst | Co-catalyst/ solvent | Oxirane | $Temp_R$ (°C) | $P_{CO2}$ (bar) | $t_R$ (h) | Conversion (%) | Yield (%) | TOF ($h^{-1}$) | Ref. |
|---|---|---|---|---|---|---|---|---|---|
| $ZnBr_2$/CIL | - | PO | 110 | 15 | 1 | - | 96 | 2886 | 204 |
| LiBr | DMAP | PO | 130 | 20 | 3 | - | 91 | 295 | 205 |
| $NbCl_5$ | DMAP | PO | 50 | 5 | 12 | - | 96 | 32 | 206 |
| $Ni(PPh_3)_2Cl_2/PPh_3/Zn$ | $Bu_4NBr$ | PO | 120 | 25 | 1 | 99 | 99 | 3544 | 207 |
| $1Cp_2TiCl_2$ / 4 KI | THF | PO | 150 | 12 | 4 | - | 99 | - | 208 |
| HCOOH/ KI | - | AGE | 120 | 9.6 | 3 | 87 | 78 | - | 209 |
| $Zn_4(OCOCF_3)_6O$ | $Bu_4NI$/- | PO | 25 | 1 | 6 | 83 | 83 | - | 210 |
| $Ni(acac)_2$ | n-$Et_4NBr$ | SO | 50 | 50 | 4.5 | 100 | 97 | - | 211 |
| Cyanocobalamin | n-$Bu_4NI$ | PO | 140 | 8 | 6 | - | >99 | 330 | 212 |
| Sn(meso-porphyrin) $Cl_2$ | DMAP | PO | 150 | 14 | 5 | - | 99 | 794 | 213 |
| NaI | $PPh_3$/PhOH | PO | 120 | 40 | 4 | 100 | 100 | - | 214 |
| $(pyridine)_2ZnBr_2$ | - | EO | 100 | 35 | 1 | - | 97 | 580 | 215 |
| PS-DFILX | - | PO | 125 | 20 | 0.5 | 99 | 97 | - | 216 |
| Tetramethylguanidine –silica | -/ Toluene | PO | 130 | 50 | 8 | 92 | 90 | - | 217 |
| SBA-15 / triazolium ionic liquids | - | PO | 110 | 20 | 2 | - | 69-72 | 30-104 | 218 |
| KI/lecithin | - | PO | 100 | 20 | 4 | - | 98 | - | 219 |
| CTF-P-HSA[a] | | ECH | 130 | 6.9 | 4 | 100 | 96 | - | 220 |
| $[DMIm]_2[Br_2Zn(Me_2PO_4)_2]$ | - | EO | 140 | 35 | 1 | - | - | 4989 | 221 |
| MCM-41-Imi/Br | - | SO | 100 | 30 | 4 | 99 | 98 | 36 | 222 |
| $SiO_2$/[BMIm]Br | - | PO | 160 | 80 | 4 | 100 | 96 | - | 223 |
| CILBr-Si | - | AGE | 110 | 18.2 | 3 | 99 | 99 | - | 224 |
| $ZnBr_2$/[BMIm]Br | - | PO | 100 | 15 | 1 | - | 98 | 5580 | 225 |
| $ZnBr_2$/[BMIm]Br | - | PO | 100 | 15 | 1 | - | 86 | 3015 | 225 |
| $ZnCl_2$/[BMIm]Br | - | PO | 100 | 15 | 1 | - | 95 | 5410 | 225 |
| $[BMIm]^+[BF_4]^-$ | - | PO | 110 | 25 | 6 | 100 | 100 | - | 226 |

**TABLE 5.5** *(Continued)*

| Catalyst | Co-catalyst/ solvent | Oxirane | $Temp_R$ (°C) | $P_{CO2}$ (bar) | $t_R$ (h) | Conversion (%) | Yield (%) | TOF ($h^{-1}$) | Ref. |
|---|---|---|---|---|---|---|---|---|---|
| Cr(III)(salen)Cl | DMAP/ DCM | PO | 100 | 7 | 1 | - | - | 916 | 227 |
| Cr(III)(pyrrolidine-salen)Cl | - | PO | 80 | 20 | 1 | - | 39 | 1936 | 228 |
| Cr(III)(pyrrolidine-salen)Cl | - | EO | 80 | 20 | 1 | - | 67 | 3338 | 228 |
| Cr(III)(salophen)$NO_3$ | DMAP | PO | 40 | 15 | 3 | - | - | 618 | 229 |
| Al(III)(salen)Cl | $Bu_4NBr$ | PO | 110 | 160 | 1 | - | - | 2220 | 230 |
| Al(III)(salen)Et | $Bu_4NI$ | PO | 35 | 6 | 8 | - | - | 84 | 231 |
| (salen)AlOAl(salen) | $Bu_4NBr$ | SO | 25 | 1 | 24 | - | 98 | - | 232 |
| (salen)AlOAl(salen) | $Bu_4NBr$ | PPO | 25 | 1 | 24 | - | 99 | - | 232 |
| (salen)AlOAl(salen) | $Bu_4NBr$ | HO | 25 | 1 | 3 | - | 88 | - | 232 |
| (salen)AlOAl(salen) | $Bu_4NBr$ | DO | 25 | 1 | 3 | - | 64 | - | 232 |
| [Co(III)(salen)] $O_2CCCl_3$ | $Bu_4NBr$ | PO | 25 | 15 | 2 | 49 | 25 | 245 | 233 |
| [Co(III)(salen)]Cl | DMAP | PO | 100 | 21 | 1.5 | - | 100 | 1200 | 234 |
| [Ru(III)(salen)]Cl | PTAT | PO | 70 | 8 | 0.33 | - | 98 | 588 | 235 |
| [Mn(III)salen]Br | - | SO | 140 | 200 | 3 | - | 92 | 183 | 236 |
| $SiO_2$-immobilized [Mn(III)salen]Br | - | SO | 140 | 200 | 3 | - | 95 | 196 | 236 |
| [Sn(IV)(salen)]$I_2$ | DMAP | PO | 120 | 7 | 5.15 | - | - | 531 | 237 |
| Zn(salphen) complex | $NBu_4I$ | SO | 45 | 80 | 3 | - | 96 | - | 238 |
| Zeolite beta | Nil | ECH | 120 | 6.9 | 3 | 100 | 85 | 27 | 239 |
| Zeolite beta | $CH_3CN$ | PO | 120 | 6.9 | 5 | 100 | 100 | 16 | 239 |
| Zeolite beta | $CH_3CN$ | SO | 120 | 6.9 | 8 | 96 | 96 | 10 | 239 |
| Zeolite beta | $CH_3CN$ | BO | 120 | 6.9 | 5 | 87 | 70 | 14 | 239 |
| AlPc-Y | DMAP/ DCM | ECH | 120 | 6.9 | 4 | 97 | 97 | 606 | 240 |
| AlPc-Y | DMAP/ DCM | PO | 120 | 6.9 | 4 | 85 | 72 | 450 | 240 |
| TS-1 | DMAP | ECH | 120 | 6.9 | 4 | 89 | 87 | 829 | 241 |
| Ti-MCM-41 | Nil | SO | 140 | 6.9 | 10 | 100 | 82 | 595 | 242 |
| Ti-SBA-15-*pr*-Ade (Si/Ti=104) | Nil | ECH | 120 | 6.9 | 4 | 94 | 89 | - | 242 |
| Ti-SBA-15-*pr*-Ade (Si/Ti=40) | $CH_3CN$ | PO | 120 | 6.9 | 6 | 96 | 95 | - | 242 |

**TABLE 5.5** *(Continued)*

| Catalyst | Co-catalyst/ solvent | Oxirane | $Temp_R$ (°C) | $P_{CO2}$ (bar) | $t_R$ (h) | Conversion (%) | Yield (%) | TOF ($h^{-1}$) | Ref. |
|---|---|---|---|---|---|---|---|---|---|
| Ti-SBA-15-*pr*-Ade (Si/Ti=40) | Nil | SO | 120 | 6.9 | 8 | 94 | 89 | - | 242 |
| ZIF-8 | Nil | ECH | 80 | 7 | 4 | 73 | 73 | - | 243 |
| SmOCl | DMF | PO | 200 | 140 | 8 | 100 | 99 | - | 244 |
| $Ca_2Al(OH)_6Cl.2H_2O$ | DMF | SO | 100 | 48 | 1 | 98 | 90 | - | 245 |
| Zn-Mg-Al-O | Nil | PO | 140 | 25 | 12 | 99 | 89 | - | 246 |
| MIL-68(In)-$NH_2$ | -/DMF | SO | 150 | 8 | 8 | 74 | - | 1.06 | 247 |
| DMC Complex | - | ECH | 140 | 1.2 | 6 | 95 | 91 | - | 248 |
| θ-ZrP/BMIMCl | - | PO | 110 | 3 | 12 | - | 79 | - | 249 |
| Cr-MIL-101 | $Bu_4NBr$ | SO | 25 | 8 | 48 | 98 | 95 | - | 250 |
| Fe-MIL-101 | $Bu_4NBr$ | PO | 25 | 8 | 24 | 99 | 95 | - | 250 |
| $[(n-C_4H_9)_4N.]_6H$ $[PW_{11}Co(H_2O)O_{39}]$ | - | PO | 150 | 30 | 12 | 99 | 99 | 83 | 251 |

DMAP = N,N-dimethylaminopyridine, DMF = N,N-dimethylformamide, DCM = dichloromethane, PTAT = phenyltrimethylammonium tribromide, EO = ethylene oxide, PO = propylene oxide, ECH = epichlrohydrin, SO = styreneoxide, AGH = allylicglycidil ether.; BO = butylene oxide, HO = n-hexene oxide, DO = N-decene oxide, PPO = 2-phenylpropylene oxide, PS-DFILX = polymer supported diol functionalized ionic liquids, acac = acetyl acetonate, CIL= carboxylic-acid functionalized imidazolium-based ILs, note: active site consideration for TOF calculation is not uniform. [a]Co-valent triazine complexes

Ionic liquids (ILs) have been studied either as homogeneous or heterogeneous catalysts, where in the latter case ILs were used in the form of a supported species. ILs show bi-functional activity as they act as solvent as well as catalyst due to the considerable solubility of $CO_2$ on it. ILs can easy be separated from the reaction mixture and reused. Imidazole salt, quaternary ammonium salt, quaternary phosphonium salt, etc. are some active ILs for this reaction [221–226]. $SiO_2$ is the most effective support for immobilizing or supporting ILs. Mesoporous materials such as SBA-15 and MCM-41 also are used to immobilize ILs [218, 222].

Metal salen complexes were found to be active for the synthesis of cyclic carbonate in the presence of quarternary ammonium salt cocatalyst. Highly selective formation of cyclic carbonate using aluminum salen complexes was due its favorability into the dissociation of growing polymer chain than the corresponding chromium complex.

#### 5.3.2.2 FROM OTHER ROUTES

One-pot synthesis of cyclic carbonate from olefin is one of the alternative approaches (Fig. 5.7). Its actually the combination of two steps where in the first-step, alkene was oxidized to epoxide and in the second step the reaction mixture containing epoxide is coupled with $CO_2$. Arresta et al. [252] examined the activity of $Nb_2O_5$ and several other metal oxide catalysts for styrene carbonate synthesis via oxidative carboxylation of styrene and $Nb_2O_5$ was found to be the best. Eghbali and Li [253] studied the combination of bromamide ion (N-bromosuccinamide, tetrabutylammonium bromide or sodium bromide) and DBU (1,8-diazabicyclo[5.4.0]undec-7-ene) or DMAP (N, N-dimethylaminopyridine) for catalyzing oxidative carboxylation at ambient reaction conditions (40–60° C and 16–30 bar pressure) with high cyclic carbonate yield. Ionic liquids and gold/silica with zinc bromide and oxidant also gave satisfactory results [254]. Cyclo-addition of $CO_2$ with 1,2-diol is more valuable and promising alternative route for cyclic carbonate synthesis even though it is limited by thermodynamics (Fig. 5.7). Tomishige et al. reported ceria-based catalysts for this reaction [255]. $CeO_2$–$ZrO_2$ solid solution was an active catalyst for propylene carbonate (PC) and ethylene carbonate (EC) formation from propylene glycol (PG) and ethylene glycol (EG), respectively. The activity of $CeO_2$–$ZrO_2$ solid solution depends on its composition and calcination temperature. $Bu_2SnO$ and $Bu_2Sn(OMe)_2$ were also reported as efficient catalysts for transformation of PG to PC under super-critical $CO_2$ condition [256]. In order to overcome the equilibrium limitations, Huang et al. [257] proposed PC synthesis from PG and $CO_2$ in the presence of modified ZnO using $CH_3CN$ as a solvent as well as a dehydrating agent. Recently, Silva et al. [258] used bezonitrile instead of acetonitrile with alkali carbonate catalyst for PC synthesis from PG, which improves the PC yield to 20%. $CO_2$ insertion into cyclic ketal under super-critical $CO_2$ medium and electrochemical addition of $CO_2$ to epoxides are the other alternative routes for cyclic carbonate synthesis. Although tremendous research work is in progress on the alternative routes, development of efficient catalysts for the synthesis of cyclic carbonates is of great importance.

### 5.3.3 POLYCARBONATES

Polycarbonates are the most important polymers with versatile applications. Excellent optical properties and specific weight index of polycarbonates make them ideal lens material in eye glasses. There are two traditional methods available for commercial production of polycarbonates from bis-phenol, that is, through solvent interfacial polymerization and transesterification [259]. An alternative way for the manufacture of polycarbonate is the reaction between $CO_2$ and high free energy compounds such as epoxides. Alternating copolymerization of epoxies with $CO_2$ (Scheme 5.6) produces biodegradable aliphatic polycarbonates. The polymers have gas-barrier (30–50 wt% $CO_2$) and degradation properties, which fabricated them for food packaging, automotive parts and electronics processing applications [260]. Aliphatic polycarbonates production through alternative copolymerization of $CO_2$ has already been commercialized. Density and thermal stability of polypropylene carbonate produced by the copolymerization of $CO_2$ and propylene oxide is lesser than that of bis-phenol-A polycarbonate [259]. In coupling reaction between carbon dioxide and epoxides, catalysts assist the reaction by activating $CO_2$ or epoxide or both at a time. $CO_2$ activation can occur through both nucleophilic and electrophilic attack. The catalyst, which is having both Lewis acidity and Lewis basicity is suitable for this reaction, where Lewis acid activates the epoxide ring opening or $CO_2$ and Lewis base assists the nucleophilic activation of $CO_2$.

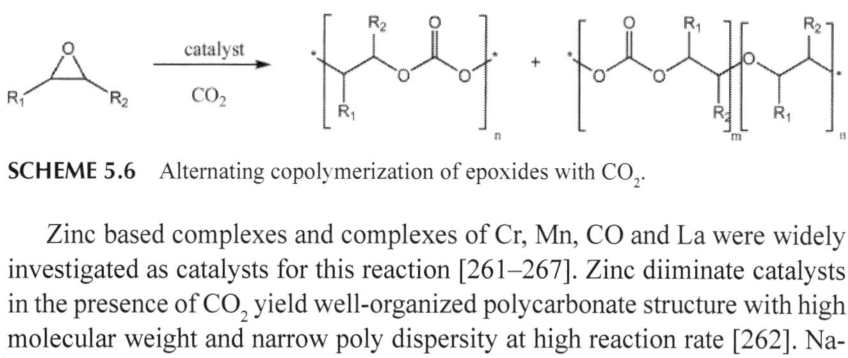


**SCHEME 5.6** Alternating copolymerization of epoxides with $CO_2$.

Zinc based complexes and complexes of Cr, Mn, CO and La were widely investigated as catalysts for this reaction [261–267]. Zinc diiminate catalysts in the presence of $CO_2$ yield well-organized polycarbonate structure with high molecular weight and narrow poly dispersity at high reaction rate [262]. Nature of the metal center plays a key role in catalytic activity. Co-based complexes are highly active for this copolymerization of ethylene and propylene oxides with $CO_2$. The high activity of Co complex is due to increase in nucleophilicity, which ultimately improves the rate of ring opening of epoxides and polycarbonate formation [263]. Functionalization of metal complex is one of the ways to improve catalytic activity of metal complexes towards copolymerization reaction. Metal–salen complex bound to quarternary ammonium unit is an example for improving the selectivity and molecular weight of polycarbonate through functionalization. The catalytic activity of metal complex for copolymerization can be significantly improved by the addition of Lewis base as a cocatalyst [268]. In addition to the homogeneous metal complex, heterogeneous catalysts were also employed for this reaction with high efficiency. Zinc dicarboxylate is the widely investigated heterogeneous catalyst system for copolymerization of epoxides and $CO_2$. Among several, zinc glutarate was found to be efficient and its activity enhancement is mainly controlled by its degree of crystallanity and crystalline size [269, 270]. Co-Zn double-metal cyanide (DMC) are the highly attractive heterogeneous catalyst for copolymerization of epoxides in an economic and sustainable way [271–274]. Catalytic activity data of double-metal cyanide catalysts for copolymerization of epoxides and $CO_2$ are listed in Table 5.6. The turn over frequency over these catalysts is higher than the homogeneous systems.

**TABLE 5.6** Catalytic activity of double-metal cyanide (DMC) complexes for copolymerization of $CO_2$ and epoxides.

| Substrate | $R_T$ (°C) | $R_t$ (h) | $P(CO_2)$ (bar) | Carbonate linkage[d] (%) | TOF[e] ($h^{-1}$) | $M_n$[f] (g/mol) | $M_w/M_n$[f] | Remarks | Ref. |
|---|---|---|---|---|---|---|---|---|---|
| CHO | 80 | 4 | 9.7 | 53 | 132 | 8900 | 1.67 | Co–Zn DMC catalyst prepared by the conventional methods used for the reaction without any solvent or additive | 275 |
| CHO | 80 | 4 | 86 | 80 | 356 | 10200 | 1.77 | | |
| VCHO | 80 | 12 | 9.7 | 73 | 38 | 10100 | 2.22 | | |
| CPO | 80 | 12 | 9.7 | 36 | 41 | 2300 | 1.55 | | |
| PO | 80 | 2 | 86 | 39 | 172 | 2600 | 1.39 | | |
| CHO | - | 0.5 | 9.7 | 74 | 3910 | 14500 | 1.5 | Micro-wave assisted reaction with Co-Zn DMC catalyst. MW power = 100 W | 271 |
| CHO | 50 | 4.6 | 40 | 80 | - | 11700 | 2.48 | THF used as a solvent along with Co-Zn DMC catalyst | 273 |
| PO | 60 | 40 | 38 | 40 | 19.3 | - | 1.93 | Co-Zn DMC catalyst prepared by ball milling $[K_4Fe(III)(CN)_6]^{3-}$ and $ZnCl_2$ used without any solvent | 276 |

**TABLE 5.6** *(Continued)*

| | | | | | | | | | |
|---|---|---|---|---|---|---|---|---|---|
| PO | 60 | 40 | 38 | 44 | 6.0 | - | 1.63 | Co-Zn DMC catalyst prepared by ball milling [$K_4Fe(III)(CN)_6]^{3-}$ and $Zn(COOCH_3)_2$ without any solvent. | |
| PO | 60 | 40 | 38 | 37 | 10.8 | - | 2.33 | Co-Zn DMC complex prepared by conventional way | |
| PO | 50 | 6.7 | 50 | 73 | - | 34700 | 2.0 | Reaction carried out using nano lamellar Zn-Co DMC catalyst; contains 98 ppm water in the system | 277 |
| CHO | 100 | 3 | 38 | 47 | 3815 | - | 1.95 | $SiO_2/Zn_3[Co(CN)_6]_2$ catalyst was used | 278 |
| PO[a] | 50 | 1 | 55 | 36 | 390 | 163000 | 5.8 | DMC catalyst was used | 279 |
| PO[b] | 90 | 24 | 55 | 43 | 18 | 25600 | 3.6 | | |
| PO[c] | 90 | 24 | 55 | 44 | 13 | 27900 | 3.7 | | |

[a]Co-Ni DMC catalyst, [b]Co-Pd DMC catalyst, [c]Co-Pt DMC catalyst, [d]Estimated by $^1H$ NMR. Carbonate linkage (%) = {carbonate linkage/[carbonate linkage + ether linkage]} × 100), [e]Turn over frequency (TOF) = moles of epoxide consumed per mole of Zn per hour, [f]Obtained from GPC analysis, PO = propylene oxide.; CHO = cyclohexene oxide, VCHO = 4-vinyl 1-cyclohexene 1,2-epoxide, CPO = Cyclopentene oxide.

### 5.3.4 UREA AND ITS DERIVATIVES

The utilization of $CO_2$ in the synthesis of urea and its derivative by carboxylation reaction of ammonia and amines has received much attention. At present, urea production is the largest $CO_2$ consuming organic synthesis and it is the most commonly used nitrogen fertilizers. Commercial production of urea is from $NH_3$-$CO_2$ technology, which actually constitutes two equilibrium reactions (Eqs. (11) and (12)).

$$2NH_3 + CO_2 \rightarrow NH_2COONH_4 \quad (11)$$

$$NH_2COONH_4 \rightarrow CO(NH_2)_2 \quad (12)$$

Reactions of ammonia and carbon dioxide forming ammonium carbamate constituted the first step, which is exothermic in nature. This is followed by the endothermic decomposition of ammonium carbamate, through dehydration, in the second step to form urea. Recently, Xiang et al. [280] reported urea synthesis at atmosphere pressure and ambient temperature.

Substituted urea derivatives are important carbonyl compounds having a versatile application in pharmaceutical, fertilizers and dye industry and as anti oxidants in gasoline and additive in the synthesis of amino-plastics. The synthesis of substituted urea received a special care because its application found in brain cancer treatment and its inhibitory effect on HIV protease enzyme [281, 282]. Current routes for the production of such compounds are non-eco-friendly in nature and expensive. The use of renewable $CO_2$ feedstock for substituted urea production is highly attractive. The formation steps of substituted urea are represented in Scheme 5.7. $CO_2$ easily combines with ammine, even at ambient conditions to form carbamic acid. The dehydration step for obtaining substituted urea requires high-energy input or the use of external agents in stoichiometric quantity; suitable catalyst is needed. $RuCl_3{\cdot}H_2O/Bu_3P$ [283] and $Ph_3SbO/P_4S_{10}$ [284] catalyze the synthesis of substituted ureas using $CO_2$. Water is formed as a by-product. Use of dehydrating agents such as $POCl_3$, propargyl alcohols, $Me_3NSO_3$, diorganophosphites and carbodiimide lift the equilibrium limitation, ultimately improving the product yield [285]. Ion et al. [286] reported the efficient synthesis of symmetric and asymmetric urea derivatives using different base catalyst, where the combination of Cs catalyst and N-methyl pyridine solvent gave high product yield even in the absence of a dehydrating agent. Hydrophilic ionic liquids are successfully demonstrated as efficient catalysts for the synthesis of dialkyl substituted ureas [285, 287]. Han et al. [285] developed [Bmim]OH ionic liquid catalyst for alkyl as well as aryl 1,3-disubstituted ureas.

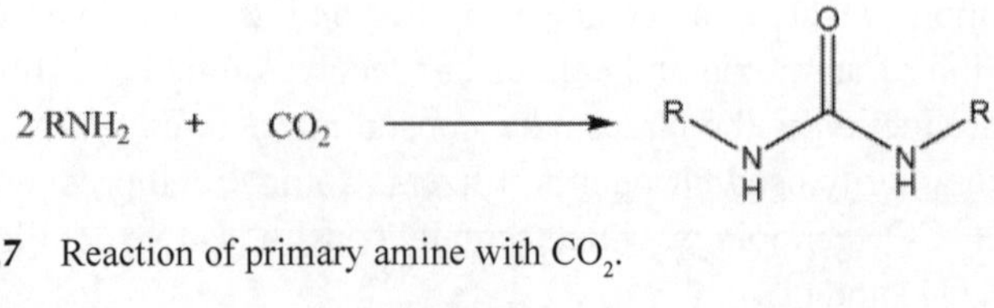


**SCHEME 5.7** Reaction of primary amine with $CO_2$.

### 5.3.5 ORGANIC CARBAMATES

Organic carbamate are stable class of compounds with unique applications in pharmaceutics, agriculture, polyurethane synthesis and more generally, for the production of intermediates for fine and commodity chemicals. The commercial process available for carbamate synthesis involves toxic reagents such as phosgene and isocyanate. The alternative non-phosgene routes like reductive carbonylation of nitro aromatics, oxidative carbonylation and methoxy carbonylation of amines have economic disadvantages. Direct reaction of $CO_2$, amine and alkyl halide is the most ecofriendly and sustainable alternative route (Scheme 5.8). In this route, carbamate anion is produced first by coupling reaction of $CO_2$ and amine (Scheme 5.8) and then, the resulting anion interacts with various alkyl halides yielding corresponding carbamates.

Several homogeneous and heterogeneous catalysts were employed for the carbamate synthesis from $CO_2$ [288]. The activity of zeolite-based catalyst for this reaction was examined by our group. Titano-silicate molecular sieves, metal pthalocyanine complexes encapsulated in zeolite-Y and zeolite-based metal-organic hybrid catalysts were explored (Table 5.7). Reactivity and stability of these catalysts is higher compared to the homogeneous catalysts like $Cs_2CO_3$ even in the absence of cocatalyst. Adenine modified mesoporous Ti–SBA-15 was a more efficient catalyst than the medium pore TS-1 catalyst for the alkyl or aryl carbamate formation [239, 242, 289]. These studies concluded that both the titanium ions (weak Lewis acid sites) and the amine moieties (basic sites) play a key role in the activity and selectivity. Recent studies on these reactions revealed that alcohols can be used as a substitute to alkyl halide, where water is formed as a byproduct instead of HX (Table 5.7) [290]. But thermodynamic and equilibrium limitation restricts the product formation, which ultimately results low carbamate yield. The use of dehydrating agent or continuous removal of water from the reaction medium is the most important way to get high carbamate yield. Synthesis of alkyl as well as aromatic carbamates with high reaction rate is the major challenge for this route. The other approaches for insertion of $CO_2$ into C-N bond are shown in Fig. 5.8.

**TABLE 5.7** Heterogeneous catalysts for carbamate synthesis from $CO_2$, amine and alkyl halide or alcohol.

| Catalyst | Amine | Alkyl halide or alcohol | Reaction conditions | Amine conversion (%) | Carbamate yield (%) | TOF | Ref. |
|---|---|---|---|---|---|---|---|
| Zeolite beta[a] | Octylamine | n-BuBr | $T_r = 80$ °C. $t_r = 4$ h, $CO_2$ pressure = 3.4 bar | 100 | 92 | 11 | 239 |

**TABLE 5.7** *(Continued)*

| | | | | | | | |
|---|---|---|---|---|---|---|---|
| SBA-15-pr-Ade | Octylamine | | | 99 | 92 | - | 242 |
| | Hexylamine | | | 97 | 85 | - | |
| | Cyclohexyl-amine | n-BuBr | | 91 | 76 | - | |
| Ti-SBA-15-pr-Ade | Aniline | n-BuBr | | 100 | 89 | - | 242 |
| | 2,4,6-tri-methyl aniline | | | 75 | 65 | - | |
| $TS\text{-}1^b$ | Aniline | | | 93 | 89 | 13 | 289 |
| | hexylamine | | | 91 | 86 | 13 | |
| | Dodecylamine | | $T_r = 80$ °C, $t_r = 4$ h, $CO_2$ pressure = 3.4 bar | 93 | 90 | 13 | |
| $MCM\text{-}41^a$ | Dodecyllamine | n-BuBr | | 100 | 90 | 36 | 291 |
| | Hexylamine | | | 90 | 85 | 33 | |
| | Cyclohexyl-amine | | | 94 | 76 | 34 | |
| | Benzylamine | | | 76 | 69 | 5.9 | |
| $CeO_2$ | Benzylamine | | | > 99 | 79 | - | 290 |
| | Cyclohexyl-methanamine | $CH_3OH$ | $T_r = 150$ °C, $t_r = 12$ h, $CO_2$ pressure = 50 bar | 72 | 61 | - | |
| | $CH_3NH_2$ | | | 48 | 48 | - | |
| | Benzylamine | | | 67 | 64 | - | |
| | Cyclohexyl-methanamine | $C_2H_5OH$ | | 44 | 42 | - | |

[a]Turnover frequency (TOF) = moles of amine converted per mole of quaternary ammonium ion (template) in as-synthesized catalyst per hour. [b]TOF = moles of substrate converted per mole of Ti per hour.

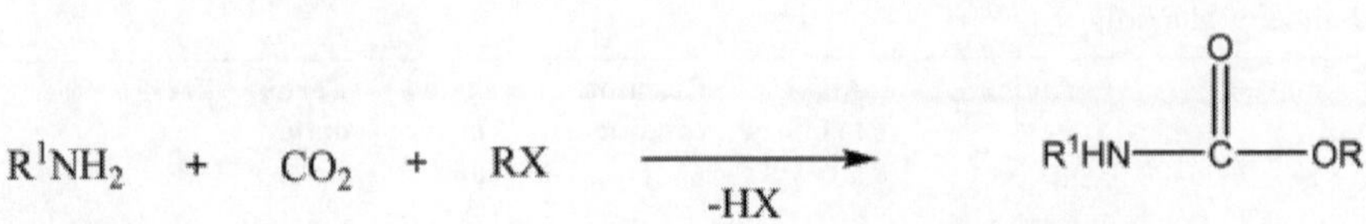


**SCHEME 5.8** Organic carbamates from a primary amine, alkyl halide and $CO_2$.

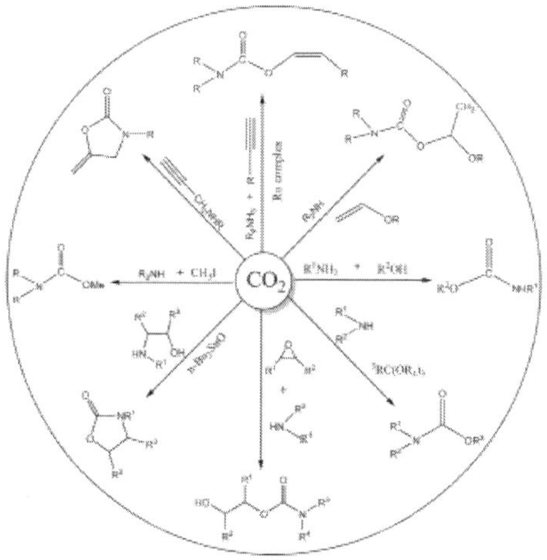


**FIGURE 5.8** Various approaches for insertion of $CO_2$ into C-N bonds.

### *5.3.6 FORMIC ACID AND ACETIC ACID*

Formic acid is the simplest carboxylic acid having versatile applications including in fuel cells. CO is used as a raw material in the commercial synthesis of formic acid. The alternative green route using $CO_2$ is relevant and advantageous to the environment. $CO_2$ hydrogenation into formic acid has thermodynamical limitations. But it can be made possible by working at lower reaction temperature using high $CO_2$-$H_2$ pressures and stoichiometric amounts of base [292–294]. Several homogeneous transition metal based catalysts, especially those of rhodium, ruthenium, and iridium were reported for the hydrogenation of $CO_2$ to formic acid. Among them, the Ru based catalyst was found to be more active (Table 5.8) [293, 295]. Tai et al. [296] reported Ru-phosphine system for this reaction and suggested that in this complex activity is related to the electronic effects and the bite angle. Development of suitable heterogeneous catalyst systems is highly desirable. Heterogeneous supported immobilized homogeneous catalysts are attractive because of they retain the activity ease of separation of formic acid from the catalyst. Amino functionalized silica immobilized Ru complexes showed TOF of 1384 $h^{-1}$ for $CO_2$ hydrogenation to HCOOH under supercritical $CO_2$ conditions [297]. Recently, Preti et al. [298] studied the heterogeneous gold catalyst activity on this reaction and found that it catalyzes the $CO_2$ hydrogenation reaction in the presence of $NEt_3$ to produce formic acid-$NEt_3$ adduct, which finally was separated with the help of high-boiling $(n\text{-}C_6H_{13})_3N$. Development of new ecofriendly and economically feasible catalyst technology is needed to overcome the current problems in this reaction.

**TABLE 5.8** $CO_2$ conversion into formic acid with different catalyst systems.

| Pre-catalyst /catalyst | Reaction conditions | TON[a] | HCOOH yield (mmol) | Ref. |
|---|---|---|---|---|
| Ru(2 wt%)/$Al_2O_3$ | $T_r$ =80 °C, 5 MPa of $H_2$, $P_{Tot}$ = 13.5 MPa, $t_R$ = 1 h, additive = $NEt_3$ (5mL), solvent = ethanol (15 mL). | 91 | 9.1 | 299 |
| Ru(2 wt%)/ activated carbon | | 10 | 1 | |
| 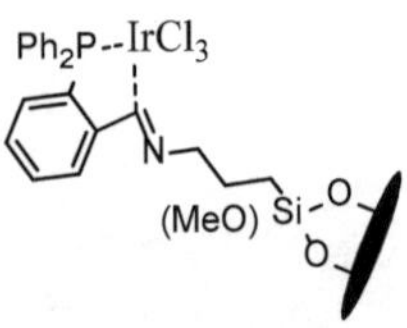 | $T_r$ = 120 °C, $P_{Tot}$ = 4.0 MPa with $H_2/CO_2$ = 1:1, $t_r$ = 2 h, additive = $NEt_3$ (2 mL), solvent = deionized water (20 mL), catalyst = 100 mmol Ir | 2300 | - | 300 |
| Si–$(CH_2)_3NH(CSCH_3)$ {$RuCl_3(PPh_3)$} | $T_r$ =80 °C, 9 MPa of $H_2$, $P_{Tot}$ =18 MPa, $t_r$ = 2 h, additive[b] = [DAMI][OTf] = 0.4 g, solvent = $H_2O$ (3 mL) | 1840 | - | 294 |
| Si–$(CH_2)_3NH(CH_2)_3CH_3$)–Ru | $T_r$ = 80 °C, 4 MPa of $H_2$, $P_{Tot}$ = 16 MPa, $t_r$ = 1 h, additive = $NEt_3$ (5 mL), solvent = ethanol (3 mL), ligand = $PPh_3$ | 1384 | - | 297 |
| ($\eta^6$-arene)Ru(bis-NHC)[c] | $T_r$ = 200 °C, 2 MPa of $H_2$, $P_{Tot}$ = 4 MPa, $t_r$ = 75 h, additive = KOH (10 mL, 1 M), solvent = $H_2O$ | 23000 | 0.46 | 301 |
| $RuCl_2(PMe_3)_4$ | $T_r$ = 50 °C, 8 MPa of $H_2$, $P_{Tot}$ = 20 MPa, $t_r$ = 75 h, additive = $NEt_3$ and $H_2O$, solvent = Sc. $CO_2$ | 7200 | - | 302 |
| $TpRu(PPh_3)(CH_3CN)H$[c] | $T_r$ = 100 °C, $P_{Tot}$ = 5 MPa with $H_2/CO_2$ = 1:1, $t_r$ = 16 h, additive = $NEt_3$ (2 mL), solvent = $CF_3CH_2OH$ (20 mL).; | 1815 | - | 303 |
| [$Ru_2$(μ-CO)$(CO)_4$(μ-$Ph_2PCH_2PPh_2)_2$] | Room temperature reaction, $P_{Tot}$ =7 MPa with $H_2/CO_2$ = 1:1, $t_r$ = 21 h, additive = $NEt_3$ (15 mL), solvent = acetone (30 mL) | 2160 | 128 | 304 |
| $RuCl(OAc)(PMe_3)_4$ | $T_e$ = 80 °C, 6.9 MPa of $H_2$, $P_{Tot}$ = 18.7 MPa, $t_r$ = 1 h, additive = $NEt_3$ and $C_6F_5OH$, solvent = Sc.$CO_2$ | 95000 | - | 305 |

**TABLE 5.8** *(Continued)*

| Pre-catalyst /catalyst | Reaction conditions | TON[a] | HCOOH yield (mmol) | Ref. |
|---|---|---|---|---|
| [Cp*Ir(phen)Cl]Cl | $T_r$ = 120 °C, $P_{Tot}$ = 6 MPa with $H_2/CO_2$ = 1:1, $t_R$ = 48 h, carried out in the presence of aq. KOH | 222000 | - | 306 |
| PNP-Ir(III) trihydride complex | $T_r$ = 120 °C, $P_{Tot}$ = 6 MPa with $H_2/CO_2$ = 1:1, $t_r$ = 48 h, carried out in presence of 1 M (5 mL) aq. KOH and 0.1 mL tetrahydrofuran | 3500000 | - | 307 |
| $RhCl(PPh_3)_3$ | $T_r$ = 25 °C, 1.96 MPa of $H_2$, $P_{Tot}$ = 3.9 MPa, $t_r$ = 20 h, reaction carried out using methanol as a solvent and $PPh_3$ as ligand | - | 2.5 | 308 |

[a]TON= moles formic acid formed per mole of metal; [b][DAMI] [TfO] = 1,3-di(N,N-dimethylaminoethyl)-2-methylimidazolium trifluoromethanesulfonate; [c]NHC = N-heterocyclic carbine, Tp=hydrotris(pyrazolyl)borate.

Acetic acid is having a great demand in plastics, textiles, paper, paints, and adhesives industries. At present, it is the produced by the Monsanto process. Alternative route for the synthesis of acetic acid is by using $CO_2$ and methane. A South African patent by Freund and Wambach [309] claimed that with one or more metals of group VIA, VIIA, and VIIIA on alumina, the direct synthesis of acetic acid from $CO_2$ and $CH_4$ at 100–600°C and 0.1–20 MPa, a selectivity of 70–95% is obtained. Cu-Co oxides [310] and Pd or Pt-based catalysts [311] are found to be effective for the activation of C-H bond in methane. Introduction of oxygen to the reactant system is another way to overcome thermodynamic limitations. Huang et al. [312] explored the catalytic activity of $V_2O_5$-$PdCl_2/Al_2O_3$ on the direct synthesis of acetic acid from $CH_4$ and $CO_2$ in the presence of $O_2$ and showed the water-vapor effect on the rate of formation of acetic acid.

### *5.3.7 DIMETHYL FORMAMIDE*

Dimethyl formamide (DMF) is a well-known organic solvent having applications in various sectors such as plastics, pharmaceutics, paint industries and peptide synthesis. Catalytic carbonylation is the commonly used procees for DMF synthesis. Replacing toxic CO by renewable $CO_2$ has attained much attention. Raney nickel was first used for this purpose and found to be an effective catalyst for the reaction between dimethyl ammine, $CO_2$ and $H_2$ in methanol or phenol [313]. In 1970, Haynes et al. [314] put forwarded the activity of $Pd(CO_3)[P(C_6H_5)_3]_2$ on DMF synthesis from $CO_2$.

Several homogeneous catalyst systems were employed for this reaction. Ru complexes showed high catalytic activity in DMF synthesis. $RuCl_2[P(CH_3)_3]_4$ exhibited TOF of 10,000 $h^{-1}$ [315]. The reaction occurs in two steps, firstly $CO_2$ is hydrogenated into formate, which subsequently dehydrated to give formamide. $RuCl_2(dppe)_2$ showed higher activity compared to $RuCl_2[P(CH_3)_3]_4$ with a TOF of 3,60,000 $h^{-1}$ [316]. Silica-immobilized Ru complexes were used as heterogeneous catalyst for this reaction, which enabled the recyclability of the catalyst and made the product separation easy [317]. Generally, water addition to the reaction mixture hindered the byproduct formation, which ultimately enhanced the DMF yield. Instead of using Ru complexes, reaction was also carried out with simple $RuCl_3$ and $Ru/Al_2O_3$ in the presence phosphine, dppe or $PPh_3$. This system was also highly active and selective [318]. Recently, Liu et al. [319] reported the use of Cu/ZnO catalyst (Cu: Zn = 3: 2) for this reaction and attained a maximum DMF yield of 97% at 140°C in 6 h. They suggested that particle size, dispersion and morphology of Cu and ZnO phases play a significant role on catalyst activity and proposed a tentative reaction mechanism involving synergistic effect between Cu and ZnO.

## 5.4 $CO_2$ AS SOLVENT AND CATALYST PRECURSOR

$CO_2$ in the form of supercritical fluid has unique properties. Supercritical $CO_2$ (Sc. $CO_2$) is a viable and ecofriendly alternative to hazardous and flammable petrochemical-derived solvent and is also beneficial for stereo-selective organic synthesis. The highlighting factors of Sc.$CO_2$ solvent in comparison with traditional solvents are its tunable physical properties by changing the strength and density of $CO_2$ through its temperature pressure changes. In addition, it is a potential operating medium for process intensification of selective separation and extraction of products by controlling their solubility in reagents. Considering safety, solubility is an important criterion of the solvent selection for the food and pharmaceutical applications. Sc.$CO_2$ is a well-known solvent in polymer manufacture, processing and modification [320] and commercially used for the production of fluoropolymers and hydrogenation and alkylation reactions [321]. Diasterio selective enhancement in sulfur oxidation of S-methyl cysteine using Sc.$CO_2$ in place of conventional solvents was reported by Rayner [322]. In those studies, they attained the corresponding trans-isomer with almost 100% selectivity at an optimum $CO_2$ density of 0.7 g/ mL. Dapurkar et al. [323] studied the selective oxidation of tetralin to 1-tetralone by molecular oxygen over mesoporous CrMCM-41 molecular sieve catalyst in Sc. $CO_2$ solvent and found that Sc. $CO_2$ improves 1-tetralone selectivity and suppresses leaching of chromium from CrMCM-41. Selva et al. [324] reported the selectivity enhancement in olefin metathesis using $CO_2$ as a solvent in place of n-heptane and toluene. Recent studies of the catalytic addition of glycerol, mono-and polyalcohols to 2,3-dimethylbuta-1,3-diene showed selective telomerization with replacing of volatile organic solvent by Sc. $CO_2$ [325]. Application of Sc. $CO_2$ as a solvent is also extended to material research.

Many chemical reactions are catalyzed by acids but on a large-scale can generate significant amounts of acid waste, which requires neutralization and/or disposal. In principle, $CO_2$ dissolved in water is mildly acidic, due to formation of carbonic acid and should be capable of catalyzing a range of reactions. Simple pressure release at the end of the reaction brings the pH back to levels, which require minimal neutralization. Hydrolysis of ketal to cyclohexanone and ethylene glycol and epoxides to diols are a few examples where $CO_2$ in presence of water have been used as a catalyst [322].

## 5.5 SOME $CO_2$-BASED NOVEL TECHNOLOGIES

Researchers at University of California Los Angeles (UCLA) have genetically engineered a microorganism (Ralstonia eutropha H16) that converts carbon dioxide into liquid fuels viz., isobutanol and 3-methyl-1-butanol, which could be used for cars or other combustion engines. This bacterium feeds on formic acid created using electricity and produces combustible alcohols [326]. This discovery leads to a possibility for a car with solar panels on the roof produces formic acid from photo-electrochemical means of $CO_2$ fixation, which is then converted into alcohol fuel source as the car drives along. In the ECFORM electrochemical process, proprietary tin-based alloys are used as cathodes to convert $CO_2$ into formate salts. Small concentration of $H_2$ and CO byproducts are also produced at the cathode. Oxygen evolution reaction takes place at the anode. However, an economically viable electrochemical technology requires optimization of parameters viz., high current density, high Farady efficiency, low specific electricity consumption and long electrode lifetime [327]. Carbon Sciences, a Californian company announced a new technique to recycle $CO_2$ into fuel [328] in, which a biochemical process converts $CO_2$ into basic hydrocarbons – methane (C1), ethane (C2) and propane (C3), which can then be used to make higher-grade fuels like gasoline and jet fuels. Mitsui Chemical Inc., Japan announced the construction of a pilot plant (33,000 gallons per annum) at the company's Osaka facility for producing methanol from industrial $CO_2$ effluent and photocatalyst produced hydrogen [329]. Highly efficient thin film, anatase titania (photon to current quantum efficiency = 60%) was proposed to use as a photocatalyst to generate $H_2$ from water and company's proprietary ultra high activity catalyst for producing methanol. Carbon Sources [330] declared developing breakthrough $CO_2$-based gas-to-liquids ($CO_2$-GTL) fuel technology. This process produces fuels such as gasoline, diesel and jet-fuel from $CO_2$ and $CH_4$. According to the company, the $CO_2$-GTL process (unlike the conventional GTL) operates at much lower temperature and pressures in the presence of a novel catalyst and produces a renewable and clean burning gasoline. The team at Sandia National Laboratory [331] is reversing the combustion of fossil fuels. Sunlight and cobalt ferrite catalyst in a solar reactor split water and carbon dioxide, producing hydrogen and carbon monoxide. These can then be used for making liquid fuels such as gasoline via FT synthesis.

## 5.6 CONCLUSION

There is a great opportunity for using and converting of $CO_2$ into a variety of industrial chemicals and fuels under CCU concept. Some of the $CO_2$-based processes are at advanced stages and the other are at infancy. Development of more such reactions would possibly have some positive impact on the environment and lead to green and sustainable technologies and novel materials. There is need to enhance market acceptability of $CO_2$ chemicals by developing innvotaive processes to bring down their cost. $C_1$ chemistry and heterogeneous catalysis will play a major role in achieving success in this direction.

## KEYWORDS

- **Acetic acid**
- **Catalytic $CO_2$ hydrogenation**
- **$CO_2$ as solvent**
- **Cyclic carbonates**
- **Dimethyl formamide**
- **Formic acid**
- **Long-chain hydrocarbons**
- **Manufacture fuels from $CO_2$**
- **Organic carbamates**
- **Supercritical $CO_2$**

## REFERENCES


1. http://scrippsco2.ucsd.edu/data/atmospheric co2.html
2. Sakakura, T.; Choi, J-C.; Yasuda, H. Transformation of Carbon dioxide. *Chem. Rev.* **2007,** 107, 2365–2387.
3. Ratnasamy, P.; Srinivas, D. Chemicals from Carbon Dioxide in *Handbook of Heterogeneous Catalysis*, 2nd Edition.; G. Ertl, H. Knözinger, F. Schüth and J. Weitkamp Eds.; Wiley: 2008; Vol. 7, Ch. 14, Section 14.17.4, p 3717–3732.
4. Hu, B.; Guild, C.; Suib, S.L. Thermal, Electrochemical, and Photochemical Conversion of $CO_2$ to Fuels and Value-added Products. *Journal of $CO_2$ Utilization.* **2013**, 1, 18–27.
5. Mikkelsen, M.; Jørgensen, M.; Krebs, F. C.; Ferey, G.; Serre, C.; Devic, T.; Maurin, G.; Jobic, H. Llewellyn, P. L.; De Weireld, G.; Vimont, A.; Daturi, M.; Chang, J. S. Why Hybrid Porous Solids Capture Greenhouse gases?. *Chem. Soc. Rev*, **2011**, 40, 550–562.
6. Hunt, A. J.; Sin, E. H. K.; Marriott, R.; Clark, J. H. Generation, Capture, and Utilization of Industrial Carbon Dioxide. *Chem Sus Chem*, **2010**, 3, 306–322.
7. Riduan, S. N.; Zhang, Y. Recent Developments in Carbon dioxide Utilization Under Mild Conditions. *Dalton Trans.* **2010**, 39, 3347–3357.
8. Olah, G. A.; Goeppert, A.; Surya Prakash, G. K. Chemical Recycling of Carbon Dioxide to Methanol and Dimethyl Ether: From Greenhouse Gas to Renewable, Environmentally Carbon Neutral Fuels and Synthetic Hydrocarbons. *J. Org. Chem.* **2009**, 74, 487–498.
10. Halmann, M.M. Chemical Fixation of Carbon Dioxide-Methods for Recycling $CO_2$ into Useful Products.; CRC Press, Boca Raton, FL, 1993.
11. Omae, I. Recent Developments in Carbon dioxide Utilization for the Production of Organic Chemicals. *Coordination Chemistry Reviews.* **2012**, 256, 1384–1405.
12. Davenport, R.E Issues Facing Global Methanol Industry (SRI consulting presented to chemical week conference) 2004.
13. Saito, M.; Murata, K. Development of High Performance Cu/ZnO-Based Catalysts for Methanol Synthesis and the Water-Gas Shift Reaction. *Catal. Surv. Asia*, **2004**, 8, 285–294.
14. Fujitani, T.; Nakamura, I.; Uchijima, T.; Nakamura, J. The kinetics and Mechanism of Methanol Synthesis by Hydrogenation of $CO_2$ over a Zn-Deposited Cu(111) Surface. *Surf. Sci.* **1997**, 383, 285–298.
15. Choi, Y.; Futagami, K.; Fujitani, T.; Nakamura, J. The role of ZnO in Cu/ZnO Methanol Synthesis Catalysts — Morphology Effect or Active Site Model?. *Appl. Catal. A. Gen.* 2001, 208, 163–167.
16. Arena, F.; Italiano, G.; Barbera, K.; Bonura, G.; Spadaro, L.; Frusteri, F. Basic Evidences for Methanol-Synthesis Catalyst Design. *Cat. Today.* **2009**, 143, 80–85.
17. Amara, A.; Bettahar, M.; Olivier, D. Methanol Synthesis on Monovalent Copper Species Stabilized on Silica and Zinc Oxide. *Appl. Catal.* **1989**, 51, 141–154.
18. Chen, H. B.; Liao, D. W.; Yu, L. J.; Lin, Y. J.; Zhang, H. B.; Tsai, K. R. Influence of Trivalent Metal ions on the Surface Structure of a Copper-Based Catalyst for methanol Synthesis. *Appl. Surf. Sci.* **1999**, 147, 85–93.

19. Liaw, J.; Chen, Y. Z. Liquid-phase Synthesis of Methanol from $CO_2/H_2$ Over Ultrafine CuB Catalysts. *Appl. Catal. A. Gen,* **2001**, 206, 245–256
20. Arena, F.; Barbera, K.; Italiano, G.; Bonura, G.; Spadaro, L.; Frusteri, F. Synthesis, Characterization and Activity Pattern of Cu–ZnO/$ZrO_2$ Catalysts in the Hydrogenation of Carbon dioxide to Methanol. *J. Catal.* **2007**, 249, 185–194.
21. Saito, M.; Takeuchi, M.; Fujitani, T.; Toyir, J.; Luo, S.; Wu, J.; Mabuse, H.; Ushikoshi, K.; Mori, K.; Watanabe, T. Advances in Joint Research Between NIRE and RITE for Developing a Novel Technology for Methanol Synthesis from $CO_2$ and $H_2$. *Appl. Organometal. Chem.* **2000**, 14, 763–772.
22. Saito, M.; Fujitani, T.; Takeuchi, M.; Watanabe, T. Development of Copper/Zinc Oxide-Based Multicomponent Catalysts for Methanol Synthesis from Carbon dioxide and Hydrogen. *Appl. Catal.A. Gen.* **1996**, 138, 311–318.
23. Wu, J.; Luo, S.; Toyir, J.; Saito, M.; Takeuchi, M.; Watanabe, T. Optimization of Preparation Conditions and Improvement of Stability of Cu/ZnO-based Multicomponent Catalysts for Methanol Synthesis from $CO_2$ and $H_2$. *Catal. Today.* **1998**, 45, 215–220.
24. Wu, J.; Saito, M.; Takeuchi, M.; Watanabe, T.; The Stability of Cu/ZnO-Based Catalysts in Methanol Synthesis from $CO_2$- Rich Feed and from CO-Rich Feed. *Appl. Catal. A: Gen.* **2001**, 218, 235–240.
25. Toyir, J.; Piscina, P.R.; Fierro, J. L. G.; Homs, N. Highly Effective Conversion of $CO_2$ to Methanol over Supported and Promoted Copper-Based Catalysts: Influence of Support and Promoter. *Appl. Catal. B,* **2001**, 29, 207–215.
26. Yoshihara, J.; Campbell, C. T. Methanol Synthesis and Reverse Water–Gas Shift Kinetics over Cu(110) Model Catalysts: Structural Sensitivity. *J. Catal.,* **1996**, 161, 776–782.
27. Ovesen, C. V.; Clausen, B. S.; Schiøtz, J.; Stoltze, P.; Topsøe, H.; Nørskov, J. K. Kinetic Implications of Dynamical Changes in Catalyst Morphology during Methanol Synthesis over Cu/ZnO Catalysts. *J. Catal.* **1997**, 168, 133–142.
28. Arena, F.; Barbera, K.; Italiano, G.; Bonura, G.; Spadaro, L.; Frusteri, F. Synthesis, Characterization and Activity Pattern of Cu–ZnO/$ZrO_2$ Catalysts in the Hydrogenation of Carbon dioxide to Methanol. *J. Catal.* **2007**, 249, 185–194.
29. Liu, J.; Shi, J.; He, D.; Zhang, Q.; Wu, X.; Liang, Y.; Zhu, Q. Surface active Structure of Ultra-fine Cu/$ZrO_2$ Catalysts used for the $CO_2$+$H_2$ to Methanol Reaction. *Appl. Catal A: Gen.* **2001**, 218, 113–119.
30. An, X.; Li, J. L.; Zuo, Y. Z.; Zhang, Q. D.; Wang, Z.; Wang, J. F. A Cu/Zn/Al/Zr Fibrous Catalyst that is an Improved $CO_2$ Hydrogenation to Methanol Catalyst. *Catal. Lett.* **2007**, 118, 264–269.
31. Raudaskoski, R.; Niemela, M. V.; Keiski, R. L. The Effect of Ageing time on Coprecipitated Cu/ZnO/$ZrO_2$ Catalysts used in Methanol Synthesis from $CO_2$ and $H_2$. *Top. Catal.* **2007**, 45, 57–60.
32. Guo, X. M.; Mao, D. S.; Wang, S.; Wu, G. S.; Lu, G. Z. Combustion Synthesis of CuO–ZnO–$ZrO_2$ Catalysts for the Hydrogenation of Carbon Dioxide to Methanol. *Catal. Commun.* **2009**, 10, 1661–1664.
33. Arena, F.; Italiano, G.; Barbera, K.; Bordiga, S.; Bonura, G.; Spadaro, L.; Frusteri, F. Solid-State Interactions, Adsorption Sites and Functionality of Cu-ZnO/$ZrO_2$ Catalysts in the $CO_2$ Hydrogenation to $CH_3OH$. *Appl. Catal A. Gen.* **2008**, 350, 16–23.

34. Arena, F.; Italiano, G.; Barbera, K.; Bonura, G.; Spadaro, L.; Frusteri, F. Basic Evidences for Methanol-Synthesis Catalyst Design *Catal. Today*. **2009**, 143, 80–85.
35. Fisher, I. A.; Bell, A. T. In-Situ Infrared Study of Methanol Synthesis from $H_2/CO_2$over $Cu/SiO_2$and $Cu/ZrO_2/SiO_2$. *J. Catal.* **1997**, 172, 222–237.
36. Jung, K. T.; Bell, A. T. Effects of Zirconia Phase on the Synthesis of Methanol over Zirconia-Supported Copper. *Catal. Lett.* **2002**, 80, 63–68.
37. Liu, X. M.; Lu, G. Q.; Yan, Z. F. Nanocrystalline Zirconia as Catalyst Support in Methanol Synthesis. *Appl. Catal A: Gen.* **2005**, 279, 241–245.
38. Sloczynski, J.; Grabowski, R.; Olszewski, P.; Kozlowska, A.; Stoch, J.; Lachowska, M.; Skrzypek, J.; Effect of Metal Oxide Additives on the Activity and Stability of Cu/ZnO/$ZrO_2$ Catalysts in the Synthesis of Methanol from $CO_2$ and $H_2$. *Appl. Catal A: Gen.* **2006**, 310, 127–137.
39. Sloczynski, J.; Grabowski, R.; Kozlowska, A.; Olszewski, P.; Lachowska, M.; Skrzypek, J.; Stoch, J. Effect of Mg and Mn Oxide Additions on Structural and Adsorptive Properties of Cu/ZnO/$ZrO_2$ Catalysts for the Methanol Synthesis from $CO_2$. *Appl. Catal., A: Gen.* **2003**, 249, 129–138.
40. Toyir, J.; Piscina, P.R.; Fierro, J. L. G.; Homs, N. Catalytic Performance for $CO_2$ Conversion to Methanol of Gallium-Promoted Copper-based Catalysts: Influence of Metallic Precursors. *Appl. Catal., B.* **2001**, 34, 255–266.
41. Fujita, S.; Moribe, S.; Kanamori, Y.; Kakudate, M.; Takezawa, N. Preparation of a Coprecipitated Cu/ZnO Catalyst for the Methanol Synthesis from $CO_2$ — Effects of the Calcination and Reduction Conditions on the Catalytic Performance. *Appl. Catal. A: Gen.* **2001**, 207, 121–128.
42. Ponce, A. A.; Klabunde, K. J. Chemical and Catalytic Activity of Copper Nano Particles Prepared via Metal Vapor Synthesis. *J. Mol. Catal. A: Chem.* **2005**, 225, 1–6.
43. Gotti, A.; Prins, R. Basic Metal Oxides as Cocatalysts for $Cu/SiO_2$Catalysts in the Conversion of Synthesis Gas to Methanol. *J. Catal.* **1998**, 178, 511–519.
44. Liu, Y.; Zhang, Y.; Wang, T. J.; Tsubaki, N. Efficient Conversion of Carbon Dioxide to Methanol Using Copper Catalyst by a New Low-temperature Hydrogenation Process. *Chem. Lett.* **2007**, 36, 1182–1183.
45. Liang, X. L.; Dong, X.; Lin, G. D.; Zhang, H-. B. Carbon Nanotube-Supported Pd–ZnO Catalyst for Hydrogenation of $CO_2$ to Methanol. *Appl. Catal.B.* **2009**, 88, 315–322.
46. Sloczynski, J.; Grabowski, R.; Kozlowska, A.; Olszewski, P.; Stoch, J.; Skrzypek, J.; Lachowska, M. Catalytic Activity of the M/(3ZnO·$ZrO_2$) System (M = Cu, Ag, Au) in the Hydrogenation of $CO_2$ to Methanol. *Appl. Catal. A: Gen.*, **2004**, 278, 11–23.
47. Solymosi, F.; Erdohelyi, A.; Lancz, M.; Surface Interaction between $H_2$ and $CO_2$ over Palladium on Various Supports. *J. Catal.* **1985**, 95, 567.
48. Tsubakia, N.; Fujimoto, K. Promotional SMSI Effect on Supported Palladium Catalysts for Methanol Synthesis. *Top. Catal.* **2003**, 22, 325–335.
49. Ashley, A. E.; Thompson, A. L.; O'Hare, D. Non-Metal-Mediated Homogeneous Hydrogenation of $CO_2$ to $CH_3OH$. *Angew. Chem.Int. Ed.* **2009**, 48, 9839–9843.
50. Riduan, S. N.; Zhang, Y.; Ying, J.Y. Conversion of Carbon Dioxide into Methanol with Silanes over N-Heterocyclic Carbene Catalysts. *Angew. Chem.* **2009**, 121, 3372–3375

51. Riduan, S. N.; Ying, J. Y.; Zhang, Y. Mechanistic Insights into the Reduction of Carbon Dioxide with Silanes over N-Heterocyclic Carbene Catalysts. *Chem Cat Chem.* **2013**, 5, 1490–1496.
52. Cabrera, I. M.-; Granados, M. L.; Fierro, J. L. G. Pd-Modified Cu–Zn Catalysts for Methanol Synthesis from $CO_2/H_2$ Mixtures: Catalytic Structures and Performance. *J. Catal.* **2002**, 210, 285–294.
53. Witoon, T.; Permsirivanich, T.; Donphai, W.; Jaree, A.; Chareonpanich, M. $CO_2$ Hydrogenation to Methanol over Cu/ZnO Nanocatalysts Prepared via a Chitosan-assisted Co-precipitation Method. *Fuel Proc. Tech.* **2013**, 116, 72–78.
54. Zhang, L.; Zhang, Y.; Chen, S. Effect of promoter $SiO_2$, $TiO_2$ or $SiO_2$-$TiO_2$ on the Performance of CuO-ZnO-$Al_2O_3$ Catalyst for Methanol Synthesis from $CO_2$ Hydrogenation. *Appl. Cat. A: Gen.* **2012**, 415–416, 118–123.
55. Cabrera, I. M.-; Granados M. L.; Fierro, J. L. G. Reverse Topotactic Transformation of a Cu–Zn–Al Catalyst during Wet Pd Impregnation: Relevance for the Performance in Methanol Synthesis from $CO_2/H_2$ Mixtures. *J. Catal.* **2002**, 210, 273–284.
56. P. Gao, F. Li, F. Xiao, N. Zhao, N. Sun, W. Wei, L. Zhongc and Y. Sun, Preparation and activity of Cu/Zn/Al/Zr catalysts via hydrotalcites containing precursors for methanol synthesis from $CO_2$ hydrogenation, *Catal. Sci. Technol.* **2012**, 2, 1447–1454.
57. Sakurai, H.; Haruta, M. Carbon Dioxide and Carbon Monoxide Hydrogenation over Gold Supported on Titanium, Iron, and Zinc Oxides. *Appl. Catal. A: Gen.* **1995**, 127, 93–105.
58. Collins, S. E.; Chiavassa, D. L.; Bonivardi, A. L.; Baltanas, M. A. Hydrogen Spillover in $Ga_2O_3$–Pd/$SiO_2$ Catalysts for Methanol Synthesis from $CO_2/H_2$. *Catal. Lett.* **2005**, 103, 83–88.
59. Cabrera, I.M.; Granados, M.L.; Terreros, P.; Fier-ro, J.L.G. $CO_2$ Hydrogenation over Pd-modified Methanol Synthesis Catalysts. *Catal. Today.* **1998**, 45, 251–256.
60. Sankaranarayanan, S.; Srinivasan, K. Carbon Dioxide- A Potential Raw Material for the Production of Fuel Additives and Bioderived Chemicals. *Ind.J.Chem Sec.A*. **2012**, 51, 1252–1262.
61. Inui, T.; Yamamoto, T. Effective Synthesis of Ethanol from $CO_2$ on Polyfunctional Composite Catalysts. *Catal. Today*. **1998**, 45, 209–214.
62. Inui, T.; Yamamoto, T.; Inoue, M.; Hara, H.; Takeguchi, T.; Kim, J. B. Highly Effective Synthesis of Ethanol by $CO_2$-Hydrogenation on Well Balanced Multi-functional FT-type Composite Catalysts. Appl. Catal. A: Gen. **1999**, 186, 395–406.
63. Takagawa, M.; Okamoto, A.; Fujimura, H.; Izawa, Y.; and Arakawa, H. Ethanol Synthesis from Carbon Dioxide and Hydrogen. *Stud. Surf. Sci. Catal.* **1998**, 114, 525–528.
64. Kusama, H.; Okabe, K.; Sayama, K.; Arakawa, H. $CO_2$ Hydrogenation to Ethanol over Promoted Rh/$SiO_2$ Catalysts. *Catal.Today*. **1996**, 28, 261–266.
65. Kusama, H.; Okabe, K.; Sayama, K.; Arakawa, H. Ethanol Synthesis by Catalytic Hydrogenation of $CO_2$ over Rh-Fe$SiO_2$ Catalysts. *Energy*, **1997**, 22, 343–348.
66. Kurakata, H.; Izumi, Y.; Aika, K. Ethanol Synthesis from Carbon Dioxide on $TiO_2$-Supported [$Rh_{10}$Se] Catalyst. *Chem. Commun.* **1996**, 389–390.
67. Izumi, Y.; Kurakata, H.; Aika, K. Ethanol Synthesis from Carbon Dioxide on [$Rh_{10}$Se]/$TiO_2$ Catalyst Characterized by X-Ray Absorption Fine Structure Spectroscopy. *J. Catal.* **1998**, 175, 236–244.

68. Tominaga, K.-i.; Sasaki, Y. Ruthenium-Catalyzed One-pot Hydroformylation of Alkenes using Carbon dioxide as a Reactant. *J. Mol. Catal. A: Chem.* **2004**, 220, 159–165.
69. Fujita, S. I.; Okamura, S.; Akiyama, Y.; Arai, M. Hydroformylation of Cyclohexene with Carbon Dioxide and Hydrogen Using Ruthenium Carbonyl Catalyst: Influence of Pressures of Gaseous Components. *Int. J. Mol. Sci.* **2007**, 8, 749–759.
70. Tominaga, K. An Environmentally Friendly Hydroformylation using Carbon Dioxide as a Reactant Catalyzed by Immobilized Ru-complex in Ionic liquids. *Catal. Today*. **2006**, 115, 70–72.
71. Kontkanen, M. L.; Oresmaa, L.; Moreno, M. A.; Janis, J.; Laurila, E.; Haukka, M. One-Dimensional Metal Atom Chain $[Ru(CO)_4]n$ as a Catalyst Precursor—Hydroformylation of 1-Hexene using Carbon Dioxide as a Reactant. *Appl. Catal. A: Gen.* **2009**, 365, 130–134.
72. Arcoumanis, C.; Bae, C.; Crookes, R.; Kinoshita, E. The Potential of Di-methyl ether (DME) as an Alternative Fuel for Compression-Ignition Engines: A Review. *Fuel*, **2008**, 87, 1014–1030.
73. Yaripour, F.; Baghaei, F.; Schmidt, I.; Perregaard, J.; Synthesis of Dimethyl Ether from Methanol over Aluminium Phosphate and Silica–Titania Catalysts. *Catal. Commun.* **2005**, 6, 542–549.
74. Aguayo, A. T.; Erena, J.; Mier, D.; Arandes, J. M.; Olazar, M.; Bilbao, J. Kinetic Modeling of Dimethyl Ether Synthesis in a Single Step on a $CuO-ZnO-Al_2O_3/\gamma\text{-}Al_2O_3$ Catalyst. *Ind. Eng. Chem. Res.* **2007**, 46, 5522–5530.
75. Wang, T.; Wang, J.; Jin, Y. Slurry Reactors for Gas-to-Liquid Processes: A Review. *Ind. Eng. Chem. Res.* **2007**, 46, 5824–5847.
76. Zhang, M.-H.; Liu, Z.-M.; Lin, G.-D. Zhang, H.-B. Pd/CNT-promoted Cu $ZrO_2$/HZSM-5 Hybrid Catalysts for Direct Synthesis of DME from $CO_2/H_2$. *Appl. Catal. A: Gen.* **2013**, 451, 28–35.
77. Bonuraa, G.; Cordarob, M.; Spadaroa, L.; Cannillaa, C.; Arenab, F.; Frusteri, F. Hybrid Cu–ZnO–$ZrO_2$/H-ZSM5 System for the Direct Synthesis of DME by $CO_2$ Hydrogenation. *Appl. Catal. B: Env.* **2013**, 140–141, 16–24.
78. García-Trenco, A.; Martínez, A. Direct Synthesis of DME from Syngas on Hybrid CuZnAl/ZSM-5 Catalysts: New Insights into the Role of Zeolite Acidity. *Appl. Catal. A: Gen.* **2012**, 411–412, 170–179.
79. Wang, S.; Maoa, D.; Guo, X.; Wua, G.; Lu, G. Dimethyl ether Synthesis via $CO_2$ Hydrogenation over CuO–$TiO_2$–$ZrO_2$/HZSM-5 Bifunctional Catalysts. *Catal. Comm.* **2009**, 10, 1367–1370.
80. Sun, K.; Lu, W.; Wang, M.; Xu, X. Low-temperature Synthesis of DME from $CO_2/H_2$ over Pd-ModifiedCuO–ZnO–$Al_2O_3$–$ZrO_2$/HZSM-5 Catalysts. *Catal. Comm.* **2004**, 5, 367–370.
81. Qi, G.-X.; Fei, J.-H.; Zheng, X.-M.; Hou, Z.-Y. DME Synthesis from Carbon Dioxide and Hydrogen over Cu-Mo/HZSM-5. *Catal. Lett..* **2001**, 72, 121–124.
82. Eren˜a, J.; Garon˜a, R.; Arandes, J. M.; Aguayo, A. T.; Bilbao, J. Effect of Operating Conditions on the Synthesis of Dimethyl ether over a CuO-ZnO-$Al_2O_3$/NaHZSM-5 Bifunctional Catalyst. *Catal. Today*. **2005**, 107–108, 467–473.

83. Zha, F.; Ding, J.; Chang, Y.; Ding, J.; Wang, J.; Ma, J. Cu–Zn–Al Oxide Cores Packed by Metal-Doped Amorphous Silica–Alumina Membrane for Catalyzing the Hydrogenation of Carbon Dioxide to Dimethyl Ether. *Ind. Eng. Chem. Res.*. **2012**, 51, 345.
84. Vance, C. K.; Bartholomew, C. H. Hydrogenation of Carbon dioxide on Group VIII Metals. III, Effects of Support on Activity/Selectivity and Adsorption Properties of Nickel. *Appl. Catal.* **1983**, 7, 169–177.
85. Mori, S.; Xu, W.-C.; Ishidzuki, T.; Ogasawara, N.; Imai, J.; Kobayashi, K. Mechanochemical Activation of Catalysts for $CO_2$ Methanation. *Appl. Catal. A: Gen.* **1996**, 137, 255–268.
86. Chang, F.-W.; Hsiao, T.-J.; Chung, S.-W.; Lo, J.-J. Nickel Supported on Rice Husk Ash–Activity and Selectivity in $CO_2$ Methanation. *Appl. Catal. A: Gen.* **1997**, 164, 225–236.
87. Chang, F.-W.; Tsay, M.-T.; S.-P. Liang. Hydrogenation of $CO_2$ over Nickel Catalysts Supported on Rice Husk ash Prepared by Ion Exchange. *Appl. Catal. A: Gen.* **2001**, 209, 217–227.
88. Chang, F. W.; Tsay, M. T.; Kuo, M. S. Effect of Thermal Treatments on Catalyst Reducibility and Activity in Nickel Supported on RHA–$Al_2O_3$ Systems. *Thermochim. Acta.***2002**, 386, 161–172.
89. Chang, F.-W.; Kuo, M.-S.; Tsay, M.-T.; Hsieh, M.-C. Hydrogenation of $CO_2$ over Nickel Catalysts on Rice Husk Ash-Alumina Prepared by Incipient Wetness Impregnation. *Appl. Catal. A: Gen.* **2003**, 247, 309–320.
90. Du, G.; Lim, S.; Yang, Y.; Wang, C.; L. Pfefferle, Haller.G.L. Methanation of Carbon Dioxide on Ni-Incorporated MCM-41 Catalysts:The Influence of Catalyst Pretreatment and Study of Steady-State Reaction. *J.Catal.* **2007**, 249, 370–379.
91. Lee, G. D.; Moon, M. J.; Park, J. H.; Park, S. S.; Hong, S. S. Raney Ni Catalysts Derived from Different Alloy Precursors Part II. CO and $CO_2$ Methanation Activity. *Korean J. Chem. Eng.* **2005**, 22, 541–546.
92. Sehested, J.; Larsen, K. E.; Kustov, A. L.; Frey, A. M.; Johannessen, T.; Bligaard, T.; Andersson, M. P.; Norskov, J. K.; Christensen, C. H. Discovery of Technical Methanation Catalysts Based on Computational Screening. *Top. Catal.* **2007**, 45, 9–13.
93. Yamasaki, M.; Habazaki, H.; Asami, K.; Izumiya, K.; Hashimoto, K. Effect of Tetragonal $ZrO_2$ on the Catalytic Activity of Ni/$ZrO_2$ Catalyst Prepared from Amorphous Ni–Zr Alloys. *Catal. Commun.* **2006**, 7, 24–28.
94. Ocampo, F.; Louis, B.; Roger, A.-C. Methanation of Carbon dioxide over Nickel-based $Ce_{0.72}Zr_{0.28}O_2$ Mixed Oxide Catalysts Prepared by Sol–gel Method. *Appl. Catal. A: Gen.* **2009**, 369, 90–96.
95. Ocampo, F.; Louis, B.; Minsker, L. K.-; Roger, A.-C. Effect of Ce/Zr composition and Noble Metal Promotion on Nickel based $Ce_xZr_{1-x}O_2$ Catalysts for Carbon Dioxide Methanation. *Appl. Catal. A: Gen.* **2011**, 392, 36–44.
96. Fischer, I.A.; Bell, A.T. A Comparative Study of CO and $CO_2$ Hydrogenation over Rh/$SiO_2$. *J. Catal.* **1996**, 162, 54–65.
97. Choe, S. J.; Kang, H. J.; Kim, S. J.; Park, S. B.; Park, D. H.; Huh, D. S. Adsorbed Carbon Formation and Carbon Hydrogenation for $CO_2$ Methanation on the Ni(111) Surface: ASED-MO Study. *Bull. Korean Chem. Soc.* **2005**, 26, 1682–1688.

98. Beuls, A.; Swalus, C.; Jacquemin, M.; Heyen, G.; Karelovic, A.; Ruiz, P. Methanation of $CO_2$: Further Insight into the Mechanism over Rh/γ-$Al_2O_3$ Catalyst. *Appl. Catal. B: Env.* **2012**, 113–114, 2–10.
99. Aldana, P.A.U; Ocampo, F.; Kobl, K.; Louis, B.; Starzyka, F. T.-; Daturi, M.; Bazin, P.; Thomas, S.; Roger, A.C. Catalytic $CO_2$ Valorization into $CH_4$ on Ni-Based Ceria-Zirconia. Reaction Mechanism by Operando Techniques, **2013**, 215, 201–207.
100. Kustov, A. L.; Frey, A. M.; Larsen, K. E.; Johannessen, T.; Norskov, J. K.; Christensen, C. H. CO Methanation over Supported Bimetallic Ni–Fe Catalysts: From Computational Studies Towards Catalyst Optimization. *Appl. Catal. A: Gen.* **2007**, 320, 98–104.
101. Kowalczyk, Z.; Stołecki, K.; Raro´g-Pilecka, W.; Mis´kiewicz, E.; Wilczkowska.E.; Karpin´ski, Z. Supported Ruthenium Catalysts for Selective Methanation of Carbon Oxides at Very Low $CO_x/H_2$ Ratios. *Appl. Catal. A: Gen.* **2008**, 342, 35–39.
102. Abe, T.; Tanizawa, M.; Watanabe, K.; Taguchi, A. $CO_2$ Methanation Property of Ru Nanoparticle-Loaded $TiO_2$ Prepared by a Polygonal Barrel-Sputtering Method Energy. *Environ. Sci.* **2009**, 2, 315–321.
103. Luo, L. T.; Li, S. J.; Zhu, Y. The Effects of Yttrium on the Hydrogenation Performance and Surface Properties of a Ruthenium-Supported Catalyst. *J. Serb. Chem. Soc.* **2005**, 70, 1419–1425.
104. Kim, H. Y.; Lee, H. M.; J-N. Park. Bifunctional Mechanism of $CO_2$ Methanation on Pd-MgO/$SiO_2$ Catalyst: Independent Roles of MgO and Pd on $CO_2$ Methanation. J. *Phys. Chem. C* **2010**, 114, 7128–7131.
105. Yu, K.-P.; Yu, W.-Y.; Kuo, M.-C.; Liou, Y.-C.; Chien, S.-H. Pt/Titania-nanotube: A Potential Catalyst for $CO_2$ Adsorption and Hydrogenation. *Appl. Catal. B: Env.* **2008**, 84, 112–118.
106. Jacquemin, M.; Beuls, A.; Ruiz, P. Catalytic Production of Methane from $CO_2$ and $H_2$ at Low Temperature : Insight on the Reaction Mechanism. *Catal. Today* **2010**, 157, 462–466.
107. Research Horizons of Nanosystems StructureRuiz, P.; Jacquemin, M.; Blangenois, N. Catalytic $CO_2$ Methanation Process. Patent WO 2010006386 A2, January 21, 2010.
108. Karelovic, A.; Ruiz, P. Mechanistic Study of Low Temperature $CO_2$ Methanation over Rh/$TiO_2$ Catalysts. *J. Catal.* **2013**, 301, 141–153.
109. Karelovic, A.; Ruiz, P. $CO_2$ Hydrogenation at Low Temperature over Rh/-$Al_2O_3$ Catalysts: Effect of the Metal Particle Size on Catalytic Performances and Reaction Mechanism. *Appl. Catal. B: Env.* **2012**, 113–114, 237–249.
110. Iizuka, T.; Tanaka, Y.; Tanabe, K. Hydrogenation of CO and $CO_2$ over Rhodium Catalysts Supported on Various Metal Oxides. *J. Catal.* **1982**, 76, 1–8.
111. Williams, K.J.; Boffa, A.B.; Salmeron, M.; Bell, A.T.; Somorjai, G.A. The Kinetics of $CO_2$ Hydrogenationon a Rh Foil Promoted by Titania Overlayers *Catal. Lett.* **1991**, 9, 415.
112. Lunde, P.J.; Kester, F.L. Carbon Dioxide Methanation on a Ruthenium Catalyst. *Ind. Eng. Chem. Proc. Des. Dev.* **1974**, 13, 27–33.
113. Weatherbee, G.D.; Bartholomew, C.H. Hydrogenation of $CO_2$ on Group VIII metals: IV. Specific Activities and Selectivities of Silica-Supported Co, Fe, and Ru. *J. Catal.* **1984**, 87, 352

114. Darensbourg, D.J.; Banch, C.G.; Ovalles, C. Mechanistic Aspects of Catalytic Carbon Dioxide Methanation. *Rev. Inorg. Chem.* **1985**, 7, 315–339.
115. Weatherbee, G.D.; Bartholomew, C.H. Hydrogenation of $CO_2$ on Group VIII Metals: I. Specific Activity of Ni/$SiO_2$. *J. Catal.* **1981**, 68, 67–76.
116. Gnanamani, M. K.; Shafer, W. D.; Sparks, D. E.; Davis, B. H. Fischer–Tropsch Synthesis: Effect of $CO_2$ Containing Syngas over Pt Promoted Co/γ-$Al_2O_3$ and K-Promoted Fe Catalysts. *Catal. Comm.* **2011**, 12, 936–939.
117. Visconti, C.G.; Lietti, L.; Tronconi, E.; Forzatti, P.; Zennaro, R.; Finocchio, E. Fischer–Tropsch Synthesis on a Co/$Al_2O_3$ Catalyst with $CO_2$ Containing Syngas. *Appl. Catal.A: Gen.* **2009**, 355, 61–68.
118. Herranz, T.; Rojas, S.; Alonso, F.J.P. -; Ojeda, M.; Terreros, P.; Fierro, J.L.G. Carbon Dioxide Hydrogenation over Silica-Supported Iron-based Catalysts Influence of the Preparation Route *Appl. Catal. A: Gen.* **2006**, 308, 19–30.
119. Dorner, R.W.; Hardy, D.R.; Williams, F.W.; Willauer, H.D. Heterogeneous Catalytic $CO_2$ Conversion to Value-Added Hydrocarbons. *Energy Environ. Sci.* **2010**, 3, 884–890.
120. Dorner, R.W.; Hardy, D.R.; Williams, F.W.; Willauer, H.D. $C_2$-$C_5$+ Olefin Production from $CO_2$ Hydrogenation using Ceria Modified Fe/Mn/K Catalysts. *Catal. Commun.* **2011**, 15, 88–92.
121. Rongxian, B.; Yisheng, T.; Yizhuo, H. Study on the Carbon Dioxide Hydrogenation to Iso-alkanes over Fe–Zn–M/Zeolite Composite Catalysts. *Fuel Process. Technol.* **2004**, 86, 293–301.
122. http://www.kit.edu/visit/pi_2013_12783.php
123. Bychkov, V.Y.; Tyulenin, Y.P.; Firsova, A.A.; Shafranovsky, E.A.; Gorenberg, A.Y.; Korchak, V.N. Carbonization of Nickel Catalysts and its Effect on Methane Dry Reforming. *Appl. Catal. A: Gen.* **2013**, 453, 71–79.
124. Jones, G.; Jakobsen, J.G.; Shim, S.S.; Kleisa, J.; Andersson, M.P.; Rossmeisl, J.; Abild-Pedersen, F.; Bligaard, T.; Helveg, S.; Hinnemann, B. J.; Rostrup-Nielsen, R.; Chorkendorff, I.; Sehested, J.; Nøskov, J.K. First Principles Calculations and Experimental Insight into Methane Steam Reforming over Transition Metal Catalysts. *J. Catal.* **2008**, 259, 147–160.
125. Rostrup-Nielsen, J.R. Aspects of $CO_2$-reforming of Methane. *Stud. Surf. Sci. Catal.* 1994, 85, 25–41.
126. Rezaei, M.; Alavi, S. M.; Sahebdelfar, S.; Yan, Z.-F. Syngas Production by Methane Reforming with Carbon Dioxide on Noble Metal Catalysts. *J. Nat. Gas Chem.* **2006**, 15, 327–334.
127. Pompeo, F.; Nichio, N. N.; Souza, M.M. V.M; Cesar, D. V.; Ferretti, O. A.; Schmal, M. Study of Ni and Pt Catalysts Supported on g-$Al_2O_3$ and $ZrO_2$ Applied in Methane Reforming with $CO_2$. *Appl. Catal. A: Gen.* **2007,** 316, 175–183.
128. Yang, R. Q.; Xing, C.; Lv, C. X.; Shi, L.; Tsubaki, N. Promotional Effect of $La_2O_3$ and $CeO_2$ on Ni/R-$Al_2O_3$ Catalysts for $CO_2$ Reforming of $CH_4$. *Appl. Catal., A. Gen.* **2010**, 385, 92–100.
129. Chen, W.; Zhao, G.; Xue, Q.; Chen, L.; Lu, Y. High Carbon-Resistance Ni/$CeAlO_3$-$Al_2O_3$ Catalyst for $CH_4$/$CO_2$ Reforming. *Appl. Catal. B: Env.* **2013**, 136–137, 260–268.

130. Kathiraser, Y.; Thitsartarn, W.; Sutthiumporn, K.; Kawi, S. Inverse $NiAl_2O_4$ on $LaAlO_3$–$Al_2O_3$: Unique Catalytic Structure for Stable $CO_2$ Reforming of Methane. *J. Phys. Chem. C* **2013**, 117, 8120–8130.
131. Gardnera, T. H.; Spivey, J.J.; Kuglera, E.L.; Pakhare, D. $CH_4$–$CO_2$ Reforming over Ni-substituted Barium Hexaaluminate Catalysts. *Appl. Catal. A: Gen.* **2013**, 455, 129–136.
132. Xie, X.; Otremba, T.; Littlewood, P.; Schomaäcker, R.; Thomas, A. One-Pot Synthesis of Supported, Nanocrystalline Nickel Manganese Oxide for Dry Reforming of Methane. *ACS Catal.* **2013**, 3, 224–229.
133. Zhu, J.; Peng, X.; Yao, L.; Deng, X.; Dong, H.; Tong, D.; Hu, C. Synthesis Gas Production from $CO_2$ Reforming of Methane over Ni-Ce/$SiO_2$ Catalyst: The Effect of Calcination Ambience. *int. J. Hydr.Energy*. **2013**, 381, 117 -126.
134. Zhang, Y.; Yoneyama, Y.; Tsubaki, N. Simultaneous Introduction of Chemical and Spatial Effects via a New Bimodal Catalyst Support Preparation Method. *Chem. Commun.* **2002**, 1216–1217.
135. Shi, L.; Yang, G.; Tao, K.; Yoneyama, Y.; Tan, Y.; Tsubaki, N. An Introduction of $CO_2$ Conversion by Dry Reforming with Methane and New Route of Low-Temperature Methanol Synthesis. *Acc.Chem. Res.* **2013**, 46, 1838–1847.
136. Tao, K.; Zhang, Y.; Terao, S.; Tsubaki, N. Development of Platinum-based Bimodal Pore Catalyst for $CO_2$ Reforming of $CH_4$. *Catal. Today* **2010**, 153, 150–155.
137. Tao, K.; Zhang, Y.; Terao, S.; Yoneyama, Y.; Kawabata, T.; Matsuda, K.; Ikeno, S.; Tsubaki, N. Chemical and Spatial Promotional Effects of Bimodal Pore Catalysts for Methane Dry Reforming. *Chem. Eng. J.* **2011**, 170, 258–263.
138. Hu, Y. H.; Solid-solution Catalysts for $CO_2$ Reforming of Methane. *Catal.Today*. **2009**, 148, 206–211.
139. Kumar, B.; Smieja, J. M.; Sasayama, A.F.; Kubiak, C.P. Tunable, Light-Assisted Co-generation of CO and $H_2$ from $CO_2$ and $H_2O$ by Re(bipy-tbu)$(CO)_3$Cl and p-Si in Non-aqueous Medium. *Chem. Commun.* **2012**, 48, 272–274.
140. Mikkelsen, M.; Jørgensen, M.; Krebs, F.C. The Teraton challenge. A Review of Fixation and Transformation of Carbon Dioxide. *Energy Environ. Sci.* **2010**, 3, 43–81.
141. Leino, E.; Maki-Arvela, P.; Eta.V.; Murzin, D.Y.; Salmi, T.; Mikkola, J.-P. Conventional Synthesis Methods of Short-chain Dialkylcarbonates and Novel Production Technology via Direct Route from Alcohol and Waste $CO_2$. *Appl. Catal.A. Gen* **2010, 383,** 1–13.
142. Shaikh, A.G.; Sivaram, S. Organic Carbonates. *Chem. Rev.* **1996**, 96, 951–976.
143. Tundo, P.; Selva, M. The Chemistry of Dimethyl Carbonate. *Acc. Chem. Res.* **2002**, 36, 706–716.
144. Ono, Y. Catalysis in the Production and Reactions of Dimethyl Carbonate, an Environmentally Benign Building Block. *Appl. Catal. A. Gen.* **1997,** 155, 133–166.
145. Delledonne, D.; Rivetti, F.; Romano, U. Developments in the Production and Application of Dimethyl Carbonate. *Appl. Catal. A*: *Gen.* **2001**, 221, 241–251.
146. Pachecoa, M.A.; Marshall, C.L. Review of Dimethyl Carbonate (DMC) Manufacture and Its Characteristics as a Fuel Additive. *Energy and Fuels* **1997**, 11, 2–29.
147. Peng, W.; Zhao, N.; Xiao, F.; Wei, W.; Sun, Y. Recent Progress in Phosgene-free Methods for Synthesis of Dimethyl Carbonate. *Pure Appl. Chem.* **2012**, 84, 603–620.

148. Keller, N.; Rebmann, G.; Keller, V. Catalysts, Mechanisms and Industrial Processes for the Dimethyl Carbonate Synthesis. *J. Mol. Catal. A: Chem.* **2010**, 317, 1–18.
149. Fang, S.; Fujimoto, K. Direct Synthesis of Dimethyl Carbonate from Carbon Dioxide and Methanol Catalyzed by Base *Applied Catal. A: Gen.* **1996**, 142, L1- L3.
150. Tomishige, K.; Ikeda, Y.; Sakaihori, T.; Fujimoto, K. Catalytic Properties and Structure of Zirconia Catalysts for Direct Synthesis of Dimethyl Carbonate from Methanol and Carbon Dioxide. *J.Catal.***2000**, 192, 355–362.
151. Xei, S.; Bell, A.T. An Raman Study of Dimethyl Carbonate Synthesis from Carbon Dioxide and Methanol over Zirconia. *Catal. Lett.* **2000**, 70, 137–143.
152. Jung, K.T.; Bell, A.T. An Infrared Study of Dimethyl Carbonate Synthesis from Carbon Dioxide and Methanol over Zirconia. *J. Catal.* **2001**, 204, 339–347.
153. Ikeda, Y.; Sakaihori, T.; Tomishige, K.; Fujimoto, K. Promoting Effect of Phosphoric Acid on Zirconia Catalysts in Selective Synthesis of Dimethyl Carbonate from Methanol and Carbon Dioxide. *Catal. Lett.* **2000**, 66, 59–62.
154. Ikeda, Y.; Asadullah, M.; Fujimoto, K.; Tomishige, K. Structure of the Active Sites on $H_3PO_4/ZrO_2$ Catalysts for Dimethyl Carbonate Synthesis from Methanol and Carbon Dioxide. *J. Phys. Chem. B* **2001**, 105, 10653–10658.
155. Jiang, C.J.; Guo, Y.H.; Wang, C.G.; Hu, C.W.; Wu.Y.; Wang, E.B. Synthesis of Dimethyl Carbonate from Methanol and Carbon Dioxide in the Presence of Polyoxometalates under Mild Conditions. *Appl. Catal. A: Gen.* **2003**, 256, 203–212.
156. Tomishige, K.; Furusawa, Y.; Ikeda, Y.; Asadullah, M.; Fujimoto, K. $CeO_2$–$ZrO_2$ Solid Solution Catalyst for Selective Synthesis of Dimethyl Carbonate from Methanol and Carbon Dioxide *Catal. Lett.* **2001**, 76, 71–74.
157. Choi, J.C.; He, L.N.; Yasuda, H.; Sakakura, T. Selective and High Yield Synthesis of Dimethyl Carbonate Directly from Carbon Dioxide and Methanol *Green Chem.* **2002**, 4, 230–234.
158. Tomishige, K.; Kunimori, K. Catalytic and Direct Synthesis of Dimethyl Carbonate Starting from Carbon Dioxide using $CeO_2$-$ZrO_2$ Solid Solution Heterogeneous Catalyst: Effect of $H_2O$ Removal from the Reaction System *Appl. Catal. A: Gen.* **2002**, 237, 103–109.
159. Eta, P.; Mäki-Arvela, J.-P.; Mikkola, K.; Kordas, Murzin, D.Y.; Salmi, T.in: *D.LMarmaduke (Ed.), Progress in Heterogeneous Catalysis, Nova Publishers, NewYork* **2008**, pp. 135–155, ISBN 978–1-60456–979–7.
160. Yoshida, Y.; Arai, Y.; Kado, S.; Kunimori, K.; Tomishige, K. Direct Synthesis of Organic Carbonates from the Reaction of $CO_2$ with Methanol and Ethanol over $CeO_2$ Catalysts. *Catal. Today* **2006**, 115, 95–101.
161. La, K.W.; Song, I.K. Direct Synthesis of Dimethyl Carbonate from $CH_3OH$ and $CO_2$ by $H_3PW_{12}O_{40}/Ce_xTi_{1-x}O_2$ Catalyst. *React. Kinet. Catal. Lett.* **2006**, 89, 303–309.
162. Lee, H. J.; Park, S.; Song, I.K.; Jung, J.C. Direct Synthesis of Dimethyl Carbonate from Methanol and Carbon Dioxide over $Ga_2O_3/Ce_{0.6}Zr_{0.4}O_2$ Catalysts: Effect of Acidity and Basicity of the Catalysts. *Catal Lett.* **2011**, 141, 531–537.
163. Wu, X.L.; Meng, Y.Z.; Xiao, M.; Lu, Y.X. Direct synthesis of Dimethyl Carbonate (DMC) using Cu-Ni/VSO as Catalyst *J. Mol. Catal. A: Chem.* **2006**, 249, 93–97.
164. Li, C.F.; Zhong, S.H. Study on Application of Membrane Reactor in Direct Synthesis DMC from $CO_2$ and $CH_3OH$ over Cu–KF/MgSiO Catalyst. *Catal. Today* **2003**, 82, 83–90.

165. Wu, X.L.; Xiao, M.; Meng, Y.Z.; Lu, Y.X. Direct Synthesis of Dimethyl Carbonate on $H_3PO_4$ Modified $V_2O_5$. *J. Mol. Catal. A: Chem.* **2005**, 238, 158–162.
166. Fan, B.B.; Zhang, J.L; Li, R.F.; Fan, W.B. Preparation of Functional Heterogeneous Organotin Catalyst Tethered on SBA-15. *Catal. Lett.* **2008**, 121, 297–302.
167. Aresta, M.; Dibenedetto, A; Nocito, F.; Pastore, C. Comparison of the Behaviour of Supported Homogeneous Catalysts in the Synthesis of Dimethyl Carbonate from methanol and Carbon Dioxide: PolyStyrene-Grafted Tin-Metallorganic Species Versus Silesquioxanes Linked Nb-Methoxo Species *Inorg. Chim. Acta* **2008**, 361, 3215–3220.
168. Bian, J.; Xiao, M.; Wang, S.J.; Lu, Y.X.; Meng, Y.Z. Carbon Nanotubes Supported Cu–Ni Bimetallic Catalysts and Their Properties for the Direct Synthesis of Dimethyl Carbonate from Methanol and Carbon Dioxide. *Applied Surf. Sci.* **2009**, 255, 7188–7196.
169. Wang, X.J.; Xiao, M.; Wang, S.J.; Lu, Y.X.; Meng, Y.Z. Direct Synthesis of Dimethyl Carbonate from Carbon Dioxide and Methanol using Supported Copper (Ni, V, O) Catalyst with Photo-Assistance. *J. Mol. Catal. A: Chem.* **2007**, 278, 92–96.
170. Wang, J.Q.; Sun, J.; Shi, C.Y.; Cheng, W.G.; Zhang, X.P.; Zhang, S.J. Synthesis of Dimethyl Carbonate from $CO_2$ and Ethylene Oxide Catalyzed by $K_2CO_3$-Based Binary Salts in the Presence of $H_2O$. *Green. Chem.* **2011**, 13, 3213–3217.
171. Bhanage, B. M.; Fujita, S.; Ikushima, Y.; Arai, M. Synthesis of Dimethyl Carbonate and Glycols from Carbon Dioxide, Epoxides, and Methanol using Heterogeneous Basic Metal Oxide Catalysts with High Activity and Selectivity. *Appl. Catal. A: Gen.* **2001**, 219, 259–266.
172. Cui, H.; Wang, T.; Wang, F.; Gu, C.; Wang, P.; Dai, Y. One-Pot Synthesis of Dimethyl Carbonate Using Ethylene Oxide, Methanol, and Carbon Dioxide under Supercritical Conditions. *Ind. Eng. Chem. Res.* **2003**, 42, 3865–3870.
173. Eta, V.; Maki-Arvela, P.; Salminen, E.; Salmi, T.; Murzin, D.Y.; Mikkoh, J.-P. The Effect of Alkoxide Ionic Liquids on the Synthesis of Dimethyl Carbonate from CO2 and Methanol over ZrO2-MgO. *Catal. Lett.* **2011**, 141, 1254–1261.
174. Chang, Y.; Jiang, T.; Han, B.; Liu, Z.; Wu, W.; Gao, L.; Li, J.; Gao, H.; Zhao, G.; Huang, J. One-pot Synthesis of Dimethyl Carbonate and Glycols from Supercritical $CO_2$, Ethylene Oxide or Propylene Oxide, and Methanol. *Appl. Catal. A: Gen.* **2004**, 263, 179–186.
175. Jiang, J.; Hua, R. Synthesis of Dimethyl Carbonate from CO2, Methanol, and Epoxides Using $Re(CO)_5Cl/K_2CO_3$ as Catalyst System. *Chem. Res. Chinese Univ.* **2007**, 23, 374–376.
176. Li, Y.; Zhao, X.; Wang, Y. Synthesis of Dimethyl Carbonate from Methanol, Propylene Oxide and Carbon Dioxide over KOH/4A Molecular SIeve Catalyst. *Appl Catal. A: Gen.* **2005**, 279, 205–208.
177. Tian, J.; Wang, J.; Chen, J.; Fan, J.; Cai, F.; He, L. One-pot Synthesis of Dimethyl Carbonate Catalyzed by $n\text{-}Bu_4NBr/n\text{-}Bu_3N$ from Methanol, Epoxides, and Supercritical $CO_2$. *Appl. Catal. A: Gen.* **2006**, 301, 215–221.
178. Chen, X.; Hu, C.; Su, J.; Yu, T.; Gao, Z. One-Pot Synthesis of Dimethyl Carbonate Catalyzed by $[bmim]BF_4/CH_3ONa$ *Chinese J. Catal.* **2006**, 27, 485–488.
179. Bhanage, B.M.; Fujita, S.; Ikushima, Y.; Torii, K.; Arai, M. Synthesis of Dimethyl Carbonate and Glycols from Carbon Dioxide, Epoxides and Methanol Using

Heterogeneous Mg Containing Smectite catalysts: Effect of Reaction Variables on Activity and Selectivity Performance. *Green. Chem.* **2003**, 5, 71–75.
180. Tian, J.; Miao, C.; Wang, J.; Cai, F.; Du, Y.; Zhao, Y.; He, L. Efficient Synthesis of Dimethyl Carbonate from Methanol, Propylene Oxide and $CO_2$ Catalyzed by Recyclable inOrganic Base/Phosphonium Halide-Functionalized Polyethylene glycol. *Green. Chem.* **2007**, 9, 566–571.
181. De, C.; Lu, B.; Lv, H.; Yu, Y.; Bai, Y.; Cai. Q. One-Pot Synthesis of Dimethyl Carbonate from Methanol, Propylene Oxide and Carbon Dioxide Over Supported Choline hydroxide/MgO. *Catal. Lett.* **2009**, 128, 459–464.
182. Li, J.; Wang, L.; Shi, F.; Liu, S.; He, Y.; Lu, L.; Ma, X.; Deng, Y. Quaternary Ammonium Ionic Liquids as Bi-functional Catalysts for One-Step Synthesis of Dimethyl Carbonate from Ethylene Oxide, Carbon Dioxide and Methanol. *Catal. Lett.* **2011**, 141, 339–346.
183. Tomishige, K.; Kunimori, K. Catalytic and Direct Synthesis of Dimethyl Carbonate Starting from Carbon Dioxide Using $CeO_2$-$ZrO_2$ Solid Solution Heterogeneous Catalyst: Effect of $H_2O$ Removal from the Reaction System. *Appl. Catal. A: Gen.* **2002**, 237, 103–109.
184. Wang, W.; Wang, S.; Ma, X.; Gong, J. Crystal Structures, Acid–Base Properties, and Reactivities of $Ce_xZr_{1-x}O_2$ Catalysts. *Catal. Today*. **2009**, 148, 323–328.
185. Tkatchenko, D.B.-; Bernard, F.; Demoisson, F.; Plasseraud, L.; Sanapureddy, S.R. Tin-Based Mesoporous Silica for the Conversion of $CO_2$ into Dimehtyl Carbonate. *Chem Sus Chem* **2011**, 4, 1316–1322.
186. La, K.W.; Jung, J.C.; Kima, H.; Baeck, S.-H.; Songa, I.K.; Effect of Acid–Base Properties of $H_3PW_{12}O_{40}/Ce_xTi_{1-x}O_2$ Catalysts on the Direct Synthesis of Dimethyl Carbonate from Methanol and Carbon Dioxide: A TPD Study of $H_3PW_{12}O_{40}/Ce_xTi_{1-x}O_2$ Catalysts. *J. Mol. Catal. A: Chem.* **2007**, 269, 41–45.
187. Tkatchenko, D.B.; Chambrey, S.; Keiski, R.; Ligabue, R.; Plasseraud, L.; Richard, P.; Turunen, H. Direct Synthesis of Dimethyl Carbonate with Supercritical Carbon Dioxide: Characterization of a Key Organotin Oxide Intermediate *Catal. Today* **2006**, 115, 80–87.
188. Fujita, S.; Bhanage, B.M.; Ikushima, Y.; Arai, M. Synthesis of Dimethyl Carbonate from Carbon Dioxide and Methanol in the Presence of Methyl Iodide and Base Catalysts Under Mild Conditions: Effect of Reaction Conditions and Reaction Mechanism *Green Chem.* **2001**, 3, 87–91.
189. Zhou, W.; Zhao, X.; Wang, Y.; Zhang, J. Synthesis of Diphenyl Carbonate by Transesterification over Lead and Zinc Double Oxide Catalyst. *Appl. Catal. A. Gen.* **2004**, 260, 19–24.
190. Fu, Z.; Ono, Y.; Two-step Synthesis of Diphenyl Carbonate from Dimethyl Carbonate and Phenol using $MoO_3SiO_2$ Catalysts. *J. Mol. Catal. A: Chem.* **1997**, 118, 293–299.
191. Lee, H.; Kim, S.J.; Ahn, B.S.; Lee, W.K.; Kim, H.S. Role of Sulfonic Acids in the Sn-Catalyzed Transesterification of Dimethyl Carbonate with Phenol. *Catal. Today.* **2003**, 87, 139–144.
192. Kim, W.B.; Lee, J.S. Gas Phase Transesterification of Dimethyl Carbonate and Phenol over Supported Titanium Dioxide. *J. Catal.* **1999**, 185, 307–313.

193. Wang, S.; Ma, X.; Gong, J.; Gao, N.; Guo, H.; Yang, X.; Xu, G. Characterization and Activity of Stannum Modified Hβ Catalysts for Transesterification of Dimethyl Oxalate with Phenol. *Catal. Today* **2004**, 93–95, 377–381.
194. Ma, X.; Wang, S.; Gong, J.; Yang, X.; Xu, G. A Comparative Study of Supported $TiO_2$ Catalysts and Activity in Ester Exchange between Dimethyl Oxalate and Phenol. *J. Mol. Catal. A: Chem.* **2004**, 222, 183–187.
195. Li, Z.; Qin, Z.; Zhu, H.; Wang, J. Synthesis of Diphenyl Carbonate from $CO_2$, Phenoxide, and $CCl_4$ with $ZnCl_2$ as Catalyst. *Chem. Lett.* **2006**, 35, 784–785.
196. Li, Z.; Qin, Z. Synthesis of Diphenyl Carbonate from Phenol and Carbon Dioxide in Carbon Tetrachloride with Zinc Halides as Catalyst. *J. Mol. Catal. A: Chem.* **2007**, 264, 255–259.
197. Fan, G.; Fujita, S.; Zou, B.; Nishiura, M.; Meng, X.; Arai, M. Synthesis of Diphenyl Carbonate from Phenol and Carbon Dioxide in the Presence of Carbon Tetrachloride and Zinc Chloride. *Catal. Lett.* **2009**, 133, 280–287.
198. Fan, G.; Zhao, H.; Duan, Z.; Fang, T.; Wana, M.; He, L. A Novel Method to Synthesize Diphenyl Carbonate from Carbon Dioxide and Phenol in the Presence of Methanol. *Catal. Sci. Technol.* **2011**, 1, 1138–1141.
199. Matsuda, H.; Ninagawa, A.; Nomura, R.; Tsuchida, T. Reaction of Carbon Dioxide with Ethylene Oxide in the Presence of Organotin Compounds. *Chem. Lett.* **1979**, 8, 573–574.
200. Nomura, R.; Ninagawa, A.; Matsuda, H. Synthesis of Cyclic Carbonates from Carbon Dioxide and Epoxides in the Presence of Organo antimony Compounds as Novel Catalysts. *J. Org. Chem.* **1980**, 45, 3735–3738.
201. Baba, A.; Nozaki, T.; Matsuda, H. Carbonate Formation from Oxiranes and Carbon Dioxide Catalyzed by Organotin Halide-Tetraalkylphosphonium Halide Complexes. *Bull. Chem. Soc. Jpn.* **1987**, 60, 1552–1554.
202. De Pasquale, R.J. Unusual Catalysis with Nickel (0) Complexes. *J. Chem. Soc. Chem. Commun.* **1973**, 157–158.
203. Tsuda, T.; Chujo, Y.; Saegusa, T. Copper(I) Cyanoacetate as a Carrier of Activated Carbon Dioxide. *J. Chem. Soc. Chem. Commun.* **1976**, 415–416.
204. Xiao, L.; Lv, D.; Wu, W. Brønsted Acidic Ionic Liquids Mediated Metallic Salts Catalytic System for the Chemical Fixation of Carbon Dioxide to Form Cyclic Carbonates. *Catal. Lett.* **2011**, 141, 1838–1844.
205. Ramidi, P.; Munshi, P.; Gartia, Y.; Pulla, S; Biris, A. S.; Paul, A.; Ghosh, A. Synergistic Effect of Alkali Halide and Lewis Base on the Catalytic Synthesis of Cyclic Carbonate from $CO_2$ and Epoxides. *Chem. Phys. Lett.* **2011**, 512, 273–277.
206. Monassier, A.; D'Elia, V.; Cokoja, M.; Dong, H.; Pelletier, J. D.A.; Basset, J.-M.; Kuhn, F.E. Synthesis of Cyclic Carbonates from Epoxides and $CO_2$ under Mild Conditions Using a Simple, Highly Efficient Niobium-Based Catalyst. *Chem Cat Chem* **2013**, 5, 1321–1324.
207. Li, F.; Xia, C.; Xu, L.; Sun, W.; Chen, G. A Novel and Effective Ni Complex Catalyst System for the Coupling Reactions of Carbon Dioxide and Epoxides. *Chem. Commun.* **2003,** 2042–2043.
208. Bai, D.; Nian, G.; Wang, G.; Wang, Z. Titanocene Dichloride/KI: An Efficient Catalytic System for Synthesis of Cyclic Carbonates from Epoxides and $CO_2$. *Appl. Organometal. Chem.* **2013**, 27, 184–187.

209. Tharun, J.; Mathai, G.; Kathalikkattil, A. C.; Roshan, R.; Kwaka, J.-Y.; Park, D.-W. Microwave-Assisted Synthesis of Cyclic Carbonates by a Formic acid/KI Catalytic System. *Green Chem.* **2013**, 15, 1673–1677.
210. Yang, Y.; Hayashi, a, Y.; Fujii, Y.; Nagano, T.; Kita, Y.; Ohshima, T.; Okuda, J.; Mashima, K. Efficient Cyclic Carbonate Synthesis Catalyzed by Zinc Cluster Systems Under Mild Conditions. *Catal. Sci. Technol.* **2012**, 2, 509–513.
211. Kumar, S.; Jain, S.L.; Sain, B. Metal Acetylacetonates as Highly Efficient and Cost Effective Catalysts for the Synthesis of Cyclic Carbonates from $CO_2$ and Epoxides. *Catal. Lett.* **2012**, 142, 615–618.
212. Baia, D.; Jing, H.; Wang, G. Cyclic Carbonate Synthesis from Epoxides and $CO_2$ over Cyanocobalamin/n-$Bu_4$NI. *Appl. Organometal. Chem.* **2012**, 26, 600–603.
213. Bai, D.; Duan, S.; Hai, L.; Jing, H. Carbon Dioxide Fixation by Cycloaddition with Epoxides, Catalyzed by Biomimetic Metalloporphyrins. *Chem Cat Chem.* **2012**, 4, 1752–1758.
214. Huang, J.W.; Shi, M. *J. Org. Chem.* Chemical Fixation of Carbon Dioxide by NaI/$PPh_3$/PhOH. **2003**, 68, 6705–6709.
215. Kim, H. S.; Kim, J. J.; Kwon, H. N.; Chung, M. J.; Lee, B. G.; Jang, H. G. Well-Defined Highly Active Heterogeneous Catalyst System for the Coupling Reactions of Carbon Dioxide and Epoxides. *J. Catal.*, **2002**, 205, 226–229.
216. Watile, R.A.; Deshmukh, K. M.; Dhake, K.P.; Bhanage, B. M. Efficient Synthesis of Cyclic Carbonate from Carbon Dioxide Using Polymer Anchored Diol functionalized Ionic Liquids as a Highly Active Heterogeneous Catalyst. *Catal. Sci. Technol.* **2012**, 2, 1051–1055.
217. Adam, F.; Batagarawa, M. S. Tetramethylguanidine–Silica Nanoparticles as an Efficient and Reusable Catalyst for the Synthesis of Cyclic Propylene Carbonate from Carbon Dioxide and Propylene Oxide. *Appl. Catal. A: Gen.* **2013**, 454, 164–171.
218. Chenga, W.; Chena, X.; Suna, J.; Wanga, J.; Zhang, S. SBA-15 Supported Triazolium-Based Ionic Liquids as Highly Efficient and Recyclable Catalysts for Fixation of $CO_2$ with Epoxides. *Catal. Today*. **2013**, 200, 117–124.
219. Song, J.; Zhang, B.; Zhang, P.; Ma, J.; Liu, J.; Fan, H.; Jiang, T.; Han, B. Highly Efficient Synthesis of Cyclic Carbonates from $CO_2$ and Epoxides Catalyzedby KI/lecithin. *Catal. Today*. **2012**, 183, 130–135.
220. Roeser, J.; Kailasam, K.; Thomas, A. Covalent Triazine Frameworks as Heterogeneous Catalysts for the Synthesis of Cyclic and Linear Carbonates from Carbon Dioxide and Epoxides *Chem Sus Chem* **2012**, 5, 1793–1799.
221. Lee, J. K.; Kim, Y. J.; Choi, Y-S.; Lee, H.; Lee, J.S.; Hong, J.; Jeong, E.-K.; Kima, H.S.; Cheong, M. Zn-Containing Ionic Liquids Bearing Dialkyl Phosphate Ligands for the Coupling Reactions of Epoxides and $CO_2$. *Appl. Catal. B: Env.* **2012**, 111–112, 621–627.
222. Appaturi, J.N.; Adam, F. A facile and Efficient Synthesis of Styrene Carbonate via Cycloaddition of $CO_2$ to Styrene Oxide over Ordered Mesoporous MCM-41-Imi/Br Catalyst. *Appl. Catal. B: Env.* **2013**, 136–137, 150–159.
223. Wang, J.-Q.; Yue, X.-D.; Cai, F.; He, L.-N. Solventless Synthesis of Cyclic Carbonates from Carbon Dioxide and Epoxides Catalyzed by Silica-Supported Ionic Liquids under Supercritical Conditions. *Catal. Comm.* **2007**, 8, 167–172.

224. Han, L.; Choi, H.-J.; Choi, S.-J.; Liub, B.; Park, D.-W. Ionic Liquids Containing Carboxyl Acid Moieties Grafted onto Silica: Synthesis and Application as Heterogeneous Catalysts for Cycloaddition Reactions of Epoxide and Carbon Dioxide. *Green Chem.* **2011**, 13, 1023–1028.
225. Li, F.; Xiao, L.; Xia, C.; Hu, B. Chemical Fixation of $CO_2$ with Highly Efficient $ZnCl_2$/[BMIm]Br Catalyst System. *Tet. Lett.* **2004**, 45, 8307–8310.
226. Peng, J.; Deng, Y. Cycloaddition of Carbon Dioxide to Propylene Oxide Catalyzed by Ionic Liquids. *New J. Chem.* **2001**, 25, 639–641
227. Paddock, R.L.; Nguyen, S.T.; Chemical $CO_2$ Fixation: Cr(III) Salen Complexes as Highly Efficient Catalysts for the Coupling of $CO_2$ and Epoxides. *J. Am. Chem. Soc.* **2001,** *123,* 11498–11499.
228. Zhang, X.; Jia, Y.-B.; Lu, X.-B.; Li, B.; Wanga, H.; Sun, L.-C. Intramolecularly Two-Centered Cooperation Catalysis for the Synthesis of Cyclic Carbonates from $CO_2$ and Epoxides. *Tetrah. Lett.* **2008**, 49, 6589–6592.
229. Niu, Y.; Zhang, W.; Li, H.; Chen, X.; Sun, J.; Zhuang, X.; Jing, X. Carbon Dioxide/Propylene Oxide Coupling Reaction Catalyzed by Chromium Salen Complexes *Polymer* 2009, 50, 441–446.
230. Lu, X.-B.; Feng, X.-J.; He, R. Catalytic Formation of Ethylene Carbonate from Supercritical Carbon Dioxide/Ethylene Oxide Mixture with Tetradentate Schiff-base Complexes as Catalyst. *Appl. Catal. A: Gen.* **2002**, 234, 25–33.
231. Lu, X.-B.; Zhang, Y.-J.; Liang, B.; Li, X.; Wang, H. Chemical Fixation of Carbon Dioxide to Cyclic Carbonates under Extremely Mild Conditions with Highly Active Bifunctional Catalysts. *J. Mol. Catal. A: Chem.* **2004**, 210, 31–34.
232. Meléndez, J.; North, M.; Pasquale, R. Synthesis of Cyclic Carbonates from Atmospheric Pressure Carbon Dioxide Using Exceptionally Active Aluminium(salen) Complexes as Catalysts. *Eur. J. Inorg. Chem.* **2007**, 3323–3326.
233. Lu, X.-B.; Liang, B.; Zhang, Y.-J.; Tian, Y.- Z.; Wang, Y.-M.; Bai, C.-X.; Wang, H.; Zhang, R. Asymmetric Catalysis with $CO_2$: Direct Synthesis of Optically Active Propylene Carbonate from Racemic Epoxides. *J. Am. Chem. Soc.* **2004**, 126, 3732–3733.
234. Paddock, R.L.; Nguyen, S.T. Chiral (salen)CoIII Catalyst for the Synthesis of Cyclic Carbonates. *Chem. Commun.* **2004**, 1622–1623.
235. Jing, H.; Chang, T.; Jin, L.; Wu, M.; Qiu, W. Ruthenium Salen/Phenyl Trimethyl Ammonium Tribromide Catalyzed Coupling Reaction of Carbon Dioxide and Epoxides. *Catal. Comm.* **2007**, 8, 1630–1634.
236. Jutz, F.; Grunwaldt, J.-D.; Baiker, A. Mn(III)(salen)-Catalyzed Synthesis of Cyclic Organic Carbonates from Propylene and Styrene Oxide in "Supercritical" $CO_2$. *J. Mol. Catal. A: Chem.* **2008**, 279, 94–103.
237. Jing, H.; Edulji, S. K.; Gibbs, J.M.; Stern, C.L.; Zhou, H.; Nguyen, S. T. (Salen)Tin Complexes: Syntheses, Characterization, Crystal Structures, and Catalytic Activity in the Formation of Propylene Carbonate from $CO_2$ and Propylene Oxide. *Inorg. Chem.* **2004**, 43, 4315–4327.
238. Taherimehr, M.; Decortes, A.; Al-Amsyar, S. M.; Lueangchaichaweng, W.; Whiteoak, C. J.; Escudero-Ada, E. C.; Kleij, A. W.; Pescarmona, P. P. A highly active Zn(salphen) Catalyst for Production of Organic Carbonates in a Green $CO_2$ Medium. *Catal. Sci. Technol.* **2012**, 2, 2231–2237.

239. Srivastava, R.; Srinivas, D.; Ratnasamy, P. Zeolite-based Organic–Inorganic Hybrid Catalysts for Phosgene-free and Solvent-free Synthesis of Cyclic Carbonates and Carbamates at Mild Conditions Utilizing $CO_2$. *Appl. Catal. A: Gen.* **2005**, 289, 128–134.
240. Srivastava, R.; Srinivas, D.; Ratnasamy, P. Synthesis of Cyclic Carbonates from Olefins and $CO_2$ over Zeolite-based Catalysts. *Catal. Lett.* **2003,** 89, 81–85.
241. Srivastava, R.; Srinivas, D.; Ratnasamy, P. Synthesis of Polycarbonate Precursors over Titanosilicate Molecular sieves. *Catal. Lett.* **2003,** 91, 133–137.
242. Srivastava, R.; Srinivas, D.; Ratnasamy, P. $CO_2$ Activation and Synthesis of Cyclic Carbonates and Alkyl/Aryl Carbamates over Adenine-Modified Ti-SBA-15 Solid Catalysts. *J. Catal.* **2005**, 233, 1–15.
243. Miralda, C. M; Macias, E.E.; Zhu, M.; Ratnasamy, P.; Carreon, M. A. Zeolitic Imidazole Framework-8 Catalysts in the Conversion of $CO_2$ to Chloropropene Carbonate. *ACS Catal.* **2012**, 2, 180–183.
244. Yasuda, H.; He, L.- N.; Sakakura, T. Cyclic Carbonate Synthesis from Supercritical Carbon Dioxide and Epoxide over Lanthanide Oxychloride. *J. Catal.* **2002**, 209, 547–550.
245. Kuwahara, Y.; Yamashita, H. A New Catalytic Opportunity for Waste Materials: Application of Waste Slag based Catalyst in $CO_2$ Fixation Reaction. *J. $CO_2$ Utilization.* **2013**, 1, 50–59.
246. Dai, W. L.; Yin, S.-F.; Guo, R.; Luo, S. L.; Du, X.; Au, C.T. Synthesis of Propylene Carbonate from Carbon Dioxide and Propylene Oxide Using Zn-Mg-Al Composite Oxide as High-efficiency Catalyst. *Catal. Lett.* **2010**, 136, 35–44.
247. Lescouet, T.; Chizallet, C.; Farrusseng, D. The Origin of the Activity of Amine-Functionalized Metal-Organic Frameworks in the Catalytic Synthesis of Cyclic Carbonates from Epoxide and $CO_2$. *Chem Cat Chem* **2012**, 4, 1725–1728.
248. Tharun, J.; Dharman, M. M.; Hwang, Y.; Roshan, R.; Park, M.S.; Park, D.-W. Tuning Double Metal Cyanide Catalysts with Complexing Agents for the Selective Production of Cyclic Carbonates over Polycarbonates. *Appl. Catal. A: Gen.* **2012,** 419–420, 178–184.
249. Hu, H.; Martin, J.C.; Zhang, M.; Southworth, C.S.; Xiao, M.; Meng, Y.; Sun, L. Immobilization of Ionic Liquids in θ-Zirconium Phosphate for Catalyzing the Coupling of $CO_2$ and Epoxides. *RSC Advances*. **2012**, 2, 3810–3815.
250. Zalomaeva, O. V.; Maksimchuk, N. V.; Chibiryaev, A. M.; Kovalenko, K.A.; Fedin, V. P.; Balzhinimaev, B.S. Synthesis of Cyclic Carbonates from Epoxides or Olefins and $CO_2$ Catalyzed by Metal-organic Frameworks and Quaternary Ammonium salts. *J. Energ. Chem.* **2013**, 22, 130–135.
251. Chen, F.; Li, X.; Wang, B.; Xu, T.; Chen, S.-L.; Liu, P.; Hu, C. Mechanism of the Cycloaddition of Carbon Dioxide and Epoxides Catalyzed by Cobalt-Substituted 12-Tungstenphosphate. *Chem. Eur. J.* **2012**, 18, 9870–9876.
252. Aresta, M.; Dibenedetto, A.; Tommasi, I. Direct Synthesis of Organic Carbonates by Oxidative Carboxylation of Olefins Catalyzed by Metal oxides: Developing Green Chemistry Based on Carbon Dioxide. *Appl. Organomet. Chem.* **2000**, 14, 799–802.
253. Eghbali, N.; Li, C.-J. Conversion of Carbon Dioxide and Olefins into Cyclic Carbonates in Water. *Green Chem.* **2007**, 9, 213–215.

254. Sun, J.; Fujita, S.-I.; Zhao, F.; Hasegawa, M.; Arai, M. A Direct Synthesis of Styrene Carbonate from Styrene with the $Au/SiO_2$–$ZnBr_2$/$Bu_4NBr$ Catalyst System. *J. Catal.* **2005**, 230, 398–405.
255. Tomishige, K.; Yasuda, H.; Yoshida, Y.; Nurunnabi, M.; Li, B.; Kunimori, K. Catalytic Performance and Properties of Ceria Based Catalysts for Cyclic Carbonate Synthesis from Glycol and Carbon Dioxide. *Green Chem.* **2004**, 6, 206–214.
256. Du, Y.; Kong, D.-L.; Yang, H.-Y.; Cai, F.; Tian, J.-S.; Wang, J.-Q.; He, L.-N. Sn-Catalyzed Synthesis of Propylene Carbonate from Propylene Glycol and $CO_2$ under Supercritical Conditions J. *Mol. Catal. A: Chem.* **2005**, 241, 233–237.
257. Huang, S.; Liu, S.; Li, J.; Zhao, N.; Wei, W.; Sun, Y. Modified Zinc Oxide for the Direct Synthesis of Propylene Carbonate from Propylene Glycol and Carbon dioxide. *Catal. Lett.* **2007**, 118, 290–294.
258. Silva, E.D.; Dayoub, W.; Mignani, G.; Raoul, Y.; Lemaire, M. Propylene Carbonate Synthesis from Propylene Glycol, Carbon dioxide and Benzonitrile by Alkali Carbonate Catalysts. *Catal. Comm.* **2012**, 29, 58–62.
259. Fukuoka, S.; Fukawa, I.; Tojo, M.; Oonishi, K.; Hachiya, H.; Aminaka, M.; Hasegawa, K.; Komiya, K.A Novel Non-Phosgene Process for Polycarbonate Production from CO2: Green and Sustainable Chemistry in Practice. *Catal. Surv. Asia.* **2010**, 14, 146–163.
260. Darensbourg. D. J. Making Plastics from Carbon Dioxide: Salen Metal Complexes as Catalysts for the Production of Polycarbonates from Epoxides and $CO_2$. *Chem. Rev.* **2007,** 107, 2388–2410.
261. Inoue, S.; Koinuma, H.; Tsuruta, T. Copolymerization of Carbon Dioxide and Epoxide *J. Polym. Sci., Part B* **1969**, *7*, 287–292.
262. Eberhardt, R.; Allmendinger, M.; Zintl, M.; Troll, C.; Luinstra, G. A.; Rieger, B. New Zinc Dicarboxylate Catalysts for the $CO_2$ Propylene Oxide Copolymerization Reaction: Activity Enhancement through Zn(II)-Ethylsulfinate Initiating Groups. *Macromol. Chem. Phys.* **2004**, 205, 42–47.
263. Lu, X.-B.; Darensbourg. D. J. Cobalt Catalysts for the Coupling of $CO_2$ and Epoxides to Provide Polycarbonates and Cyclic Carbonates. *Chem. Soc. Rev.,* **2012**, 41, 1462–1484.
264. Darensbourg, D. J.; Mackiewicz, R. M. Role of the Cocatalyst in the Copolymerization of $CO_2$ and Cyclohexene Oxide Utilizing Chromium Salen Complexes. *J. Am. Chem. Soc.* **2005**, 127, 14026–14038.
265. Darensbourg, D. J.; Yarbrough, J. C. Mechanistic Aspects of the Copolymerization Reaction of Carbon Dioxide and Epoxides, Using a Chiral Salen Chromium Chloride Catalyst. *J. Am. Chem. Soc.* **2002**, 124, 6335–6342.
266. Cui, D. M.; Nishiura, M.; Hou, Z. M. Alternating Copolymerization of Cyclohexene Oxide and Carbon Dioxide Catalyzed by Organo Rare Earth Metal Complexes. *Macromolecules* **2005**, 38, 4089–4095.
267. Sugimoto, H.; Ohshima, H.; Inoue, S. Alternating Copolymerization of Carbon Dioxide and Epoxide by Manganese Porphyrin: The First Example of Polycarbonate Synthesis from 1-atm Carbon Dioxide. *J. Polym. Sci., Part A: Polym. Chem.* **2003**, 41, 3549–3555.

268. Pescarmona, P.P.; Taherimehr, M. Challenges in the Catalytic Synthesis of Cyclic and Polymeric Carbonates from Epoxides and $CO_2$. *Catal. Sci. Technol.,* **2012**, 2, 2169–2187.
269. Klaus, S.; Lehenmeier, M.W.; Herdtweck, E.; Deglmann, P.; Ott, A.K.; Rieger, B. Mechanistic Insights into Heterogeneous Zinc Dicarboxylates and Theoretical Considerations for $CO_2$-Epoxide Copolymerization. *J. Am. Chem. Soc.* **2011**, 133, 13151–13161.
270. Kember, M.R.; Buchard, B.; Williams, C.K. Catalysts for $CO_2$/Epoxide Copolymerization. *Chem. Commun.* **2011**, 47, 141–163.
271. Dharman, M. M.; Ahn, J.-Y.; Lee M.-K.; Shim, H.-L.; Kim, K.-H.; Kim, Il.; Park, D.-W. Moderate Route for the Utilization of $CO_2$-Microwave Induced Copolymerization with Cyclohexene Oxide using Highly Efficient Double Metal Cyanide Complex catalysts Based on $Zn_3[Co(CN)_6]$. *Green Chem.* **2008**, 10, 678–684.
272. Dong, Y.; Wang, X.; Zhao, X.; Wang, F. Facile Synthesis of Poly(ether carbonate)s via Copolymerization of $CO_2$ and Propylene Oxide Under Combinatorial Catalyst of Rare Earth Ternary Complex and Double Metal Cyanide Complex. *J. Polymer Sci. Part A: Polymer Chem.* **2012**, 50, 362–370.
273. Sun, X.-K.; Zhang, X.-H.; Wei, R.-J.; Du, B.-Y.; Wang, Q.; Fan, Z.-Q.; Qi, G.-R. Mechanistic Insight into Initiation and Chain Transfer Reaction of $CO_2$/Cyclohexene Oxide Copolymerization Catalyzed by Zinc-Cobalt Double Metal Cyanide Complex Catalysts. *J. Polymer Sci. Part A: Polymer Chem.* **2012**, 50, 2924–2934.
274. Gao, Y.; Qin, Y.; Zhao, X.; Wang, F.; Wang, X. Selective Synthesis of Oligo(Carbonate-Ether) Diols from Copolymerization of $CO_2$ and Propylene Oxide under Zinc-Cobalt Double metal Cyanide Complex. *J Polym Res* **2012**, 19, 9878–9886.
275. Kim, Il.; Yi, M. J.; Byun, S. H.; Park, D. W.; Kim, B.U.; Ha, C. S. Biodegradable Polycarbonate Synthesis by Copolymerization of Carbon Dioxide with Epoxides Using a Heterogeneous Zinc Complex. *Macromol. Symp.* **2005**, 224, 181–192.
276. Zhang, W.; Lin, Q.; Cheng, Y.; Lu, L.; Lin, B.; Pan, L.; Xu, N. Double Metal Cyanide Complexes Synthesized by Solvent-Free Grinding Method for Copolymerization of $CO_2$ and Propylene Oxide. *J. Appl. Polym. Sci.* **2012**, 123, 977–985.
277. Zhang, X.-H.; Wei, R.-J.; Sun, X.-K.; Zhang, J.- F.; Du, B.-Y.; Fan, Z.-Q.; Qi, G.-R. Selective Copolymerization of Carbon dioxide with Propylene Oxide Catalyzed by a Nanolamellar Double Metal Cyanide Complex Catalyst at Low Polymerization Temperature s. *Polymer.* **2011**, 52, 5494–5502.
278. Sun, X.- K.; Zhang, X.-H.; Liu, F.; Chen, S.; Du, B.-Y.; Wang, Q.; Fan, Z.-Q.; Qi, G.-R. Alternating Copolymerization of Carbon Dioxide and Cyclohexene Oxide Catalyzed by Silicon Dioxide/Zn-CoIII Double Metal Cyanide Complex Hybrid Catalysts with a Nanolamellar Structure. *J. Polym. Sci. Part A: Polym. Chem.* **2008**, 46, 3128–3139.
279. Robertson, N. J.; Qin, Z.; Dallinger, G. C.; Lobkovsky, E. B.; Lee, S.; Coates, G. W. Two-dimensional Double Metal Cyanide Complexes: Highly Active Catalysts for the Homopolymerization of Propylene Oxide and Copolymerization of Propylene Oxide and Carbon Dioxide. *Dalton Trans.* **2006**, 5390–5395.
280. Xiang, X.; Guo, L.; Wu, X.; Ma, X.; Xia, Y. Urea Formation from Carbon dioxide and Ammonia at Atmospheric Pressure. *Environ. Chem. Lett.* **2012**, 10, 295–300.

281. Gabriele, B.; Salerno, G.; Mancuso, R.; Costa, M. Efficient Synthesis of Ureas by Direct Palladium-Catalyzed Oxidative Carbonylation of Amines. *J. Org.Chem.* **2004**, 69, 4741–4750.
282. Esteves-Souza, A.; Pissinate, K.; Nascimento, M.G.; Grynberg, N.F.; Echevarria, A. Synthesis, Cytotoxicity, and DNA-Topoisomerase Inhibitory Activity of New Asymmetric Ureas and Thioureas. *Bioorg. Med. Chem.* **2006**, 14, 492–499.
283. Fournier, J.; Bruneau, C.; Dixneuf, P.H.; L´ecolier, S. Ruthenium-Catalyzed Synthesis of Symmetrical N, N'-Dialkylureas Directly from Carbon Dioxide and Amines. *J. Org. Chem.* **1991**, 56, 4456–4458.
284. Nomura, R.; Hasegawa, Y.; Ishimoto, M.; Toyosaki, T.; Matsuda, H. Carbonylation of Amines by Carbon Dioxide in the Presence of an Organoantimony Catalyst. *J. Org. Chem.* **1992,** 57, 7339–7342.
285. Jiang, T.; Ma, X.; Zhou, Y.; Liang, S.; Zhang, J.; Han, B. Solvent-free Synthesis of Substituted Ureas from $CO_2$ and Amines with a Functional Ionic Liquid as the Catalyst. *Green Chem.*, **2008**, 10, 465–469.
286. Ion, A.; Parvulescu, V.; Jacobsa, P.; De Vos, D. Synthesis of Symmetrical or Asymmetrical Urea Compounds from $CO_2$ via Base Catalysis. *Green Chem.* **2007**, 9, 158–161.
287. Shim, Y. N.; Lee, J.K.; Im, J. K.; Mukherjee, D.K.; Nguyen, D. Q.; Cheong, M.; Kim, H. S.Ionic Liquid-Assisted Carboxylation of Amines by $CO_2$: A Mechanistic Consideration. *Phys. Chem. Chem. Phys.* **2011**, 13, 6197–6204.
288. Chaturvedi, D. Perspectives on the Synthesis of Organic Carbamates. *Tetrahedron.* **2012**, 68, 15–45.
289. Srivastava, R.; Manju, M.D.; Srinivas, D.; Ratnasamy, P. Phosgene-Free Synthesis of Carbamates over Zeolite-based Catalysts. *Catal. Lett.* **2004**, 97, 41–47.
290. Honda, M.; Sonehara, S.; Yasuda, H.; Nakagawa, Y.; Tomishige, K. Heterogeneous $CeO_2$ Catalyst for the One-Pot Synthesis of Organic Carbamates from Amines, $CO_2$ and Alcohols. *Green Chem.* **2011**, 13, 3406–3413.
291. Srivastava, R.; Srinivas, D.; Ratnasamy, P. Syntheses of Polycarbonate and Polyurethane Precursors Utilizing $CO_2$ over Highly Efficient, Solid As-synthesized MCM-41 Catalyst. *Tetrah. Lett.* **2006**, 47, 4213–4217.
292. Hayashi, H.; Ogo, S.; Fukuzumi, S.Aqueous hydrogenation of carbon dioxide catalysed by water-soluble Ruthenium Aqua Complexes under Acidic Conditions. *Chem. Comm.* **2004**, 2714–715.
293. Jessop, P.G. Homogeneous Hydrogenation of Carbon Dioxide. In The Handbook of Homogenous Hydrogenation, de Vries, J. G., Elsevier, C. J., Eds.; Wiley-VCH: Weinheim, 2007; Vol.1; P. 489–511.
294. Zhang, Z. F.; Hu, S. Q.; Song, J. L.; Li, W. J.; Yang, G. Y.; Han, B.X. Hydrogenation of $CO_2$ to Formic Acid Promoted by a Diamine-Functionalized Ionic Liquid. *Chem Sus Chem.* **2009**, 2, 234–238.
295. Wang, W.; Wang, S.; Ma, X.; Gong, J. Recent Advances in Catalytic Hydrogenation of Carbon Dioxide. *Chem. Soc. Rev.* **2011**, 40, 3703–3727.
296. Tai, C. C.; Pitts, J.; Linehan, J.C.; Main, A. D.; Munshi, P.; Jessop, P G. Formation of Ruthenium Catalysts for the Homogeneous Hydrogenation of Carbon Dioxide. *Inorg. Chem.* **2002**, 41, 1606–1614.

297. Zhang, Y.; Fei, J.; Yu, Y.; Zheng, X. Silica Immobilized Ruthenium Catalyst used for Carbon Dioxide Hydrogenation to Formic Acid (I): The Effect of Functionalizing Group and Additive on the Catalyst Performance Catal. *Commun.* **2004**, 5, 643–646.
298. Preti, D.; Resta, C.; Squarcialupi, S.; Fachinetti, G. Carbon Dioxide Hydrogenation to Formic Acid by Using a Heterogeneous Gold Catalyst. *Angew. Chem. Int. Ed.* **2011**, 50, 12551–12554.
299. Hao, C.; Wang, S.; Li, M.; Kang, L.; Ma, X. Hydrogenation of $CO_2$ to Formic Acid on Supported Ruthenium Catalysts. *Catal. Today.* **2011**, 160, 184–190.
300. Xu, Z.; McNamara, N. D.; Neumann, G. T.; Schneider, W. F.; Hicks, J. C. Catalytic Hydrogenation of $CO_2$ to Formic Acid with Silica-Tethered Iridium Catalysts. *Chem Cat Chem.* **2013**, 5, 1–3.
301. Sanz, S.; Azua, A.; Peris, E. '(η6-arene)Ru(bis-NHC)' Complexes for the Reduction of $CO_2$ to Formate with Hydrogen and by Transfer Hydrogenation with iPrOH. *Dalton Trans.* **2010**, 39, 6339–6343.
302. Jessop, P. G.; Hsiao, Y.; Ikariya, T.; Noyori, R. Homogeneous Catalysis in Supercritical Fluids: Hydrogenation of Supercritical Carbon Dioxide to Formic Acid, Alkyl Formates, and Formamides. *J. Am. Chem. Soc.* **1996,** *118,* 344–355.
303. Ng, S. M.; Yin, C.; Yeung, C. H.; Chan, T. C.; Lau, C. P. Ruthenium-Catalyzed Hydrogenation of Carbon Dioxide to Formic Acid in Alcohols. *Eur. J. Inorg. Chem.* **2004**, 1788–1793.
304. Gao, Y.; Kuncheria, J. K.; Jenkins, H. A.; Puddephatt, R. J.; Yap, G. P. A. The Interconversion of Formic Acid and Hydrogen/Carbon Dioxide using a Binuclear Ruthenium Complex Catalyst. *J. Chem. Soc., Dalton Trans.*, **2000**, 3212–3217.
305. Munshi, P.; Main, A. D.; Linehan, J. C.; Tai, C.-C.; Jessop, P. G. Hydrogenation of Carbon Dioxide Catalyzed by Ruthenium Trimethylphosphine Complexes: The Accelerating Effect of Certain Alcohols and Amines. *J. Am. Chem. Soc.* **2002**, 124, 7963–7971.
306. Himeda, Y.; Komatsuzaki, N.O-; Sugihara, H.; Kasuga, K. Simultaneous Tuning of Activity and Water Solubility of Complex Catalysts by Acid-Base Equilibrium of Ligands for Conversion of Carbon Dioxide. *Organometallics.* **2007**, 26, 702–712.
307. Tanaka, R.; Yamashita, M.; Nozaki, K. Catalytic Hydrogenation of Carbon Dioxide Using Ir(III)-Pincer Complexes. *J. Am. Chem. Soc.* **2009**, 131, 14168–14169.
308. Ezhova, N. N.; Kolesnichenko, N. V.; Bulygin, A. V.; Slivinskii, E. V.; Han, S. Hydrogenation of $CO_2$ to Formic Acid in the Presence of the Wilkinson Complex. *Russ. Chem. Bull. Int. Ed.* **2002**, 51, 2165–2169.
309. Freund, H. J.; Wambach, J. Process for the Preparation of Acetic Acid, Republic of South Africa, Patent No. 1995/06606.
310. Huang, W.; Xie, K.C.; Wang, J. P.; Gao, Z. H.; Yin, L. H.; Zhu, Q. M. Possibility of Direct Conversion of $CH_4$ and $CO_2$ to High-Value Products. *J. Catal.* **2001**, 201, 100–104.
311. Wilcox, E. M.; Roberts, G. W.; Spivey, J. J. Direct Catalytic Formation of Acetic Acid from $CO_2$ and Methane. *Catal. Today.* **2003**, 88, 83–90.
312. Huang, W.; Zhang, C.; Yin, L.; Xie, K. Direct Synthesis of Acetic Acid from $CH_4$ and $CO_2$ in the Presence of $O_2$ over a $V_2O_5$-$PdCl_2/Al_2O_3$ Catalyst. *J. Nat. Gas Chem.* **2004**, 13, 113–115.

313. Farlow, M. W.; Adkins, H. The Hydrogenation of Carbon Dioxide and a Correction of the Reported Synthesis of Urethanes *J. Am. Chem. Soc.,* 1935, 57, 2222–2223.
314. Haynes, P.; Slaugh, L.H.; Kohnle, J.F. Formamides from Carbon Dioxide, Amines and Hydrogen in the Presence of Metal Complexes.*Tetrahedron Lett.* **1970**, 11, 365–368.
315. Jessop, P. G.; Hsiao, Y.; Ikariya, T.; Noyori, R. Catalytic Production of Dimethyl Formamide from Supercritical Carbon Dioxide. *J. Am. Chem. Soc.* **1994**, 116, 8851–8852.
316. Krocher, O.; Koppel, R.A.; Baiker, A. Highly Active Ruthenium Complexes with Bidentate Phosphine Ligands for the Solvent-Free Catalytic Synthesis of N, N-Dimethyl Formamide and Methyl Formate. *Chem. Commun.* **1997**, 453–454.
317. (a) Krocher, O.; Koppel, R.A.; Baiker, A. Silica Hybrid Gel Catalysts Containing Ruthenium Complexes: Influence of Reaction Parameters on the Catalytic Behaviour in the Synthesis of N, N-Dimethyl formamide from Carbon Dioxide. *J. Mol. Catal. A: Chem.* **1999**, 140, 185–193. (b) Krocher, O.; Koppel, R.A.; Froba, M.; Baiker, A. Silica Hybrid Gel Catalysts Containing Group(VIII) Transition Metal Complexes: Preparation, Structural, and Catalytic Properties in the Synthesis of N, N-Dimethylformamide and Methyl Formate from Supercritical Carbon Dioxide. *J. Catal.* **1998**, 178, 284–298.
318. (a). Rohr, M.; Grunwaldt, J. D.; Baiker, A. A Simple route to Highly Active Ruthenium Catalysts for Formylation Reactions with Hydrogen and Carbon Dioxide. *J. Mol. Catal. A: Chem.* **2005**, 226, 253–257. (b) Rohr, M.; Grunwaldt, J. D.; Baiker, A. Formylation with Supercritical Carbon Dioxide over $Ru/Al_2O_3$ Modified by Phosphines: Heterogeneous or Homogeneous Catalysis? *J. Catal.* **2005**, 229, 144–153.
319. Liu, J.; Guo, C.; Zhang, Z.; Jiang, T.; Liu, H.; Song, J.; Fan, H.; Han, B. Synthesis of Dimethylformamide from $CO_2$, $H_2$ and Dimethylamine over Cu/ZnO. *Chem. Commun.* **2010**, 46, 5770–5772.
320. Nalawade, S. P.; Picchioni, F.; Janssen, L.P.B.M. Supercritical Carbon Dioxide as a Green Solvent for Processing Polymer Melts: Processing Aspects and Applications. *Prog. Polym. Sci.* **2006**, 31, 19–43.
321. Beckman, E. J. Supercritical and Near-Critical $CO_2$ in Green Chemical Synthesis and Processing. *Supercrit. Fluids* **2004**, 28, 121–191.
322. Rayner, C.M. The Potential of Carbon Dioxide in Synthetic Organic Chemistry. *Organic Process Research and Development.* **2007**, 11, 121–132.
323. Dapurkar, S. E.; Kawanami, H.; Yokoyama, T.; Ikushima, Y. Highly Selective Oxidation of Tetralin to 1-Tetralone over Mesoporous CrMCM-41 Molecular Sieve Catalyst using Supercritical Carbon Dioxide. *New J. Chem.* **2009**, 33, 538–544.
324. Selva, M.; Perosa, A.; Fabrisa, M.; Canton, P. The Metathesis of $\alpha$-Olefins over Supported Re-Catalysts in Supercritical $CO_2$. *Green Chem.* **2009**, 11, 229–238.
325. Conceiçao, L.; Łukasika, R.B-; and Łukasik, E. B-Supercritical $CO_2$ as an Effective Medium for a Novel Conversion of Glycerol and Alcohols in the Heterogeneous Telomerisation of Butadiene. *Green Chem.* **2012**, 14, 673–681.
326. http://www.extremetech.com/extreme/124383-bacteria-converts-carbon-dioxide-into-liquid-fuel/
327. http://www.ship-technology.com/features/featureelectrically-efficient-biofuel-photosynthesis/featureelectrically-efficient-biofuel-photosynthesis-4.html

328. http://articles.cnn.com/2008–10–08/tech/co2.fuel_1_recycling-projects-fuel-co2?_s=PM:TECH
329. http://newenergyandfuel.com/http:/newenergyandfuel/com/2008/08/29/a-new-leading-process-for-$CO_2$-to-methanol/.
330. http://www.carbonsciences.com/01/technology.html.
331. Service, R. F. Sunlight in Your Tank. *Science*. **2009**, 326, 1472–1475.
332. Song, C., and Pan, W., Tri reforming of methane: A Novel Concept for Synthesis of Industrially Useful Synthesis Gas with Desired $H_2$/CO Ratios using $CO_2$ in Flue Gas over Power Plants without $CO_2$ Seperation. *Prepr. Pap. Am. Chem. Soc. Div. Fuel Chem.*, **2004**, 49, 128–131.

CHAPTER 6

# SUPERCRITICAL $CO_2$ IN ORGANIC SYNTHESIS

S. MAYADEVI

## CONTENTS

## 6.1 INTRODUCTION

Supercritical fluids (SCFs) are characterized by gas like viscosity and solvating properties of a wide range of organic solvents. Two principal roles are played by them while participating in chemical transformations viz., to dissolve and transport reactant species to coreactants and catalyst particles and impart activation energy to the reactants via solute-solvent collisions. It has been established that the chemical reaction rates within supercritical media are enhanced resulting in reduced energy demand on the chemical process. More frequent higher energy solute-solvent collisions are stated to be responsible for enhanced reaction rates. In addition, heat transfer, micromixing phenomena and catalyst activity are promoted. Supercritical fluids as reaction media also eliminate other energy demanding auxiliary operations that are normally associated with organic solvents. Catalysis in SCF media is endowed with other advantages viz., extended catalyst life due to less coke formation, continuous rinsing of catalyst internal porous structure, switchable or tunable selectivity and safer process. All these advantages are highly relevant to industrial process development. Environmentally benign $CO_2$ and $H_2O$ offer significant potential to replace current reliance on a range of hazardous, relatively expensive and environmentally damaging organic solvents that are used on an extensive global basis. An attempt is made in this chapter to highlight the potential applications of supercritical $CO_2$ as sustainble alternatives to the existing technologies of important industrial catalytic processes, like hydrogenation, hydroformylation, oxidation and C-C bond formation, their theoretical aspects and reported process intensification efforts.

Supercritical fluids exist at a state above their critical temperature ($T_c$) and critical pressure ($P_c$), but below the pressure required to condense them into a solid [1]. Typical phase diagram of carbon dioxide, illustrating the different domains is presented in Fig. 6.1 [2]. It can be seen that the gas-liquid phase boundary moves up with increase in temperature and pressure. The liquid becomes less dense as the temperature increases and the gas becomes denser as the pressure increases. At critical point, the densities of both gas and liquid phases become equal and the phase difference disappears. The isothermal compressibility of any pure fluid, at critical point is infinite and SCFs exhibit interesting properties in the region near the critical point ($1.0 < T_r < 1.1$ and $1 < P_r < 2$), where they exist as a single phase. In this region SCFs possess both liquid and gaseous properties. Liquid like densities and dissolution characteristics make them efficient solvents while gas like viscosities and diffusivities help in reducing the mass transfer resistances. Further, small changes in temperature and pressure can result in up to 100-fold changes in solubility [3]. This behavior of SCFs is being exploited effectively in many commercial separation processes. The critical properties of some fluids are presented in Table 6.1.

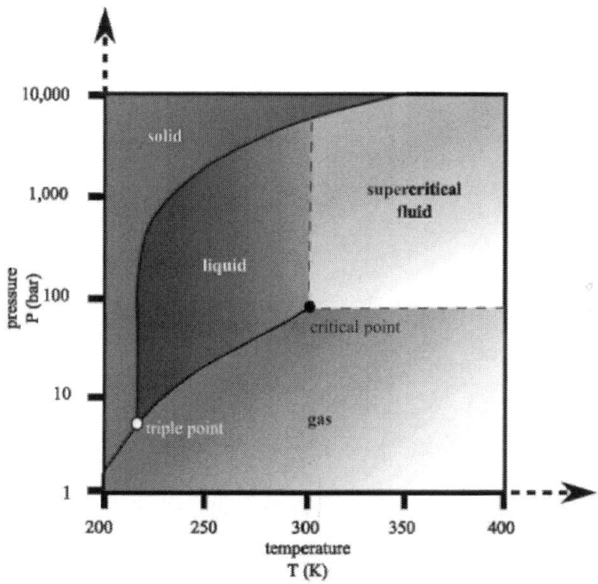


**FIGURE 6.1** Phase diagram of Carbon Dioxide (Source: http://upload.wikimedia.org/wikipedia/commons/1/13/Carbon-dioxide-pressuretemperature phase-diagram.svg).

Among the SCFs, supercritical carbon dioxide ($ScCO_2$) is a highly attractive medium for organic reactions as it is inexpensive, readily available, non-toxic and environmentally benign. Further, its properties (viz. density, viscosity, diffusivity and heat capacity) can be easily tuned between those of liquid and gaseous $CO_2$ by varying its temperature and/or pressure. Chemicals dissolved in $ScCO_2$ have diffusivities higher and viscosities lower than those of conventional organic solvents. Near critical point, large changes in fluid density and related properties are observed with small changes in temperature and pressure, providing opportunities for engineering the reaction environment through small variations in temperature and pressure [4]. Increase in pressure leads to an increase in solvent density and solubility of the solute. The temperature effect on solubility is pressure dependent. In the region around the gas-liquid phase boundary in Fig. 6.1, that is, below/above a certain pressure, called the crossover pressure, changes in temperature leads to sudden changes in the density and the dissolution capacity of $CO_2$ due to change of phase. The first major industrial application of $ScCO_2$ was for decaffeination of roasted coffee beans [5]. The focus later shifted to chemical reactions in SCF and a special issue of Chemical Reviews was dedicated to reactions in supercritical media [4, 6–8].

**TABLE 6.1** Critical properties of some fluids [4].

| SCF | Critical Temp. (°C) | Critical Pressure (MPa) | Critical Density ($kg.m^{-3}$) |
|---|---|---|---|
| Water ($H_2O$) | 373.9 | 22.06 | 322 |
| 1-Propanol ($C_3H_7OH$) | 263.6 | 5.17 | 275 |
| 2-Propanol ($C_2H_5(OH)CH_3$) | 235.1 | 4.762 | 273 |
| Ethanol ($C_2H_5OH$) | 240.7 | 6.137 | 276 |
| Methanol ($CH_3OH$) | 239.4 | 8.092 | 272 |
| Propane ($C_3H_8$) | 96.6 | 4.25 | 217 |
| Propene ($C_3H_6$) | 91.6 | 4.601 | 233 |
| Ethane ($C_3H_8$) | 32.2 | 4.884 | 203 |
| Carbon dioxide ($CO_2$) | 30.9 | 7.375 | 468 |
| Ethene ($C_2H_4$) | 9.1 | 5.041 | 214 |

## 6.2 $ScCO_2$ IN ORGANIC SYNTHESIS

Carbon dioxide is noninflammable, nontoxic, inert towards oxidation, and readily available. Like all SCFs, near the critical point, $CO_2$ has dissolving power comparable to that of liquids, is more compressible than dilute gases and has transport properties intermediate between those in gas and liquid phase. This combination of physical properties lead to high reaction rates, improved selectivity and elimination of mass transfer problems for reactions in $ScCO_2$. Increased reaction rates and selectivity in the critical region result due to several factors, viz. (i) increased solubility of reactant gases in $ScCO_2$ (ii) rapid diffusion of solutes into and within the supercritical phase (iii) weakening of solvation around reacting species (iv) local clustering of reactants and/or solvent (v) large negative activation volume, and (vi) reduction in cage effect in radical reactions. Further, the properties of $ScCO_2$ vary with density, which is very sensitive to changes in temperature and pressure in the critical region.

Supercritical carbon dioxide has significant safety advantage over organic solvents as it is noninflammable. Since it is a nonreactive solvent in most cases, side reactions and by-product formation are rare in $ScCO_2$ mediated reactions. However, there are also instances where $ScCO_2$ is used as a reactant. As the solubility of reactants and products in $ScCO_2$ is a strong function of temperature and pressure, it is possible to effect *in-situ* product separation by suitable temperature and pressure manipulations.

### 6.2.1 HOMOGENEOUS CATALYTIC REACTIONS

Homogeneous reactions using transition metal catalysts can be engineered in $ScCO_2$ medium for tuneable reaction rates and selectivity. It is possible to achieve deliberate control of the phase behavior, dissolution of reactants, recovery of products and catalyst by moderately changing the reaction temperature and pressure.

### 6.2.2 ENHANCEMENT IN GAS SOLUBILITY

Many homogeneous reactions (e.g., hydrogenation, hydroformylation, oxidation, etc.) are limited by the solubility of the gaseous reactant ($H_2$ and $O_2$) in the reacting medium. Gaseous reactants have limited solubility in liquid solvents, but high solubility in $ScCO_2$. As the reaction rate is dependent on the concentration of the reacting gas in the homogeneous phase, the use of $ScCO_2$ results in increased reaction rates and better selectivity.

### 6.2.3 ENHANCEMENT IN CATALYST SOLUBILITY

Metal-organic complexes and amorphous fluoropolymers used as catalysts in many organic syntheses exhibit excellent solubility in $ScCO_2$ [9]. Fluoropolymers such as poly(chlorotrifluoroethylene), fluorinated polyacrylates, copolymers of fluorinated acrylates with methacrylate, *n*-butyl acrylate, styrene, ethylene, etc. which are used to enhance the solubility of the catalyst are also reported to be highly soluble in $ScCO_2$ [9]. Temperature and pressure dependent catalyst solubility not only enhances its activity, but also permits easy recovery and recycle of the catalyst at the end of the reaction. It is possible to precipitate the catalyst and sometimes the product by suitable adjustments in temperature and pressure, and integrate reaction and separation.

### 6.2.4 CARBON DIOXIDE-EXPANDED SOLVENT MEDIA

Dense $CO_2$ (below critical pressure) is also used as a cosolvent along with organic solvents. The use of $CO_2$ along with $CO_2$ soluble solvents expands its application domain. Compared to neat organic solvent, $CO_2$ expanded solvents exhibit decreased solubility for solutes [10–12] and increased solubility for gases (e.g., $H_2$ and $O_2$). Part replacement of organic solvent with dense $CO_2$ can ensure the catalyst solubility in the reaction medium, enhance the solubility of reactant gas while preserving the advantages of traditional solvent. Thus, the advantages of traditional solvent are optimally combined with the environmentally benign nature of $CO_2$ leading to several advantages viz. flexibility in solvent selection, increased reaction rate and selectivity at subcritical operating pressures. However, the problems associated with the use of organic solvents is not eliminated in this method of operation.

### 6.2.5 HETEROGENEOUS CATALYTIC REACTIONS

Unlike homogeneous catalysis, reactants and catalyst are present in different phases in heterogeneous catalysis. This necessitates effective mass and heat transfer through the interface an essential prerequisite for the reaction. The sequence of different steps in a heterogeneous catalytic reaction are: (1) diffusion of gaseous reactant from bulk gas phase to gas-liquid interface; (2) absorption and diffusion of gaseous reactants from the interface to bulk liquid phase; (3) diffusion of both gas and liquid reactants to the external surface of the catalyst through the stagnant film surrounding the catalyst particle; (4) internal diffusion of the reactants from the catalyst surface to the pores in case of porous catalysts; (5) adsorption of reactants on the catalyst; (6) surface reaction; (7) desorption of products from the catalyst surface; (8) internal diffusion of the products through catalyst pores; and (9) external diffusion of the products through the stagnant liquid layer covering the catalyst to the bulk liquid phase. Overall reaction rate and the product distribution are significantly affected by mass transfer resistance at each step, which is eliminated under supercritical conditions. In principle, each of the above steps is affected either directly or indirectly by change in pressure of the reaction medium from subcritical to supercritical [4].

### 6.2.6 LOCAL DENSITY ENHANCEMENT: CLUSTERING/ MOLECULAR CHARISMA

The reactant molecules and $ScCO_2$ molecules generally vary in size, shape, interaction strength and polarity. Further, due to the high compressibility and large free volume of $ScCO_2$, the attractive forces move the molecules to energetically favorable locations, giving rise to nonuniform spatial distribution of solvent molecules about a solute molecule. This is termed as local density enhancement/clustering/molecular charisma and gives rise to unusual solvation effects [4] and affects the rates and selectivities of chemical reactions.

### 6.2.7 PHASE BEHAVIOUR AND SOLUBILITY

All reactants under supercritical conditions exist as a single homogeneous fluid phase leading to elimination of interface mass transfer resistances observed in multiphase systems. This feature along with pressure tuneable solvent properties and possibility of easier products separation can be exploited only for operation in supercritical region. While it is easy to identify the SCF region of pure fluids, the location of phase boundaries for multicomponent systems may be complicated. Critical point in a multicomponent reacting system is composition dependent and changes with the extent of reaction. It means that the critical point of the reacting system will change with time for a batch reactor, and with location in a continuous fixed bed reactor. An understanding of the phase behavior is necessary for fully exploiting the advantages of SCF properties in heterogeneous catalytic reactions. However, this information is somewhat scanty.

### 6.2.8 TRANSPORT PROPERTIES

High diffusion coefficients and elimination of interfaces results in significantly enhanced mass transfer in $ScCO_2$. Theoretical studies on diffusivity of some organic compounds in $CO_2$ and $CO_2$/methanol and $CO_2$/acetone mixed solvents indicated enhanced local solvent density around the solute for pure solvent and preferential attraction of methanol/acetone (cosolvent) around the solute in mixed solvent systems, significantly influencing the solute diffusivity. $ScCO_2$ has viscosity comparable to and density considerably higher than gaseous $CO_2$ at similar temperature. Hence the kinematic viscosity is very low in the critical region. This has beneficial influence on mass transfer as natural convection effects are inversely proportional to the square of kinematic viscosity [13].

## 6.3 THEORETICAL ASPECTS

In the supercritical region, the liquid and vapor phases are indistinguishable. Supercritical fluid has properties (diffusivity, viscosity, thermal conductivity, heat capacity) intermediate between liquid and gas phase. The isothermal compressibility $\kappa_T$ of pure fluid is given by:

$$\kappa_T = [(\partial \varrho/\partial P)_T]/\varrho \quad (1)$$

Here $\varrho$ is the density, T is the temperature and P is the pressure. At critical point $\kappa_T$ is very large and approaches infinity.

### 6.3.1 SOLUBILITY OF SOLUTES IN $ScCO_2$

Supercritical $CO_2$ has been highly favored in many extraction processes due to its beneficial properties (noninflammable, nontoxic and ambient critical temperature ). SCF based extraction and separation require the determination of its selectivity towards the individual components of the mixture to be separated. Although pure component solubility of a large number of compounds in $ScCO_2$ is available, data on multicomponent solubility is scarce [14].

Solubility of solids in sub and supercritical fluids has been reviewed by Škerget and co-workers [15]. Modelling of multicomponent solubilities in $ScCO_2$ correlating pure component solubility based on equation of state (EOS), usually Peng-Robinson equation of state (PR-EOS), and suitable mixing rule (usually the van der Waals one-fluid mixing rule) are available in literature [16–17]. Data on critical properties and sublimation pressure required by the EOS models are not available for many compounds and are estimated by the group contribution methods, leading to certain amount of errors in estimation.

Density based semi empirical models have also been proposed for correlating the solubility of solids in $ScCO_2$. These models are based on the concept of association of molecules. They do not account for cosolute concentration and cannot be applied directly to correlate the solubility of solid mixtures in $ScCO_2$. Hence, like the models using EOS, density based models also use separate correlations (mixing rules) that use pure component solubility, to obtain the selectivity as a function of temperature and pressure [14]. Rao and Madras [14] have also proposed models that use the association of molecules as well as the activity coefficients for estimating selectivity of $ScCO_2$ for solid mixtures.

### 6.3.2 REACTION EQUILIBRIUM AND KINETICS

Carbon dioxide has critical temperature ($T_c$) 30.9°C and critical pressure ($P_c$) 7.375 MPa. As the critical pressure of $CO_2$ is relatively high, pressure effects on chemical equilibrium and reaction rates can be important. The effect of pressure on equilibrium constant $K_x$ (based on mole fraction $x$) and $K_C$ (based on concentration, C), at constant T, $x$ and C [4] are given by equations (2) and (3) respectively:

$$[(\ln K_x)/\partial P]_{T,x} = -(\Delta V_r/RT) \tag{2}$$

$$[(\ln K_C)/\partial P]_{T,C} = -(\Delta V_r/RT) + \kappa_T \Sigma \nu_i \tag{3}$$

In these equations, $\Delta V_r$ is the change in the reaction volume, given by the difference between the partial molar volumes of the product(s) and the reactant(s) and R is the universal gas constant. The equation for the pressure effect on concentration-based equilibrium constant (Eq. (3)) has an additional term that accounts for the volume effects, that is, the isothermal compressibility $\boldsymbol{\kappa}_T$ of the reaction mixture and the stoichiometric coefficients $\nu_i$. According to transition-state theory [4], for elementary reactions, the variation of the magnitude and direction of the mole fraction based rate constant $k_x$ with change in pressure is given by:

$$[(\ln k_x)/\partial P]_{T,x} = -(\Delta V^*/RT) \tag{4}$$

where $\Delta V^*$ is the difference between the partial molar volume of the activated complex and the sum of molar volume of reactants. The pressure dependent variation of concentration based rate constant, $k_C$ [4, 18] is given by:

$$[((\ln k_C)/\partial P]_{T,C} = -(\Delta V^*/RT) + [(1-n)/P]\{1-[\partial(\ln \kappa_T)/\boldsymbol{\partial}(\ln P)]_{T,C}\} \tag{5}$$

Equations (4) and (5) assume that the transition-state transmission coefficient is independent of pressure and is equal to unity. The above equations imply that a negative $\Delta V^*$ leads to an acceleration in rate with increase in P while a positive $\Delta V^*$ results in a retardation in rate. The value of $\Delta V^*$ for high pressure liquid phase reactions range between −30 and 30 $cm^3$ $mol^{-1}$. This represents approximately equal contribution from solvation and volume changes in the activated complex as a result of bond formation/breakage/other mechanistic features [4, 19] and corresponds to an acceleration/retardation in the rate by a factor of 3.5 between atmospheric pressure and a pressure of 100 MPa [20]. The knowledge of $\Delta V^*$ is useful in understanding the reaction mechanism of elementary reactions, in the absence of solvent contribution. However, solvation contribution is dominant near the critical region of the solvent where the large negative partial molar volumes of the reactants and transition states give rise to extremely large values of $\Delta V^*$. Hence it is not possible to extract mechanistic information from $\Delta V^*$ values in the critical region [4].

Most of the kinetic analysis of organic reactions in $ScCO_2$ presented in literature is based on the normal reaction mechanism (in the absence of $ScCO_2$) and use mole fraction for the treatment of data, instead of concentration. The general simplified assumption used implies that $CO_2$ modifies the phase equilibrium and does not influence the other typical steps in a multiphase reaction system.

## 6.4 REACTIONS INVOLVING $SCCO_2$ AS REACTANT

Although $ScCO_2$ is a preferred solvent in many organic syntheses due to its inert nature, it is a Lewis acid and reacts with strong bases (amines, phosphines, alkyl anions) [21, 22]. Initial studies reported on $CO_2$ activation in $ScCO_2$ involved the hydrogenation of $ScCO_2$ to formic acid and its esters [6, 23–25]. Hydrogenation reactions often use palladium catalyst, which can be poisoned by CO generated during $CO_2$ reduction above 343 K [6]. Reaction products and product distribution in high temperature catalytic hydrogenation vary with reaction conditions and constitute CO, $CH_4$, $CH_3OH$, formic acid derivatives [26]. Some of the reactions in $ScCO_2$ where $CO_2$ is the medium as well as the reactant include synthesis of styrene carbonate with 100% selectivity and yield using $ZnBr_2$ and the quaternary salt $n$-$Bu_4NI$ as catalyst [27], synthesis of poly(ether-carbonate) by copolymerization of $CO_2$ with oxiranes, which is considered to be a green alternative to the traditional phosgene route [28, 29], and formation methyl formate and formamide derivatives by $ScCO_2$ hydrogenation of methanol, ammonia and amines respectively [26, 30].

## 6.5 ORGANIC SYNTHESIS IN $SCCO_2$

The use of organic solvents as medium in chemical processes often generates disproportionately large amount of waste and lead to complex purification steps. Although the solvent is partly recycled, it is eventually either lost to environment or require disposal as waste. In addition to being a benign solvent, $ScCO_2$ enhances the solubility of gaseous reactants in the reaction mixture, increases catalyst activity and selectivity by eliminating interphase transport resistances. Reactions in $ScCO_2$ can be pressure tuned to optimize the catalyst activity, recovery and recycle and to maximize the product yield. This is true for both homogeneous and heterogeneous reactions [21]. Typical reactions include hydrogenation, hydroformylation, oxidation, alkylation, disproportionation and esterification [26, 30].

### 6.5.1 HYDROGENATION

Hydrogenation is one of the important reactions in organic synthesis and is carried out in gas or liquid phase. It is a catalytic reaction, where fluid/gas phase reactions use solid catalysts in a two-phase reaction medium. Liquid phase reactions are carried out in two (gas-liquid or liquid-solid) or three phase (gas-liquid-solid) reaction medium depending on whether the catalyst is homogeneous or heterogeneous. Hydrogenation reactions in $ScCO_2$ medium are carried out in batch/continuous mode and homogeneous/heterogeneous phase. The benefits of hydrogenations in $ScCO_2$ medium are enhanced reaction rates and product yields, higher catalyst efficiency, safer process and tuneable selectivity. The hydrogenation of *o*-chloronitrobenzene to *o*-chloroaniline [31] is presented as an example in Scheme 6.1.

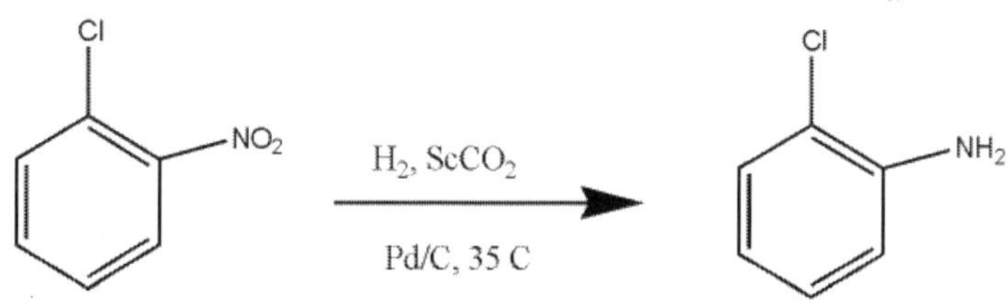


*o*-Chloronitrobenzene *o*-Chloroaniline

**SCHEME 6.1** Hydrogenation of *o*-chloronitrobenzene to *o*-chloroaniline.

**TABLE 6.2** Hydrogenation reactions in $ScCO_2$ using rhodium-based catalyst [5].

| Sr. No. | Reaction | Catalyst | Benefit of $ScCO_2$ | Ref. |
|---|---|---|---|---|
| 1. | Hydrogenation of *tert*-butylphenols | Rh/C | Higher *cis* ratio (0.95) | 33 |
| 2. | Hydrogenation of Tetralin | Rh/γ-$Al_2O_3$, Rh/C | High rate and high *cis*-decalin yield with Rh/C catalyst | 34 |
| 3. | Hydrogenation of 4-alkylphenols | Rh/C | High selectivity to *cis*-4-alkylcyclohexanols | 35 |
| 4. | Hydrogenation of 4-isopropylphenol | HCl-Rh/C and $H_3PO_4$-Rh/C | High yield of *cis*-4-isopropylcyclohexanols | 36 |
| 5. | Hydrogenation of biphenyl | Rh/C, Ru/C | >99% yield, TON = 1490 for Rh/C | 37 |

**TABLE 6.2** *(Continued)*

| Sr. No. | Reaction | Catalyst | Benefit of $ScCO_2$ | Ref. |
|---|---|---|---|---|
| 6. | Hydrogenation of isophorone | Rh, Pt, Pd on Charcoal, | High activity for Rh/Charcoal, Pt/Charcoal catalysts | 38 |
| 7. | Ring Hydrogenation of Phenol and cresol | Rh/Charcoal | Increase in catalyst activity, enhanced mass transfer | 39 |
| 8. | Ring hydrogenation of naphthalene and 1-naphthol | Rh, Pd, Pt, Ru supported on C or $Al_2O_3$ | High selectivity to decalin over Rh/C | 40 |
| 9. | Ring hydrogenation of phenyl ethanols | Rh, Pd, Pt, Ru supported on C or $Al_2O_3$ | High selectivity (>80%) to Cyclohexyl ethanols over supported Ru catalysts | 41 |
| 10. | Hydrogenation of naphthalene | Rh, Pd, Pt, Ru supported on C or $Al_2O_3$ | High activity for Rh/C catalyst, high TOF | 42, 43 |
| 11. | Tetrahydrofurfural alcohol to 1,5-pentanediol | Rh/MCM-41 | High conversion, selectivity under mild conditions | 44 |
| 12. | Stereoselective hydrogenation of itaconic acid esters | Rhodium-organic catalyst | Improved selectivity, high rate | 45 |

The concept of switchable selectivity in continuous hydrogenation of furfural in $ScCO_2$ by changing the operating conditions was demonstrated by Poliakoff and group [32]. The product tuneability based on reaction temperature for this reaction is presented in Scheme 6.2. Literature reveals that Rhodium supported on carbon has been extensively employed as catalyst in hydrogenation reactions in $ScCO_2$. This is followed by palladium, platinum, ruthenium, nickel, iridium and bifunctional catalysts viz. palladium-copper. Most of the supported catalysts use carbon in different forms as support. Alumina, calcium carbonate and MCM-41 have also been used as supports for rhodium based catalysts. Some hydrogenation reactions in $ScCO_2$ using rhodium catalyst are presented in Table 6.2. Examples of hydrogenation reactions in $ScCO_2$ using platinum, palladium and other catalysts are presented in Table 6.3.

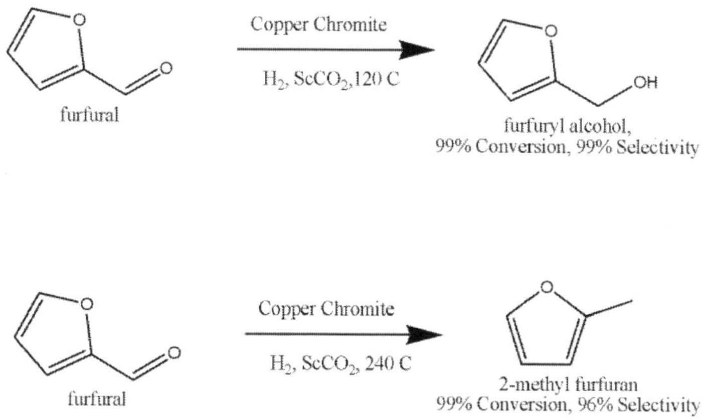


**SCHEME 6.2** Switchable selectivity in functional hydrogenation.

**TABLE 6.3** Hydrogenation reactions in $ScCO_2$ using platinum, palladium, and other catalysts [5].

| Sr. No. | Reaction | Catalyst | Benefit of $ScCO_2$ | Ref. |
|---|---|---|---|---|
| 1. | Hydrogenation of Sertraline | $Pd/CaCO_3$ | High selectivity towards *Cis* products | 46 |
| 2. | Isophorone to trimethylcyclohexanone | Pd/alumina | Efficient reaction, high rate | 47 |
| 3. | Hydrogenation of Furfural | Pd/C | Switchable selectivity to 5 different products by varying the reaction conditions | 32 |
| 4. | Hydrogenation of Pinene | Pd/C | Significantly faster reaction rates, safer operation | 48 |
| 5. | Cyclohexene to cyclohexane | Pd/C | High catalyst activity and catalyst life | 49 |
| 6. | *o*-Chloronitrobenzene to *o*-chloroaniline | Pd/C | High reaction rate and product selectivity | 31 |
| 7. | Hydrogenation of 2-butyne-1,4-diol | 5 wt% Pd/C | High rate, 100% selectivity to butane-1,4-diol | 50 |

**TABLE 6.3** *(Continued)*

| Sr. No. | Reaction | Catalyst | Benefit of $ScCO_2$ | Ref. |
|---|---|---|---|---|
| 8. | Styrene oxide to 2-phenyl ethanol | Pd-Cu catalyst encapsulated with polyurea | 100% conversion, 100% selectivity | 51 |
| 9. | Aldol condensation followed by hydrogenation in a single reactor | Bifunctional Pd-acid catalyst | High activity, selectivity, long catalyst life | 52–55 |
| 10. | Hydrogenation of isophorone | Rh, Pt, Pd on Charcoal, | High activity for Rh/Charcoal, Pt/Charcoal catalysts | 38 |
| 11. | Ring hydrogenation of naphthalene and 1-naphthol | Rh, Pd, Pt, Ru supported on C or $Al_2O_3$ | High selectivity to decalin over Rh/C | 40 |
| 12. | Ring hydrogenation of phenyl ethanols | Rh, Pd, Pt, Ru supported on C or $Al_2O_3$ | High selectivity (>80%) to Cyclohexyl ethanols over supported Ru catalysts | 41 |
| 13. | Hydrogenation of naphthalene | Rh, Pd, Pt, Ru supported on C or $Al_2O_3$ | High activity for Rh/C catalyst, high TOF | 42, 43 |
| 14. | Hydrogenation of 1,2-Epoxy-ethylbenzene | Pd/C, Pt/C | Suppression of dehydroxylated product | 56 |
| 15. | Hydrogenation of unsaturated aldehydes | $Pt/Al_2O_3$ | High selectivity for unsaturated alcohol | 57 |
| 16. | Nitrostyrene to aminostyrene | $Pt/TiO_2$ | Increase in reaction rate, selectivity | 58 |
| 17. | Hydrogenation of vegetable oil and fatty acid | Nickel based catalyst | High yield | 59, 60 |
| 18. | Enantioselective hydrogenation of *N*-(1-phenylethylidene) aniline | Chiral Iridium | High rate (20 times faster than in organic solvent) | 61 |

The rate of hydrogenation of liquid phase hydrogenation reactions is generally proportional to the liquid phase $H_2$ concentration, which is often limited by the rate of diffusion of $H_2$ from gas to liquid phase near ambient pressure. Operation under supercritical conditions ensures complete miscibility between $H_2$ and $ScCO_2$ and has beneficial effect on the hydrogenation reactions. Platinum, palladium, rhodium, ruthenium and iridium complexes have been used as homogeneous catalysts in hydrogenation reactions. The reaction mechanism in transition metal catalyzed hydrogenations is either by insertion or by hydrogen transfer (radical mechanism). Insertion is favored in hydric metal complexes of rhodium, ruthenium, etc. while radical mechanism is common in acidic metal hydride catalysts such as $MnH(CO)_5$ and $CoH(CO)_4$ [6]. In the case of hydric metal complexes in $ScCO_2$, the insertion of $CO_2$ into the metal-hydride bond of the catalyst can lead to the formation of formate complexes [24, 25] that can inhibit the catalyst activity, unless they rapidly revert to hydride and $CO_2$ [6].

Homogeneous hydrogenation reactions are carried out using neutral transition metal complexes or highly selective positively charged cationic complexes [62]. While neutral transition metal complexes have high solubility in $ScCO_2$, cationic complexes are not readily soluble. The solubility of cationic complexes can be considerably enhanced by its perfluorination with moieties such as $C_6F_{13}$, with no compromise on the performance [45]. The substrate often acts as a cosolvent enhancing the catalyst solubility in $ScCO_2$ and the reaction rate [62]. Liquid phase hydrogenation in $ScCO_2$ eliminates transport resistances leading to high reaction rate and product yield. Further, it is possible to have kinetic and selectivity control by changing the gas phase $CO_2/H_2$ ratio as well as the operating conditions like pressure and temperature.

Carrying out catalytic reactions in heterogeneous phase using solid catalyst has advantages like ease of catalyst separation, recovery, regeneration and recycle. Hence the soluble metal complex catalysts are being increasingly replaced by supported metal catalysts in hydrogenation reactions in the presence of $ScCO_2$. Tables 6.2 and 6.3 reveal that Pt, Pd, Rh, Ru, Ni supported on different inorganic materials have been used for the hydrogenation reactions in $ScCO_2$ with charcoal, active carbon, graphite, alumina, calcium carbonate, titanium dioxide, MCM-41 and polyurea have been used as supports for hydrogenation catalysts [5].

Stereoselective hydrogenation reactions in $ScCO_2$ lead to the formation of the *cis*-products with high selectivity and yield [33–36, 46, 61]. Control of hydrogen concentration in the highly exothermic hydrogenation of alkenes and alkynes by use of $CO_2$ is an important aspect for safe operation as well as product selectivity [6, 48]. Carbon dioxide inventory and its compression add to major operating costs for reactions in $ScCO_2$. The inventory and compression costs can be reduced by *in-situ* generation of $CO_2$ through decomposition of formic acid. This has been demonstrated in some hydrogenation reactions [63] without compromising on the product yield. Another development is the use of reactors for multiple reactions in sequence as part of process intensification effort. This has been carried out for Aldol condensation followed by hydrogenation in a single reactor using bi-functional palladium catalyst with the advantages of high activity, selectivity and long catalyst life [52–55].

Kinetics of hydrogenation are analyzed typically as multiphase reactions using Langmuir-Hinshelwood, power law, Langmuir-Hinshelwood-Hougen-Watson, etc. models [51, 64]. Although there are several advantages of carrying out hydrogenation reactions in $ScCO_2$, the CO formation and associated poisoning of noble metal catalysts is reported to result in loss of catalyst activity at temperature above 70 C and long reaction times [49].

### *6.5.2 HYDROFORMYLATION*

Addition of a hydrogen atom and formyl group (CHO) to a C=C bond occurs in hydroformylation or oxo synthesis. It is an important industrial process for the production of linear/branched aldehydes from alkenes. The other reactants are CO and $H_2$ and the reactions are usually carried out in liquid phase under homogeneous conditions. Aldehydes are important intermediates used for the production of a variety of organic chemicals viz., alcohols, carboxylic acids, aldol products, diols, cetals, ethers, acroliens and esters [65]. Examples of hydoformylation reactions are presented in Scheme 6.3(a) and (b). Commercial hydroformylation reactions use cobalt and rhodium complexes as catalysts and require an organic or aqueous solvent for catalyst recovery leading to significant post processing, additional expenses and environmental implications. Homogeneous hydroformylation reactions in $ScCO_2$ solvent provides ease of catalyst dissolution and recovery, control of phase behavior and tuning of reaction through small changes in reaction temperature and pressure [65]. In heterogeneous hydroformylations, it is possible to obtain enhancement in reaction rate, catalyst life and to control product distribution by using $ScCO_2$ as the medium of reaction. Recent research in hydroformylation reactions in $ScCO_2$ medium are presented in Table 6.4.

**TABLE 6.4** Typical examples of hydroformylation reactions.

| Sr. No. | Reaction | Catalyst | Benefit of $ScCO_2$ | Ref. |
|---|---|---|---|---|
| 1. | Hydroformylation of olefins | Rhodium complexes | Enhancement in reaction rate | 66 |
| 2. | Hexene to heptanal | Rhodium complex | Improved catalyst stability | 67 |

**TABLE 6.4** *(Continued)*

| Sr. No. | Reaction | Catalyst | Benefit of $ScCO_2$ | Ref. |
|---|---|---|---|---|
| 3 | 1-octene | Rhodium complex | Improved reaction rate and regioselectivity with increase in ligand concentration, High TOF, similar to that in organic solvents on use of $ScCO_2$. | 68, 69 |
| 4 | Hydroformylation of long chain alkenes | Rhodium complex/ homogeneous and immobilized | High activity for homogeneous catalyst | 70 |
| 5 | Styrene | Rh/(R, S)-3-$H_2F_6$-BINAPHOS | High conversion, selectivity, catalyst life | 71 |

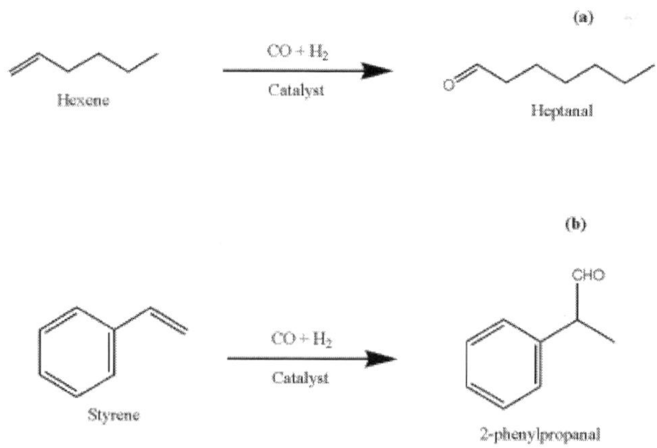


**SCHEME 6.3** Hydroformylation of hexane and styrene.

Problems in the recovery of expensive metal complex catalyst in homogeneous hydroformylation reactions directed the research to heterogenization of the soluble complex catalysts. This has been achieved by chemically bonding the active catalyst to inorganic or polymer supports, for example, silica, carbon, MCM-41. As heterogeneous hydroformylation catalysts are very expensive, they are used where the product selectivity is critical as in the case of pharmaceutical products [65]. The loss of active catalysts by leaching from the support to the reaction medium is a major problem faced by heterogeneous catalysts. Developments in hydroformylation reactions in supercritical fluids and the application of Nuclear Magnetic resonance (NMR) spectroscopy was reviewed by Niesse and Woelk [62]. Hydroformylation reactions have been conducted in a combination of $ScCO_2$ and ionic liquids [72, 73]. These systems work well for gas-phase reactions. Haumann and Riisagar [74] recently reviewed hydroformylation employing room temperature IL (RTIL). *In-situ* production of CO from $CO_2$ during the reaction is also reported in this review.

The kinetics of hydroformylaion in $ScCO_2$ has been studied using empirical/semiempirical rate equations. The kinetics of hydroformylation of propylene in $ScCO_2$ (homogeneous catalyst) was studied by Guo and Akgerman [75] using the empirical equation proposed by Cornils [76]

$$r = k\, C_p\, W_{cat}\, (P_{H2}/P_{CO}) \tag{6}$$

$$r = k'\, C_p \tag{7}$$

where

$$k' = k\, W_{cat}\, (P_{H2}/P_{CO}) \tag{8}$$

In the above equations, k is the rate constant, k' is the modified rate constant given by Eq. (8) for reactions where the ratio of $P_{H2}/P_{CO}$ is constant, $C_p$ is the concentration of propylene, $W_{cat}$ is the catalyst weight, and $P_{H2}$ and $P_{CO}$ are the partial pressures of $H_2$ and CO respectively. They reported a pressure dependent variation in rate constant.

Tadd and co-workers [67] examined the kinetics of hydroformylation of 1-hexene in $ScCO_2$, taking into account the complete reaction network for isomerization and aldehyde formation, considering each step to be a first order reaction in parallel or series. The complex set of rate expressions were numerically solved using a MATLAB program. Kinetics of hydroformylation in $ScCO_2$ for both homogeneous and heterogeneous catalysts was reviewed by Bektesevic [65]. The kinetic model for homogeneously catalyzed hydroformylation of 1-octene in $ScCO_2$ using a complex rhodium catalyst proposed by Palo and Erkey [77] is presented below:

$$r = k\, (C_{H2})^{0.48}\, (C_{cat})^{0.84}\, (C_{1\text{-octene}})^{0.5} / [1 + 0.69\, (C_{CO})^{2.2}] \tag{9}$$

Here $C_{H2}$, $C_{cat}$, $C_{1\text{-octene}}$, $C_{CO}$ are the concentrations in mol/L of $H_2$, catalyst, 1-octene and CO respectively and k is the reaction rate constant. Kinetic studies of 1-octene hydroformylation on arylphosphine modified rhodium complex indicated that the hydrogen concentration dependence on the reaction rate was nearly first order [78, 79]. In general, the kinetic studies using different solvents revealed that the reaction rate was five times faster in $ScCO_2$ than in toluene [65]. Tadd et al. studied the hydroformylation of 1-hexene in $ScCO_2$ using rhodium complex anchored on phosphinated silica [80]. The initial rate expression proposed by them, presented as Eq. (10), suggested that the immobilized system was behaving like a tethered homogeneous system [65].

$$r_0 = K_A\, C_{H2,\,0}\, C_{CO,\,0}\, C_{1\text{-hexene},\,0}/[1 + K_C\, C_{1\text{-hexene},\,0} + (1 + K_B\, C_{CO,\,0})^2] \quad (10)$$

In Eq. (10), $K_A$, $K_B$ and $K_C$ are model parameters, and $C_{H2,\,0}$ $C_{CO,\,0}$ $C_{1\text{-hexene},\,0}$ are the initial concentrations of $H_2$, CO and 1-hexene, respectively. Lin and Akgerman [81] studied the hydroformylation of styrene over rhodium catalyst bound to a fluoroacrylate copolymer backbone through phosphine ligands and developed a rate expression for the reaction. This is given by:

$$r = k\, C_S\, W_{cat}/(K_1 + K_2\, C_S) \quad (11)$$

$C_S$ is the concentration of styrene in the reactor, $W_{cat}$ is the weight of the catalyst, k is the reaction rate constant and $K_1$ and $K_2$ are functions that include the concentrations of $H_2$ and CO.

### 6.5.3 OXIDATION

Oxidation reactions in $ScCO_2$ offer safer operation as it cannot be over oxidized like the organic solvents. Therefore side product formation due to the oxidation of organic solvents is eliminated for reactions in $ScCO_2$ [49]. In addition to the safety aspect, $O_2$ and $CO_2$ are completely miscible under supercritical conditions at all concentrations [9]. $CO_2$ acts as an inert diluent for molecular oxygen, improving the operational safety for oxidation reactions. There are minimum four critical components for the transition metal catalyzed oxidation reactions. These are the catalyst, the substrate, terminal oxidant and the medium. Some benign, inexpensive and popular terminal oxidants are molecular oxygen from air, hydrogen peroxide and hypochlorite salts. Some of the expensive oxidants used are iodobenzene and its variants, peroxy acids (e.g., chloroperoxybenzoic acid, monoperoxysulfuric acid), alkyl peroxides (e.g., *t*-BuOOH) and amine oxides. Manganese based catalysts are active with terminal oxidants like $H_2O_2$, $O_2$ and NaOCl. $H_2O_2$ and its derivatives are used as terminal oxidants in case of rhenium, titanium, vanadium and molybdenum complex catalysts. Cobalt, ruthenium, iron and chromium catalysts use oxygen as terminal oxidant while iron and chromium complexes use aqueous NaOCl as terminal oxidant along with a cosolvent. The advantages of these systems have been discussed by Musie et al. [9] $ScCO_2$-expanded solvent media offer higher oxidant solubility without compromising on the catalyst solubility. Hence an optimized mixture of solvent and $ScCO_2$ is found to give a higher reaction performance compared to the use of either the solvent or $ScCO_2$ in isolation [5].

Catalytic oxidation of organic substrates in $ScCO_2$ both in homogeneous and heterogeneous phase has been reviewed by Seki and Baiker [81]. Continuous/semi continuous oxidation reactions in heterogeneous phase are typically carried out at temperature greater than or equal to 200 C and use a range of oxide and mixed oxide catalysts (Co, Mo, W, Ni, Si, Cu, Mn, etc.) supported on $Al_2O_3$ or carbon, $Pt/TiO_2$ etc. Partial oxidation of aromatic/aliphatic hydrocarbons to alcohols/aldehydes, olefin epoxidation, olefin to alcohol and ether, etc., are some examples of oxidation reactions in $ScCO_2$. There is an increasing interest on photocatalytic reaction in $ScCO_2$, which is a comparatively new research effort. The oxidation of geraniol to citral is presented in Scheme 6.4. Some research on organic oxidation reactions in $ScCO_2$ is presented as Table 6.5. The superier performance of $ScCO_2$ is evident from the presented data.

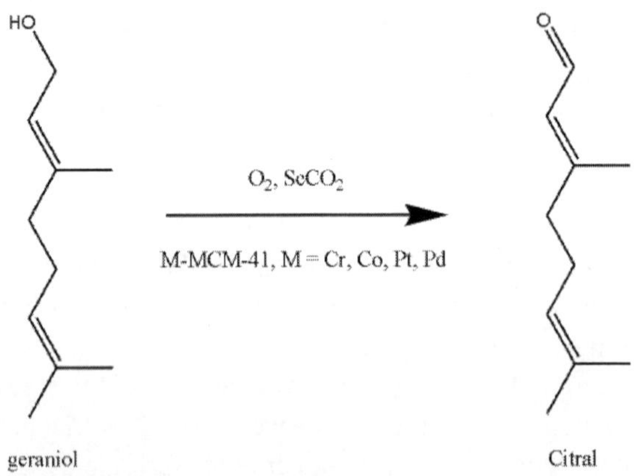


**SCHEME 6.4** Oxidation of geraniol to citral.

**TABLE 6.5** Examples of organic oxidation reactions in $ScCO_2$ [5].

| Sr. No. | Reaction | Catalyst | Advantage of $ScCO_2$ | Ref. |
|---|---|---|---|---|
| 1. | Oxidation of benzyl alcohol | 0.5 wt% Pd/ Alumina or Pd/C | High rate, 93–97% selectivity to benzaldehyde, ease of catalyst recycle | 82 |
| 2. | Oxidation of alcohols to aldehydes and ketones (benzyl alcohol to benzaldehyde) | $Au/TiO_2$ | 97% conversion and 95% selectivity to aldehyde, ease of catalyst recycle | 83 |

**TABLE 6.5** *(Continued)*

| Sr. No. | Reaction | Catalyst | Advantage of $ScCO_2$ | Ref. |
|---|---|---|---|---|
| 3. | Oxidation of alcohols to aldehydes | $Au/Pd/TiO_2$, Pt/C, Pd/C | High selectivity to aldehyde, ease of catalyst recyle | 83, 84 |
| 4. | Oxidation of 1-and 2-propanol | Pt/C | Selective formation of ketone at mild temperature, longer catalyst life | 85 |
| 5. | Oxidation of primary and secondary alcohols to corresponding carbonyl compounds | Chromium trioxide /$SiO_2$ | No solvent, increase in catalyst stability, ease of catalyst recycle | 86 |
| 6. | Oxidation of geraniol | M-MCM-41, M=Cr, Co, Pt, Pd | Enhancement in conversion and yield of citral, ease of catalyst recycle. | 87 |

### *6.5.4 C-C BOND FORMATION, ALKYLATION, ACETYLATION, ESTERIFICATION*

C-C bond formation reactions in $ScCO_2$ include Heck-Mizoroki olefin formation from aryl halide, cyclotrimerization of alkynes, Suzuki-, Sonogashira- and Stille-coupling reactions using Co or Pd complexes and have been reviewed by Ikariya and Kayaki [88]. The mechanism of C-C bond formation is similar in $ScCO_2$ and in organic solvents, but higher conversion and selectivity are usually observed in $ScCO_2$. *tert*-Butylation of naphthalene is presented in Scheme 6.5 as an example. The scheme indicates the possibility of formation of a variety of primary and secondary products in the alkylation reaction. High conversion and selectivity to 2,6-dialkylated product was observed when the reaction was carried out in $ScCO_2$ over ReY catalyst [89]. Esterification of oleic acid to methyl oleate is presented as an example in Scheme 6.6. Esterification of oleic acid and ethanol over lipozyme IM 60 immobilized on duolite is reported to give similar or higher rates in $ScCO_2$ compared to the organic solvent [90].

Esterification/transesterification of fatty acids/triglycerides in $ScCO_2$ for biodiesel production is an area of increasing research interest. Vegetable oils/

fats are hydrolyzed to give fatty acids, which on esterification with methanol produce the methyl esters for the generation of energy (biodiesel). An enhancement in catalyst activity and reaction rate in $ScCO_2$ has been reported in the enzyme catalyzed/noncatalytic transesterification of several vegetable oils, for example, castor and linseed oil [91], sesame and mustard oil [92], corn oil [93], sunflower oil [94], menhaden oil [95], rapeseed oil [96], etc., with methanol/ethanol. Transesterification of seasame and mustard oil are reported to proceed to 70% conversion in $ScCO_2$ while 100% conversion could be obtained when supercritical methanol/ethanol was used as the medium [92]. High yields and faster kinetics in $ScCO_2$ make the enzyme catalyzed (immobilized Novozym 435 catalyst) continuous synthesis of fatty acid methyl esters a potential process for biodiesel production [93, 94].

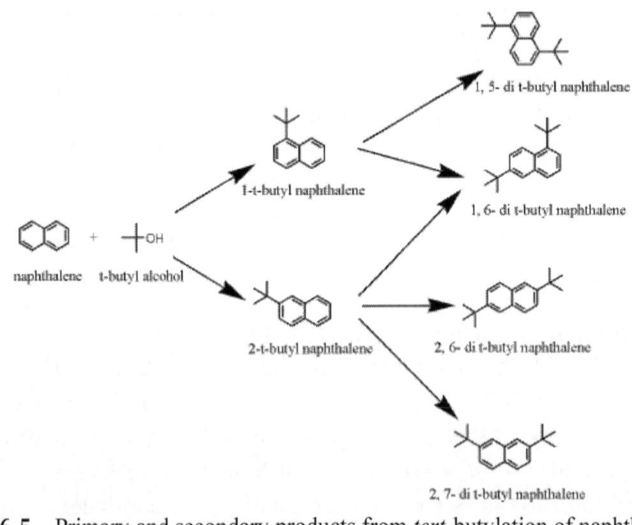


**SCHEME 6.5** Primary and secondary products from *tert*-butylation of naphthalene.

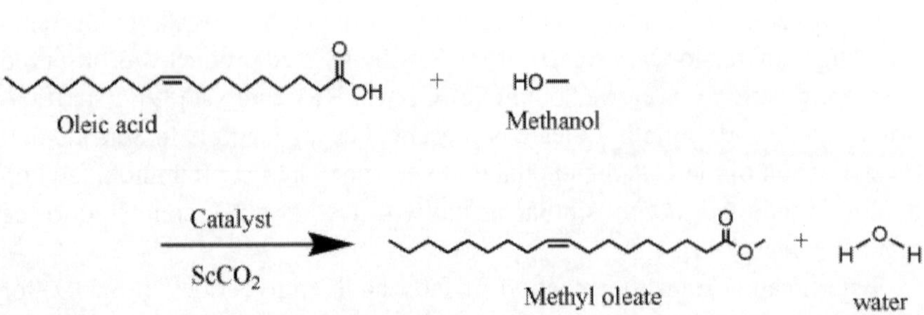


**SCHEME 6.6** Esterification of oleic acid and methanol to methyl oleate.

## 6.6 PROCESS INTENSIFICATION AND SCALE-UP ASPECTS

The most successful application of $ScCO_2$ is the decaffeination of coffee beans [55]. The main reason for its success is the saving in energy costs, as caffeine can be removed from compressed $CO_2$ without release of pressure, by extraction with water. $ScCO_2$ separation is used in food industry for other processes too as $CO_2$ is nontoxic.

There has been considerable research in organic reactions in $ScCO_2$. Initial research in this area was related to replacement of conventional organic solvents, elimination of interfacial mass and heat transport resistance, and enhancment in solubility of homogeneous catalyst in gas-liquid organic reactions. The focus soon shifted to heterogenization of homogeneous catalyst by immobilization on different supports and use of metal oxide catalysts in heterogeneous phase. Leaching of the active catalyst to the solution was a serious issue in the case of immobilized metal complex catalysts.

Although there is huge accumulation of research findings, technology development and commercial deployment has not kept pace with the advance in research. Most of the research work has been carried out in batch high-pressure reactors (autoclaves). Scale up of these autoclaves to commercial volumes involves very high capital investment. This is further complicated by the fact that the selectivities observed in small-scale operations (research) are often not reproducible on scale up as it is difficult to precisely control the phase changes at higher scale of operation. The product recovery at the end of reaction is by the depressurization of $CO_2$, which has to be recompressed and recycled. This tremendously increases the energy cost for recompression, in case a side stream of the required pressure is not already available elsewhere in the plant, making the process economically unattractive.

Continuous flow reactions are considered to be more sustainable in terms of quality, safety and economic perspective [97]. Continuous reactions in $ScCO_2$ have been critically reviewed by Hans and Poliakoff [55]. The major energy burden of a packed bed supercritical flow reactor comes from the $CO_2$ compression to supercritical conditions. This can be reduced if a $CO_2$ stream of similar pressure is available elsewhere in the plant. Although this is not realized at present, this may become possible as the $CO_2$ Capture and Storage (CCS) technology matures.

An innovative way of facing this challenge is sequencing of multiple reactions in the same reactor without release of pressure, so that the energy cost is distributed. Several examples are presented in the review by Hans and Poliakoff [55]. An example suggested is the dehydration of cyclohexanol to cyclohexene and its reaction with phenol to form the alkylate, cyclohexyl phenol using a common catalyst, $Al_2O_3$ in a series of two reactors with a water trap in-between to prevent the catalyst deactivation due to water [98].

Controlled phase changes by manipulation of the reactor conditions for product separation, without release of pressure has been demonstrated by Bourne et al. [99] for the separation of pure $\gamma$ -valerolactone (product) in the hydrogenation of levulinic acid. They used water as a cosolvent to effect the phase separation. In another example, density based change in solvent properties of $CO_2$ for the catalyst was used in a reactor with two temperature zones [100]. At lower temperature zone, the $CO_2$ density was sufficient to dissolve the catalyst, substrate and the reaction takes place in the homogeneous phase with the formation of products. As the stream flows towards the higher temperature zone, $CO_2$ density decreases sufficiently to precipitate the catalyst; it falls down to the cooler reaction zone and gets dissolved. The product solubility is not affected by the change in temperature and is carried off along with $CO_2$. Combination of Ionic Liquid and $ScCO_2$ for hydrogenation reactions and $ScCO_2$ expanded solvents for oxidation reactions are also reported to have synergistic effect compared to the use of either in isolation.

## 6.7 CONCLUSIONS

$ScCO_2$ is a highly attractive medium for organic synthesis since it is nontoxic, inert towards oxidation, inexpensive, readily available and environmentally benign. Near critical point $CO_2$ has dissolving power comparable to that of liquids, is more compressible than dilute gases, and has transport properties intermediate between that of gases and liquids. This combination of physical properties is exploited in organic syntheses in $ScCO_2$ to achieve high reaction rates, improved selectivity and elimination of interphase mass transfer resistances. Organic reactions using $ScCO_2$ as a reactant and/or reaction medium are discussed in this chapter. $ScCO_2$ is used either as a replacement of organic solvent medium or in combination with organic solvents/ionic liquids in gas-liquid and liquid-liquid systems. Specific reactions discussed are hydrogenation, hydroformylation, oxidation, carbon-carbon coupling, alkylation, acetylation and esterification with evidence of enhanced fluid and catalyst solubilities, minimum interface mass transfer resistances and improved transport phenomena. An analysis of reaction equillibria and kinetics of these chemical processes mediated by $ScCO_2$ has demonstrated its ability to modify the phase equilibrium phenomenon without influencing other typical systems in the multiphase process systems. The reported literature on the process intensification and scale up of $ScCO_2$ mediated reactions has shown that their technology development and commercial deployment have not kept pace with the research advances.

## KEYWORDS


- **Supercritical fluids**
- **Spercritical $CO_2$**
- **Organic synthesis**
- **Homogeneous catalytic reactions**
- **Heterogeneous catalytic reactions**
- **Supercritical $CO_2$ as reactant**
- **Hydroformylation**
- **Process intensification**


## REFERENCES


1. Jessop, P. G.; Leitner, W. E. Chemical Synthesis Using Supercritical Fluids; Wiley-VCH: Weinheim, 1999.
2. http://upload.wikimedia.org/wikipedia/commons/1/13/Carbon_dioxide_pressuretemperature_phase_diagram.svg.
3. Randolph, T. W.; Blanch, H. W.; Prausnitz, J. M.; Wilke, C. R. Biotechnol. Lett., 7, 325 (1985).
4. Baiker, A. Chem. Rev., 99, 453 (1999).
5. Mayadevi, S. Ind. J. Chem. 51A 1298 (2012).
6. Jessop, P. G.; Ikaria, T.; Noyori, R. Chem. Rev. 1999, 99, 475 (1999).
7. Kendall, J. L.; Canelas, D. A.; Young, J. L.; DeSimone, J. M. Chem. Rev., 99, 543 (1999).
8. Mesiono, A. J.; Beckman, E. J.; Russel, A. J. Chem. Rev., 99, 623 (1999).
9. Musie, G.; Wei, M.; Subramaniam, B.; Busch, D.H. Coord. Chem. Rev., 2–19–221, 789 (2001).
10. Gallagher, P. M.; Coffey, M. P.; Krukonis, V. J.; Klasutis, N. American Chemical Symposium Series, vol. 406, American Chemical Society, Washington, DC, 1989.
11. De la Fuente Badilla, J. C.; Peters, C. J.; Arons, J.; de Swaan. J. Supercrit. Fluids, 17 (1), 13 (2000).
12. Tai, C. Y.; Cheng, C.-S. AIChE J., 44 (4), 989 (1998).
13. Debenedetti, P. G.; Reid, R. C. AIChE J., 32, 2034 (1986).
14. Reddy, S. N.; Madras, G. Sep. Purif. Technol., 89, 181 (2012).
15. Škerget, M.; Knez, Z.; Hrncic, M. K. J. Chem. Eng. Data, 56, 694 (2011).

16. Cho, D. W.; Shin, J.; Shin, M. S.; Bae, W.; Kim, H. J. Chem. Thermodynamics, 47, 177 (2012).
17. Cho, D. W.; Shin, J.; Bae, W.; Kim, H.; Lee, J. H.; Shin, M. S. Fluid Phase Equilib, 319, 37 (2012).
18. Dooley, K. M.; Brodt, S. R.; Knopf, F. C. Ind. Eng. Chem. Res., 26, 1267 (1987).
19. van Eldik, R.; Asano, T.; le Noble, W. J. Chem. Rev., 89, 549 (1989).
20. Asano, T.; le Noble, W. J. Chem. Rev., 78, 407 (1978).
21. Inoue, S.; Yamazaki, N. Organic and Bioorganic Chemistry of Carbon Dioxide; Wiley and sons, New York, 1982.
22. West, K. N.; Wheeler, C.; McCarney, J. P.; Griffith, K. N.; Bush, D.; Liotta, C. L.; Eckert, C. A.; J. Phys. Chem. A, 105, 3947 (2001).
23. Niessen, H. G.; Woelk, K.; Top. Curr. Chem., 276, 69 (2007).
24. Jessop, P. G.; Ikaria, T.; Noyori, R. Nature, 368, 231 (1994).
25. Jessop, P. G.; Hsiao, Y.; Ikaria, T.; Noyori, R. J. Am. Chem. Soc., 118, 344 (1996).
26. Jessop, P. G.; Ikaria, T; Noyori, R. Chem. Rev., 95, 259 (1995).
27. Sun, J.; Fujita, S-I.; Zhao, F.; Arai, M.; Appl. Catal. A: General, 287, 221 (2005).
28. Kuran, W. Prog. Polym. Sci., 23, 919 (1998).
29. Cheng, M.; Moore, D. R.; Reczek, J. J.; Chamberlain, B. M.; Lobokovsky, E. B.; Coates, G. W. J. Am. Chem. Soc., 123, 8783 (2001).
30. Jessop, P. G.; Hsiao, Y.; Ikaria, T.; Noyori, R. J. Chem. Soc. Chem. Commun., 453 (1997).
31. Xi, C.; Cheng, H.; Hao, J.; Cai, S.; Zhao, F. J. Mol. Catal. A: Chemical, 282, 80 (2008).
32. Stevens, J. G.; Bourne, R. A.; Twigg, M. A.; Poliakoff, M. Angew. Chem. Int. Ed. 49, 8856 (2010).
33. Hiyoshi, N.; Rode, C. V.; Sato, O.; Tetsuka, H.; Shirai, M. J. Catal., 252,57 (2007).
34. Hiyoshi, N.; Mine, E.; Rode, C. V.; Sato, O.; Shirai, M. Appl. Catal. A: General, 310, 194 (2006).
35. Hiyoshi, N.; Bando, K. K.; Sato, O.; Yamaguchi, A.; Rode, C. V.; Shirai, M. Catal. Commun., 10, 1702 (2009).
36. Hiyoshi, N.; Sato, O.; Yamaguchi, A; Rode, C. V. Green Chem., 14, 633 (2012).
37. Hiyoshi, N.; Rode, C. V.; Sato, O.; Shirai, M. Appl. Catal. A: General, 288, 4 (2005).
38. Sato, T.; Rode, C. V.; Sato, O.; Shirai, M. Appl. Catal. B: Environmental, 49, 181 (2004).
39. Rode, C. V.; Joshi, U. D.; Sato, O.; Shirai, M. Chem. Commun., 1960 (2003).
40. Shirai, M.; Rode, C. V.; Mine, E.; Sasaki, A.; Sato, O.; Hiyoshi, N. Catal. Today, 115, 248 (2006).
41. Sato, T.; Sato, O.; Arai, K.; Mine, E.; Hiyo, N.; Rode, C. V.; Shirai, M. J. Supercrit. Fluids, 37, 87 (2006).
42. Hiyoshi, N.; Osada, M.; Rode, C. V.; Sato, O.; Shirai, M. Appl. Catal. A: General, 331, 1 (2007).
43. Hiyoshi, N.; Yamaguchi, A.; Rode, C. V.; Sato, O.; Shirai, M. Catal. Commun., 10, 1681 (2009).
44. Chatterjee, M.; Kawanami, H.; Ishizaka, T.; Sato, M.; Suzuki, T.; Suzuki, A. Cat. Sci. Technol., 1, 1466 (2011).
45. Lange, S.; Brinkmann, A.; Trautner, P.; Woelk, K.; Bargon, J.; Leitner, W. Chirality 12, 450 (2000).

46. Clark, P.; Poliakoff, M.; Wells, A. Adv. Synth. Catal., 349, 2655 (2007).
47. Devetta, L.; Giovanzana, A.; Canu, P.; Bertucco, A.; Minder, B. J. Catal. Today, 48, 337 (1999).
48. Chouchi, D.; Gourgouillon, Courel, M.; Vital, J.; da Ponte, M. N. Ind. Eng. Chem. Res., 40, 2551 (2001).
49. Beckman, E. J. J. Supercrit. Fluids, 28, 121 (2004).
50. Kriaa, K.; Serin, J-P.; Contamine, F.; Cezac, P.; Mercadier, J. J. Supercrit. Fluids, 49, 227 (2009).
51. Yadav, G. D.; Lawate, Y. S. J. Supercrit. Fluids, 59, 78 (2011).
52. Seki, T.; Grundwaldt, D.; Baiker, A.; Chem. Commun., 3562 (2007).
53. Seki, T.; Grundwaldt, D.; van Vegten, N.; Baiker, A. Adv. Synth. Catal., 350, 691 (2008).
54. Stevens, J. G.; Bourne, R. A.; Poliakoff, M. Green Chem., 11, 409 (2009).
55. Han, X.; Poliakoff, M. Chem. Soc. Rev., 41, 1428 (2012).
56. Rode, C. V.; Hiyoshi, N.; Mine, E.; Shirai, M. Ind. Eng. Chem. Res., 48, 9457 (2009).
57. Bhanage, B. M.; Ikushima, Y.; Shirai, M.; Arai, M. Cat. Lett., 62, 175 (1999).
58. Fujita, S-I.; Yoshida, H.; Asai, K.; Meng, X.; Arai, M. J. Supercrit. Fluids, 60, 106 (2011).
59. Andersson, M. B. O.; King, J. W.; Blomberg, L. G.; Green Chem., 2, 230 (2000).
60. King, J. W.; Holliday, R. L.; List, G. R.; Snyder, J. M. J. Am. Oil Chem. Soc, 78, 107 (2001).
61. Kainz, S.; Brinkmann, A.; Leitner, W.; Pfaltz, A. J. Am. Chem. Soc., 121, 6421 (1999).
62. Neissen, H. G.; Woelk, K. Top. Curr. Chem., 276, 69 (2007).
63. Hitzler, M. G.; Smail, F. R.; Ross, S. K.; Poliakoff, M. Org. Proc. Res. Dev., 7, 357 (2005).
64. Phiong, H-S; Lucien, F. P.; Adesina, A. A. J. Supercrit. Fluids, 25, 155 (2003).
65. Bektesevic, S.; Kleman, A. M.; Marteel-Parrish, A. E.; Abraham, M. A. J. Supercrit. Fluids, 38, 232 (2006).
66. Koch, D.; Leitner, W. J. Am. Chem. Soc., 120, 13398 (1998).
67. Tadd, A. R.; Marteel, A.; Mason, M. R.; Davies, J. A.; Abraham, M. A. J. Supercrit. Fluids, 25, 183 (2003).
68. Koeken, A. C. J.; van Den Broeke, L. J. P.; Benes, N. E.; Keurentjes, J. T. F.; J. Mol. Catal. A: Chemical, 346, 1 (2011).
69. Koeken, A. C. J.; van Den Broeke, L. J. P.; Benes, N. E.; Keurentjes, J. T. F.; J. Mol. Catal. A: Chemical, 346, 94 (2011).
70. Disser, C.; Muennich, C.; Luft, G. Appl. Catal. A: General, 296, 201 (2005).
71. Francio, G.; Wittmann, K.; Leitner, W. J. Organomet. Chem., 621, 130 (2001).
72. Sellin, M. F.; Webb, P. B.; Cole-Hamilton, D. J. Chem. Commun., 781 (2001).
73. Webb, P. B.; Sellin, M. F.; Kunene, T. E.; Williamson, S.; Slawin, A. M. Z.; Cole-Hamilton, D. J. J. Am. Chem. Soc., 125, 15577 (2003).
74. Haumann, M.; Riisager, A. Chem. Rev., 108, 1474 (2008).
75. Guo, Y.; Akgerman, A. Ind. Eng. Chem. Res., 36, 4581 (1999).
76. Cornils, B. Hydroformylation, Oxo Synthesis, Roelen Reaction. In *New Synthesis with Carbon Monoxide*; Falbe, J., Ed.; Springer-Verlag: Berlin, pp 16 and 17 (1980).
77. Palo, D. R.; Erkey, C. Ind. Eng. Chem. Res., 38, 3786 (1999).
78. Davis, T.; Erkey, C. Ind. Eng. Chem. Res., 39, 3671 (2000).

79. Banet Osuna, A. M.; Chen, W.; Hope, E. G.; Kemmitt, R. D. W.; Paige, D. R.; Stuart, A. M.; Xiao, J.; Xu, L. J. Chem. Soc., Dalton Trans., 4052 (2000).
80. Tadd, A. R.; Marteel, A.; Mason, M. R.; Davies, J. A.; Abraham, M. A. Ind. Eng. Chem. Res., 41, 4514 (2002).
81. Seki, T.; Baiker, A. Chem. Rev., 109, 2409 (2009).
82. Jenzer, G.; Mallat, T.; Baiker, A. Catal. Lett., 73, 5 (2001).
83. Wang, X.; Kawanami, H.; Dapurkar, S. E.; Venkataramanan, N. S.; Chatterjee, M.; Yokoyama, T.; Ikushima, Y. Appl Catal A: General, 349, 86 (2008).
84. Steele, A. M.; Zhu, J.; Tsang, S. C. Catal. Lett., 73, 9 (2001).
85. Glaser, R.; Josl, R.; Willardt, J. Topics Catal., 22, 31 (2003).
86. Gonzalez-Nunez, M. E.; Mello, R.; Olmos, A.; Aceete, R.; Asesio, G. J. Org. Chem., 71, 1039 (2006).
87. Dapurkar, S. E.; Kawanami, H.; Chatterjee, M.; Rode, C. V.; Yokoyama, T.; Ikushima, Y. Appl. Catal. A: General, 394, 209 (2011).
88. Ikariya, T.; Kayaki, Y. Catal. Surveys from Japan, 4, 38 (2000).
89. Marathe, R. P.; Mayadevi, S.; Pardhy, S. A.; Sabne, S. M.; Sivasanker, S. J. Mol. Cat., 181, 201 (2002).
90. Goddard, R.; Bosley, J.; Al-Duri, B. J. Chem. Technol. Biotechnol., 75, 715 (2000).
91. Varma, M. N.; Madras, G. Ind. Eng. Chem. Res., 46, 1 (2007).
92. Varma, M. N.; Deshpande, P. A.; Madras, G. Fuel, 89, 1641 (2010).
93. Ciftci, O. N.; Temelli, F. J. Supercrit. Fluids, 58, 79 (2011).
94. Rodrigues, A. R.; Paiva, A.; da Silva, M. G.; Simoes, P.; Barreiros, S. J. Supercrit. Fluids, 56, 259 (2011).
95. Shin, S-K.; Sim, J-E.; Kishimura, H.; Chun, B-S. J. Ind. Eng. Chem., 18, 546 (2012).
96. Galia, A.; Scialdone, O.; Tortorici, E. J. Supercrit. Fluids, 56, 186 (2011).
97. Wiles, C.; Watts, P. Green Chem., 14, 38 (2012).
98. Amandi, R.; Scovell, K.; Licence, P.; Lotz, T. J.; Poliakoff, M. Green Chem., 9, 797 (2007).
99. Bourne, R.A.; Stevens, J. G.; Ke, J.; Poliakoff, M. Chem. Commun., 4632 (2007).
100. Harwardt, T.; Francio, G.; Leitner, W. Chem. Commun., 46, 6669 (2010).

# PART IV

## INTENSIFICATION OF ORGANIC SYNTHESES

# CHAPTER 7

# AMMOXIDATION OF HETERO AROMATIC COMPOUNDS TO THE CORRESPONDING NITRILES

ANDREAS MARTIN and NARAYANA V. KALEVARU

## CONTENTS

## 7.1 INTRODUCTION

The conversion of various olefins, aromatic and hetero-aromatic compounds in the presence of air (oxygen) and ammonia towards their corresponding nitriles in a continuous process is termed as ammoxidation or ammonoxidation or oxidative ammonolysis [1]. It is a rather simple one-pot reaction leading from hydrocarbons or hetero-aromatics to nitriles that are mainly used as intermediates for the production of a wide variety of industrially important chemicals. Outstanding examples are the manufacture of acrylonitrile from propene, xylene dinitriles from *p*- and *m*-xylene, 2-cyanopyrazine from 2-methylpyrazine, nicotinonitrile from 3-picoline, etc. Scheme 7.1 shows the general stoichiometry of an olefinic and aromatic or hetero-aromatic reactant to their corresponding nitriles, respectively. The one-step reaction runs after a first H-abstraction via an allyl- and benzyl-like species formed in an activation step. Subsequently, an aldehyde stage is formed that reacts with an activated ammonia surface species towards a N-containing intermediate (imine [2, 3] or amine/amide [4, 5]) that is further oxidatively dehydrogenated to form the desired nitrile.

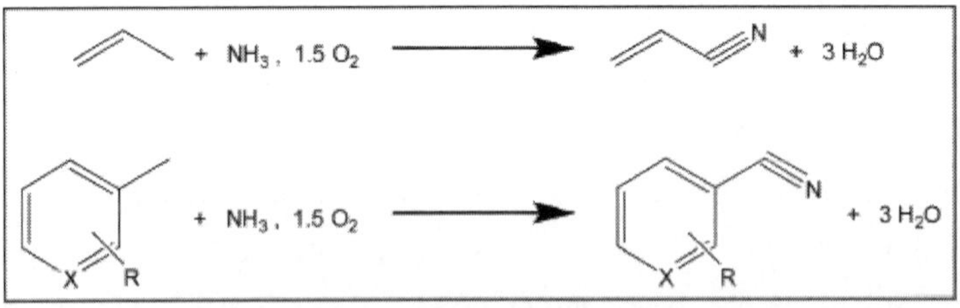


**SCHEME 7.1** Stoichiometry of the ammoxidation of an olefinic and aromatic or hetero-aromatic compound, respectively (R = –$CH_3$, –Cl, –$OCH_3$ and others, X = C, N, S).

Obviously, the only by-product is water making the reaction very "green." In general, conversions above 90% and product selectivity of 85–95% make the ammoxidation process very efficient and economical. However, carbon oxides can be formed as side-products due to overoxidation. The amount of these by-products however depends upon the reaction conditions and reactants to be converted [1]. Thus, the position of a substituent, its size and/or its electronic properties strongly influences the result. For example, the conversion of 2,6-dichloro toluene suffers from the presence of two large chlorine substituents on both the ortho positions of the ring, narrowing the accessibility of the methyl group being the reaction center. In addition, the formation of hydrogen cyanide is sometimes reported but the concentrations are rather insignificant, particularly in the case of aromatics and hetero-aromatics. However, larger amounts of HCN are only reported in case of conversion of olefins or alkanes.

In recent years, alkanes (e.g., ethane or propane) are subjected to ammoxidation leading to acetonitrile and acrylonitrile, respectively [6, 7]. In case of propane, the reactant activation runs via an oxidative dehydrogenation step by using bifunctional catalysts, for instance MoVNbTeO, GaVNbSbO and VPO solids [8, 9]. Pilot scale ACN manufacture from propane was achieved in early 2000. Asahi Kasei, Mitsubishi and BP Chemicals (INEOS) are reported to be active in developing such a process for ACN from propane in the past several years. In fact, Asahi Kasei has already run a plant with 70 thousand metric tons/annum capacity in 2006 in South Korea on a trial basis using propane as a feedstock but this plant was stopped after one year. They also planned to commission another plant in Thailand in 2009–2010 with a capacity of 200 thousand metric tons/annum. The best yield of ACN reported so far from propane feedstock is ca. 60% [10] while it is ≥80% from propylene [11].

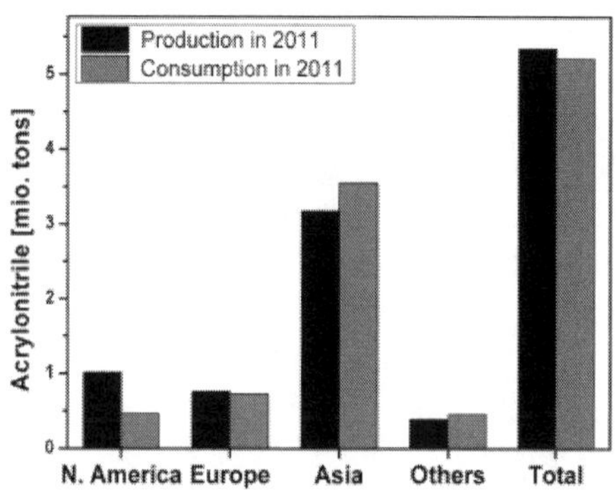


**FIGURE 7.1** Worldwide production and consumption of acrylonitrile in 2011.

Among all nitriles produced through ammoxidation, acrylonitrile occupies the most commanding position due to its high demand and large-scale production. ACN is used primarily to produce resins, fibers, acrylamide, adiponitrile, nitrile rubbers, etc. Figure 7.1 shows the worldwide production as well as consumption of acrylonitrile in 2011 amounting to over 5 million metric tons. An average annual consumption growth rate of acrylonitrile for the next four years is anticipated to be around 3%. Among all regions, Asia is the highest producer (3.2 million metric tons) and also the highest consumer (3.6 million metric tons) of this nitrile. Again, in Asia, China's contribution (1.3 million metric tons) is high followed by Japan. It is also obvious from the Fig. 7.1

that roughly 60% of world production capacity and close to 70% world's consumption is located in Asia. North America is the second largest producer of ACN, which accounts to ca. 20%. INEOS is the largest producer (~20%) of ACN followed by Asahi Kasei Corporation (>10%). Other producers are SOLUTIA and CNPC. Furthermore, some new plants with several hundred thousand metric tons capacity are also planned to be commissioned by SINOPEC, SABIC, CNPC and other companies.

In general, the ammoxidation reactions are carried out in the presence of a heterogeneous catalysts at elevated temperatures, slightly increased pressure and in the gas phase. Multicomponent oxidic catalysts, mainly containing transition metal oxides are in use. Most catalysts for olefin conversions consist of bismuth and molybdenum oxides. They can be doped by a wide variety of further transition metals. In addition, oxide components like silica, alumina and titania are incorporated as inert diluents or active constituents. The catalysts for the reactions of aromatics and hetero-aromatics mainly contain vanadium, molybdenum and other transition metals. In addition, the reaction can also be carried out in liquid phase. The reported literature on this subject is rather scanty.

The reaction follows the Mars-van Krevelen mechanism [12, 13, 14], that is, the oxygen needed for the oxidation reaction is liberated from the oxidic framework of the catalyst as $O^{2-}$. After adsorption of the reactant on the catalyst surface with the help of Lewis acidic sites, H-abstraction occurs and allyl or benzyl-like species are formed. The $O^{2-}$ species lead to further intermediate oxidation due to nucleophilic substitution into the activated reactant. The generated oxygen vacancies have to be refilled with gas phase oxygen (Scheme 7.2). Due to such transformations, an active catalyst must have good oxygen and electron mobility in the bulk. Otherwise, these properties have to be well balanced to avoid overoxidation and increased total oxidation.

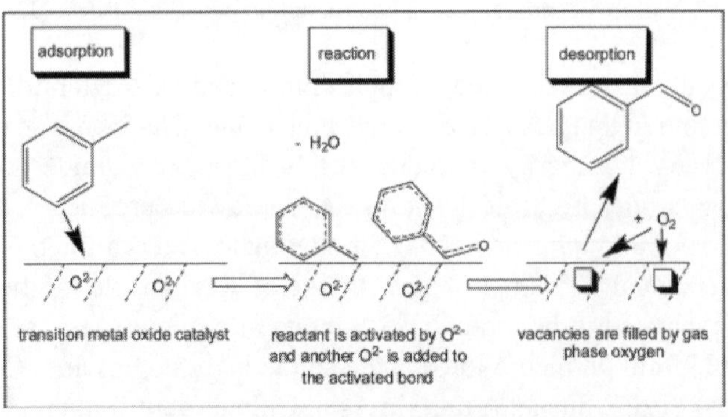


**SCHEME 7.2** Schematic description of the Mars-van Krevelen oxidation mechanism.

In addition, it was proven that the nitrogen of the nitrile group is also inserted from surface activated ammonia. For instance, pulse ammoxidation experiments using $^{15}N$-containing ammonia feed and $^{14}N$-containing ammonium vanadium phosphate catalyst showed the formation of $^{14}N$-containing nitrile during the first pulses. However, an increase of $^{15}N$-containing proportion in the nitrile product could be observed in subsequent runs with raising pulse numbers (Scheme 7.3) [15].

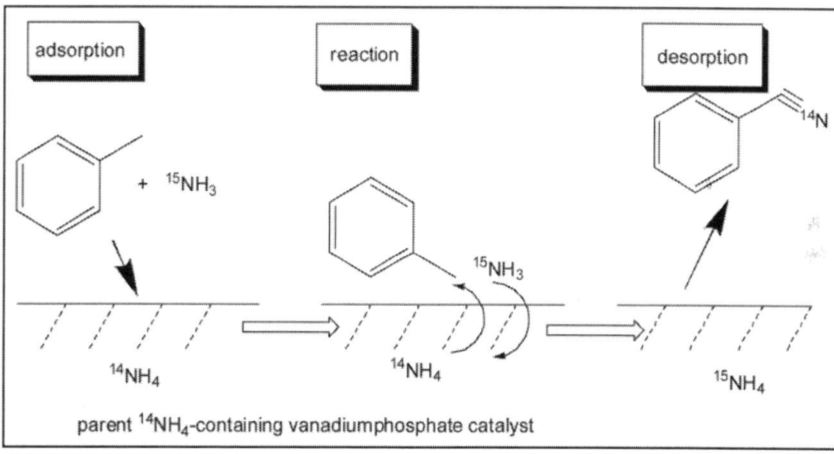


**SCHEME 7.3** Schematic description of the insertion of nitrogen according to the mechanism of Mars and van Krevelen [12] for oxygen insertion proven by $^{14}N$ vs. $^{15}N$ isotope experiments [1, 15].

The solid catalysts are mostly applied as bulk materials, mixtures of active constituents and inert diluent and eggshell catalysts using more or less inert cores of alumina or silica. These compositions are available as tablets, spheres, extrudates or spray-dried finer particles depending on the reactors to be used. Fixed bed multitubular and fluidized bed reactors are used. Outstanding industrially operated examples are the conversions of propene to acrylonitrile or 3-picoline to nicotinonitrile, for example. Scheme 7.4 depicts various examples on reactants, nitrile products and their subsequent uses.

The heterogeneously catalyzed ammoxidation is a reaction used by the industry for the last 60 years. A recent review provides a short historic survey [1]. Leonid Andrussow was the first to prepare nitriles and amides by gas phase oxidations in the presence of ammonia. He is also known for his contribution to Andrussow reaction, that is, the formation of hydrogen cyanide from methane, ammonia and oxygen [16, 17]. Another milestone was the invention of the ammoxidation of propene to acrylonitrile and its transfer to the industrial scale by Standard Oil of Ohio (SOHIO, USA) scientists in the mid-1950s [18]. Today, this reaction is processed in multimillion-ton scale as already stated above. Production of large-scale industrial hetero aromatic nitriles started in the early 1960s in the U.S. with first producers such as Nepera Chemical Co. and Reilly Industries. The latter opened their first industrial plant at Indianapolis, USA. Later, in 1980s Degussa AG (Germany) erected a first plant for the manufacture of nicotinonitrile in Europe at Antwerp (Belgium) [19]. Reilly Industries (USA), today part of Vertellus Specialties, a leading manufacturer of synthetic pyridines and final products, took over this plant 20 years later.

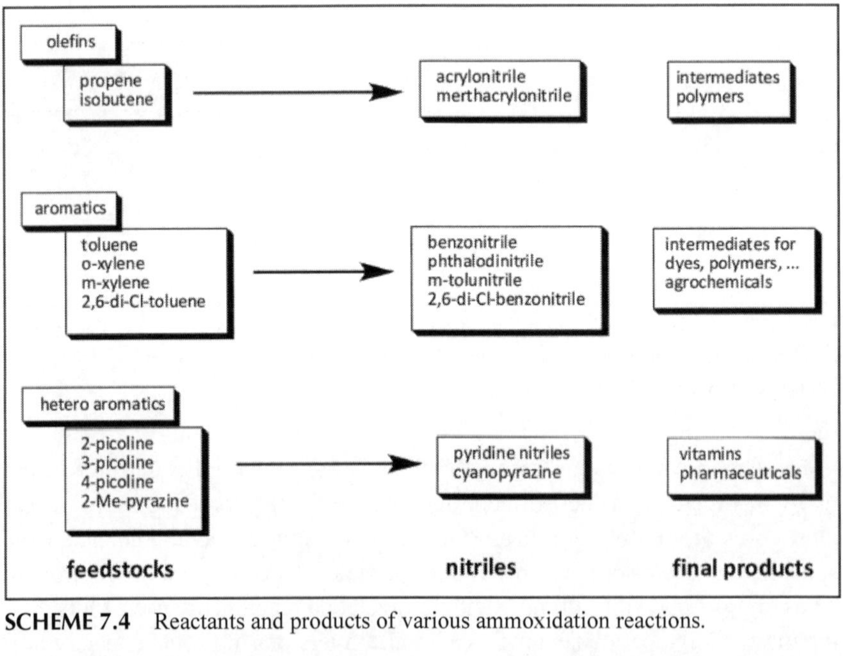


**SCHEME 7.4** Reactants and products of various ammoxidation reactions.

## 7.2 RELEVANCE OF HETERO AROMATICS AND THEIR NITRILES IN THE CHEMICAL INDUSTRY

The ammoxidation of hetero aromatics (and substituted methyl aromatics) opens the way for the synthesis of intermediates, specialties and fine chemicals. Various pharmaceuticals, agrochemicals, dyestuffs and other specialty products are produced from substituted aromatic or hetero aromatic nitrile intermediates [e.g., 1, 20]. Some of such reactions are already commercialized and run with a capacity of several thousand tons scale. Examples are the conversion of 2-, 3-, and 4-picolines and 2-methylpyrazine to the corresponding nitriles to be discussed below more in detail.

Before going to discuss the importance of hetero aromatic nitriles and their relevance to chemical industry, we would like to briefly describe some useful aspects of the feedstocks and their global consumption statistics. For instance, pyridines are good examples of hetero aromatics and their alkyl derivatives (e.g., alkyl pyridines, picoline isomers and 2-methyl-5-ethylpyridine) are used as feedstocks in the ammoxidation processes to produce different hetero aromatic nitriles that have good commercial significance. World consumption of pyridine derivatives increased from ~120 thousand metric tons in 2007 to 150 thousand metric tons in 2010, which is expected to increase further to >170 thousand metric tons by 2015 (Fig. 7.2a). It is also evident that the major production and consumption is concentrated in Asia followed by Europe. To our knowledge, no production is reported from Africa, Middle East and Oceania regions. The majority of the new capacity additions were installed in Asia in recent times to meet the high demand for niacinamide/niacin and paraquat (1,1′-dimethyl-4,4′-bipyridinium dichloride) in this region. In fact, such new capacity additions in India and China have led to clear enhancement of global production. Therefore, the consumption has grown remarkably in both India and China since 2007, while the U.S. has experienced only a marginal growth. VERTELLUS Specialties (USA), JUBILANT Life Sciences (India), RED SUN (China), and LONZA AG (Switzerland & China) are some of the major producers of these chemicals.

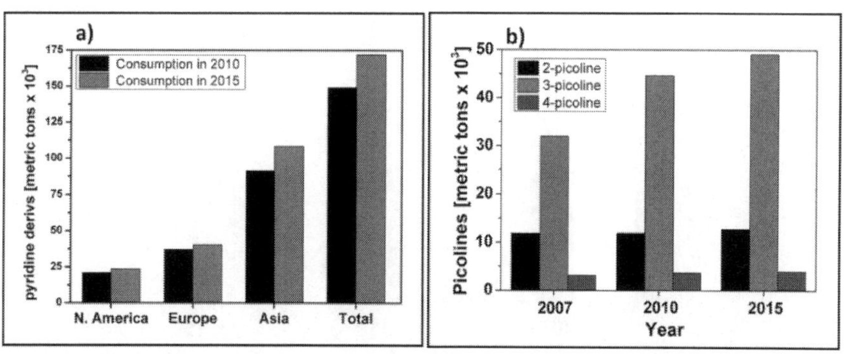


**FIGURE 7.2** Regional consumption of pyridine derivatives in 2010 & 2015 (a) and worldwide consumption of 2-, 3-, and 4-picolines in the period from 2007 to 2015 (b).

Among the pyridine derivatives, ca. 40% contribution comes from the picoline isomers and hence the worldwide consumption of all the three isomers of picolines (2-, 3- and 4-) is shown in Fig. 7.2b. Among all, 3-picoline (3-methylpyridine) occupies the dominating position and its contribution accounts to roughly 70–75% of total production of all the three picoline isomers. It is obvious from Fig. 7.2b that the demand for 3-picoline is significantly increasing with a clear rise in its production from 32 thousand metric tons in 2007 to 45 thousand metric tons in 2010 and then to ca. 50 thousand metric tons by 2015. Such increase is expected from the high demand of niacin particularly in Asian region. On the other hand, the demand for 2- and 4-picolines is not that high and is maintained more or less constant over a period from 2007 to 2015.

Among all pyridine nitriles, nicotinonitrile or 3-cyanopyridine (3CP) is the high-volume product and is widely used for various commercial applications. The majority of this nitrile is used for the production of niacin (nicotinic acid) and/or niacinamide (nicotinamide). However, certain amounts of niacin are also produced from other sources. Niacin (also known as vitamin $B_3$) is a derivative of pyridine, which is a color less and water-soluble solid. It is one of the main essential vitamins for metabolic activities and found in food stuff such as meat, fish, milk, eggs, cereals and green vegetables. It is also used as memory enhancer, to regulate cholesterol levels, to treat pellagra, to decrease the risk of cardiovascular issues as well as animal feed, etc. In addition, small amounts of 3CP are also used for preparing agricultural chemicals.

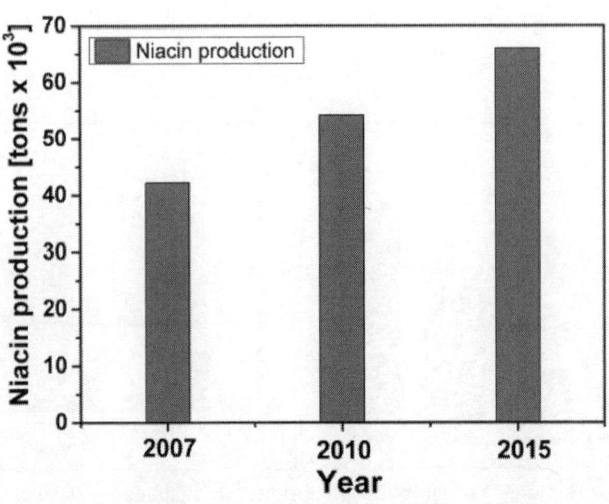


**FIGURE 7.3** Worldwide production of niacin during 2007 to 2015.

Considering the importance of niacin and its major production from 3CP through ammoxidation, the statistics on its worldwide production is illustrated in Fig. 7.3. The worldwide production of niacin and nicotinamide, mainly based on nicotinonitrile, amounted to 22 thousand metric tons in 1995. The production in 2007 was nearly doubled reaching 35 to 40 thousand metric tons [21]. A nice overview is given on the developments in nicotinate technology over the past 10–15 years by R. Chuck from Lonza in 2005 [22]. It is clear from Fig. 7.3 that the demand for niacin is growing continuously with time. The amount of niacin is increased from ca. 40 thousand metric tons in 2007 to 55 thousand metric tons in 2010, which is estimated to increase further to over 65 thousand metric tons by 2015. An average annual growth rate of 4% is expected for niacin until 2015.

The U.S. alone consumes about 10 thousand metric tons of niacin per annum, of which 70% is consumed for animal feed (feed supplements for dairy cattle, poultry, pet foods, etc.), ca. 20% for food (e.g., rice, flour, beverages, baby food), and the balance of 10% for cosmetics and pharmaceutical applications. Indian consumption of niacin/niacinamide was also high, which is expected to reach ca. 15 thousand metric tons by 2015.

In the United States, niacin is produced by the ammoxidation of 3P to 3CP, which upon hydrolysis gives rise to niacin. In this route, yields up to ~90% could be obtained. Vertellus is the main producer of niacin/niacinamide in the U.S. with its annual production capacity of ca. 6 thousand metric tons. Antwerp site of the Vertellus is the main producer of niacin/niacinamide in Western Europe. Additional amounts required will be imported into the U.S. from companies like Jubilant and Lonza. U.S. imports increased from ca. 2.5 thousand metric tons in 1980s to roughly six thousand metric tons by now. At the same time, U.S. exports about two thousand metric tons of niacin to various parts of the world. The price of niacin in the U.S. was ca. 6 US$ per kg in 1980 s, which is now increased to ~9 US$ per kg. On the other hand, U.S. exports of 3CP have declined now due to increased production of 3CP in Asian regions (particularly in China) in recent times.

Oxidation of MEP with nitric acid is another route for niacin manufacture, which is practiced mainly in Western Europe. Furthermore, Western European exports of niacin are also facing strong competition from the lower price offered by China and India. In Asia, China produces more than 20 thousand metric tons of niacinamide where Lonza Guangzhou Nansha Ltd. is the major producer of the country and is likely to expand further its production capacity soon. Vertellus has planned to expand the production in China and to construct a plant with a capacity of 10 thousand metric tons at Nantong site (China) to produce niacinamide. This unit has been built and has been running since December 2011. On the whole, the use of pyridine derivatives for the production of niacin is forecast to grow considerably in the years to come.

Isonicotinonitrile (4-cyanopyridine, 4CP) can be obtained by the ammoxidation of 4-picoline. It is mainly used in the preparation of Isoniazid, an antitubercular drug and for the production of 4-dimethylaminopyridine (**DMAP**). The latter is generally used as an acylation catalyst **for** the synthesis of agrochemicals, pharmaceuticals and polymers. Vertellus is the largest producer of 4CP. The Chengdu XiYa Chemical Technology Co. and Shanghai Boyle Chemical Co. China, Tokyo Chemical Industry Co. Japan, DKSH, Switzerland, Vertellus Specialties UK (formerly Seal Sands Chemicals Ltd.) are major global suppliers of 4CP. 2-Cyanopyridine (2CP) can be manufactured in the same way by ammoxidation of 2-picoline. However its application is limited and thus, production amounts are very low only.

Much of the Vertellus picoline ammoxidation technology was developed by its precursor companies (Reilly, Nepera, and Degussa) based on fixed and fluid-bed reactors. In addition to 3CP and 4CP, the ammoxidation of 2-picoline to 2-cyanopyridine (2CP) has been practiced since 1970s.

Besides ammoxidation, 2- and 3Ps are also oxidized to the corresponding aldehydes using fixed-bed reactors. These reactions run on a much smaller commercial scale (Raschig GmbH, Germany). In addition, picolines can be oxidized further to the corresponding acids as already mentioned above [e.g., 23]. Several companies have made announcements of their intentions to build plants for the catalytic vapor-phase conversion of 3-picoline towards niacin.

Another good example for industrial manufacture of nitriles via ammoxidation is the conversion of methylpyrazines to their corresponding nitriles [24]. The ammoxidation of 2-methylpyrazine to cyanopyrazine was commercialized by Koei Seiyaku Co. in 1985 [25]. In addition, the ammoxidation of 2-methylpyrazine is produced in India by two companies. Pyrazinonitrile (2-cyanopyrazine, 2CPy) is commonly used as the upstream product of pyrazinamide manufacture. Pyrazinamide is an effective antitubercular drug. It is often used in combination with Rifampicin® (bactericidal antibiotic drug) and very effective in reducing the treatment period to ca. 2 months from 9 months (in absence of pyrazinamide). It is available in a wide variety of compositions. Pyrazinamide, is of course the pyrazine analog of nicotinamide, which is a white crystalline powder and sparingly soluble in water. In addition, it is used as an antiuric drug and acts on urate transporter in the kidney. World Health Organization (WHO) report of 2012 [26] reveals that the tuberculosis (TB) still remains an enormous global burden. There were an estimated 8.7

million new cases of TB in 2011 and 1.4 million deaths across the globe. Geographically, the burden of TB is the highest in Asia and Africa. India and China together account for ca. 40% of the world's TB cases. However, the new data show a considerable decrease (45%) in TB prevalence since 2002. The Millennium Development Goal of WHO is to prevent and reverse the TB epidemic by 2015. In view of this, there is a need for developing new and efficient drugs and drug intermediates (for for example, pyrazinamide and cyanopyrazine) to handle TB problem across the globe in an effective manner.

## 7.3 MANUFACTURE OF PYRIDINE NITRILES

In this section, we summarise the literature results on the hetero aromatic nitriles obtained by ammoxidation processes, in particular, the details on 2CP, 3CP, 4CP and 2CPy will be discussed. The type of catalysts applied so far, the conversion (X) of educts and yield (Y) of desired nitriles achieved (state-of-the-art) and other important issues related to the catalytic performance will be briefly illustrated in the following sub-sections.

### *7.3.1 AMMOXIDATION OF 2P TO 2CP*

The reaction scheme for the ammoxidation of 2-picoline to 2-cyanopyridine is shown in Scheme 7.5. In general, this reaction can be carried out in fixed bed or fluidized bed reactors in the temperature range from 300 to 450°C. Among the three isomers, 2-picoline ammoxidation is less investigated. The first studies on this reaction appeared in the literature in the late 1960s and early 1970s by Russian researchers [e.g., 27, 28] using V-W-O and $V_2O_5/TiO_2$ catalysts at reaction temperatures of 380–405°C.

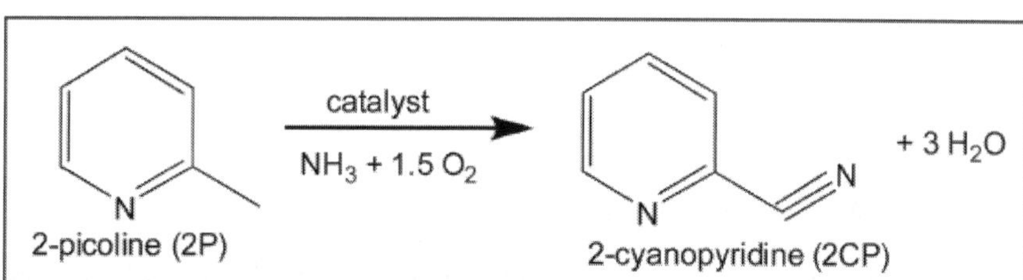


**SCHEME 7.5** Ammoxidation of 2P.

Very recently, Chinese researchers claimed 85% yield of 2CP at a 2-picoline conversion of >97% by vapor phase ammoxidation using alumina supported oxide catalyst at reaction temperatures in the range from 300–450°C [29]. Ammoxidation of 2-picoline and the other two isomers over molybdenum phosphate catalyst was studied by Lee et al. [30, 31], and reported varying reactivity of these three isomers. The authors state that the conversion of picolines follows the order: 2-picoline > 4-picoline > 3-picoline. This order is somewhat surprising because of the expected steric hindrance with the position of methyl group and its distance from the N-atom of the ring. Based on this principle, 2-picoline should be less reactive, whereas 4-picoline should be highly reactive. They also claimed almost 100% conversions and >90% selectivity of nitriles irrespective of picoline isomer used. In contrast, Vorobev et al. [32] reported different order of reactivity of these three picolines obtained over vanadium oxide-containing catalysts which follows: 4-picoline > 2-picoline > 3-picoline. They found some correlations between the mentioned rate constants and the gas-phase basicities of the starting compounds, characterizing electron-acceptor properties of the N-atoms, as well as the deprotonation constants of the methyl groups of picolines in the gas phase or bound via the nitrogen atoms with clusters modeling Brønsted acid centers of the vanadium oxide surface. Their explanation on the observed variations related to the reactivity properties of these three isomers seem to be logical. On the other hand, 10 wt% $V_2O_5$/silicated γ-alumina commercial catalyst [33] showed the highest activity for ammoxidation of 4-picoline and the activity decreases in the order, 4-picoline > 3-picoline > 2-picoline. The activity results were correlated with the oxygen chemisorption capacities of the catalyst at different temperatures.

In addition, some kinetic studies on the oxidative ammonolysis of 2-picoline using V-Sn/$TiO_2$ catalysts were reported [34]. The majority of work on this particular reaction was carried out by Russian researchers and in particular Suvorov's group [35]. It is also noteworthy that most of their results were published in Russian journals [e.g., 36]. Moreover, VPO catalysts (e.g., a-$VOPO_4$ phase) were also applied by our group for the ammoxidation of three isomeric picolines to their corresponding cyanopyridines [3].

### 7.3.2 AMMOXIDATION OF 3P TO 3CP

The stoichiometric reaction of ammoxidation of 3P to 3CP is shown in Scheme 7.6. As mentioned earlier, among the three structural isomers, 3P ammoxidation was extensively investigated. This process is also available on commercial scale using different catalyst compositions by different industries across the globe. Here, we describe some important issues related to the activity and/ or selectivity properties and the nature of catalysts applied for this reaction.

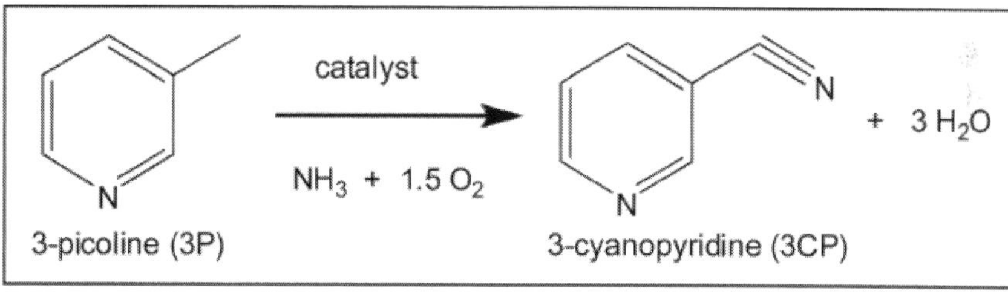


**SCHEME 7.6** Ammoxidation of 3-picoline.

### *7.3.3 VANADIUM PHOSPHATE BASED CATALYSTS*

Recently, a variety of VPO-based catalysts, for instance VPO/$TiO_2$ [37], Sb-VPO/$Al_2O_3$ [38], industrial VPOs [39], V-Mo-P-O/support [40], V-Cr-P-O/$SiO_2$ [41] were reported. They gave a conversion of 3P close to 100% and a yield of 3CP over 80%, respectively. The ammoxidation of 3P to nicotinonitrile over a VPO/$SiO_2$ catalyst system was investigated with respect to the influence of the catalyst components, reaction temperature, the ratio of air to ammonia to water. Using an optimum catalyst and optimum reaction conditions the highest molar yield is around 85% at 100% conversion [42].

We have also applied bulk and supported VPOs for the ammoxidation of 3P to 3CP under varying reaction conditions and studied their performance. It has been observed from our own investigations that all the VPO catalysts consist of both Lewis (L-Py) and Brønsted ($PyH^+$) sites in different proportions, which however depend upon the content of VPO (Fig. 7.4). It is also noted that the anatase supported VPO catalysts exhibit intense bands compared to their parent materials such as bulk VPO and pure $TiO_2$. This result implies that the supported VPOs are more acidic than the parent (pure) samples. Such enhancement in the acidity of supported catalysts is due to improved interaction between the support and active phase, which in turn displayed a considerable influence on the catalytic performance.

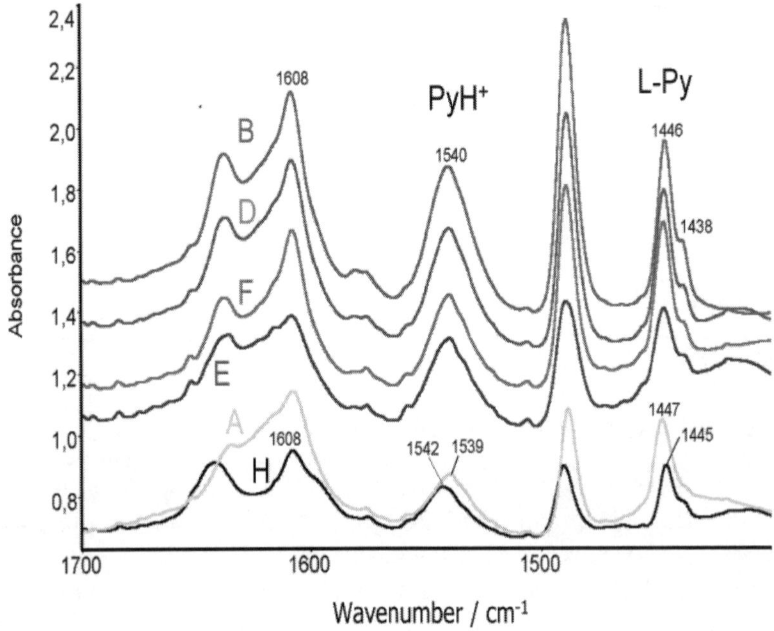


**FIGURE 7.4** Pyridine-FTIR spectra of bulk and $TiO_2$ supported VPO solids (A: bulk VPO; B: 5% VPO/$TiO_2$; D: 15% VPO/$TiO_2$; E: 20% VPO/$TiO_2$; F: 25% VPO/$TiO_2$; H: pure $TiO_2$) [36].

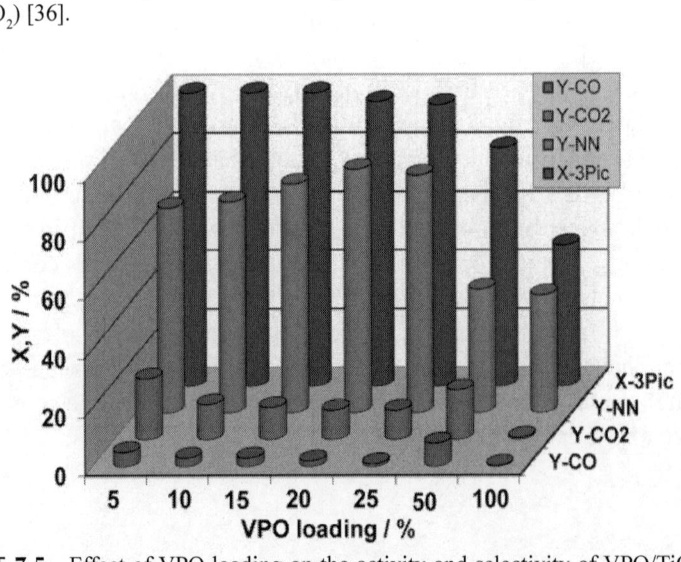


**FIGURE 7.5** Effect of VPO loading on the activity and selectivity of VPO/$TiO_2$ solids (T=360°C; (molar ratio of 3-pic: $H_2O$: $NH_3$: air = 1: 9: 6: 30 (3-pic = 10 mmol/h); 3 g (4.0 mL) catalyst; dilution (catalyst to glass beads) = 1:2 (by wt.)).

Furthermore, the activity tests on the performance of VPOs clearly demonstrate that the supported VPO catalysts displayed better performance than the pure VPO solid. In addition, the content of VPO has also shown considerable influence on the activity and selectivity behavior of the catalysts. Bulk VPO showed very low conversion (<30%) compared to supported catalysts (~100%). As expected, 3CP is the major product, while CO and $CO_2$ are the main by-products. However, small amounts of pyridine and isonicotinonitrile are also observed as side products depending upon the catalyst employed. The selectivity of 3CP is increased from 70 to >85% with rise in VPO content from 5 to 20 wt% and then decreased with further rise in VPO content. Among all, better performance is obtained over 20 wt% VPO/$TiO_2$ catalyst (conversion of 98% and yield of 83%) (Fig. 7.5). Such increase in selectivity of 3CP could be attributed to the changes in acidic characteristics of the catalysts. The correlation of acidity against activity is also described elsewhere [36]. It appears that when the VPO loading is increased beyond a certain level, the catalytic behavior of the catalysts seems to be moving in the direction of bulk VPO and thereby reduced activity of the catalysts. The bulk VPO solid gave only 48% conversion of 3P with 83% selectivity of 3CP. The results revealed that an increase in VPO loading beyond 20 wt% is not effective in improving the performance. On the whole, it can be stated that 20 wt% VPO/$TiO_2$ has exhibited superior performance in terms of obtaining higher yield of 3CP compared to others. In order to further improve the yield of 3CP some additional catalyst compositions based on VPO were also tried. Such investigations on the usage of promoted and supported VPOs (e.g., VPFeO/$TiO_2$) ultimately led to extremely high yields of 3CP up to 99% at total conversion of 3P. These results are however guarded by various patents [43, 44, 45, 46, 47]. Reddy et al. [48, 49] also tried supported VPO catalysts and claimed ca. 80% yield of 3CP from the ammoxidation of 3P.

### *7.3.4 OTHER CATALYST COMPOSITIONS*

A series of $V_2O_5$/$TiO_2$ catalysts were used in the selective ammoxidation of 3P to 3CP by various research groups [50, 51, 52]. Jiang et al. investigated VCrO/$SiO_2$ catalysts with varying Cr/V molar ratios (0.2–1.0) for the ammoxidation of 3P [53] and found that the active components on the silica surface were mainly amorphous $V_2O_5$ and $CrVO_4$. Their $NH_3$-TPD results showed that the acidity of the catalysts decreased with an increase in Cr/V ratio, which in turn revealed a clear influence on the catalytic performance. Low acidity gave low conversion but high 3CP selectivity. Among all, the catalyst with

Cr/V=0.6 exhibited the highest yield of 3CP. Direct synthesis of 3CP by vapor phase ammoxidation of 3P was carried out on highly dispersed vanadia catalysts supported on zirconia [54]. In 1980s, Andersson et al. contributed significantly to the ammoxidation of picoline with focus on the development of catalysts, structural dynamics of V-oxides, and mechanistic aspects [55, 56, 57, 58]. Efforts to apply alumina-supported $V_2O_5$ and bismuth molybdate catalysts led to obtain relatively poor yields of 3CP close to 60% [59]. Chary et al. [60] studied a series of $V_2O_5/Nb_2O_5$ catalysts with varying vanadia loadings (2–12 wt%) for ammoxidation of 3-picoline and claimed that the activity increases up to 6 wt% $V_2O_5$, which corresponds to monolayer coverage and then remains constant at higher vanadia loadings.

Freshly prepared and spent VTiSbSiOx ammoxidation catalysts for producing 3CP from 3-picoline were extensively studied in our group [61]. Catalytic runs with spent samples (after a few years of tests) revealed a significant drop in catalytic activity compared to their fresh ones due to occurrence of some structural alterations during long-term industrial use. Characterization by UV/VIS spectroscopy revealed a partial reduction of V(V) and V(IV) to V(III). EPR spectroscopy was effectively showed structural changes leading to a more pronounced formation of antiferromagnetically interacting vanadium(IV) oxide clusters, which in turn caused a loss in vanadium dispersion. XRD showed a slight increase in crystallinity of the spent samples. Consequently, all these effects collectively led to the observed drop in activity. However, the original activity could be restored by reoxidation under airflow by short-term heating to 600°C. By means of such regeneration, the reduced vanadium species (mainly V(III)) were partly reoxidized and the dispersion of the vanadium(IV) oxide clusters could be restored again (as evidenced by EPR spectroscopy). Ultimately, the catalytic activity/selectivity of regenerated samples was improved again to a comparable level to that of fresh solids.

Kinetics and mechanism of the vapor-phase ammoxidation of 3P was reported in mid 1970s by Prasad and Kar [62]. Radheshyam et al. [63] also reported kinetic studies and mechanism of the ammoxidation of 3P over $V_2O_5/ZrO_2$ catalysts in a differential flow reactor in the temperature range from 300–410°C. They claimed from their kinetics studies that the mechanism of the reaction is of the "Redox" type and the rate equation deduced assuming a steady state involves a three-stage oxidation-reduction process during the conversion of 3-picoline to 3CP. Based on that a tentative mechanism of the process has been proposed.

In addition to the usage of conventional oxide supports (e.g., $TiO_2$, $Al_2O_3$, $SnO_2$, $ZrO_2$ etc.), some selected metal fluorides were also used as efficient and novel catalyst supports for vanadia catalysts. For instance, $V_2O_5$ supported on $MgF_2$ [64, 65, 66], $AlF_3$ [67, 68], $CeF_3$ [69, 70] have shown good catalytic performance. The results obtained on metal fluoride supported $V_2O_5$ catalysts are quite comparable with those of commercial catalysts. Conversion of 3P (90–100%) and yield of 3CP (80–85%) were successfully achieved on these fluoride supported catalyst compositions.

Recently, vanadium-containing oxynitrides were also used as effective new class of catalysts for the ammoxidation of 3P [71, 72]. Furthermore, SAPO and modified SAPOs, such as Sb-V-SAPO-5, Mo-V-SAPO-5 and Mn-V-SAPO-5, were used for the ammoxidation of 3 and 4Ps picolines and claimed 85% and 90% yields of their corresponding nitriles, respectively [e.g., 73]. Best results on this reaction were reported in various patents [e.g., 42, 74, 75]. The yield of 3CP achieved was over 90% at almost total conversion of picoline. More details on the ammoxidation processes in general and 3P to 3CP in particular can be found in a recent review of Cavani et al. [76]. Roy et al. [77] reported the application of Ti-pillared clay based vanadia catalysts for producing hetero aromatic nitriles from picolines at 350–450°C.

Besides the use of vanadium-based catalysts, a wide variety of other catalyst compositions were reported. A recent review focussed on FeSbO based catalysts promoted by appropriate additives as suitable for the ammoxidation of alkyl-substituted aromatics and hetero aromatic compounds. A unique preparation method of a fluid-bed catalyst is presented using nitric acid oxidation of antimony trioxide catalyzed with iron ions. The catalysts thus prepared have superior catalytic and physical properties. [78]. In addition, some unique compositions were reported by different research groups. For instance, new ammoxidation catalysts based on rhenium carbonyl cluster complexes containing antimony and bismuth ligands were reported by Adam et al. [79]. Single-site multifunctional catalysts based on $[Cu_4Ru_{12}C_2]$ nanocluster anchored to inner walls of mesoporous silica were also used in the ammoxidation of 3P [80].

Moreover, efforts were also made on producing 3CP and niacin/niacinamide directly from 3P using Re-Bi catalysts by liquid-phase ammoxidation [e.g., 81]. Eventhough, these compositions based on Re-Sb (e.g., $Re_2Sb_2$) or Re-Bi ($Re_2Bi_2$) are rare and new, they gave relatively poor performance compared to the state-of-the-art, that is, low conversion of 3P (40–60%) and low selectivities (~75%) and niacin/niacinamide (<15%).

### 7.3.5 AMMOXIDATION OF 4P TO 4CP

The reaction of the ammoxidation of 4P is shown in Scheme 7.7. The first studies on this reaction appeared in late 1960s [82], where only low yields of 4CP (45–50%) could be obtained at reaction temperatures of 375–415°C. Very recently, Mikahailovskaya et al. [83] employed V-Ti-Sn-O catalysts of various compositions in the ammoxidation of both 3P and 4P. They claim that these solids exhibit high activity under mild conditions with minimum excess of ammonia. The best catalytic activity in oxidative ammonolysis of methyl pyridines are exhibited by the catalysts having $V_2O_5$: $TiO_2$: $SnO_2$= 1: 4: 4 (where $SnO_2$ = 54.6%). A conversion of 4P close to 100% could be achieved at relatively low temperature (i.e., 250–290°C) with the yield of 4CP between 85–90%. In addition, the application of V-oxides promoted by $Cr_2O_3$, $SnO_2$ and $ZrO_2$ additives were investigated by Vorobev et al. [84].

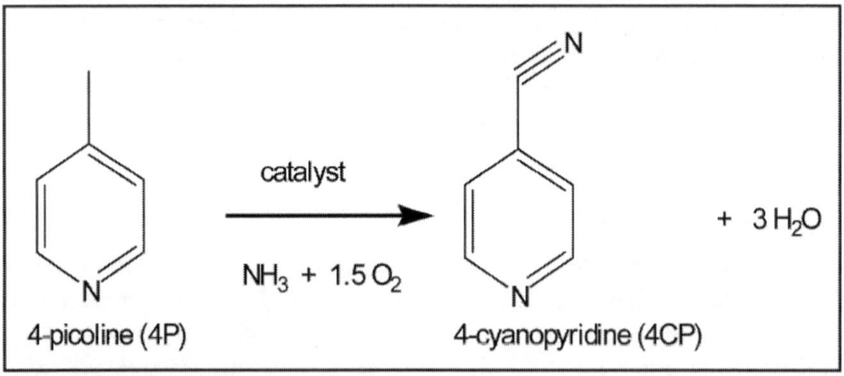


**SCHEME 7.7** Ammoxidation of 4-picoline.

Our own investigations [85] on the ammoxidation of 3P and 4P revealed that their reactivity is strongly dependent on the position of methyl group on the ring. We have used $CeO_2/TiO_2$ catalysts and observed the following trend: 4P > 3P. More over the TOFs obtained on 4P are more than double the TOFs observed with 3P. It can be inferred that the two structural isomers with methyl group at different positions from nitrogen atom showed marked steric effect on them. 4-picoline with methyl group farther away from nitrogen atom in the ring seems to be more favorably activated than 3P on the catalyst surface for participation in the reaction. Verma et al. [86] claimed 4CP yield of 85% with 4P conversion of 90% over the V-Sb-O catalyst. Suvurov et al. [87] made some interesting observations on the ammoxidation activity of different pyridine derivatives with varying side-chain lengths. Ammoxidation reactivities of different pyridine-based substrates decreased in the following order: 4-ethylpyridine > 4-picoline > 2-picoline > 4-acetylpyridine > 3-picoline > 4-vinylpyridine > toluene. The authors claimed that the reaction was less selective in the case of 2-carbon side chains. Correlations of reactivity with the energies of proton loss were discussed.

Silica/zeolite based catalysts [88], SAPO [89], VSAPO [90], modified SAPOs with Sb, Mo, Mn [91], VAPO [92] were used for the synthesis of 4CP by vapor phase ammoxidation to obtain 90% yield of 4CP and 85% yield of 3CP from their corresponding picoline isomers. Verma et al. [85] reported 85% yield of 4CP using VSbO-based catalysts with 90% conversion of 4P [85]. Furthermore, catalytic ammoxidation of 4P to 4CP in the presence of modified forms of gumbrin and synthetic X-type zeolite was cited as a rare example [93]. However, this study led to much lower yields (~40%) of 4CP. The ammoxidation of 4P of all the three isomeric picolines to their corresponding cyanopyridines on crystalline VPO catalysts was also investigated about two decades ago by our group [3].

Das et al. [94, 95] correlated the catalytic properties of selected metal oxides (e.g., $Co_3O_4$, $WO_3$, etc.) in the ammoxidation of 4P to understand the mechanism of the reaction. They correlated their oxidation kinetics with the heat of formation of metal oxides. The rate-determining step was found to be the reaction between the adsorbed picoline and oxygen. They carried out the ammoxidation of 4P in an integral fixed-bed reactor over $V_2O_5$-$CrO_3$-$Al_2O_3$ catalyst at 300–400°C. Different rate equations based on 2- and 3-stage mechanism were proposed. They assumed steady state involving simultaneous reduction of catalyst by picoline and its oxidation with oxygen similar to that considered in the Mars-van Krevelen mechanism. Rate equations were also derived by the Langmuir-Hinshelwood mechanism assuming the surface reaction between adsorbed 4P and oxygen atoms as the rate-determining step. The rate expressions corresponding to the best fit for experimental data reflected a first order kinetics with respect to 4P and 0.5 order with respect to oxygen [94]. In addition, they investigated various other catalyst relevant properties such as the heat of formation and *d*-electron characteristics.

Gut and Dirr [96] carried out kinetic studies on the ammoxidation of 4P in a fixed-bed integral reactor using $VO_x/Al_2O_3$ catalyst at 290–400°C. They reported that the ammoxidation reaction was controlled by pore-diffusion phenomena. The activation energy for 4P conversion was 12 kcal/mol in the diffusion region, and 21 kcal/mol for the chemically controlled reaction. Besides the main reaction, two consecutive reactions were also reported, such as the reaction of $NH_3$ with isonicotinonitrile to isonicotinamidine, and iso-nicotinaldehyde to 4-pyridyl-N-(4-picolinealdimino) carbinol. To our knowledge, no other studies claimed the formation of these products.

## 7.4 MANUFACTURE OF PYRAZINE NITRILES

### 7.4.1 AMMOXIDATION OF 2MP TO 2CPy

The ammoxidation of 2MP to 2CPy is an important reaction from both industrial and academic points of view (Scheme 7.8).

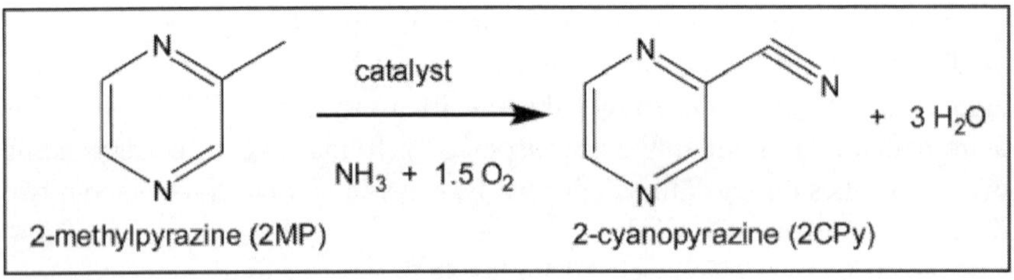


**SCHEME 7.8** Ammoxidation of 2 MP.

Very recently, we have developed novel metal vanadates as efficient catalysts for synthesizing 2CPy by ammoxidation of 2MP [97, 98, 99]. At the first instance, we have checked the potentiality of various bulk metal vanadates such as $AlVO_4$, $FeVO_4$, $CrVO_4$, $NbVO_5$, $LaVO_4$ and $BiVO_4$. Among them, $NbVO_5$ displayed the best performance followed by $LaVO_4$ solid [98]. Considering this fact, two series of catalysts were further prepared with varying Nb:V and La:V ratios and checked the influence of such M/V ratios on the catalytic performance.

The correlation between surface V:M ratios and the catalytic properties of different metal vanadate catalysts is illustrated in Fig. 7.6. These investigations clearly demonstrated that the phase composition and enrichment of vanadium in the near- surface- region (investigated by X-ray photoelectron spectroscopy (XPS)) strongly depend on the nature of the metal used in the metal vanadates. Furthermore, these two properties play an important role on the activity and selectivity behavior of the catalysts [99]. It was observed that higher surface V:M ratios enhanced the selectivity of CPy but decreased the conversion of 2MP.

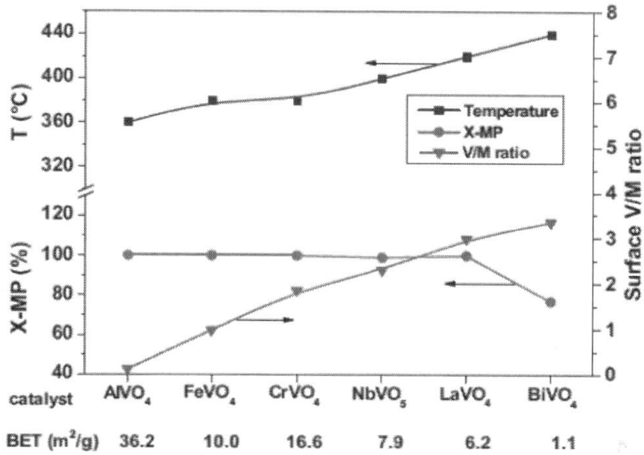


**FIGURE 7.6** Correlation between the maximum conversion at given temperature and the near-surface-region V: M molar ratio of various metal vanadate catalysts (reaction conditions: catalyst wt. = 1 g; reactant molar ratio MP: $H_2O$: $NH_3$: air: $N_2$ = 1:13:7:26:22; total flow = 9506 mL/h; GHSV = 9500 $h^{-1}$) [96, 98].

The influence of the La:V ratio in LaVOx solids on the catalytic performance is depicted in Fig. 7.7. As expected, pure $La_2O_3$ is almost inactive while the combination of La and V in an appropriate ratio substantially improved both activity and selectivity. This had clearly demonstrated that the catalytic performance strongly depends on the La:V ratio. The conversion of 2MP is increased progressively with rise in vanadium content and reached almost 100% with simultaneous enhancement in the yield of 2CPy. Among different $LaVO_x$ solids examined, the catalyst with optimum La:V ratio (i.e., $La_{0.1}V_{0.9}O_x$) exhibited the best yield of 2CPy (>85%) at almost total conversion with high space-time-yields (STY) exceeding 525 $g_{CPy}/kg_{cat}/h$ [100] along with good long-term stability during time-on-stream. To the best of our knowledge, this is the best result reported so far in the literature.

Besides metal vanadates, other catalyst compositions were also tried in the past several decades. For instance, VPO based solids [101], $FePO_4$ [102], MoVPO [103], heteropolyacids [104, 105], $V_2O_5/TiO_2$ [106, 107, 108], V-$MOO_x$ [109] are widely used. Two isomorphous $VOPO_4$ solids (α– and β–) were prepared and tested for the ammoxidation of 2MP to 2CPy [103, 110]. The reaction was carried out in the temperature range of 360–420°C. Between the two mono-phosphates tested, the α-$VOPO_4$ showed better activity compared to β-VOPO4. Interestingly, in situ synthesized 12-molybdophosphoric acid (AMPA) catalysts displayed much better performance compared to their parent $VOPO_4$ solids. Among all catalysts tested, AMPA-α-$VOPO_4$ exhibited the best performance (X-2MP ≈ 90%, S-2CPy = 65% at T = 420°C). The differences in catalytic performance of the tested catalysts are ascribed to the structural differences of the used $VOPO_4$ solids. Bondareva et al. [106] investigated the mechanism of 2MP ammoxidation on a $V_2O_5/TiO_2$ catalyst by in situ-FTIR spectroscopy. The structure of surface intermediates was identified and the sequence of their transformation was ascertained. The interaction of 2MP with catalyst surface results in the transformation of coordinatively bound 2MP into oxygenated surface compounds, for example, aldehyde-like complex and asymmetric carboxylate species. The main reaction product, amidopyrazine, is formed through the interaction of the surface oxy-intermediates with adsorbed ammonia species. Another study by Hong et al. [103] using MoVPO solid, a 2CPy selectivity of 93% and yield of 86% could be achieved at a reaction temperature of 380°C under molar ratio of 2MP:$H_2O$:$O_2$:$NH_3$ = 1:8:8:8.

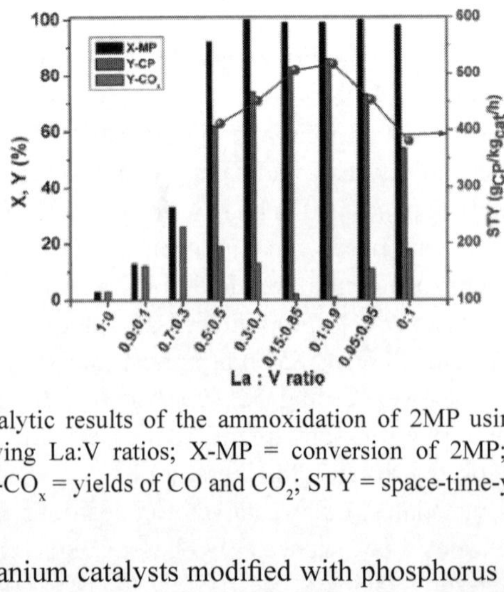


**FIGURE 7.7** Catalytic results of the ammoxidation of 2MP using various LaVO x catalysts with varying La:V ratios; X-MP = conversion of 2MP; Y-CPy = yield of 2-cyanopyrazine; Y-$CO_x$ = yields of CO and $CO_2$; STY = space-time-yield [96, 98].

Vanadium-titanium catalysts modified with phosphorus (acidic) additives were studied in the ammoxidation of 2MP by Bondareva et al. [107, 108]. The authors claim that the introduction of high amounts of the acid additives resulted in decreased activity and selectivity due to formation of low active ternary compounds. The nature of active sites of the samples modified with 1–5 wt% of $P_2O_5$ showed that the $V^{5+}$ cations are strongly bound with $TiO_2$ and located in the distorted octahedral oxygen environment. On the other hand, V-Ti-O catalysts modified with basic additives (1–15 wt% Na or K) have also caused a decrease in the activity and selectivity of the catalysts due to formation of low-active bronzes (e.g., $MeV_6O_{15}$) and vanadates ($NaVO_3$, $KVO_3$ and $K_3V_5O_{14}$) [108]. In addition, the authors claimed that the active centers of binary catalysts include $V^{5+}$ cations, with distorted octahedral coordination strongly bonded with titania apparently owing to formation of V-O-Ti bonds. In another study, Bondareva et al. [111] using V-Sb oxide catalysts of varying compositions reported that the vanadium antimonate had the lowest selectivity in the series of binary compounds. At Sb/V ratio >1, a solid solution of the tetrahedrally coordinated $V^{5+}$ ions in $Sb_6O_{13}$ was the most selective, while the solid with Sb/V <1 was the active one.

Some investigations were focused on the kinetic studies of the ammoxidation of 2MP using Sb-V-Mn mixed oxides [112]. Apparent partial reaction orders for the title reaction in excess of steam at 310–370°C and ambient pressure were evaluated. A pseudo-first order model was proposed, which fitted well with the experimental data. The authors also reported low activation energy values (i.e., 50 kJ/mol or less), which indicated that the kinetics are probably controlled by internal diffusion. The same research group in other studies examined many different mixtures of transition metal oxides, different binders and supports both in the presence and in the absence of steam in the ammoxidation of 2MP. The best results were claimed over Sb-V-Mn oxides supported on $\alpha$-$Al_2O_3$ at 375–425°C. Under these conditions, >90% conversion and 75% selectivity to 2CPy were obtained [113].

## 7.5 MECHANISTIC ASPECTS

In general, the ammoxidation reaction of methyl aromatics and/or hetero aromatics runs via redox mechanism as proposed by Mars and van Krevelen [12]. Most of the catalysts used so far contain transition metal oxides with easily changing valence states (e.g., V, Mo, etc.). Essential steps of the reaction mechanism are: (i) chemisorption of the methyl aromatic or hetero aromatic reactant on the catalyst surface followed by H-abstraction (i.e., C-H bond disassociation) to form a benzylic intermediate, (ii) insertion of nitrogen into a surface bonded partially oxidized intermediate and (iii) desorption of the formed nitrile and iv) reoxidation of the catalyst by gas-phase oxygen. Literature survey [114, 115] revealed that the H-abstraction occurs via C-H bond dissociation in three different possible ways, such as (i) heterolytic with the abstraction of hydrogen atom in an anionic form followed by carbocation,

(ii) homolytic with the formation of a radical of hydrocarbon, (iii) heterolytic, with protonation of hydrogen and formation of a carbanion. The type of C-H bond dissociation however depends upon the nature of the catalyst used. Additionally, the position, size and electronic effects of substituents also matters a lot [e.g., 116]. Over vanadium-tin oxide [117] and vanadium phosphate catalysts [118], the reactivity of substituted toluenes with electron-withdrawing substituents is enhanced compared to the un-substituted toluene. On the other hand, electron-donating substituents also show a rather decreasing effect on activity and selectivity [118] due to stronger chemisorption of substrate, intermediates and products on the electrophilic catalyst surface. Such situation leads to a delayed desorption of products, which can then easily undergo consecutive oxidation. Nonetheless, these issues can be investigated in detail using various spectroscopic techniques. For instance, the electronic interaction of substituted toluenes with vanadium phosphates can be observed by insitu ESR spectroscopy. Similarly, formation of different reaction intermediates during the course of the reaction can be identified by insitu FTIR spectroscopy. It is known that the interaction of the substrate with the catalyst surface is stronger in the presence of electron-donating substituents, while such interaction is distinctly lower when electron-withdrawing substituents are present [119].

Until now, $V_2O_5$, VPO, Mo, and Sb based catalysts are widely used for various ammoxidation reactions. Though the steps involved in almost all ammoxidation reactions are same irrespective of the catalyst system used, the mode of adsorption of substrate, the nature of active species involved differ from catalyst to catalyst. Although recent studies have thrown some light on the mechanism of ammoxidation reaction on different oxide catalysts, a clear understanding of various reaction steps involved is still under debate. The mechanism of ammoxidation of different substrates over different catalysts is discussed below.

### *7.5.1 MECHANISM OF VAPOR PHASE AMMOXIDATION OVER $V_2O_5$-CONTAINING CATALYSTS*

Andersson and Lundin [58] proposed a mechanism for the vapor phase ammoxidation of 3P on V-oxide catalysts. They state that the adsorption of 3P on the vanadium surface takes place in the first step followed by H-abstraction that leads to transformation of $(V{=}O)^{3+}$ surface groups into $(V\text{-}O)^{2+}$ groups. In the second step, the hydrogen atom from methyl group reacts with neighboring –OH group with the formation of water as by-product. If there is no ammonia in the feed gas, then the surface compound can be desorbed in the form of aldehyde-like species. However, they have not found any aldehyde species in their product analysis in presence of $NH_3$ and hence they proposed that ammonia can react with the aldehyde complex before it is desorbed, forming either amide species or imine species depending on hydrogen rearrangement and reaction conditions with the formation of $H_2O$ as a by-product. The resulting amide or imine then transforms into nitrile and then the catalyst is reoxidized by gas phase oxygen according to a redox mechanism. Controversy exists till today in the literature on the formation of amide or imine as an intermediate.

In fact, three moles of water are produced for every mole of 3P reacted. Generally, the formation of water during the reaction or additionally supplied water is of great importance in changing the acid-base properties of catalysts and also on the formation of nicotinic acid. However, Prasad et al. [62] have reported that water vapor had no influence on the yield of nicotinonitrile in the ammoxidation of 3P. Andersson and Lundin [58] suggest that the water does not take part in the formation of 3CP and probably may not be able to hydrolyze 3CP to nicotinic acid due to presence of $NH_3$ in the reactant feed gas mixture and its high nucleophilic character than the water. Based on this property, they explained why the yield of nicotinonitrile was high and the yield of nicotinic acid was low despite additional supply of water vapor along with the reactant feed gas. They observed small amounts of pyridine, carbon oxides and tar as by-products in this reaction. They reported that the formation of pyridine probably occurs from the destruction of the intermediate complexes, which increases with temperature leading to an increase in the amount of carbon oxides. The formation of high boiling compounds like tar might be due to the polymerization of intermediate imine-like species.

Murakami et al. [120] and Niwa et al. [121] investigated the reaction mechanism of ammoxidation by FTIR spectroscopy using $V_2O_5/Al_2O_3$ catalysts. They claim that the reaction proceeds via interaction of ammonium ions with surface benzoate ions. Cavalli et al. [5, 122] and Busca et al. [123] proposed benzyl amine and benzaldehyde species as reaction intermediates in the ammoxidation of toluene over $V_2O_5/TiO_2$ catalysts. Otamiri and Andersson [124] proposed vanadium imido species (V=NH) and vanadium hydroxylamino species (V–NH–OH) as nitrogen insertion sites in $V_2O_5$ catalysts. These species in turn react with adsorbed toluene to form an amine (R–$CH_2$–$NH_2$) and/or imine (R–CH=NH) surface intermediate, which subsequently transform into nitrile as a final product. A mechanism proposed by Sanati and Andersson [125] includes $(C_6H_5)CH(NH_2)O$– and $(C_6H_5)CH(NH_2)(O–)_2$ species as reaction intermediates. Ramis et al. [126] reported the formation of amido groups in addition to protonated and coordinated ammonia species during the adsorption of ammonia. The formation of vanadium imido species (V=NH) and benzylamine surface species were also postulated by Azimov et al. [127] in the ammoxidation of toluene over V-Sb-Bi-O catalysts.

### 7.5.2 MECHANISM OF TOLUENE AMMOXIDATION OVER VPO CATALYSTS

The mechanism of formation of surface intermediates during the interaction of vanadyl pyrophosphate, $(VO)_2P_2O_7$, with the feed components of the toluene ammoxidation was first studied by in situ-FTIR spectroscopy [2]. The interaction of toluene and possible reaction intermediates like benzaldehyde and benzyl amine were studied by us using in situ-FTIR spectroscopy too. The investigations revealed that the ammoxidation of toluene probably proceeds via benzaldehyde intermediate followed by benzyl imine like species from $NH_4^+$ surface species, which then leads to the formation of final nitrile product by subsequent oxidative dehydrogenation. However, no benzamide or benzylamine species were detected. In addition, it is noticed that the surface species generated upon adsorption of benzaldehyde were similar to those formed from toluene, indicating the role of the benzaldehyde as an intermediate in the nitrile formation path. On the other hand, the adsorption of benzyl amine in presence of oxygen did not lead to the formation of benzonitrile. This fact shows that an ammoxidation mechanism of toluene through a benzaldehyde intermediate is a more preferred path while the reaction pathways via benzamide or benzyl amine intermediates seem to be more unlikely. We have carried out some investigations [2] by the individual adsorption of all reactants (e.g., toluene, ammonia and air) individually and also in combination to identify the reaction intermediates and to propose the reaction mechanism. Such experiments were also performed by the adsorption of possible reaction intermediates like benzaldehyde, benzyl amine, etc. under working conditions and to evaluate the adsorption behavior of reaction intermediates and possible reaction pathways by in situ-FTIR combined with other methods. Based on these results, a more probable reaction mechanism of toluene ammoxidation over $(VO)_2P_2O_7$ catalyst has been proposed [2]. This study also considered the interaction of surface vanadyl groups with oxygen and ammonia. Such FTIR investigations on the ammonia adsorption on $(VO)_2P_2O_7$ revealed the existence of three types of N-containing surface species such as i) protonated ions ($NH_4^+$), ii) coordinatively adsorbed $NH_3$ and iii) -$NH_2$ groups. All of them may act as the so-called nitrogen insertion species. However, their concentration depends on the extent of ammonia interaction with the pyrophosphate catalyst surface [128].

It was suggested that the ammonia adsorption is accompanied by a redox interaction with the $(VO)_2P_2O_7$, resulting in a partial reduction of $V^{4+}$ to $V^{3+}$ and generation of hydroxyl groups by hydrolysis or ammonolysis of P-O-V bonds forming ammonium ions and amido species. The formation of methylene-like species generated upon the adsorption of toluene on the surface proceeds probably via oxidative H-abstraction from the methyl group of toluene adsorbed through p-complex of the aromatic ring with the adjacent $V^{IV}$=O group. Miyata et al. [129] also reported benzyl and benzaldehyde species as intermediates and their formation occurs through an interaction of the toluene methyl group and surface V=O entities of V-oxides on $ZrO_2$ and $TiO_2$ supports.

On the other hand, a portion of benzaldehyde species might be oxidized to benzoate ions and consecutively the 'total oxidation products' becoming the main side reaction products of the ammoxidation. The reaction steps of the nitrile formation were reasonably assigned to one edge sharing $VO_6$ octahedra unit of the pyrophosphate surface. The coordinatively unsaturated part (Lewis acidic site) of the $VO_6$ dimer could bind the aromatic substrate via π-complex formation whereas the reaction sequence of the partial oxidation could run on the other site according to a V....V distance of the vanadyl groups of ca. 3.4 Å [130] and $C\text{-}C_{ring}$ and $C\text{-}CH_3$ distances of toluene molecule of 1.4 Å and 1.52 Å, respectively [131] that is, the distance of π-complex center to the carbon atom of the methyl group amounts to ca. 3 Å.

It was claimed elsewhere [132] that at least two adjacent VO groups located closely are necessary for the chemisorption step and the subsequent catalytic reaction over vanadyl-containing catalysts. One VO site of edge-shared $VO_6$ octahedra, existing for example in vanadium phosphate catalysts, acts in the chemisorption of a toluene reactant as a coordinatively unsaturated (Lewis) site via pi-interaction, the other one as an oxidation site via the V=O group. Additionally, high catalytic activity requires effective exchange pathways for the electron transport. In VPO catalysts, the electron density alters due to effective spin-spin exchange interactions (detected by in situ-ESR spectroscopy) at a discrete surface vanadyl ion that can easily be delocalized by the overlapping d-orbitals of the exchange-coupled centers [133]. This mechanism was also supported in further studies based on temporal-analysis-of-products (TAP) reactor use, which gave an evidence for the formation of shortlived benzaldehyde species as reaction intermediate during interaction of ammoxidation feed pulses with the $(VO)_2P_2O_7$ catalyst [134]. Ammonia, the source of N-insertion, would be chemisorbed and activated by the catalyst under reaction conditions [122, 135, 136] but does not directly involve from the gas phase as proven by the use of $(^{14}NH_4)_2(VO)_3(P_2O_7)_2$ catalysts as shown above (see Scheme 7.3 [15]). Special investigations using transient experiments resulted in $^{15}N$-containing nitrile product during first pulses of $^{15}NH_3$ feed on $^{14}N$-containing $(NH_4)_2(VO)_3(P_2O_7)_2$ catalysts supported this idea [15]. As in case of vanadium phosphate catalysts, $NH_4^+$ ions may act as potential N-insertion species in the ammoxidation cycle and suitable chemisorption sites (OH-groups) will be formed during the catalytic reaction owing to the hydrolytic break up of V-O-P and P-O-P bonds [127]. Otherwise, the formation of imido species (=NH) is often discussed as active site for the N-insertion of catalysts, containing for example, molybdenum instead of vanadium [124, 137, 138]. Ammonia not only acts as a reaction partner but also involves in the reduction of catalyst [127, 131] or may block weakly bound oxygen sites, which cause an increase in the selectivity of the partial oxidation products [122, 123, 139]. Reaction steps of the ammoxidation of toluene over a vanadium phosphate catalyst have also been investigated by XRD, TPR, TPSR, transient experiments and FTIR-, Raman-, and ESR-spectroscopy to achieve a comprehensive mechanistic scheme [132, 133, 140].

## 7.6 SUMMARY AND OUTLOOK

The ammoxidation of methyl aromatic and hetero aromatic compounds provides a convenient route to produce large number of industrially important nitrilcs in a single step. Herein we summarized the present state of knowledge on the usage of various catalysts, reaction mechanisms, latest developments, and the key issues related to catalytic performance of the production of various hetero aromatic nitriles by vapor phase ammoxidation. It is known that the ammoxidation occurs through a redox mechanism and hence the catalyst compositions employed must contain some redox components (e.g., V. Mo, Cr, W, etc.). However, other parameters such as acidity, reducibility, surface area, surface composition, etc. also play a vital role on the catalytic properties. Generally, high activity and selectivity should be engineered by selecting appropriate chemical components, using suitable preparation methods to achieve the desired properties (surface area, porosity, dispersion, etc.) and formulating the pellet to ensure good access to active sites. Moreover, one should also note that achieving the activity alone is not sufficient enough unless it is balanced with the other requirements such as lifetime of the catalyst, thermal stability, mechanical strength, poisoning, fouling, etc.

In this chapter, the application of a variety of catalysts for different ammoxidation reactions was discussed in detail. Special focus was given to two industrially important reactions; such as i) ammoxidation of picolines (all the three isomers) and ii) ammoxidation of 2MP. The nitrile products of these two specified ammoxidation reactions are not only important from commercial point of view but also present opportunities for academic research. This report also deals with pharmaceutical chemicals (e.g., TB treatment using pyrazinamide).

The mechanistic principles could also serve as design rules for preparing novel compositions with desired architecture and unique properties. The main issue in all industrial ammoxidation processes is productivity based on activity and selectivity and these two properties depend on stability and durability of catalysts employed. Such issues will certainly facilitate the selection and testing of new catalyst formulations/compositions with good stability. This will also entail developing and identifying much-needed tasks to derive correlations between structure and activity that will assist in the selection of the next generation catalysts with more efficiency.

The ammoxidation of 3P to 3CP and 2MP to 2CPy have already been commercialized by different industries. Almost total conversion of educts and close to 90% yields of desired nitriles have been successfully achieved. Metal vanadates developed very recently in our group could also be potential ammoxidation catalysts of the future. In view of innovative and academic work that is being carried out at various academic and industrial laboratories all over the world, more scientific breakthroughs can be expected in the near future. The future course of work on new developments has yet to be figured out with novel strategies. Mechanical failure due to attrition is one of the serious problems in many commercial catalytic processes including oxidation and ammoxidation. Therefore, the future efforts by various researchers/engineers should aim at developing effective methods to overcome such problems. Scientists should also take advantage of highly sophisticated surface spectroscopic studies and computer modeling for enhanced understanding of catalyst properties and catalytic reactions at the molecular level more effective, which in turn enable designs for optimization purposes.

To design and develop novel catalysts for various oxidation and ammoxidation processes, the focal point of all research should aim at understanding the reaction mechanisms and deriving structure activity relationships. However, choice of active components, type of supports and promoters, method of preparation and formulation of catalyst must be related to mechanism, kinetics, and adsorption.

Microstructured reactors may help to master exothermicity, probably membranes can enhance oxygen mobility in a selective direction according to Mars-van Krevelen mechanism in such a way that total and nonselective oxidation paths can be avoided. The ammoxidation is not only important from commercial point of view but also present opportunities for fundamental research and also contribute to the well-being of mankind. Eventhough great deal has already been achieved in the field of ammoxidation, new approaches and new attempts are certainly necessary. Another research direction is the synthesis of bifunctional and/or multifunctional catalysts, which might open new perspectives for future experiments. On the whole, we are hopeful that the future of the ammoxidation processes looks brighter and new revolutionary ideas are expected to come in the coming future.

## ACKNOWLEDGMENTS

Special thanks to Dr. Colin McAteer (PhD) from Vertellus Specialties Inc. (Indianapolis, USA) and Dr. Achim Fischer from Evonik Industries AG (Hanau, Germany) for their support and also for providing various valuable information.

## KEYWORDS

- **2-Cyanopyrazine**
- **2-Methylpyrazine**
- **3-Picoline**
- **Acetonitrile**
- **Acrylonitrile**
- **Ammoxidation**
- **Hetero-aromatic compounds**
- **Nicotinonitrile**
- **Pyrazinamide**

## REFERENCES

1. Martin, A. and Kalevaru, V. N. Heterogeneously catalyzed ammoxidation: a valuable tool for one-step synthesis of nitriles. *Chem Cat Chem* **2,** 1504–1522 (2010).
2. Zhang, Y., Martin, A., Berndt, H., Lücke, B., and Meisel, M. FTIR investigation of surface intermediates formed during the ammoxidation of toluene over vanadyl pyrophosphate. *J Mol Catal A* **118,** 205–214 (1997).

3. Martin, A., Luecke, B., Seeboth, H., and Ladwig, G. Ammoxidation of picolines on vanadium phosphate catalysts. *Appl Catal A* **49,** 205–211 (1989).
4. Centi, G., Cavani, F., and Trifirò, F. Selective oxidation by heterogeneous catalysis-fundamental and applied catalysis, Springer, pp. 363–495 (2001).
5. Cavalli, P., Cavani, F., El-Sawi, M., Manenti, I., and Trifirò, F. Kinetic and mechanistic analysis of toluene ammoxidation to benzonitrile on vanadium-titanium oxides. *Ind Eng Chem Res* **26,** 804–810 (1987).
6. Rojas, E., Guerrero-Perez, M. O., and Banares, M. A. Niobia-supported nanoscaled bulk-nio catalysts for the ammoxidation of ethane into acetonitrile. *Catal Lett* **143,** 31–42 (2013).
7. Nguyen, T. T., Burel, L., Nguyen, D. L., Pham-Huu, C., and Millet, J. M. M. Catalytic performance of MoVTeNbO catalyst supported on SiC foam in oxidative dehydrogenation of ethane and ammoxidation of propane. *Appl Catal A* **433–434,** 41–48 (2012).
8. Lopez-Medina, R., Rojas, E., Banares, M. A., and Guerrero-Perez, M. O. Highly active and selective supported bulk nanostructured MoVNbTeO catalysts for the propane ammoxidation process. *Catal Today* **192,** 67–71 (2012).
9. Castelli, A., Ballarini, N., and Cavani, F. New rutile Ga/V/Nb/Sb mixed oxides as catalysts for propane ammoxidation to acrylonitrile. *Catal Today* **192,** 72–79 (2012).
10. Ushikubo, T., Oshima, Y.-K., Kayo, A., Umezawa, T., Kiyono, K.-I., Sawaki, I., Nakamura, H. US 5 472 925 (1995) *(Mitsubishi Chemical Corporation, Japan)*
11. Suresh, D. D., Friedrich, M. S., Seely, M. J. US 5 212 137 (1993) *(Standard Oil Co. Ohio, USA).*
12. Mars, P. and van Krevelen, D. W. Oxidations carried out by means of vanadium oxide catalysts. *Chem Eng Sci* **3,** 41–59 (1954).
13. Kaichev, V. V., Gladky, A. Yu., Prosvirin, I. P., Saraev, A. A., Hävecker, M., Knop-Gericke, A., Schlögl, R., and Bukhtiyarov, V. I. In situ XPS study of self-sustained oscillations in catalytic oxidation of propane over nickel. *Surf Sci* **609,** 113–118 (2013).
14. Efremenko, I. and Neumann, R. Computational insight into the initial steps of the mars–van krevelen mechanism: electron transfer and surface defects in the reduction of polyoxometalates. *J Am Chem Soc* **134,** 20669–20680 (2012).
15. Martin, A., Zhang, Y., Zanthoff, H.-W., Meisel, M., and Baerns, M. The role of ammonium ions during toluene ammoxidation on α-$(NH_4)_2[(VO)_3(P_2O_7)_2]$ used as catalyst. *Appl Catal A* **139,** L11–L16 (1996).
16. http://en.wikipedia.org/wiki/Andrussow_process
17. *Andrussow, L. Angew Chem* **48,** *593 (1935).*
18. http://acswebcontent.acs.org/landmarks/landmarks/soh/index.html
19. Weissermel, K., Arpe, H. J., Industrielle Organische Chemie, Wiley-VCH, Weinheim, pp. 211 (1998).
20. Pollak, P., Romeder, G., Hagedorn, F., Gelbke, H. P., Ullmann's encyclopedia of industrial chemistry, Vol. 17A pp. 371 (Eds.: B. Elvers, S. Hawkins, G. Schulz), VCH, Weinheim, (1991).
21. Chuck, R., Lonza, A. G. Proc. 1st Int. IUPAC conference on green sustainable chemistry, Sept 2006.
22. Chuck, R. Technology development in nicotinate production. *Appl Catal A* **280,** 75–82 (2005).

23. Heinz, D., Hölderich, W.F., Krill, S., Böck, W., and Huthmacher. K. $V_2O_5/TiO_2$ catalysts for the vapor-phase oxidation of β-picoline: influence of the $TiO_2$-carrier. *J Catal* **192,** 1–10 (2000).
24. Shimizu, S., Shoji, T., Kono, K., Nakaishi T., EP 301540 *(Koei Chemical Co., Ltd., Japan)* (1987).
25. http://www.koeichem.com/english/company/history.htm
26. Global Tuberculosis Report of WHO (2012).
27. Afanas'eva, T. A., Kagarlitskii, A. D., Kan, I. I.; Suvorov, B. V. Khimiya Geterotsiklicheskikh Soedinenii **4,** 675 (1969).
28. Suvorov, B. V., Kagarlitskii, A. D., Sembaev, D. Kh., Kolodina, I. S., Loiko, A. I. GB 1317064 A *(Institut Khimicheskikh Nauk, Alma-Ata, USSR)* (1973).
29. Ding, C., Zhu, X., Liu, F., Zhou, X., Xue, J. CN 101602720 A (2009) *(Nantong Acetic Acid Chemical Co., China)*.
30. Lee, D. -K., Chang, T. -S., Cho, D. -H., Hong, S. B., Shin, C. -H. Ammoxidation of isomeric picolines on molybdenum phosphate catalyst. *Korean J Chem Eng* **20,** 279–283 (2003).
31. Lee, Y. K., Shin, C. -H., Chang, T. -S., Cho, D. -H., Lee, D. -K. EP 684074 A1 *(Korea Research Institute of Chemical Technology, Taejon, Korea)* (1995).
32. Vorob'ev, P. B., Gabdrakipov, V. Z., Mikhailovskaya, T. P., and Sembaev, D. Kh. Reactivity of Isomeric Picolines in Oxidative Ammonolysis on a Vanadium Oxide Catalyst. *Russian J Gen Chem* (Translation of Zhurnal Obshchei Khimii) **71,** 605–607 (2001).
33. Masthan, S. K., Narayana, K. V., Venkat Rao, V., and Kanta Rao, P. Characterization and evaluation of a commercial vanadia catalyst for ammoxidation of picolines. *Ind J Chem Technol* **3,** 313 (1996).
34. Vorob'ev, P. B., Glubokovskikh, L. K., and Sembaev, D. Kh. Izvestiya Ministerstva Obrazovaniya i Nauki Respubliki Kazakhstan, Seriya Khimicheskaya **3,** 34 (2002).
35. Suvorov, B. V., Sembaev, D. Kh., Afanas'eva, T. A., Tolmacheva, T. P., Saurambaeva, L. I., Gostev, V. I., Kan, I. I., and Glubokovskikh, L. K. SU 1014228 A1 *(Institut Khimicheskikh Nauk Akademia Nauk KazSSR, Alma-Ata, USSR)* (1991).
36. Suvorov, B. V., Glubokovskikh, L. K., Kan, I. I., Berstenev, S. V. Zhurnal Prikladnoi Khimii 64 (1991) 2451.
37. Kalevaru, V. N., Madaan, N., and Martin, A. Synthesis, characterization and catalytic performance of titania supported VPO catalysts for the ammoxidation of 3-picoline. *Appl Catal A* **391,** 52–62 (2011).
38. Basude, M. Ammoxidation of 3-picoline over antimony promoted VPO catalysts and alumina supported Sb-VPO catalyst. *J Chem Pharm Res* **3,** 311–322 (2011).
39. Martin, A., Janke, C., and Kalevaru, V. N. Ammoxidation of 3-picoline to nicotinonitrile over VPO catalysts. *Appl Catal A* **376,** 13–18 (2010).
40. Roy, S. K., Dutta, P., Nandi, L.,N., Yadav, S.,N., Mondal, T.,K., Ray, S.,C., Mitra, S., and Samuel, P. The influence of support on ammoxidation of 3-picoline over vanadia catalyst. *J Mol Catal A* **223,** 211–215 (2004).
41. Ma, Y.; Zhou, X.; Yang, Z.; Zhang, Y.; and Feng, X. Shiyou Huagong **33,** 407 (2004).
42. Yu, Peng., Huang, Chi., Li, Xiao-Yun., Zheng, Qiong, Wuhan Daxue Xuebao, Lixueban **49,** 179 (2003).
43. Fischer, A., Martin, A., Narayana, K. V., Lücke, B., Weckbecker, C., Huthmacher, K. DE 10 335 454 A1, *(Reilly Industries, Inc., Indianapolis, USA)* (2005).

44. Fischer, A., Martin, A., Lücke, B., Narayana, K. V., Weckbecker, C., Huthmacher, K. WO 002005016505 A2 *(Reilly Industries, Inc., Indianapolis, USA)* (2005).
45. Fischer, A., Martin, A., Lücke, B., Narayana, K. V., Weckbecker, C., Huthmacher, K. EP 1654233 A2 *(Reilly Industries, Inc., Indianapolis, USA)* (2006).
46. Fischer, A., Martin, A., Lücke, B., Narayana, K.V., Weckbecker, C., Huthmacher, K. US 2007/ 01179 84 A1 *(Reilly Industries, Inc., Indianapolis, USA)* (2007).
47. Fischer, A., Martin, A., Lücke, B., Narayana, K. V., Weckbecker, C., Huthmacher, K., Narayana, K. V. CN 100545149 *(Reilly Industries, Inc., Indianapolis, USA)* (2009).
48. Reddy, B. M., Kumar, M. V., Manohar, B. *Chem Ind* (Dekker) **62,** 487 (1995).
49. Reddy, B. M., Manohar, B. *Chem Ind* (London, UK) (1992) 182-3.
50. Narayana, K. V., Venugopal, A., Rama Rao, K. S., Khaja Masthan, S., Venkat Rao, V., and Kanta Rao, P. Ammoxidation of 3-picoline over $V_2O_5/TiO_2$ (anatase) system. II. Characterisation of the catalysts by DTA, SEM, FTIR, ESR and oxygen and ammonia chemisorptions. *Appl Catal A* **167,** (1998) 11–22.
51. Narayana, K. V., Venugopal, A., Rama Rao, K. S., Venkat Rao, V., Khaja Masthan, S., and Kanta Rao, P. Ammoxidation of 3-picoline over $V_2O_5TiO_2$ (anatase) system I. Relationship between ammoxidation activity and oxidation state of vanadium. *Appl Catal A* **150,** 269–278 (1997).
52. Zhang, F., Chen, F., Xiao, Q., Zhong, Y., and Zhu, W. *Adv Mater Res* **396–398,** 791 (2012).
53. Jiang, F., Wei, R., Gao, L., Xiao, G., and Niu, L., Ammoxidation of 3-picoline to nicotinonitrile using silica-supported VCrO catalysts. *Res Chem Intermed* **39,** 1353–1361 (2013).
54. Chary, K. V. R., Kishan, G., Narayana, K. V., and Bhaskar, T. Ammoxidation of 3-Picoline to Nicotinonitrile over Highly Dispersed V2O5/ZrO2 Catalysts. *J Chem Res* 314 (1998).
55. Andersson, A., Bovin, J. O., and Walter, P. Ammoxidation of 3-picoline: An activity and high-resolution electron microscopic investigation of vanadium oxide catalysts. *J Catal* **98,** 204–220 (1986).
56. Andersson, A. Activities of V-Ti-O catalysts in the ammoxidation of 3-picoline: II. Acid-base properties and infrared spectra. *J Catal* **76,** 144–156 (1982).
57. Andersson, A. Structural dynamics of a $V_2O_5SnO_2$ catalyst in the ammoxidation of 3-picoline. *J Catal* **69,** 465–474 (1981).
58. Andersson, A. and Lundin, S. T. Activities of vanadium oxides in ammoxidation of 3-picoline. *J Catal* **58,** 383–395 (1979).
59. Prasad, R. and Kar, A. K. *Chem. Age Ind.* **26,** 375 (1975).
60. Chary, K. V. R., Kishan, G., Kumar, C. P., Sagar, G. V., and Niemantsverdriet, J. W. Characterization and reactivity of vanadium oxide catalysts supported on niobia. *Appl Catal A* **245,** 303–316 (2003).
61. Martin, A., Kalevaru, V. N., Lücke, B., and Brückner, A. Deactivation and oxidative regeneration of $VTiSbSiO_x$ catalyst for ammoxidation of 3-picoline to nicotinonitrile. *Appl Catal A* **335,** 196–203 (2008).
62. Prasad, R. And Kar, A. K. Kinetics and Mechanism of the Vapor-Phase Ammoxidation of 3-Picoline. *Ind Eng Chem Proc Des Dev* **15,** 170–175 (1976).

63. Radheshyam, A., Reddy, V. S., Dwivedi, R., and Prasad, R. Kinetic studies and mechanism evolution of the ammoxidation of 3-picoline over $V_2O_5/ZrO_2$ catalyst. *Can J Chem Eng* **83,** 274–280 (2005).
64. Kalevaru, V. N., Raju, B. D., Venkat Rao, V., and Martin, A. Preparation, characterisation and catalytic evaluation of $MgF_2$ supported $V_2O_5$ catalysts for ammoxidation of 3-picoline. *Appl Catal A* **352,** 223–233 (2009).
65. Kalevaru, V. N., Raju, B. D., Venkat Rao, V., and Martin, A. Ammoxidation of 3-picoline over $V_2O_5/MgF_2$ catalysts: Correlations between nicotinonitrile yield and $O_2$ and $NH_3$ chemisorption properties. *Catal Commun* **9,** 715–720 (2008).
66. Narayana, K. V., David Raju, B., Khaja Masthan, S., Venkat Rao, V., and Kanta Rao, P. Reactivity of $V_2O_5/MgF_2$ catalysts for the selective ammoxidation of 3-Picoline. *Catal Lett* **84,** 27–30 (2002).
67. Narayana, K. V., Masthan, S. K., Venkat Rao, V., Raju, B. D., and Kanta Rao, P. Influence of $V_2O_5$ content on ammoxidation of 3-picoline over $V_2O_5/AlF_3$ catalysts. *Catal Commun* **3,** 173–178 (2002).
68. Narayana, K. V., Raju, B. D., Masthan, S. K., Rao, V. V., Rao, P. K., Subrahmanian, R., and Martin, A. ESR spectroscopic characterization of $V_2O_5/AlF_3$ ammoxidation catalysts. *Catal Commun* **5,** 457–462 (2004).
69. Narayana, K. V., Masthan, S. K., Venkat Rao, V., and Kanta Rao, P. $CeF_3$ as a novel support for preparing highlyactive and selective vanadia catalysts for the ammoxidation of 3-picoline. *J Chem Res* 328 (1997).
70. Narayana, K. V.; Raju, B. D., Masthan, S. K., Venkat Rao, V., Kanta Rao, P., and Martin, A. Cerium fluoride supported $V_2O_5$ catalysts: physico-chemical characterization and 3-picoline ammoxidation activity. *J Mol Catal* **223,** 321–328 (2004).
71. Janke, C., Radnik, J., Bentrup, U., Martin, A., and Brückner, A. Vanadium-containing oxynitrides: effective catalysts for the ammoxidation of 3-picoline. *ChemCatChem* **1,** 485–491 (2009).
72. Janke, C., Schneider, M., Bentrup, U., Radnik, J., Martin, A., Scholz, G., and Brückner, A. Impact of phosphorus and nitrogen on structure and catalytic performance of VZrPON oxynitrides in the ammoxidation of 3-picoline. *J Catal* **277,** 196–207 (2011).
73. Srinivas, N., Radha Kishan, M., Kulkarni, S. J., and Raghavan, K. V. Ammoxidation of picolines over modified silicoaluminophosphate molecular sieves. *Microporous Mesoporous Mater* **39,** (2000) 125–134.
74. Kanta Rao, P., Rama Rao, K. S., Venkat Rao, V., Narayana, K. V., Venugopal, A. Rama Rao, A. V., and Subrahmanyam, M. US 5910465 A *(CSIR, New Delhi, India)* (1999).
75. Kanta Rao, P., Venkat Rao, V., Rama Rao, K. S., Venugopal, A., Narayana, K. V., Subrahmanyam, M., and Rama Rao, A. V. IN 187528 *(CSIR, New Delhi, India)* (2000).
76. Cavani, F., Centi, G., Marion, P., Eds.: Jackson, S. D., and Hargreaves, J. S. J. Metal oxide catalysis, Wiley-VCH, pp. 771 (2009).
77. Roy, S. K., Roy, S. C., Dutta, P., Nandi, L., and Yadav, S. WO 2004087310 A1 *(CSIR, New Delhi, India)* (2004).
78. Sasari, Y., Yamamoto, H., Mori, K., and Miyaki, K. *Stud Surf Sci Catal* **121,** (1999) 335–338.
79. Adams, R. D. and Pearl, W. C. Pacifichem 2010, *Int Chem Congr Pacific Basin Soc* Honolulu, USA (2010).

80. Thomas, J. M., Raja, R., Gai, P. L., Groenbeck, H., and Hernandez-Garrido, J. C. Exceptionally Active Single-Site Nanocluster Multifunctional Catalysts for Cascade Reactions. *ChemCatChem* **2,** (2010) 402–406.
81. Raja, R., Adams, R. D., Blom, D. A., Pearl, W. C., Jr., Gianotti, E., and Thomas, J.-M. New catalytic liquid-phase ammoxidation approach to the preparation of niacin (Vitamin $B_3$). *Langmuir* **25,** (2009) 7200–7204.
82. Afanas'eva, T. A., Kagarlitskii, A. D., Serazetdinova, V. A., Saltybaeva, L. S., and Suvorov, B. V. Khimiya Geterotsiklicheskikh Soedinenii 672 (1969).
83. Mikhailovskaya, T. P., Yugay, O. K., Chukhno, N. I., and Sembaev, D. Kh. Oxidative ammonolysis of 3- and 4-methylpyridines on vanadium oxide catalysts modified with titanium and tin oxides. *Russ J Appl Chem* **85,** 191–195 (2012).
84. Vorob'ev, P. B., Serebryanskaya, A. P., and Sembaev, D. Kh. Izvestiya Natsional'noi Akademii Nauk Respubliki Kazakhstan, Seriya Khimicheskaya 46 (2009).
85. Kanta Rao, P., Rama Rao, K. S., Khaja Masthan, S., Narayana, K. V., Rajiah, T., and Venkat Rao, V. Highly active titania supported ceria catalysts for ammoxidation of picolines. *Appl Catal A* **163,** 123–127 (1997).
86. Verma, P.K. and Agarwal, A. WO 2003022819 A1 *(Jubilant Organosys Ltd., Noida, India)* (2003).
87. Vorob'ev, P. B., Suvorov, B. V., Shlyugina, I. A., Gabdrakipov, V. Z., and Mikhailovskaya, T. P. *Zhurnal Obshchei Khimii* **63,** 1332 (1993).
88. Mondal, T. K., Jha, Y., and Roy, S. K. J. Ind. Chem. Soc. **83,** 1183 (2006).
89. Kulkarni, S. J., Rao, R. R., Subrahmanyam, M., Farsinavis, S., Kanta Rao, P., and Rama Rao, A.V. IN 190432 A1 *(CSIR, New Delhi, India)* (2003).
90. Farsinavis, S., Kanta Rao, P., Rama Rao, A.V., Kulkarni, S.J., Rao, Revur, R., and Subrahmanyam, M. IN 190431 A1 *(CSIR, New Delhi, India)* (2003).
91. Srinivas, N., Radha Kishan, M., Kulkarni, S.J., and Raghavan, K. V. Ammoxidation of picolines over modified silicoaluminophosphate molecular sieves. *Microporous Mesoporous Mater* **39,** 125–134 (2000).
92. Kulkarni, S. J., Rao, R. R., Subrahmanyam, M., Farsinavis, S., Kanata Rao, P., and Rama Rao, A. V. IN 190238 A1 *(CSIR, New Delhi, India)* (2003).
93. Naskidashvili, T. I., Ioseliani, D. K., Kvantaliani, E. K., Saradzhishvili, K. G. Izvestiya Akademii Nauk Gruzii, Seriya Khimicheskaya **26,** 60 (2000).
94. Das, A., Kar, A. K., Ind. J. Technol. 18 (1980) 267.
95. Das, A. and Kar, A. Kinetics of the Vapor-Phase Ammoxidation of 4-Picoline. *Ind Eng Chem Proc Des Dev* **19,** 689–693 (1980).
96. Gut, G. and Dirr, G. Reaktionstechnische untersuchungen über die Katalytische ammoxydation von 4-picolin. *Chem Eng Sci* **29,** 443–449 (1974).
97. Dhachapally, N., Kalevaru, V.N., Brückner, A., and Martin, A. Metal vanadate catalysts for the ammoxidation of 2-methylpyrazine to 2-cyanopyrazine. *Appl Catal A* **443–444,** 111–118 (2012).
98. Martin, A., Dhachapally, N., Brückner, A., Kalevaru, V. N., American Chemical Society, Division of Petroleum Chemistry **57,** 257– (2012).
99. Dhachapally, N., Kalevaru, V.N., Radnik, J., and Martin, A. Tuning the surface composition of novel metal vanadates and its effect on the catalytic performance. *Chem Commun* **47,** 8394–8396 (2011).

100. Dhachapally, N., Kalevaru, V. N., and Martin, A. EP 2428267 A1 *(LIKAT, Rostock, Germany)* (2012).
101. Manohar, B. Ammoxidation of 2-methylpyrazine to 2-cyanopyrazine over promoted VPO catalysts and alumina supported Sb-VPO catalyst. *J Chem Pharmaceut Res* **4,** 2781–2788 (2012).
102. Nagaraju, P., Srilakshmi, Ch., Pasha, N., Lingaiah, N., Suryanarayana, I., and Saiprasad, P. S. Effect of P/Fe ratio on the structure and ammoxidation functionality of Fe-P-O catalysts. *Appl Catal A* **334,** 10–19 (2008).
103. Hong, C. and Li, Y. 2-Cyanopyrazine Prepared from 2-methylpyrazine by catalytic ammoxidation on MoVPO catalyst. *Chinese J Chem Eng* **14,** 670–675 (2006).
104. Srilakshmi, Ch., Nagaraju, P., Sreedhar, B., Saiprasad, P. S., Kalevaru, V. N., Lücke, B., and Martin, A. Ammoxidation activity of in situ synthesized ammonium salt of molybdophosphoric acid on VOPO4 catalysts. *Catal Today* **141,** 337–343 (2009).
105. Mohan Reddy, K., Lingaiah, N., Rao, P. S. N., Nagaraju, P., Saiprasad, P. S., and Suryanarayana, I. Studies on ceria supported vanadium incorporated ammonium salt of 12-molybdophosphoric acid catalysts. *Catal Lett* **130,** 154–160 (2009).
106. Bondareva, V. M., Andrushkevich, T. V., Paukshtis, E. A., Paukshtis, N. A., Budneva, A. A., and Parmon, V. N. Investigation of the reaction mechanism of methylpyrazine ammoxidation on vanadia–titania catalyst by FTIR in situ. *J Mol Catal A* **269,** 240–245 (2007).
107. Bondareva, V. M., Andrushkevich, T. V., Lapina, O. B., Vlasov, A. A., and Dovlitova, L. S. Ammoxidation of methylpyrazine over phosphorus-modified vanadium-titanium catalysts. *React Kinet Catal Lett* **79,** 165–173 (2003).
108. Bondareva, V. M., Andrushkevich, T. V., Lapina, O. B., Vlasov, A. A., Dovlitova, L. S., and Burgina, E. B. Ammoxidation of methylpyrazine over vanadium-titanium catalysts modified by alkali additives. *React Kinet Catal Lett* **78,** 355–363 (2003).
109. Reddy, B. M., Manohar, B., Reddy, E. P., Patil, K. S., Rao, and Ramarao, A.V. IN 182185 A1 *(CSIR, New Delhi, India)* (1999).
110. Srilaxmi, Ch., Lingaiah, N., Hussain, A., Saiprasad, P.S., Narayana, K.V., Martin, A., and Lücke, B. In situ synthesis and ammoxidation activity of ammonium salt of molybdophosphoric acid on $VOPO_4$ catalysts. *Catal Commun* **5,** 199–203 (2004).
111. Bondareva, V. M., Andrushkevich, T. V., Plyasova, L. M., Burgina, E. B., Lapina, O. B., and Altynnikov, A.A. *Kinet Catal* (Translation of Kinetika i Kataliz) **38,** 662 (1997).
112. Forni, L. Kinetic study of the ammoxidation of methylpyrazine to cyanopyrazine. *Appl Catal A* **37,** 305–314 (1988).
113. Forni, L. Ammoxidation of 2-methylpyrazine over oxide catalysts. *Appl Catal A* **20,** 219–230 (1986).
114. Sokolovskii, V. D. in Mechanism and kinetics of catalytic reactions, Institute of Catalysis, Novosibirsk 41 (1977).
115. Reutov, O. A., Beletskaya, I. P., and Butin, K. P. C-H acids, Crompton, T.R. (Ed.), Pergamon, New York (1978).
116. Martin, A., Lücke, B., Wolf, G.-U., and Meisel, M. Effect of the position of Cl-substituents on the nitrile formation during the ammoxidation of chlorotoluenes. *Catal Lett* **33,** 349–355 (1995).

117. Chmyr, I. M., Bukeikhanov, N. R., Suvorov, B. V., Izv. Akad. Nauk Kaz. SSR, Ser. Khim. 36m 43 (1983).
118. Martin, A. and Lücke, B. Ammoxidation of methylaromatics over vanadium phosphate catalysts. I. Effect of the size, position and electronic properties of substituents on the catalytic conversion of methylaromatics to the corresponding nitriles. *Catal Today* **32,** 279–283 (1996).
119. Brückner, A., Martin, A., Lücke, B., and Hannour, F. K. Coupled vanadyl centres in vanadium phosphorus oxide catalysts: Essential structural units for effective catalytic performance in the ammoxidation of methylaromatics. *Stud Surf Sci Catal* **110,** 919–928 (1997).
120. Murakami, Y., Niwa, M., Hattori, T., Osawa, S., Igushi, S., and Ando, H. Reaction mechanism of ammoxidation of toluene: I. Kinetic studies of reaction steps by pulse and flow techniques. *J Catal* **49,** 83–91 (1977).
121. Niwa, M., Ando, H., and Murakami, H. Reaction mechanism of ammoxidation of toluene: II. Identification of reaction intermediate adsorbed on $V_2O_5Al_2O_3$ by infrared spectroscopy. *J Catal* **49,** 92–96 (1977).
122. Cavalli, P., Cavani, F., Manenti, F., and Trifirò, F. Ammoxidation of toluene to benzonitrile on vanadium-titanium oxides catalysts prepared by precipitation. The role of catalyst composition. *Ind Eng Chem Res* **26,** 639–647 (1987).
123. Busca, G., Cavani, F., and Trifirò, F. Oxidation and ammoxidation of toluene over vanadium-titanium oxide catalysts: A Fourier transform infrared and flow reactor study. *J Catal* **106,** 471–482 (1987).
124. Otamiri, J. and Andersson, A. Kinetics and mechanisms in the ammoxidation of toluene over a $V_2O_5$ catalyst. Part 1: Selective reactions. *Catal Today* **3,** 211–222 (1998).
125. Sanati, M. and Andersson, A. DRIFT study of the oxidation and the ammoxidation of toluene over a TiO2(B)-supported vanadia catalyst. *J Mol Catal* **81,** 51–62 (1993).
126. Ramis, G., Busca, G., Bregani, F., and Forzatti, P. Fourier transform-infrared study of the adsorption and coadsorption of nitric oxide, nitrogen dioxide and ammonia on vanadia-titania and mechanism of selective catalytic reduction. *Appl Catal A* **64,** 259–278 (1990).
127. Azimov, A. B., Vislovskii, V. P., Mamedov, E. A., and Rizayev, R. G. "Dehydrogenation" mechanism for ammoxidation of alkylaromatic hydrocarbons. *J Catal* **127,** 354–365 (1991).
128. Berndt, H., Büker, K., Martin, A., Brückner, A., and Lücke, B. Redox interaction of ammonia with $(VO)_2P_2O_7$. *J Chem Soc Faraday Trans* **91,** 725–731 (1995).
129. Miyata, H., Ohno, T., and Hatayama, F. FTIR studies of the interaction of aromatic hydrocarbons with vanadium oxide layered on $ZrO_2$ and $TiO_2$. *J Chem Soc Faraday Trans* **91,** 3505–3510 (1995).
130. Thompson, M. R., Hess, A. C., Nicholas, J. B., White, J. C., Anchell, J., and Ebner, J. R. A concise description of the bulk structure of vanadyl pyrophosphate and implications for n-butane oxidation. *Stud Surf Sci Catal* **82,** 167–181 (1994).
131. Handbook of Chemistry and Physics, 75th Ed. (CRC Press, Boca Raton) (1995).
132. Martin, A. Brückner, A., Zhang, Y., and Lücke, B. Ammoxidation of methylaromatics over NH4+-containing vanadium phosphate catalysts—New mechanistic insights. *Stud Surf Sci Catal* **108,** 377–384 (1997).

133. Brückner, A., Martin, A., Steinfeldt, N., Wolf, G.-U., and Lücke, B. Global potentials for calcium and strontium solids. *J Chem Soc Faraday Trans* **92,** 425–432 (1996).
134. Martin, A., Berndt, H., Lücke, B., and Meisel, M. Reaction pathway of benzonitrile formation during toluene ammoxidation on vanadium phosphate catalysts. *Topics Catal* **3,** 377–386 (1996).
135. Belokopytov, Yu. V., Kholyavenko, K. M., and Gerei, S. V. An infrared study of the surface properties of metal oxides: 2. The interaction of ammonia with the surface of $Fe_2O_3$, ZnO, $MoO_3$, and $V_2O_5$. *J Catal* **60,** 1–7 (1979).
136. Takagi-Kawai, M., Soma, M., Onishi, T., and Tamaru, K. The adsorption and the reaction of NH3 and NOx on supported V2O5 catalysts: effect of supporting materials. *Can J Chem* **58,** 2132–2137 (1980).
137. Grasselli, R. K., Callahan, J. C. (Standard Oil Co (Ohio)), Ger Offen DE 2 314 151 (Standard Oil Co., Cleveland, USA) (1973).
138. Burrington, J. D., Kartisek, T., and Grasselli, R. Surface intermediates in selective propylene oxidation and ammoxidation over heterogeneous molybdate and antimonate catalysts. *J Catal* **87,** 363–380 (1984).
139. Grasselli, R.K. Selective oxidation and ammoxidation of olefins by heterogeneous catalysis. *J Chem Ed* **63,** 216 (1986).
140. Zhang, Y., Meisel, M., Martin, A., Lücke, B., Witke, K., and Brzezinka, K.-W. In Situ Raman Investigation on the Structural Transformation of Oxovanadium Hydrogenphosphate Hemihydrate in the Presence of Ammonia. *Chem Mater* **9,** 1086–1091 (1997).

# CHAPTER 8

# CRACKING AND OXIDATIVE DEHYDROGENATION OF ETHANE TO ETHYLENE: PROCESS AND INTENSIFICATION OPTIONS

A. SRI HARI KUMAR and P. S. SAI PRASAD

## CONTENTS

## 8.1 INTRODUCTION

### *8.1.1 IMPORTANCE OF ETHYLENE*

Ethylene is an important building block of the chemical industry. It tops the list of organic chemicals produced in the world in terms of volume. It is a colorless, flammable gas with a slightly sweet odor. Ethylene is used in several major industrial reactions such as polymerization, oxidation, halogenation, hydrohalogenation, alkylation, hydration, oligomerization, and hydroformylation. Many chemical products are obtained from these reactions. Approximately 90% of ethylene produced goes into the manufacture of chemicals like ethyl benzene, styrene, ethylene oxide, ethylene dichloride, ethylene glycols, synthetic lubricants, surfactants, plasticizers, detergents, and a variety of polyethylene compounds. Polyethylenes of various densities account for more than 50% of world ethylene demand, out of, which the films have wide applications in packaging, carrier bags and trash liners. In smaller quantities, ethylene is used as an anesthetic agent as well as a welding gas. Figure 8.1 illustrates various chemicals for, which ethylene is the raw material.

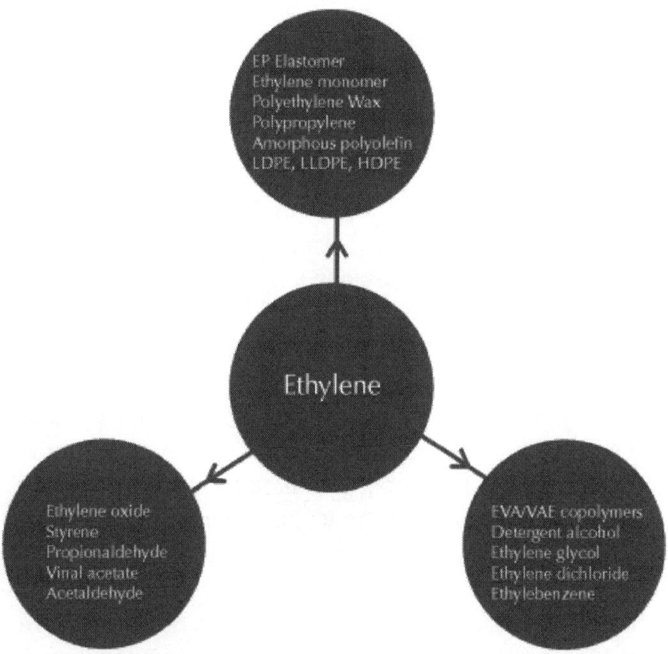


**FIGURE 8.1** Industrial importance of ethylene [1].

### 8.1.2 PRODUCTION AND CONSUMPTION PATTERNS OF ETHYLENE

The annual global production of ethylene during the past few years is depicted in Table 8.1. Ethylene production has shown a gradual increase over the past few years. During 2012 and 2013, it is postulated that approximately another 8.5 million tpa capacity will be added to the existing capacities. During 2009, about 60% of ethylene produced went into polyethylene production, 15% to ethylene oxide, about 12% to ethylene dichloride (polyvinyl chloride), styrene consumed 6% of ethylene, 1% went towards vinyl acetate monomer and about 6% was used for other products.

**TABLE 8.1** Yearly production of ethylene.

| Year | Global ethylene production capacity (million tones) |
|---|---|
| 2007 | 120 |
| 2008 | 127 |
| 2009 | 130 |
| 2010 | 138 |
| 2012 | 141 |
| 2015 (Projected) | 160 |

### 8.1.3 FEEDSTOCKS FOR ETHYLENE PRODUCTION

Various liquid and gaseous feedstocks are used for the production of ethylene. The principal feedstock is naphtha. The light gas phase feedstocks include ethane, propane, n-butane, and i-butane. The current statistics reveal that globally, around 50% of ethylene is produced from naphtha due to the ease of its transport into regions where there is a large ethylene demand. Of late, the raw materials of ethylene have become more and more diversified and lighter. Gas oils, which are the crude oil fractions, are used as feedstocks in some parts of the world. Ethylene production from ethane accounts for about 30%. Ethane is the second major component of natural gas. Thus, the production of ethylene from ethane has become a potential route for the subsequent synthesis of oxygenates and aromatic hydrocarbons. Due to the rising oil price and the development of natural gas industry, the proportion of natural gas as a raw material to ethylene production increased substantially, while that of naphtha decreased. The recent technical advancements in exploration and production of natural gas have greatly increased its market production. As a result, ethane-based ethylene production regained its competitive cost edge relative to naphtha. Tables 8.2 and 8.3 depict the world ethylene capacity (continent-wise) and the upcoming ethylene production plants, respectively [2–3].

**TABLE 8.2** Continent-wise ethylene capacity as of 1st January 2012 [3].

| Continent | Mtpa |
|---|---|
| Asia | 69.0 |
| Europe | 29.0 |
| Africa | 1.6 |
| South America | 6.4 |
| North America | 34.5 |
| Australia and Oceania | 0.5 |
| Total | 141.0 |

In Western Europe and Japan, over 80% of ethylene is produced from naphtha, the principal ethylene raw material. As per the world consumption of feedstocks for ethylene production, the order of priority is like naphtha (53%) > ethane (29%) > propane (8%) > gas oil (6%) > butane (3%) > others (1%) [5]. A shift in feedstocks occurred for the period from 1980 to 1991. In the United States and Europe larger amounts of light feedstocks (liquefied petroleum gas LPG: propane + butanes) and NGL (natural gas liquids: ethane, propane, butane) are used for ethylene production, whereas in Japan more naphtha was used in 1995 compared to 1981. The use of gas oils for ethylene production decreased slightly during the 1980 s. Since 2000 a huge amount of new capacities have been installed in the Middle East based on ethane as feedstock, increasing the share of ethane-based production significantly. This trend will continue in the coming years due to favorable production costs for ethane-based crackers.

**TABLE 8.3** Upcoming major ethylene production plants for the years 2012 to 2014 [4].

| Location | Company | Ethylene capacity, Mtpa | | |
|---|---|---|---|---|
| | | 2012 | 2013 | 2014 |
| Arzew, Algeria | Total and Sonatrach | 1.1 | | |
| Heilongjiang province, China | Daqing Petroleum and Chemical Corporation | 0.6 | | |

**TABLE 8.3** *(Continued)*

| Location | Company | Ethylene capacity, Mtpa | | |
|---|---|---|---|---|
| | | 2012 | 2013 | 2014 |
| Dahej, Gujarat, India | ONGC Petro-additions Ltd. | 1.1 | | |
| RassLaffan Industrial City, Qatar | Qatar Petroleum/ExxonMobil Corp. | 1.6 | | |
| Jurong Island, Singapore | ExxonMobil Corp. | | 1.0 | |
| Kachsiung, Taiwan | CPC Corp. | | 1.0 | |
| Lin Yuan, Kaohsiung, Taiwan | CPC Corp. | | 0.6 | |
| Riwais, Abu Dhabi | Abu, Dhabi Polymers Co. Ltd. | | 1.5 | |
| Jose, Anzoategui, Venezuela | Pequiven | 1.1 | | |
| EI Tablazo, Venezuela | Pointer | 1.0 | | |
| Port Arthur, Tex. | BASF Fina Petrochemicals | | | 0.1 |
| Total | | 5.8 | 2.6 | 0.16 |

### 8.1.4 ETHANE FROM SHALE GAS AS A FEEDSTOCK

Shale is the common name for fine-grained sedimentary rock made of clay or mud layers. The gas, which is trapped in this rock is called shale gas. The composition of shale gas is similar to that of natural gas. The typical composition of natural gas is given in Table 8.4. It contains mainly methane (70–90%) with small proportions of nitrogen, ethane, propane, the noble gasses, oxygen and carbon oxide [6].

**TABLE 8.4** Natural gas composition.

| Gas | Mole % |
|---|---|
| $CH_4$ | 70–90% |

**TABLE 8.4** *(Continued)*

| Gas | Mole % |
|---|---|
| $C_2H_6$ | |
| $C_3H_8$ | 0–20% |
| $C_4H_{10}$ | |
| $CO_2$ | 0–8% |
| $N_2$ | 0.5% |
| $O_2$ | 0–0.2% |
| $H_2S$ | 0.5% |
| Ar, He, Ne, Xe | Trace |

Since several years, the US has been recovering shale gas. The reserves are estimated at 60.6 and 827 trillion cubic feet in the years 2009 and 2011, respectively, which reveals an exponential rise. The country's gas production is predictable to increase from 23% of total gas production in 2010 to 49% by 2035 [6]. China is also looking at shale gas reserves for their energy needs [7]. The gas compositions obtained from some typical locations are provided in Table 8.5 [6, 8]. Thus, there seems to be a considerable scope for the availability of ethane for further processing.

**TABLE 8.5** Shale gas compositions of some typical locations in the world [6, 8].

| Place<br>Gas | Marcellus well (USA) | | | | Barnett well (USA) | | | | Cambay basin (India) | | Haynesville (USA) |
|---|---|---|---|---|---|---|---|---|---|---|---|
| | **11** | **2** | **3** | **4** | | **2** | **3** | **4** | **D-A** | **D-B** | |
| $C_1$ (Methane) | 79.4 | 82.1 | 83.8 | 95.5 | 93.7 | 80.3 | 81.2 | 91.8 | 60.44 | 69.99 | 93.7 |
| $C_2$ (Ethane) | 16.1 | 14.0 | 12.0 | 3.0 | 2.6 | 8.1 | 11.8 | 4.4 | 8.33 | 9.98 | 2.6 |
| $C_3$ (Propane) | 4.0 | 3.5 | 3.0 | 1.0 | 0.0 | 2.3 | 5.2 | 0.4 | 13.09 | 6.76 | 0.0 |
| $CO_2$ (Carbon dioxide) | 0.1 | 0.1 | 0.9 | 0.3 | 2.7 | 1.4 | 0.3 | 2.3 | 1.02 | 9.40 | 2.7 |
| $N_2$ (Nitrogen) | 0.4 | 0.3 | 0.3 | 0.2 | 1.0 | 7.9 | 1.5 | 1.1 | 0.09 | 0.19 | 1.0 |

## 8.2 CONVENTIONAL PROCESS TECHNOLOGIES FOR ETHYLENE PRODUCTION

The ethylene production technologies can be divided into the following two broad categories.

(i) High temperature processes: Thermal and steam cracking form the two most important high temperature processes.

(ii) Low temperature process: Oxidative dehydrogenation is the main low temperature process, which is undergoing several improvements.

The following paragraphs deal first with the high temperature processes (Section 2) and next, the low temperature process (Section 3)

### 8.2.1 THERMAL CRACKING

William Merriam Burton developed thermal cracking processes in 1912 which operated at 370–400°C and an absolute pressure of about 6 atm. It was known as the Burton process [9]. Now high temperature (450–750°C) and pressures (up to 70 atm.) are used to break the large hydrocarbons into smaller ones. In thermal cracking, the carbon-carbon (C-C) bonds are broken so that each carbon atom ends up with a single electron. In other words, free radicals are formed. In thermal cracking the feed, for example oil, is heated in the tubes of the cracking furnace to the cracking temperature. Thermal cracking usually occurs by two procedures. In one of them, cracking takes place partly in the heating tubes of the furnace and partly in subsequent pipelines. The delay times of the cracking procedure are not exactly known, but they are relatively short (< 1 min). The pressure varies greatly, going down from the furnace entrance to the exit. In the other cracking procedure, the hydrocarbon feed is first heated in the cracking furnace to a suitable reaction temperature, and the actual cracking reaction takes place in a separate reaction zone, where the delay time is considerably longer than in the preceding procedure, that is, of the order of 10 to 30 min. No heat is introduced into the reaction zone. At the other end of the reaction zone, the extraction of a mixture of liquid and gas takes place. The mixture further goes to the refining steps, for instance to distillation.

In thermal cracking of hydrocarbons, two types of reactions take place; viz., cracking where the long-chain molecules are split into smaller ones, causing reduction of viscosity and polycondensation, wherein the molecules combine and produce pitch and coke with the liberation of hydrogen. The latter is undesirable as it results in the formation of greater quantities of asphaltenes. Since the polycondensation is favored at higher temperatures, lower reaction temperature and correspondingly longer delay times are used to minimize asphaltene formation. Maintenance of an appropriate delay time is very important for obtaining better product yields. The thermal cracking has several disadvantages. The reaction is highly endothermic ($\Delta H_{590°C}$ =+143 kJ/mol), the operating pressure is high and it produces large amounts of solid, unwanted coke.

## 8.2.2 CATALYTIC CRACKING OF ALKANES TO OLEFINS

Catalytic dehydrogenation of alkanes produces light alkenes as well as hydrogen. An important advantage of catalytic cracking is that it occurs at lower temperature s, 700–800°C. However, catalytic cracking produces methane as a byproduct and also results in coking. Alkane dehydrogenation is catalyzed by Pt, mostly promoted by other elements, such as Sn, Zn, Ge, Ga, or In, to enhance alkene selectivity. Pt interacts with these elements forming bimetallic alloys. The novel trend is to use bimetallic nanoparticles of Pt and an alloying element to reduce coke formation.

The stability of the catalyst is improved by selecting a metal oxide as a support. The dehydrogenation reaction proceeds on small ensembles of surface Pt atoms. Other side reactions like hydrogenolysis, isomerization, and coking proceed through the formation of highly dehydrogenated species over multiple adsorption sites. Thus, support materials are important as they play a key role in the modification of the selectivity by catalyzing the lateral reactions. Quite often $\gamma$ -$Al_2O_3$ and $SiO_2$ are used as supports. However, nonacidic supports are found to offer less cracking. Magnesium aluminate ($MgAl_2O_4$) has been extensively studied recently and is used in the Phillips Star process for dehydrogenation of low-molecular-weight paraffins [10].

Basic supports such as calcined hydrotalcite-like materials are also employed for this purpose. Particularly, Sun et al. proposed Mg(Al)O type hydrotalcites to be the ideal catalysts. The Al cations present in the support stabilize the dispersed metal particles against sintering. Most of the reports in this direction suggest Sn as the best element for promotion. The size of Pt ensembles and the bond strength of chemisorbed hydrocarbons are modified by the addition of Sn. Thus, Pt-Sn combination performs better than Pt alone. Sn forms an alloy with platinum. The performance of the catalyst was found to be dependent on the composition of the Pt and Sn precursors and the catalyst preparation procedure. PtGa bimetallic nanoparticles were found to exhibit not only high activity and selectivity for ethane dehydrogenation, they also reduce carbon deposition. An improvement in this direction is the hydrotalcite supported PtIn catalysts, in, which the dependence on the composition of the Pt and In precursors and the procedure used for catalyst preparation were reduced. Consequently, PtIn was claimed to be effective catalyst for dehydrogenation of light alkanes. Maximum activity was achieved when the catalyst contained a bulk In/Pt ratio of 0.48, corresponding to a PtIn alloy of 12% In. The selectivity to ethane was nearly 100% for $Pt/Mg(In)(Al)_{0.48}$ [11].

### 8.2.3 STEAM CRACKING

The bulk of worldwide annual commercial production of ethylene is based on steam cracking. In this petrochemical process saturated hydrocarbons are broken down into smaller, often unsaturated, hydrocarbons. It is the principal industrial method for producing lighter alkenes, mainly ethylene and propylene. The steam cracking operation involves heating the hydrocarbon (ethane, propane, butane, naphtha or gas oil) in radiant coils in the presence of steam in a furnace. The heat is transferred to the coils by radiation. This technology has been commercially practiced since early 1940s. Hydrocarbon feed is heated with steam up to ~1050°C and fed to Cr-Ni reactor tubes. Cracked products exit at ~850°C and are rapidly quenched to 300°C to prevent secondary reactions. The product is scrubbed to remove $H_2S$ and $CO_2$, and then dried. $C_2$ and $C_3$ components are separated by low temperature fractional distillation. $C_4$ components are separated by chemical means, because their boiling points are too similar. Since the cracking is highly endothermic, high-energy input is necessary. The reaction products leaving the radiant tube are cooled within 0.02–0.1 (s) to prevent degradation of the highly reactive products by secondary reactions.

Cracking of ethane can be visualized as a free radical dehydrogenation reaction, where hydrogen is a coproduct:

$$C_2H_6 \rightarrow CH_2{=}CH_2 + H_2; \quad \Delta H_{590\,°C} = +143 \text{ KJ/mol} \tag{1}$$

The highly endothermic reaction is favored at high temperature and low pressures. Superheated steam is used to reduce the partial pressure of the reacting ethane. Superheated steam also reduces carbon deposits that are formed by the pyrolysis of hydrocarbons at high temperature s. Pyrolysis of ethane produces carbon and hydrogen.

$$C_2H_6 \rightarrow 2C + 3H_2 \tag{2}$$

Ethylene pyrolysis can also take place in the same way. Additionally, the presence of steam as a diluent reduces the chances of hydrocarbons coming in contact with the reactor tube-wall. The carbon deposits reduce heat transfer through the reactor tubes, but steam reduces this effect by reacting with them [12] (steam reforming reaction).

$$C + H_2O \rightarrow CO + H_2 \qquad (3)$$

Many side reactions occur when ethane is cracked. A probable sequence of reactions between ethylene and a formed methyl or an ethyl free radical could be represented by the following reactions.

$$CH_2 = CH_2 + \cdot CH_3 \rightarrow CH_3CH_2\cdot CH \rightarrow CH_3CH{=}CH_2 + \cdot H \qquad (4)$$

$$CH_2 = CH_2 + CH_3\cdot CH_2 \rightarrow CH_3CH_2CH_2\cdot CH \rightarrow CH_3CH_2CH{=}CH_2 + \cdot H \qquad (5)$$

Propene and l-butene, respectively, are produced in this free radical reaction. A newly formed free radical may get terminated by the abstraction of a hydrogen atom. The steam/hydrocarbon ratio, cracking temperature and furnace residence time are the important process variables in the cracking reaction. A higher steam/hydrocarbon ratio favors olefin formation. Steam reduces the partial pressure of the hydrocarbon mixture and increases the yield of olefins [12]. Steam/hydrocarbon weight ratio varies in the range of 0.2–1 for ethane cracking. Increasing temperature enhances the formation of lighter olefins, whereas low temperature cracking favors high molecular weight olefins, and aromatics. Higher temperature are usually selected to maximize olefin production and minimize the formation of carbon deposits. A typical furnace outlet temperature for cracking of ethane is approximately 800°C. In the steam cracking processes, olefins are formed as primary products. Aromatics and higher hydrocarbon compounds result from secondary reactions of the formed olefins. Accordingly, short residence times are required for high olefin yield. When ethane is the feed, shorter residence times are used to maximize olefin production and minimize benzene, toluene, xylene (BTX) and liquid yields. Residence times of 0.5–1.2 (s) are typical [12]. The effect of residence time on the ethylene yield obtained from ethane cracking is presented in Fig. 8.2 [12]. With the decrease in residence time, ethylene yield is found to increase. By and large, ethane as a feed gives mostly ethylene and less of other products [12].

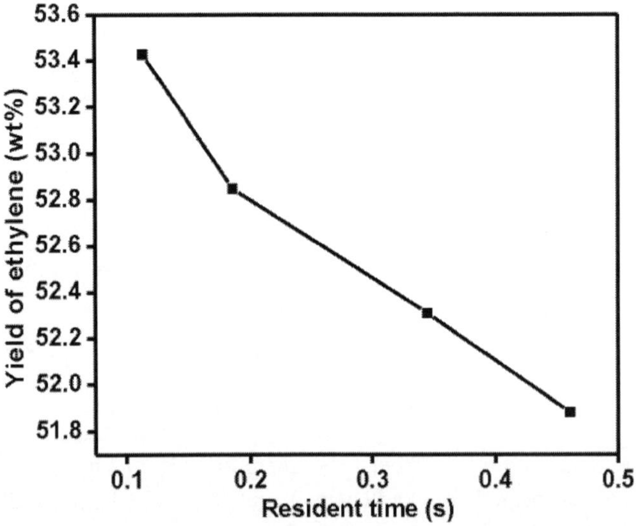


**FIGURE 8.2** Effect of resident time (conversion: 65%, steam dilution: 0.30) [12].

Steam cracking has several disadvantages. The reaction is highly endothermic. The residence time is in the order of milliseconds, which causes high-pressure drop. Coking is a significant problem in the furnace and in the transfer line exchanger. Catalyst activity decreases due to coke formation. Catalyst regeneration step is required. Coke formation reduces product yields, increases energy consumption and shortens coil service life.

### *8.2.4 CATALYTIC STEAM CRACKING IN PRESENCE OF LIMITED OXYGEN*

In the case of thermal cracking the heat transfer from the reactor walls is considerably reduced due to coke deposition on the internal walls of the tubes. This necessitates higher wall temperature (up to 1100°C). Therefore, more energy is required; consequently a reduction in the life of reactor tubes is expected. There would be frequent process shut downs. In order to overcome the problems associated with conventional steam reforming, a few alternative methodologies were adopted. One among them was performing the reaction in the presence of catalysts. The endothermic nature of the reaction was also reduced by adding limited oxygen to the feed. Thus, oxidative conversion of ethane to ethylene, using alkali metal-containing catalysts and alkaline earth metal-promoted $La_2O_3$ or other rare earth oxide catalysts came into existence.

The alkali metal-containing catalysts had poor thermal/hydrothermal stability because of the presence of low melting or volatile active components. Alkaline earth-metal promoted catalysts were found to be useful in this context. Since water was one of the products during the oxidative reforming, the catalysts had to be stable both thermally and hydrothermally and possess high mechanical strength. In order to achieve the last property, the catalytic active component was supported on a suitable carrier. The following paragraphs describe the efforts made in this direction.

Oxidative dehydrogenation (ODH) of ethane to ethylene was studied in the presence of steam and limited $O_2$. Sr promoted $La_2O_3$ (Sr/La=0.1) catalyst showed the best performance with high selectivity (>80%) for $C_{2+}$ olefins at 50–70% conversion of ethane at 800–850°C and contact times less than 10 ms. There was no coke deposition on the catalyst. At higher temperature and $C_2H_6/O_2$ ratios the process was nearly thermo-neutral or mildly exothermic/endothermic, depending upon the process conditions [13]. Auto thermal oxidative dehydrogenation of ethane was performed in the absence and presence of 15% steam on $Sr_{0.1}La_{1.0}Nd_{1.0}O_x$ catalyst. A maximum ethylene yield of 56% (S=71%) could be achieved at GHSV=$8.3{\cdot}10^4$ $h^{-1}$ ($\tau_{STP}$= –45 ms), $C_2H_6/O_2/N_2$=2/1/1 and the reaction temperature of 940° C. The presence of steam decreased the maximal temperature leading to a decrease in the ethylene yield by about 2%. No significant effect of the steam addition on product distribution was observed [14]. An energy efficient conversion of ethane to ethylene involving simultaneous oxidative conversion and thermal cracking of ethane was performed in the presence of steam and limited $O_2$ over a BaO-promoted $La_2O_3$ supported on low surface area macroporous silica-alumina. High selectivity (≥ 85%) for $C_{2+}$ olefins (at 50–60% conversion) were obtained in the process at a low contact time (<10 ms), particularly for the higher $C_2H_6/O_2$ ratios (≥ 6.0) and temperature (≥ 800°C). No coke deposition was observed. The process was claimed to be highly energy efficient and nonhazardous [15].

On a $Sr_{1.0}La_{1.0}Nd_{1.0}O_x$ catalyst, the catalytic and noncatalytic oxidative dehydrogenation of ethane to ethylene was carried out between 600 and 800°C in the presence of 30% steam. When the dead volume of the reactor was low, homogeneous gas-phase reactions were not significant up to 700°C, as compared to the catalytic ones. Above 700°C, the noncatalytic ODE reaction occurred to a large extent resulting in ethylene yields of 46% (S=66%) and 53% (S=65%) at 750 and 800°C, respectively. The addition of steam to the feed gas affected the catalytic as well as the noncatalytic oxidative dehydrogenation of ethane (ODE) slightly [16]. SrO – promoted $Sm_2O_3$, supported on low surface area macroporous silica-alumina also gave good performance (selectivity (≥80%) of $C_{2+}$ olefins at 50–70% conversion of ethane) during the ODE in presence of steam and limited $O_2$ [16]. The effect of replacing $Sm_2O_3$ with $Er_2O_3$, was been thoroughly investigated. The effects of different process parameters such as temperature (700–850°C), $C_2H_6/O_2$ feed ratio (4.0–8.0), and space velocity (50,000–200,000 $cm^3\ g^{-1}\ h^{-1}$) on the conversion of ethane and product selectivity were examined. High selectivity (≥80%) of $C_{2+}$ olefins at 50–70% conversion of ethane could be obtained at very low contact time without coke deposition on the catalyst at 800–850°C.

The oxidative steam reforming of ethane involves a complex network of both heterogeneous and homogeneous reactions such as catalytic and non-catalytic oxidative conversion of ethane reactions (exothermic) and thermal cracking of ethane (endothermic). The catalytic reactions are more predominant at temperature below 750°C. At temperature above 750°C, the homogeneous (noncatalytic) and surface-originated homogeneous reactions become increasingly important. The homogeneous and heterogeneous reactions take place simultaneously. In order to reduce the contribution due to the homogenous reaction, low dead volume reactor packed with inert support was used.

Mulla et al. [17] proposed the following mechanism for the oxidative steam reforming. Ethane is activated by molecular oxygen leading to the formation of ethyl radicals and their subsequent decomposition to ethylene [Reactions 1.6 and 1.7], the conversion of ethane purely by its thermal cracking is drastically increased.

$$C_2H_6 + O_2 \rightarrow C_2H_5 + HO_2 \text{ (homogeneous reaction)} \tag{6}$$

$$C_2H_5 \rightarrow C_2H_4 + H \text{ (homogeneous reaction)} \tag{7}$$

Ethane is expected to be activated on the catalyst surface by the surface oxygen as follows.

$$C_2H_6 + [O]_s \rightarrow C_2H_5 + [HO]_s \text{ (surface reaction)} \tag{8}$$

and the $[O]_s$ sites are regenerated by the dehydration reaction,

$$2[OH] \rightarrow [O]_s + H_2O \text{ (surface reaction)} \tag{9}$$

### 8.2.5 THE DEVELOPMENT OF THERMAL AND STEAM CRACKING METHODOLOGIES

In 1913, Standard Oil patented thermal cracking process. During 1930–1940, ethylene was first separated from coke oven gas and the first commercial plant for the production of ethylene was built by Linde at that time. The development of the world's first steam cracker took place at Baton Rouge by Standard Jersey (subsequently ExxonMobil) in 1941. In 1950s ethylene emerged as a large-volume intermediate, replacing acetylene as prime material for synthesis. After the emergence of steam cracking, all the thermal cracking plants have been replaced with that of steam cracking. Currently, ethylene is primarily produced by steam cracking of hydrocarbons with plant capacities up to 1–1.5 Mtpa. The product composition from ethane cracking are presented in Figs. 8.1–8.3 [12]. The ethylene yield obtained from ethane cracking (~53.4%) is always higher than that of the naphtha cracking (~32.4%) making ethane a more advantageous raw material.

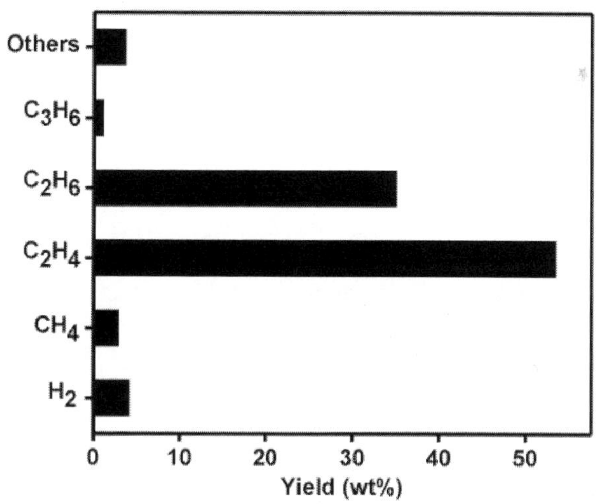


**FIGURE 8.3** Yields of individual components from ethane steam cracking (ethane conversion: 65.01%, steam dilution: 0.30, resident time: 0.1133) [12].

The following are some of the ethane based cracker plants, which came into existence in the recent years:

1. National Petrochemical Company (Iran) 2007: 1 Mtpa ethane cracker established in the 9th olefin complex at Assaluyeh.
2. National Petrochemical Company (Iran) 2008: Ethylene technology with liquid and gas furnaces. The plant came on stream in early 2008. This unit delivers the largest output of ethylene to-date in the world and is one of the large capacity steam-cracker using both gas and liquid as feedstocks to produce ethylene and propylene.

3. Yansab (Kingdom of Saudi Arabia), 2009: One of the world largest steam cracker with ultimate capacity of 1.7 Mtpa at term using ethane and propane feedstocks.
4. Rabigh(Kingdom of Saudi Arabia), 2009: World largest ethane cracker operating at 1.5 Mtpa ethylene employing deep catalytic cracking process to produce large quantities of propylene (1 Mtpa).
5. Sadara(Kingdom of Saudi Arabia): World largest mixed cracker (ethane, LPG, naphtha) operating at 1.5 Mtpa ethylene and 0.5 Mtpa propylene.
6. RasLaffan (Qatar), 2010: Large ethylene plant with a capacity of 1.3 Mtpa based on ethane cracking.
7. Sharq (Kingdom of Saudi Arabia), 2010: large ethylene plant with a capacity of 1.3 Mtpa based on ethane cracking.
8. Map Ta Phut (Thailand), 2010: Ethylene plant of mixed (ethane, naphtha) cracker technology with 0.175 Mtpa.
9. Etileno XXI (Mexico): A complex of 1.05 Mtpa ethylene cracker based on ethane feed is under construction and expected to be ready by 2014.
10. Port Arthur Steam Cracker, Texas (USA) 2013: The Port Arthur steam cracker is one of the biggest in the world, with a capacity of 1 Mtpa of ethylene. The project was commissioned in 2001 to process naphtha distiled from petroleum. In response to petroleum product price hike and the emergence of abundant gas resources, they adapted the steam cracker to give it flexibility and maintain its competitiveness. It can now use as a feedstock ethane, which costs around $30 per barrel of oil equivalent (boe) versus around $100/boe for naphtha and liquefied petroleum gases such as butane and propane, which are also cheaper.

## 8.3 OXIDATIVE DEHYDROGENATION OF ETHANE

Oxidative dehydrogenation of ethane to the ethylene is considered as a greener alternative, mainly due to its reduced demand for energy, lower temperature, no need for steam, and the absence of any equilibrium constraint on the conversion. Moreover, less coke formation is expected. ODH of lower alkanes, such as ethane, propane and butane is attractive because of the interest in exploiting alkanes more efficiently. With successful development of ODH, high yields of olefins will be possible through the conversion of much smaller volumes of alkanes. Several reviews have been reported stating the significant work done on ethane ODH since the year 1995 [18–25]. Compared with the conventional steam-cracking method of dehydrogenating alkanes to olefins and the current catalytic dehydrogenation processes require temperature in the order of 350 to 600°C. Operational efficiencies can be improved by eliminating the need for a furnace and for decoking shutdowns, lowering operating temperature s, lessening material demands, conducting fewer maintenance operations, and using a greater proportion of the alkanes in the olefin conversion process. It does not need a catalyst regeneration step, because the regeneration occurs in situ by the oxygen feed. Higher conversions can be achieved at much lower temperature s.

The catalytic ODH may completely change the way some of the most important organic chemicals are manufactured. Low molecular weight alkenes, such as ethylene and propylene, can be formed via nonoxidative dehydrogenation of the corresponding alkane, as well. However, nonoxidative dehydrogenation reactions are endothermic and lead to the concurrent formation of carbon and of lower molecular weight alkanes, both of, which decrease alkene yields. Thus, ODH of light alkanes offers a potentially attractive route to alkenes, since the reaction is exothermic and avoids the thermodynamic constraints of nonoxidative routes by forming water as a byproduct.

As shown in the following reactions, the ODH not only leads to the formation of alkene, [Eq. (1.10)] but it can proceed with the nonselective oxidation to carbon dioxide [Eq. (1.11)] or other side products like carbon monoxide and coke.

$$C_nH_{2n+2} + \frac{1}{2}O_2 \rightarrow C_nH_{2n} + H_2O \quad (10)$$

$$C_nH_{2n+2} + \frac{1}{2}(3n+1)O_2 \rightarrow nCO_2 + (n+1)H_2O \quad (11)$$

As described in several recent reviews of ODH of lower alkane on supported metal oxides [20, 24, 26–27], the activity and selectivity for ODH depends on the alkane reactant, the metal oxide catalyst used and the presence of modifiers. This may be complex, with the same catalyst showing different activities with alkanes of different chain length.

### 8.3.1 ETHANE ODH USING $O_2$ AS OXIDANT

Oxidative dehydrogenation of ethane can be carried out using oxidants like oxygen, carbon dioxide and nitrous oxide. When oxygen is the reactant, ethane is reacted with oxygen over a suitable catalyst to produce ethylene and water.

$$C_2H_6 + 0.5O_2 \rightarrow C_2H_4 + H_2 \quad (12)$$

$$C_2H_6 + 3.5O_2 \rightarrow 2CO_2 + 3H_2O \quad (13)$$

Water formation in the reaction removes the equilibrium constraint on conversion. The challenge here is to avoid undesirable partial and/or total oxidation of ethane with oxygen to carbon oxides. Selective catalysts are necessary for oxidation of ethane to ethylene.

Despite these advantages there are many challenges to be met in the ODH of alkanes since it produces considerable amount of carbon oxides. Thermal run-away of the reaction could occur due to the exothermic nature of this reaction. In case of ethane poses great difficulty in controlling the selectivity to ethylene.

The catalytic ethane oxidative dehydrogenation for ethylene production has not been commercialized yet. Even though some methodologies could be considered for commercial exploitation technologically, the requirement for air separation to obtain pure oxygen is an economic constraint. However, the advantages of oxidative dehydrogenation process made it an attractive research area in recent years. The main target in this research is to achieve suitable ethane conversion and ethylene selectivity at lower temperature s. It is well known that the carbon oxides are thermodynamically more stable than ethylene. Therefore, the catalyst must be able to arrest the reaction at ethylene formation step and prevent the formation of the carbon oxides. The success of oxidative dehydrogenation depends on the development of a catalyst with high product selectivity and yield.

#### *8.3.1.1 SUPPORTED VANADIUM CATALYSTS*

The first reports on propane and n-butane ODH on vanadium supported on MgO [28–29] were published in 1988. Since then many researchers have attempted to have a deep understanding on the vanadium oxide catalyst system [30–31]. Even though well debated, the general agreement was on the establishment of tetrahedral V species as the active site for alkane ODH. The isolated $VO_4$ tetrahedra [32–33] and $V_2O_7$ units [34] were assumed to be the main active structures for alkene selectivity. However, the $V_2O_7$ containing the labile oxygen was reported to promote the oxidation reactions to produce oxygenated products. Nieto and co-workers [34–35] reported that the catalytic activity of $V_2O_5/Al_2O_3$ increased with the V loading in the region of 0 to 6.3 wt%, whereas the selectivity to alkene showed its maximum at 3 to 4 wt% V loading. The octahedral $V^{5+}$ species showed higher catalytic activity, whereas the tetrahedral $V^{5+}$ produced a higher yield of alkene.

In ethane ODH, the activity of $V_2O_5$ catalysts increases with the $V_2O_5$ loading. The rates per V atom reach a maximum on $VO_x$ domains of intermediate size. There exists a balance between the activity of surface $VO_x$ species and their accessibility to reactants. The selectivity to ethylene reaches its maximum at intermediate values of $V_2O_5$ loading corresponding to the formation of the monolayer structure. The ethylene formation rate per V atom depends on the vanadia surface density and reaches its maximum value at intermediate surface densities [36–37]. Surface oxygen, O–H groups and especially oxygen vacancies are considered as the most prevalent reactive intermediates during ODH on active $VO_x$ domains [38–39]. Klose et al. observed that the adsorbed O species [40] mainly contribute to $CO_x$ formation.

The theoretically calculated reduction of $V^{5+}$ to $V^{4+}$ or $V^{3+}$ is always higher than the amount of vanadium reduced during the ODH [41–46]. Again, only a portion of the reduced centers are active in the catalytic reaction. The reoxidation occurs for the rest of sites in time scales longer than turnover times. Ethylene selectivity is directly proportional to the extent of surface reduction and in supported vanadia systems it follows the order $V_2O_5$-$ZrO_2$> $V_2O_5$-$Al_2O_3$> $V_2O_5$-$SiO_2$ [42–43]. Our recent work has revealed that enrichment of vanadium in the near-surface-region and formation of optimum amount of monomeric/oligomeric VO*x* species is the key reason for the better activity of $V_2O_5/Nb_2O_5$ catalyst for ethane ODH [2].

The reactivity of vanadium oxide in ethane ODH is highly dependent on the type of support [18, 37, 42, 47–51]. The performance of the catalysts depends on the characteristics of the supports and the nature of the active species formed on these supports. At a given surface density, the vanadia-support interactions determine the type of $VO_x$ structure formed. Bañares and co-workers [52–53] found that in ethane ODH, the turnover frequency values vary with the specific oxide support for both the isolated and polymeric surface $VO_x$ species. Thus, the support has a significant influence on the reaction parameters. The support cation directly affects the reactivity of the bridging V–O-support bond.

The formation of binary oxides like $VO_x$-$MoO_x$ helps the $VO_x$ species in its reduction and catalytic activity. In the case of $VO_{x-}MoO_x$ mixed oxides higher ethane ODH selectivity was observed than in the case of $VO_x$ species [54]. Solsona et al. found that the V-containing catalysts are about four times more active than pure Mo-catalyst, whereas Mo-V-containing catalysts resulted to be the most selective ones, which indicates that a synergetic effect between Mo and V takes place [54]. It was also reported that mesoporous alumina was more active than its other forms like $\gamma$ -$Al_2O_3$.

### *8.3.1.2 SUPPORTED MOLYBDENUM OXIDE CATALYSTS*

The vanadium and molybdenum oxides [55–71] based catalysts possess several similarities. Heracleous et al. [55] observed that molybdenum based catalysts were less active than vanadia-based catalysts, irrespective of the support used. Heracleous and Lemonidou [59] extensively studied $MoO_3/Al_2O_3$ catalytic system and they reported the formation of ethylene on the redox sites of the catalysts whereas, the alumina contributes to the primary oxidation and dehydrogenation of ethane to CO*x* and coke, respectively. However, higher loadings of MoO*x* (the redox sites) were found to unselectively activate the over oxidation of the olefin to carbon oxides due to the growth of $Al_2MoO_4$ crystallites. Highly dispersed two-dimensional molybdenum species, which fully encapsulate the alumina surface could result in high ethylene yields [60]. Heracleous et al. have also studied the role of preparation parameters on the structureselective properties of alumina supported molybdena catalysts [71]. Two Catalysts with different preparation conditions pH=3.9, $C_{Mo}$=1.1×10$^{-3}$ mol L$^{-1}$ and pH=6.3, $C_{Mo}$=2.7×10$^{-2}$ mol L$^{-1}$, respectively demonstrated similar ethane conversions in the temperature range of 450–600°C. However, selectivity to ethylene was 15% higher for the catalyst prepared at relatively low pH than for the sample prepared at a higher pH. The difference in the structural characteristics of molybdena phase influenced the ethylene selectivity [71].

Watson and Ozkan [72] studied the Mo containing catalysts with single oxide supports of silica and titania along with their mixed supports. The ethylene yield increased with Mo loading over Si:Ti=1:1 to a maximum of 10% Mo/Si: Ti=1:1, where an optimum coverage of MoO*x* existed and that shared several types of support ligands from both silica and titania domains of the support. MoO*x* increasingly interacted with the silica phase of the support as titania increasingly segregated into larger anatase domains. Support effect on structure and activity of molybdenum oxide catalysts for the ODH of ethane was studied by Tsilomelekis et al. [73]. The Mo surface density was varied in the range of 0.5 to 18.5 Mo/nm$^2$ over the surface of $ZrO_2$, $Al_2O_3$, $TiO_2$ and $SiO_2$ oxides and the order of the catalytic activity was found as $ZrO_2 > Al_2O_3 > TiO_2 > SiO_2$. The nature of the oxide support material and that of the Mo-O-support bond have a significant influence on the catalytic behavior of the molybdena catalysts with monolayer coverage [73]. Christodoulakis and Boghosian studied the importance of molecular structure and its impact on the catalytic activity of molybdena catalysts supported on zirconia for ethane ODH at temperature varying from 420–540°C [74]. The formation of bulk $Zr(MoO_4)_2$ crystals were observed at surface densities exceeding the monolayer (>7 Mo/nm$^2$).

The surface composition and the structural properties of the dispersed species responded to changes in the catalyst gas environment. The selectivity to ethylene was found to increase up to the monolayer coverage, with high ethylene selectivity persistent at high conversions. There existed an agreement between the apparent activity per Mo atom and the number of Mo-O-Zr bonds per Mo [74]. Similar study related to the molecular structure and reactivity of molybdenum (VI) oxide supported on titania was reported by Tsilomelekis and Boghosian [75]. The molybdenum oxide on the support was varied in the range of 1.8–17.0 Mo per $nm^2$ over $TiO_2$ (anatase) and the molecular structure of the deposited $(MoO_x)_n$ species was studied for ethane ODH at temperature of 410–480°C. Upto the monolayer coverage the selectivity to ethylene was found to increase, where the primary steps of ethane activation follow selective reaction pathways leading to ~100% $C_2H_4$ selectivity. The terminal Mo=O sites were involved in nonselective reaction turnovers [75].

### 8.3.1.3 PEROVSKITES AS CATALYSTS

Application of perovskites for ethane ODH was reported in 1990s. Velle et al. [76] carried out the reaction on $SrCe_{0.5}Yb_{0.5}O_{2.75}$ perovskite catalyst and obtained an yield of 49% of ethylene at 700°C. However, the catalyst was not stable in the presence of $CO_2$ as it decomposed to $SrCO_3$ and $Ce_{1-x}Yb_xO_{2-0.5x}$. Two different reaction mechanisms were suggested for the perovskite catalyst; the free radical mechanism at higher temperature and a redox mechanism at lower temperature and Yb contents [76]. $La_{1-x}Sr_xFeO_{3-\Delta}$ $(0 \leq x \leq 1)$ were tested in the temperature range of 300–800°C by Guanghua et al. [77] and obtained a maximum yield of 37% ethylene at 650° C. $La_{0.6}Sr_{0.4}FeO_{3-0.103}Cl_{0.164}$ catalyst was more active with $C_2H_4$ yield of 57.6% [78].

Dai et al. reported that the presence of F-or Cl-ions in the perovskites promotes lattice oxygen mobility in $La_{1-x}Sr_xFe_{O3-\Delta}$ thereby reducing the deep oxidation of $C_2H_4$ [79]. $YBa_2Cu_3O_{7-0.18}Cl_{0.13}$ catalysts exhibited 68.8% $C_2H_4$ yield at a $C_2H_6$ conversion of 92.5%. An ethylene yield of 55% was observed in the case of a short-contact-time reactor consisting of a $LaMnO_3$ based monolithic catalyst below 400°C [80]. Donsì et al., observed that the formation of ethylene was catalyzed by both monolith surfaces and the through gas-phase radical reactions [80]. Alifanti et al. [81] observed that $SmCoO_3$ and $PrCoO_3$ perovskites were active for ethane ODH even at 300°C i.e., much below the temperature previously reported for perovskites.

#### 8.3.1.4 OTHER CATALYSTS

Solsona and coauthors [82] found that the addition of $CeO_2$ to NiO had significant influence on the mixed oxide catalytic performance. The ethylene productivity was enhanced by a factor of 7 with the incorporation of low amounts of $CeO_2$. The increase in surface area and the change in NiO crystal size were the main reasons for the improved activity [82]. The impact of adding W to NiO on the activity and reducibility of the catalyst was reported [83]. In W-rich catalysts, $NiWO_4$ and $WO_3$ phases were predominant and had a strong influence of ethane conversion. The nature of Ni sites, number of acid sites and their characteristics regulate the catalytic behavior of Ni-W-O catalysts [83]. Ni-Nb-O catalyst containing 15% of Nb was found to be the best performing one with an optimum ethane conversion of 33% with 78% of selectivity to ethylene at 350°C at a W/F ratio of 0.54 gs/mL [84]. The activity of niobium phosphates for ODH of ethane were reported by Weng et al. [85]. They found that $Nb_{1.91}P_{2.82}O_{12}$ was the best performing catalyst among the different niobium phosphates ($Nb_2P_4O_{15}$, $NbOPO_4$ and $Nb_{1.91}P_{2.82}O_{12}$) studied. $Nb_{1.91}P_{2.82}O_{12}$ catalyst was highly selective to ethylene (ca. 90% selectivity at 15% conversion) and the catalyst activity was found to increase with time on stream because of the increase of $Nb_{1.91}P_{2.82}O_{12}$ concentration on the surface.

Despite all the vanadium and molybdenum containing catalysts mentioned in the previous sections, mixed oxide catalyst derived from heteropolyacids as precursors exhibited special behavior compared to that of oxides derived from normal ammonium salts. Kumar et al. worked on the ODH of ethane over $MoO_3/V_2O_5–Al_2O_3$ catalysts, where heteropoly molybdate (MPA) was taken as the precursor for $MoO_3$ [86]. The Mo species derived from heteropoly molybdate was highly active and selective compared to the Mo species derived from the routine precursor (i.e., ammonium heptamolybdate). Among the 0–20 wt.% $V_2O_5/Al_2O_3$ catalysts for ODH, 17.5 wt.% $V_2O_5/Al_2O_3$ was found to be the most active one with 28% ethane conversion ($X(C_{2-})$) and 35% ethylene selectivity ($S(C_2^=)$) with an overall yield of about 10%. The decomposition of Keggin structure (heteropoly molybdate) and the formation of Mo-V mixed oxide phases on $Al_2O_3$ appeared to be responsible for this improved performance. The catalyst also displayed long-term stability over a period of 36 h [86]. Figures 8.4 and 8.5 depict the activity obtained over (5–20 wt%) $V_2O_5$-$Al_2O_3$ (5–20 wt%) MPA/$V_2O_5$-$Al_2O_3$ catalysts, respectively. There is a clear enhancement in the ethylene selectivity with the addition MPA to the vanadium containing catalysts.

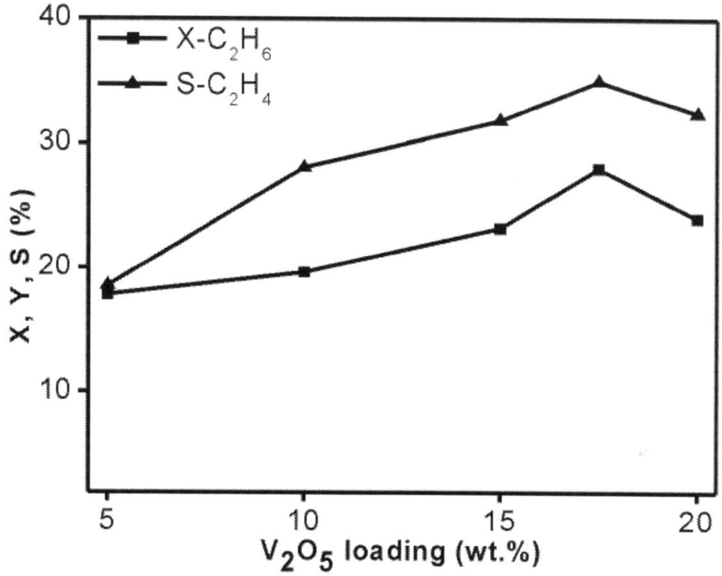


**FIGURE 8.4** Effect of $V_2O_5$ loading on –gamma-$Al_2O_3$ on conversion of ethane and selectivity of ethylene. (Reaction conditions: mole ratio of $C_2H_6/O_2/He=1/0.67/2.3$; T=600°C; catalyst weight: 1 g; GHSV=3600 $h^{-1}$; $\tau$ =1 s) [86].

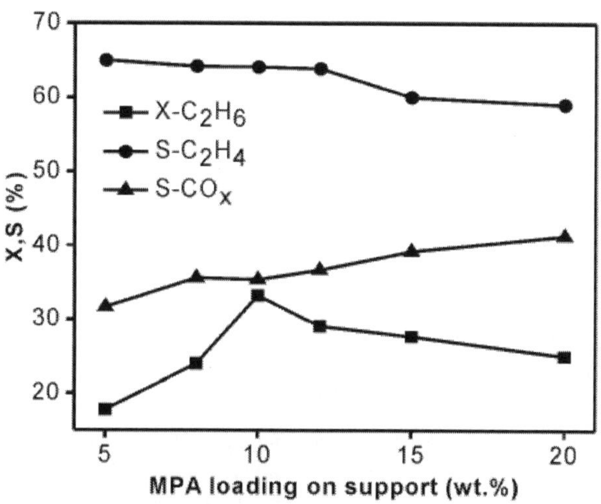


**FIGURE 8.5** Effect of MPA content on the activity and selectivity of $V_2O_5$-$Al_2O_3$ supported molybdophosphoric acid catalysts with different MPA loadings (Reaction conditions: mole ratio of $C_2H_6/O_2/He=1/0.67/2.3$; T=600°C; catalyst weight: 1 g; GHSV=3600 $h^{-1}$; $\tau$ =1 s) [86].

The precursor for the catalyst material plays a vital role in enhancing or deminishing the catalytic activity. Figure 8.6 showns the influence of the precursor on ethane ODH. The activity obtained over 10% MPA/$V_2O_5$-$Al_2O_3$ catalyst was compared with two additional catalysts with similar composition prepared using two different methods such as i) physical mixing (i.e., addition of individual oxides $MoO_3$, $V_2O_5$, $P_2O_5$ and $Al_2O_3$); and ii) impregnation of ammonia salts (ammonium salts of Mo, V, P) on $Al_2O_3$ instead of direct usage of MPA with Keggin structure. Despite having the same catalyst composition, the MPA/$V_2O_5$-$Al_2O_3$ exhibited superior performance compared to the other two. The ethylene yields obtained over these three catalysts prepared in different ways were in the following order: decomposed 10% MPA/$V_2O_5$-$Al_2O_3$ ($X(C_{2-})$=33.2% and $S(C_2^=)$=64.1%) > impregnated catalyst of ammonium salts ($X(C_{2-})$=35.1% and $S(C_2^-)$=42.4%) > physical mixing ($X(C_{2-})$=38.0% and $S(C_2^=)$=23.4%). This result clearly indicates that the method of preparation plays a key role on the catalytic performance and the decomposed MPA/$V_2O_5$-$Al_2O_3$ reveals improved performance [86].

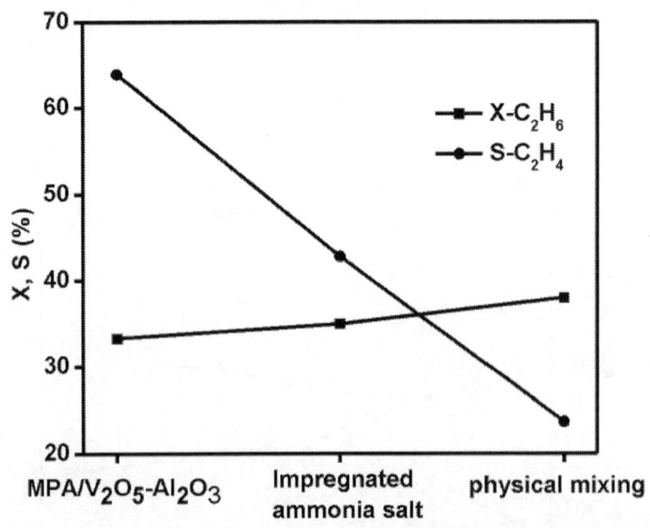


**FIGURE 8.6** Influence of precursor on the catalytic activity. (Reaction conditions: mole ratio of $C_2H_6/O_2$/He=1/0.67/2.3; T=600° C; catalyst weight: 1 g; GHSV=3600 $h^{-1}$; $\tau$ =1 s) [86].

MPA/$Al_2O_3$ catalysts also displayed significant activity and selectivity in the ODH of ethane [87]. The conversion of ethane and the selectivity towards ethylene were found to depend strongly on MPA loading (5–25 wt%) and varied in the range of 12–24% and 30–70%, respectively. Ethylene was the main product of the reaction, while $CO_2$ was the major by-product due to total oxidation. The conversion of ethane was observed to increase progressively up to 20 wt% MPA loading and then decreased, while the selectivity of ethylene increased continuously (30 to 70%) at the expense of $CO_2$ formation (selectivity decreased from >65 to 30%). Consequently, the 20 wt% MPA/$Al_2O_3$ catalyst (at 550°C) displayed its superior activity and selectivity and hence it became the highest performing catalyst among the series with an ethane conversion of 24% and an ethylene and $CO_2$ yields of 15.4% and 8.5%, respectively. Interestingly, pure MPA after decomposition gave very poor performance giving <5% conversion of ethane with ca. 10% selectivity of ethylene and ca. 90% selectivity to $CO_2$. In other words, the total oxidation was the predominant reaction on pure MPA solid. Two catalysts (prepared using two different Mo precursors such as MPA and ammonium heptamolybdate (AHM)) with the same composition of the best catalyst (20 wt% MPA) were tested for ODH. The results revealed that the nature of precursor source had considerable influence on the catalytic performance. Between the two, the catalyst prepared with MPA precursor displayed the superior performance. In other words, the catalyst prepared from AHM precursor, though exhibited comparable conversion of ethane (X=22%), its ethylene selectivity (44.5%) was significantly lower. However, its selectivity towards the total oxidation product, $CO_2$ (S-$CO_2$=55%) was considerably high in comparison with the one prepared using MPA precursor (X=24%, S-$C_2H_4$=ca. 65% and S-$CO_2$=35%). This result indicated that the precursor source is an important parameter that needs to be selected carefully [87].

### *8.3.2 ETHANE ODH WITH $N_2O$ AS OXIDANT*

This technology is still under development. It is carried out essentially on Fe-or Mo-modified, zeolites. Very few investigators reported their work on vanadium and heteropolyacid based catalysts. Investigations using CaO catalysts also appear in the literature. The monomeric and dimeric ferric ions were found to be active in ODH of ethane with nitrous oxide giving a product selectivity of ~40–60% at 350°C on Fe-modified zeolites (ferrierite and MFI). On the other hand, Fe oxide nanoclusters resulted in overoxidation of ethane and/or ethylene to C, CO and $CO_2$ [88]. Iron modified zeolites of different structures viz., ZSM-5, zeolite Y and mordenite were tested in the ethane and propane ODH. The nature of zeolite dictated the catalytic activity. The best catalytic activity was observed on iron-modified zeolites of MFI structure. Fe-Y and Fe-mordenite mainly catalyzed total oxidation [89]. The activity and selectivity of Fe-HZSM-5 catalysts in the oxidative dehydrogenation of ethane was studied at 250–350°C. The oxidized catalysts free from condensation products were active in complete oxidation of ethane. At low temperature coke formation took place, simultaneously affecting the coordination state of iron ions on the zeolite and improving the ODE selectivity [90]. Zeolite ZSM-5 with different Si/Al ratios, H-Y, mordenite modified with iron cations by means of ionic exchange were employed as catalysts. Comparison of the activity was made with those of amorphous oxide supports and crystalline neutral silicalite modified with iron cations. Only iron modified zeolites showed activity. The best catalytic performance was found on iron-modified zeolites of MFI structure. The Si/Al ratio of the ZSM-5 did not show much influence on the activity. $N_2O$ was partly used for ethane oxidation to form ethylene or carbon oxides, while a portion of the oxidant was decomposed to nitrogen and oxygen. At 450°C the reaction gave a high ethylene yield and complete $N_2O$ removal [91]. The effect of calcination temperature of 0.4% Fe-ZSM-5 zeolites with a Si/Al ratio was investigated for the dehydrogenation of $C_2$-$C_4$ hydrocarbons. It was found that the calcination temperature (600 or 900°C) slightly influenced hydrocarbon conversion, while it significantly affected the selectivity towards respective olefins.

The modification of acidic centers present in Fe-ZSM-5 samples was found to be responsible for the change in performance [92]. Metal ion species introduced into the zeolite framework seem to provide attractive redox activity. Efforts were also made to convert light paraffins into valuable olefinic chemicals. Cobalt modified zeolites of BEA type exhibited the highest activity and selectivity in ODH of ethane. The catalytic activity increased with the increasing cobalt concentration up to molar ratio Co/Al 0.5. At higher level of ion exchange, there was a dramatic decrease in activity, due to formation of oxide clusters of cobalt. Cobalt oxides caused the nonselective oxidation of ethane to carbon oxides [93]. The ODH was performed on 10% Mo/Si-Ti catalyst using either $O_2$ or $N_2O$ as the oxidant at contact times varying from 0.46 to 1.0 mg min/cm$^3$. $N_2O$ gave superior ethylene selectivity for a given ethane conversion out-performing oxygen at all contact times tested. Ethylene selectivities decreased with contact time at a far slower rate when $N_2O$ was used as the oxidant. Mo was found to be in its fully oxidized state as Mo(VI) during ethane ODH using $O_2$, whereas the reduction state decreased to +5.8 using $N_2O$ as the oxidant. The activation energy of reoxidation when using $N_2O$ was 98 kJ/mol, whereas that of $O_2$ was 41 kJ/mol leading to a reoxidation rate at least 1700 times faster when $O_2$ was the oxidant. This difference in rates accounted for the less oxidized state of molybdenum during ethane ODH with $N_2O$ and explained the behavior observed during the reaction [94]. The reaction of $C_2H_6$ with lattice oxygen, $O^{2-}$ (in the absence of gaseous oxygen), or "adsorbed" oxygen over $NiMoO_4$ catalysts was studied. The adsorbed oxygen exhibited a higher reactivity to $C_2H_6$, while lattice oxygen was more reactive towards formation of ethylene [95]. Over nano-sized VO*x* species, the olefin selectivity was significantly improved upon replacing $O_2$ with $N_2O$, due to suppressing $CO_x$ formation in the presence of $N_2O$ [96]. Adsorbed O species were formed on CaO by the decomposition of $N_2O$. The O species formed on the sites with coordination number of '4' were found to be active for ODH of ethane. Ethylene was produced with high selectivity [97].

### 8.3.3 PROCESS INTENSIFICATION OPTIONS REPORTED

Some pioneering efforts have been reported in the literature to make ethane conversion to ethylene more efficient, environmentally and energy demand wise more sound, separations are shaper and more proactive to use $CO_2$. This section highlights these efforts.

### 8.3.4 USE OF CARBON DIOXIDE IN CRACKING PROCESSES

It is possible to use $CO_2$ as an oxidant, instead of oxygen or with limited oxygen, in the noncatalytic (thermal) hydrocarbon cracking processes for the production of ethylene and other olefins. Choudhary et al. investigated the noncatalytic cracking of ethane at atmospheric pressure in the presence of $CO_2$ at different $CO_2$/ethane ratios (0–2>0) with or without oxygen ($O_2$/ethane ratio=0 to 0<3) at different temperature (750–900°C) and space velocities (1500–9000 $h^{-1}$) in the absence of steam in the feed [98].

From the above studies, the following important conclusions were drawn. (i) The presence of $CO_2$ in both thermal and oxy-cracking of ethane had a beneficial effect. (ii) In both cases, ethane conversion increased with increasing $CO_2/C_2H_6$ ratio in the feed; ethane was activated in the presence of $CO_2$. (iii) At different process conditions, ethane conversion was higher but ethylene selectivity was lower when limited $O_2$ was added to the ethane-$CO_2$ feed because of the formation of CO due to partial combustion of ethane. However, in the presence of both $CO_2$ and $O_2$, the process was expected to become highly energy efficient with large energy saving for coke-free ethylene production from ethane [98].

### 8.3.4 STRUCTURED CATALYSTS

Different compositions of cerium promoted Ni-catalysts were incorporated onto ceramic papers and were tested in the reaction. They contained silica-alumina fibers to provide a better environment for the deposition of the catalyst. The catalysts with the atomic ratio of Ce/Ni = 0.17 showed the highest ethylene productivity, due to the formation of smallest NiO crystals and due to the partial incorporation of the Ni cation into the ceria lattice [99]. MoVTeNbO catalysts were supported on preoxidized SiC foam. The coating with a two-layered structure contained significant quantities of the active phase. The structured catalyst thus obtained with two layers appeared highly efficient and stable in both reactions between 380 and 420°C [100]. ODH of ethane with $CO_2$ was carried out over the Cr-based monolithic catalysts with FeCrAl alloy foil as support. High catalytic activity and stability was obtained on these monolithic catalysts. 66.5% ethane conversion and 99.5% ethylene selectivity was achieved on the 5.0% Cr loaded monolithic catalyst at 750°C. The catalyst was stable for 1130 h [101]. Microstructured reactors were used for ethylene production via ODH of ethane. The process involving the exothermic hydrogen combustion in close contact with the endothermic dehydrogenation reaction showed significant improvement in the economics of ethylene production. The reactor walls were coated with a catalyst for dehydrogenation. 75% ethane conversion and 84% ethylene selectivity were reported on this catalyst. The operation of was stable for more than 100 h when conducted at 950°C and a contact time of <50 ms. The per pass ethylene yield exceeded that of the conventional stream cracker [102].

### 8.3.5 MEMBRANES

As an intensification option aimed at obtaining high selectivity, a catalytic reactor was integrated with an oxygen separation membrane. The membranes made of a mixed ionic-electronic conducting $Ba_{0.5}Sr_{0.5}Co_{0.8}Fe_{0.2}O_{3-\Delta}$ mixed oxides were activated using different porous catalytic layers based on rare earth-doped cerias (fluorite structure). The porous catalytic coating was deposited by screen printing. The ODH reaction was performed at 850°C. High ethylene yields (up to ≈84%) were obtained due to the high catalytic activity, controlled oxygen concentration in the gas phase and the appropriate fluid dynamics [103]. Another process intensification effort was made by using multilayer ceramic and sintered metal membranes in a dead-end configuration; the latter allowing a controlled distributed reactant feeding. $VO_x/Al_2O_3$ was chosen as the active catalyst with different dopents. Pilot scale evaluations were carried out to determine the potential of distributed dosing via membranes and compatibility of reaction and membrane properties. Back diffusion of reactants out of the catalytic zone was avoided. The metal membranes possessed good mechanical stability, the cost of production was also low. The rate of reaction and the mass transfer in the membrane could be adjusted by the transmembrane pressure and the catalyst activity [104].

The status and future trends in the development and application of novel reactors equipped with solid electrolyte based membrane materials was analyzed in a review article [105]. $BaCo_xFe_yZr_zO_3$ perovskite hollow-fiber membrane reactor was used with permeable and passivated surface segments to enable controlled oxygen insertion into the reactor. The reduced oxygen concentration offered higher ethene selectivity. At low and moderate ethane conversion, the performance of the membrane reactor was on par with the best catalysts used in cofeed mode. The ethene yield was comparable with that of the industrial steam cracking process. However, the reaction temperature was 100°C lower [106].

A membrane multitubular reactor was designed for industrial-scale ethylene production of ethane by ODH over Ni-Nb-O catalyst. The application of membrane helped in lowering the $O_2$ partial pressures inside the catalyst tubes when compared with a conventional multitubular reactor and thereby increasing the ethylene selectivity and good temperature [107]. Phase inversion spinning followed by sintering was used to develop a dense perovskite hollow fiber membrane made of $BaCo_xFe_yZr_zO_{3-\Delta}$(BCFZ, $x+y+z=1$). The oxygen partial pressure on the shell and core sides were adjusted by varying the air flow rates on the shell side (25–300 mL/min) and He flow rates on the core side. However, the ethylene selectivity (64%) was lower than the BCFZ disk-shaped membrane (79%) [108].

### 8.3.6 OPERATIONAL MODIFICATIONS

Various aspects of multistage dosing concepts were discussed in the case of membrane reactors. $VO_x/Al_2O_3$ was chosen as the catalyst. The molar $O_2/C_nH_m$ ratio was varied between 0.5–8. The analysis helped in identifying the operation range to attain higher ethylene selectivity and a simultaneously higher conversion in membrane reactors [109]. PtSn and $LaMnO_3$ monolith reactor was tested at 700–800°C with short contact times. PtSn displayed higher selectivity to ethene than $LaMnO_3$ when sufficient amount of hydrogen was added to the feed [110]. Pt-Sn impregnated monoliths were tested at 1 bar,

600–900°C and with different contents of oxygen, hydrogen and steam in the feed gas. Decrease in oxygen in the feed led to a decrease in the conversion due to lower reaction temperature. Steam had no effect on the ethane conversion or the ethene selectivity. Hydrogen/ethane ratio in feed influenced reaction considerably. At 700°C the selectivity increased from ≈50% to 93% with only a small decrease in the conversion of ethane [111].

### 8.3.7 USE OF $CO_2$ AS AN OXIDANT

For all the reactions of alkanes with oxygen, the abstracted hydrogen is oxidized; releasing the heat of reaction, and the conversion becomes significant at a much lower reaction temperature. However, because of the exothermic characteristics of these reactions in the presence of oxygen, it is necessary to remove heat to avoid the overoxidation of ethylene to carbon oxides and to provide high selectivity toward the ethylene. To overcome these difficulties, usage of $CO_2$ as a mild oxidant instead of $O_2$ is an advantage in the ODH. The ODH of alkanes in the presence of $CO_2$ is represented in Eq. (14). CO is the major byproduct in this reaction in addition to alkene and water.

$$C_nH_{2n+2} + CO_2 \rightarrow C_nH_{2n} + CO + H_2O \tag{14}$$

Oxidative dehydrogenation of ethane in presence of $CO_2$ can be described as follows:

$$C_2H_6 + CO_2 \rightarrow C_2H_4 + CO + H_2O\ (\Delta H^\circ \quad =134\ \text{kJ/mol}) \tag{15}$$

Other side reactions can also occur:

$$C_2H_6 \rightarrow C_2H_4 + H_2\ (\Delta H^\circ \quad =137\ \text{kJ/mol}) \tag{16}$$

$$C_2H_6 + 2CO_2 \rightarrow 4CO + 3H_2\ (\Delta H^\circ \quad =431\ \text{kJ/mol}) \tag{17}$$

$$H_2 + CO_2 \rightarrow CO + H_2O\ (\Delta H^\circ \quad =7\ \text{kJ/mol}) \tag{18}$$

All these reactions are observed to be endothermic reactions, although, in comparison with the side reactions, $CO_2$ ODH is more favorable from thermodynamic considerations. The following advantages were found in case of ethane ODH using $CO_2$:

- Used as a diluent to obtain high equilibrium conversion of light alkanes.
- Results in high ethylene selectivity.
- Reduce the coke formation on the catalyst.
- Maintain catalytic activity for a longer period of time.

The promoting effect of $CO_2$ in the ODH of $C_2H_6$ was first reported by Wang and his colleagues [112]. They found that $CO_2$, either formed or added to the system, increased the selectivity for the desired hydrocarbon products during the ODH of $C_2H_6$ by oxygen over $Li^+$/MgO catalysts. The improved selectivities were attributed to the poisoning effect that $CO_2$ had on the secondary reactions of alkyl radicals with the surface. Krylov et al. [113–114] conducted investigations on the catalytic ODH of $C_{1-}C_7$ alkanes by $CO_2$ on several metal oxide catalysts like $Fe_2O_3$, $Cr_2O_3$, $MnO_2$, and on multicomponent systems. They found that the $MnO_2$ based catalysts exhibited high activities at 800°C for ODH of $C_2H_6$ using $CO_2$. Further improvement could be achieved when the catalyst was modified by promoters like Hong et al. [115] investigated the ODH of ethane using activated-carbon-supported metal oxide iron, manganese, molybdenum, tungsten, and chromium oxide catalysts. Among these catalysts $Cr_2O_3$ was found to be the best catalyst. Activated-carbon-supported $Cr_2O_3$ catalyst exhibited 29.2% ethane conversion with 87.5% ethylene selectivity at 650°C. $CO_2$ facilitates the dehydrogenation of $C_2H_6$ and enhances $C_2H_6$ conversion and $C_2H_4$ yield. Moreover, coke deposition on the catalyst was significantly retarded. They have also attempted the catalyst regeneration step after dehydrogenation reactions with $CO_2$ at 700°C for 1h under an atmosphere of $CO_2$. The activity of the catalyst could only be restored to a certain extent, but the original activity of the catalyst could not be recovered completely. Coke deposition and changes in the surface state of the catalyst are the main reasons for catalyst deactivation.

Recently, another approach has been found with carbon dioxide as a mild oxidant for catalytic dehydrogenation of hydrocarbons [113, 116]. $CO_2$ is expected to increase equilibrium conversion by diluting light alkanes, and to maintain the activity of the catalyst over a long period of time by removing coke formed on the catalyst [117]. Wang et al. [116] studied the catalytic activity of supported ($Al_2O_3$, $SiO_2$, $TiO_2$ and $ZrO_2$) chromium catalysts. Among all the catalysts $Cr_2O_3/SiO_2$ catalysts exhibited higher conversion and selectivity to ethylene. The optimum $Cr_2O_3$ loading was found as 8 wt%. At 650°C, 8 wt% $Cr_2O_3/SiO_2$ catalyst exhibited 55.5% ethylene yield with 61% ethane conversion. Mimura et al. [117] reported the effects of $CO_2$ on the dehydrogenation of ethane over Cr/H-ZSM-5 catalysts. Cr/H-ZSM-5 ($SiO_2/Al_2O_3 \geq$ 190) catalyst showed the highest performance. At 650°C, the Cr/H-ZSM-5 ($SiO_2/Al_2O_3 \geq 190$) catalyst resulted in 51.6% conversion of ethane and 79.1% selectivity to ethylene. It was stated that $CO_2$ maintained the activity of the catalyst by removing the coke from surface of the catalyst. Wang et al. [118]

reported the oxidative dehydrogenation of ethane into ethylene by carbon dioxide over unsupported $Cr_2O_3$, $Cr_2O_3/SiO_2$ and a series of $Cr_2O_3/SiO_2$ catalysts modified by sulfate ions. The results showed that $Cr_2O_3/SiO_2$ is an effective catalyst for dehydrogenation of ethane and $CO_2$ in the feed promotes the catalytic activity. Sulfation of silica influenced the catalytic behavior of $Cr_2O_3/SiO_2$ depending on the amount of sulfate. $Cr_2O_{3-}6$ wt% $SO_4^{2-}$–$SiO_2$ catalysts exhibited an excellent performance, given an ethylene yield of 55% at 67% ethane conversion at 650°C. Characterizations indicated that the addition of sulfate changed the bulk and surface properties of $Cr_2O_3/SiO_2$, promoting the reduction of $Cr^{6+}$ to $Cr^{3+}$, in turn favoring the catalytic activity.

$Mo_2C$ prepared on $SiO_2$ was found to be an effective catalyst by Solymosi et al. [119] for the dehydrogenation of ethane to produce ethylene in the presence of $CO_2$. The selectivity to ethylene at 650°C was 95% at an ethane conversion of 30%. With the increase of the temperature the dry reforming of ethane became significant. It was assumed that the Mo oxycarbide formed in the reaction between $CO_2$ and $Mo_2C$ played an important role in the activation of ethane. Valenzuela et al. [120] studied the nano-structured cerium oxides, pure or doped with CaO, prepared from amorphous acetate precursors and adopting the freeze-drying method. The catalysts exhibited high surface areas and were active and selective for ODH of ethane using $CO_2$ as an oxidant. The incorporation of Ca into a solid solution of ceria framework reduced the activity but improved the selectivity to ethylene. Under reaction conditions, the pure ceria was stable even at 750°C, while the surface area of the Ca-containing sample decreased considerably. This change was accompanied by a drastic increase in the selectivity to ethylene, allowing yields of ethylene up to 22% with 91% selectivity.

Qiao et al. studied the role of promoters (i.e., Cr, Mo, W) on Ni-Nb-O catalyst for ethane ODH using $CO_2$ admixture [121]. All the promoters tested belonged to the same group of elements showing different d-characters. The activity results of the doped solids revealed a clear reduced activity compared to the original Ni-Nb-O catalyst (ethylene yield of 32%) during ODH. Among the three auxiliaries used, Cr displayed relatively superior catalytic performance compared to the other two yielding an ethane conversion of 26% and an ethylene selectivity of about 65% only. In addition, the introduction of $CO_2$ into the reactant feed mixture was found to improve the selectivity to ethylene in ODH. Ni-Nb-Cr-O solid revealed an ethylene selectivity of 85% at slightly less ethane conversion [121].

## 8.4 CONCLUSIONS

In the past, thermal and steam cracking of ethane to ethylene received considerable attention in view of their commercial importance. Currently, efforts are being made worldwide to develop new process intensification options to adequately address emerging environmental, energy and alternative feedstock resourcing issues. The importance of ethane as a feedstock for ethylene production has been increasing ever since natural gas has become the most preferred feed stock for the ethylene production. Ethane is a major secondary constituent of natural gas for using as a feedstock. Since last few years, significant research efforts have, accordingly, been made to develop ethane ODH to produce ethylene in view of its reduced energy demand, minimal thermodynamic equilibrium constraints and relatively better environmental acceptability.

Vanadium, molybdenum and perovskite based catalysts have been extensively studied for ethane ODH. The reported ethylene yields are comparable and even superior to those achieved in the case of steam and catalytic cracking. However, the high reactivity of ethylene and the resulting tendency to undergo partial or total oxidation have been limiting the achievement of high selectivity for its formation. Added to thus, relatively less attention has been given to achieve long-term stability of ODH catalysts. Finding a versatile catalyst capable of providing industrially acceptable ethane conversion/productivity coupled with very high specificity to ethylene has still remained a scientific challenge. Well known research groups in the world have been making serious efforts to understand the ODH reaction mechanism, its kinetics and catalytic properties to make it commercially more competitive than the traditional cracking and steam reforming technologies. An interesting aspect of these studies is to achieve process intensification through measures like membrane assisted reactive separations, use of $CO_2$ as oxidant in place of $O_2$ and better management of catalyst activity and specificity. Several scientific challenges including high temperature (>650°C) operation, enhanced coke formation and overcoming $CO_2$ stability barriers have to be surmounted to achieve process intensification in ethane ODH. There is tremendous scope for further innovations.

## ACKNOWLEDGMENTS

The authors thank the financial support from DST, New Delhi and DAAD-Germany for a joint project on ODH of ethane between Indian Institute of

Chemical Technology (IICT), Hyderabad and Leibniz Institute for Catalysis at University Rostock (LICAT), Germany. The authors are grateful to Dr. V. Narayana Kalevaru and Dr. Habil Andreas Martin, LICAT, for their cooperation. Thanks are also due to the Director, IICT, Hyderabad, for encouraging writing this report.

## KEYWORDS


- **Catalytic cracking**
- **Ethane to ethylene**
- **Membranes**
- **Oxidative dehydrogenation**
- **Shale gas**
- **Steam cracking**
- **Support molybdenum oxide**
- **Support vanadium oxide**
- **Thermal cracking**


## REFERENCES


1. Chemical Market Associates. Inc Alaskan Petrochemical Investment Opportunities. https://gas4alaska.info/uploads/CMAI_Petrochem_Inv_Opportunities_-_Final_11–10.pdf.
2. Qiao, A., Kalevaru, V. N., Radnik, J., Srihari Kumar, A., Lingaiah, N., Sai Prasad, P. S., and Martin, A. Oxidative dehydrogenation of ethane to ethylene over $V_2O_5/Nb_2O_5$ catalysts. *Catal. Commun.* **30**, 45–50 (2013).
3. True, W. R. Global ethylene capacity continues advance in 2011. *Oil Gas J* **110** (2012).
4. Koottungal, L. Survey shows increase in refining construction. *Oil Gas J* **110**, 32 (2012).
5. Eramo, M. Emerging Trends in Global Light Olefins. CMAI Paper presented at Australia Pacific Technical College, Seoul, 2002.
6. Bullin, K. A., and Krouskop, P. E. Compositional Variety Complicate Processing Plans for US Shale Gas. *Oil Gas J* **107**, 50–55 (2009).
7. Ramana, K. V. India holds 527 tcf of shale gas reserves. http://www.dnaindia.com/money/1685334/report-india-holds-527-tcf-of-shale-gas-reserves.
8. Sharma, S. S., Kulkarni, P. K., Kumar, A., Piyush Pankaj, Ramanathan, V., and Susanta, P. Shale energy: Developing international plays-Design and execution of effective hydraulic fracturing taps first shale gas reserves in India. *World oil* **231** (2010).
9. Universal Oil Products Co, and Globe Oil and Refining Co **1944**.

10. Bednarova, L., Lyman, C. E., Rytter, E., and Holmen, A. Effect of support on the size and composition of highly dispersed Pt–Sn particles. *J. Catal.* **211**, 335–346 (2002).
11. Sun, P., Siddiqi, G., Vining, W. C., Chi, M., and Bell, A. T. Novel Pt/Mg(In)(Al)O catalysts for ethane and propane dehydrogenation. *J. Catal.* **282**, 165–174 (2011).
12. Matar, S., and Hatch, L. F. *Chemistry of Petrochemical Processes*. Second ed.; Gulf Professional Publishing: Woburn, **2001**.
13. Choudhary, V. R., Mulla, S. A. R., and Rane, V. H. Coupling of exothermic and endothermic reactions in oxidative conversion of ethane to ethylene over alkaline earth promoted $La_2O_3$ catalysts in presence of limited $O_2$. *Appl. Energy* **66**, 51–62 (2000).
14. Mulla, S. A. R., Buyevskaya, O. V., and Baerns, M. Autothermal oxidative dehydrogenation of ethane to ethylene using $Sr_xLa_{1.0}Nd_{1.0}O_y$ catalysts as ignitors. *J. Catal.* **197**, 43–48 (2001).
15. Choudhary, V. R., and Mulla, S. A. R. Energy efficient simultaneous oxidative conversion and thermal cracking of ethane to ethylene using supported $BaO/La_2O_3$ catalyst in the presence of limited $O_2$. *Appl. Energy* **68**, 377–386 (2001).
16. Mulla, S. A. R., Buyevskaya, O. V., and Baerns, M. A comparative study on noncatalytic and catalytic oxidative dehydrogenation of ethane to ethylene. *Appl. Catal., A* **226**, 73–78 (2002).
17. Mulla, S. A. R., and Choudhary, V. R. Oxidative conversion of ethane to ethylene over supported SrO-promoted $Er_2O_3$ catalyst. *J. Mol. Catal. A: Chem.* **223**, 259–262 (2004).
18. Mamedov, E. A., and Cortés Corberán, V. Oxidative dehydrogenation of lower alkanes on vanadium oxide-based catalysts. The present state-of-the-art and outlooks. *Appl. Catal., A* **127**, 1–40 (1995).
19. Moffat, J. B. Conversion of $C_{2-}C_3$ alkanes on heteropoly oxometalates. *Appl. Catal., A* **146**, 65–86 (1996).
20. Bañares, M. A. Supported metal oxide and other catalysts for ethane conversion: a review. *Catal. Today* **51**, 319–348 (1999).
21. Bhasin, M. M., McCain, J. H., Vora, B. V., Imai, T., and Pujadó, P. R. Dehydrogenation and oxydehydrogenation of paraffins to olefins. *Appl. Catal., A* **221**, 397–419 (2001).
22. Wang, S., and Zhu, Z. H. Catalytic conversion of alkanes to olefins by carbon dioxide oxidative dehydrogenation a review. *Energy Fuels* **18**, 1126–1139 (2004).
23. Grabowski, R. Kinetics of oxidative dehydrogenation of $C_{2-}C_3$ alkanes on oxide catalysts. *Chem. Rev.* **48**, 199–268 (2006).
24. Cavani, F., Ballarini, N., and Cericola, A. Oxidative dehydrogenation of ethane and propane: How far from commercial implementation? *Catal. Today* **127**, 113–131 (2007).
25. Gärtner, C. A., van Veen, A. C., and Lercher, J. A. Oxidative dehydrogenation of ethane: Common principles and mechanistic aspects. *ChemCatChem*, n/a-n/a (2011).
26. Madeira, L. M., and Portela, M. F. Catalytic oxidative dehydrogenation of n-butane. *Cat. Rev. – Sci. Eng.* **44**, 247–286 (2002).
27. Grabowski, R. Kinetics of Oxidative Dehydrogenation of $C_{2-}C_3$ Alkanes on Oxide Catalysts. *Catal. Rev. – Sci. Eng.* **48**, 199–268 (2006).
28. Chaar, M. D., Patel, D., Kung, M. C., and Kung, H. H. Selective oxidative dehydrogenation of butane over V-Mg-O catalysts. *J. Catal.* **105**, 483–498 (1987).
29. Chaar, M. A., Patel, D., and Kung, H. H. Selective oxidative dehydrogenation of propane over V-Mg-O catalysts. *J. Catal.* **109**, 463–467 (1988).

30. Kung, H. H., and Kung, M. C. Oxidative dehydrogenation of alkanes over vanadium-magnesium-oxides. *Appl. Catal., A* **157**, 105–116 (1997).
31. López Nieto, J. M., Coenraads, R., Dejoz, A., and Vazquez, M. I. The role of metal oxides as promoters of $V_2O_5/\gamma\text{-}Al_2O_3$ catalysts in the oxidative dehydrogenation of propane. *Stud. Surf. Sci. Catal.* **110**, 443–452 (1997).
32. Corma, A., Nieto, J. M. L., and Paredes, N. Influence of the preparation methods of V-Mg-O catalysts on their catalytic properties for the oxidative dehydrogenation of propane. *J. Catal.* **144**, 425–438 (1993).
33. Michalakos, P. M., Kung, M. C., Jahan, I., and Kung, H. Selectivity patterns in alkane oxidation over $Mg_3(VO_4)_2$-MgO, $Mg_2V_2O_7$, and $(VO)_2P_2O_7$. *J. Catal.* **140**, 226–242 (1993).
34. Blasco, T., Gali, A., Lopez Nieto, J. M., and Triffiro, F. Oxidative dehydrogenation of ethane and n-butane on $VO_x/Al_2O_3$ catalysts. *J. Catal.* **169**, 203–211 (1997).
35. Concepción, P., Navarro, M. T., Blasco, T., López Nieto, J. M., Panzacchi, B., and Rey, F. Vanadium oxide supported on mesoporous $Al_2O_3$: Preparation, characterization and reactivity. *Catal. Today* **96**, 179–186 (2004).
36. Blasco, T., Galli, A., López Nieto, J. M., and Trifiró, F. Oxidative dehydrogenation of ethane and n-butane on $VO_x/Al_2O_3$ catalysts. *J. Catal.* **169**, 203–211 (1997).
37. Argyle, M. D., Chen, K., Bell, A. T., and Iglesia, E. Effect of catalyst structure on oxidative dehydrogenation of ethane and propane on alumina-supported vanadia. *J. Catal.* **208**, 139–149 (2002).
38. Chen, K., Khodakov, A., Yang, J., Bell, A. T., and Iglesia, E. Isotopic tracer and kinetic studies of oxidative dehydrogenation pathways on vanadium oxide catalysts. *J. Catal.* **186**, 325–333 (1999).
39. Argyle, M. D., Chen, K., Bell, A. T., and Iglesia, E. Ethane oxidative dehydrogenation pathways on vanadium oxide catalysts. *J. Phys. Chem. B* **106**, 5421–5427 (2002).
40. Klose, F., Joshi, M., Hamel, C., and Seidel-Morgenstern, A. Selective oxidation of ethane over a $VO_x/g\text{-}Al_2O_3$ catalyst – investigation of the reaction network. *Appl. Catal., A* **260**, 101–110 (2004).
41. Argyle, M. D., Chen, K., Resini, C., Krebs, C., Bell, A. T., and Iglesia, E. Extent of reduction of vanadium oxides during catalytic oxidation of alkanes measured by in-situ UV−visible spectroscopy. *J. Phys. Chem. B* **108**, 2345–2353 (2004).
42. Gao, X., Bañares, M. A., and Wachs, I. E. Ethane and n-butane oxidation over supported vanadium oxide catalysts: An UV–visible diffuse reflectance spectroscopic investigation. *J. Catal.* **188**, 325–331 (1999).
43. Bañares, M. A., Martínez-Huerta, M. V., Gao, X., Fierro, J. L. G., and Wachs, I. E. Dynamic behavior of supported vanadia catalysts in the selective oxidation of ethane: Raman, UV-Vis DRS and reactivity studies. *Catal. Today* **61**, 295–301 (2000).
44. Gao, X., Jehng, J.-M., and Wachs, I. E. UV–vis–NIR diffuse reflectance and raman spectroscopic studies of propane oxidation over $ZrO_2$-supported vanadium oxide catalysts. *J. Catal.* **209**, 43–50 (2002).
45. Mul, G., Banares, M. A., Garcia Cortez, G., van der Linden, B., Khatib, S. J., and Moulijn, J. A. MultiTRACK and operando Raman-GC study of oxidative dehydrogenation of propane over alumina-supported vanadium oxide catalysts. *Phys. Chem. Chem. Phys.* **5**, 4378–4383 (2003).

46. Ballarini, N., Battisti, A., Cavani, F., Cericola, A., Lucarelli, C., Racioppi, S., and Arpentinier, P. The oxygen-assisted transformation of propane to $CO_x/H_2$ through combined oxidation and WGS reactions catalyzed by vanadium oxide-based catalysts. *Catal. Today* **116**, 313–323 (2006).
47. Le Bars, J., Auroux, A., Forissier, M., and Vedrine, J. C. Active sites of $V_2O_5$/g-$Al_2O_3$ catalysts in the oxidative dehydrogenation of ethane. *J. Catal.* **162**, 250–259 (1996).
48. Khodakov, A., Olthof, B., Bell, A. T., and Iglesia, E. Structure and catalytic properties of supported vanadium oxides: Support effects on oxidative dehydrogenation reactions. *J. Catal.* **181**, 205–216 (1999).
49. Solsona, B., Blasco, T., López Nieto, J. M., Peña, M. L., Rey, F., and Vidal-Moya, A. Vanadium oxide supported on mesoporous MCM-41 as selective catalysts in the oxidative dehydrogenation of alkanes. *J. Catal.* **203**, 443–452 (2001).
50. Martínez-Huerta, M. V., Fierro, J. L. G., and Bañares, M. A. Monitoring the states of vanadium oxide during the transformation of $TiO_2$ anatase-to-rutile under reactive environments: $H_2$ reduction and oxidative dehydrogenation of ethane. *Catal. Commun.* **11**, 15–19 (2009).
51. Habel, D., Görke, O., Tovar, M., Willinger, M., Ziemann, M., Schwarz, O., Schomäcker, R., and Schubert, H. Microstraining in titania-, alumina-and silica-supported $V_2O_5$- catalysts. *J. Eur. Ceram. Soc.* **29**, 1093–1099 (2009).
52. Martínez-Huerta, M. V., Gao, X., Tian, H., Wachs, I. E., Fierro, J. L. G., and Bañares, M. A. Oxidative dehydrogenation of ethane to ethylene over alumina-supported vanadium oxide catalysts: Relationship between molecular structures and chemical reactivity. *Catal. Today* **118**, 279–287 (2006).
53. Bañares, M. A., and Wachs, I. E. Molecular structures of supported metal oxide catalysts under different environments. *J. Raman Spectrosc.* **33**, 359–380 (2002).
54. Solsona, B., Dejoz, A., Garcia, T., Concepción, P., Nieto, J. M. L., Vázquez, M. I., and Navarro, M. T. Molybdenum-vanadium supported on mesoporous alumina catalysts for the oxidative dehydrogenation of ethane. *Catal. Today* **117**, 228–233 (2006).
55. Heracleous, E., Machli, M., Lemonidou, A. A., and Vasalos, I. A. Oxidative dehydrogenation of ethane and propane over vanadia and molybdena supported catalysts. *J. Mol. Catal. A: Chem.* **232**, 29–39 (2005).
56. Bañares, M. A., and Khatib, S. J. Structure–activity relationships in alumina-supported molybdena–vanadia catalysts for propane oxidative dehydrogenation. *Catal. Today* **96**, 251–257 (2004).
57. Chen, K., Xie, S., Bell, A. T., and Iglesia, E. Structure and properties of oxidative dehydrogenation catalysts based on $MoO_3/Al_2O_3$. *J. Catal.* **198**, 232–242 (2001).
58. Chen, K., Xie, S., Iglesia, E., and Bell, A. T. Structure and properties of zirconia-supported molybdenum oxide catalysts for oxidative dehydrogenation of propane. *J. Catal.* **189**, 421–430 (2000).
59. Heracleous, E., and Lemonidou, A. A. Reaction pathways of ethane oxidative and non-oxidative dehydrogenation on g-$Al_2O_3$ studied by temperature programmed reaction (TP-reaction). *Catal. Today* **112**, 23–27 (2006).
60. Heracleous, E., Lee, A. F., Vasalos, I. A., and Lemonidou, A. A. Surface properties and reactivity of $Al_2O_3$-supported $MoO_3$ catalysts in ethane oxidative dehydrogenation. *Catal Lett* **88**, 47–53 (2003).

61. Chen, K., Iglesia, E., and Bell, A. T. Isotopic tracer studies of reaction pathways for propane oxidative dehydrogenation on molybdenum oxide catalysts. *J. Phys. Chem. B* **105**, 646–653 (2000).
62. Heracleous, E., Lemonidou, A. A., and Lercher, J. A. Mechanistic features of the ethane oxidative dehydrogenation by FTIR spectroscopy over a $MoO_3/Al_2O_3$ catalyst. *Appl. Catal., A* **264**, 73–80 (2004).
63. Heracleous, E., and Lemonidou, A. A. Homogeneous and heterogeneous pathways of ethane oxidative and nonoxidative dehydrogenation studied by temperature programmed reaction. *Appl. Catal., A* **269**, 123–135 (2004).
64. Abello, M. C., Gomez, M. F., Casella, M., Ferretti, O. A., Bañares, M. A., and Fierro, J. L. G. Characterization and performance for propane oxidative dehydrogenation of Li-modified $MoO_3/Al_2O_3$ catalysts. *Appl. Catal., A* **251**, 435–447 (2003).
65. Chen, K., Xie, S., Bell, A. T., and Iglesia, E. Alkali effects on molybdenum oxide catalysts for the oxidative dehydrogenation of propane. *J. Catal.* **195**, 244–252 (2000).
66. Liu, C., and Ozkan, U. S. Spectroscopic and structural characterization of chlorine loading effects on Mo/Si:Ti catalysts in oxidative dehydrogenation of ethane. *J. Phys. Chem. A* **109**, 1260–1268 (2005).
67. Koc, S. N., Gurdag, G., Geissler, S., Guraya, M., Orbay, M., and Muhler, M. The oxidative dehydrogenation of propane over potassium-promoted molybdenum oxide/sol–gel zirconia catalysts. *J. Mol. Catal. A: Chem.* **225**, 197–202 (2005).
68. Abello, M. C., Gomez, M. F., and Ferretti, O. Mo/$\gamma$-$Al_2O_3$ catalysts for the oxidative dehydrogenation of propane.: Effect of Mo loading. *Appl. Catal., A* **207**, 421–431 (2001).
69. Samson, K., Klisińska, A., Gressel, I., and Grzybowska, B. Effect of the rhenium additive to $VO_x/TiO_2$ and $MoO_x/TiO_2$ catalysts on their properties in oxidative dehydrogenation of propane. *React. Kinet. Catal. Lett.* **77**, 309–315 (2002).
70. Abello, M. C., Gomez, M. F., and Ferretti, O. Oxidative conversion of propane over $Al_2O_{3-}$supported molybdenum and chromium oxides. *Catal Lett* **87**, 43–49 (2003).
71. Heracleous, E., Vakros, J., Lemonidou, A. A., and Kordulis, C. Role of preparation parameters on the structureselectivity properties of $MoO_3/Al_2O_3$ catalysts for the oxidative dehydrogenation of ethane. *Catal. Today* **91–92**, 289–292 (2004).
72. Watson, R. B., and Ozkan, U. S. Mo loading effects over Mo/Si:Ti catalysts in the oxidative dehydrogenation of ethane. *J. Catal.* **208**, 124–138 (2002).
73. Tsilomelekis, G., Christodoulakis, A., and Boghosian, S. Support effects on structure and activity of molybdenum oxide catalysts for the oxidative dehydrogenation of ethane. *Catal. Today* **127**, 139–147 (2007).
74. Christodoulakis, A., and Boghosian, S. Molecular structure and activity of molybdena catalysts supported on zirconia for ethane oxidative dehydrogenation studied by operando Raman spectroscopy. *J. Catal.* **260**, 178–187 (2008).
75. Tsilomelekis, G., and Boghosian, S. An operando Raman study of molecular structure and reactivity of molybdenum(vi) oxide supported on anatase for the oxidative dehydrogenation of ethane. *Phys. Chem. Chem. Phys.* **14**, 2216–2228 (2012).
76. Velle, O. J., Andersen, A., and Jens, K. J. The oxidative dehydrogenation of ethane by perovskite type catalysts containing oxides of strontium, cerium and ytterbium. *Catal. Today* **6**, 567–574 (1990).

77. Yi, G., Hayakawa, T., Andersen, A. G., Suzuki, K., Hamakawa, S., York, A. P. E., Shimizu, M., and Takehira, K. Oxidative dehydrogenation of ethane over $La_{1-x}Sr_{x}$-$FeO_{3-\Delta}$perovskite oxides. *Catal Lett* **38**, 189–195 (1996).
78. Dai, H. X., Ng, C. F., and Au, C. T. Perovskite-type halo-oxide $La_{1-x}Sr_{x}FeO_{3-\delta}X_{\sigma}$ (X=F, Cl) catalysts selective for the oxidation of ethane to ethene. *J. Catal.* **189**, 52–62 (2000).
79. Dai, H. X., Ng, C. F., and Au, C. T. $YBa_2Cu_3O_{7-\delta}X_{\sigma}$ (X=F and Cl): Highly active and durable catalysts for the selective oxidation of ethane to ethene. *J. Catal.* **193**, 65–79 (2000).
80. Donsì, F., Pirone, R., and Russo, G. Oxidative dehydrogenation of ethane over a perovskite-based monolithic reactor. *J. Catal.* **209**, 51–61 (2002).
81. Alifanti, M., Bueno, G., Parvulescu, V., Parvulescu, V. I., and Cortés Corberán, V. Oxidation of ethane on high specific surface $SmCoO_3$ and $PrCoO_3$ perovskites. *Catal. Today* **143**, 309–314 (2009).
82. Solsona, B., Concepción, P., Hernández, S., Demicol, B., and Nieto, J. M. L. Oxidative dehydrogenation of ethane over NiO–$CeO_2$ mixed oxides catalysts. *Catal. Today* **180**, 51–58 (2012).
83. Solsona, B., López Nieto, J. M., Concepción, P., Dejoz, A., Ivars, F., and Vázquez, M. I. Oxidative dehydrogenation of ethane over Ni–W–O mixed metal oxide catalysts. *J. Catal.* **280**, 28–39 (2011).
84. Savova, B., Loridant, S., Filkova, D., and Millet, J. M. M. Ni–Nb–O catalysts for ethane oxidative dehydrogenation. *Appl. Catal., A* **390**, 148–157 (2010).
85. Weng, W., Davies, M., Whiting, G., Solsona, B., Kiely, C. J., Carley, A. F., and Taylor, S. H. Niobium phosphates as new highly selective catalysts for the oxidative dehydrogenation of ethane. *Phys. Chem. Chem. Phys.* **13**, 17395–17404 (2011).
86. Kumar, A. S. H., Upendar, K., Qiao, A., Rao, P. S. N., Lingaiah, N., Kalevaru, V. N., Martin, A., Sailu, C., and Prasad, P. S. S. Selective oxidative dehydrogenation of ethane over $MoO_3/V_2O_5$–$Al_2O_3$ catalysts: Heteropolymolybdate as a precursor for $MoO_3$. *Catal. Commun.* **33**, 76–79 (2013).
87. Kalevaru, V., Sri Hari Kumar, A., Qiao, A., Lingaiah, N., Alshammari, A., Sailu, C., Sai Prasad, P. S., and Martin, A. Catalytic behavior of decomposed molybdophosphoric acid supported on alumina for oxidative dehydrogenation of ethane to ethylene. In *Mechanisms of Catalytic Reacctions*, St. Petersburg, 2012; (in press).
88. Smoláková, L., Grygar, T., Čapek, L., Schneeweiss, O., and Zbořil, R. Speciation of Fe in Fe-modified zeolite catalysts. *J. Electroanal. Chem.* **647**, 8–19 (2010).
89. Held, A., Kowalska, J., Tuchorska, A., and Nowińska, K. Effect of zeolite structure on oxidative dehydrogenation of light paraffins. In *Stud. Surf. Sci. Catal.*, Elsevier: Volume 170, 1267–1274 (2007).
90. Kucherov, A. V., Nissenbaum, V. D., Kucherova, T. N., and Kustov, L. M. Peculiarities of catalytic oxidation of ethane by nitrous oxide on ZSM-5 type zeolites containing iron ions in different localization sites. *Kinet. Catal.* **43**, 99–106 (2002).
91. Held, A., Kowalska, J., and Nowińska, K. Nitrous oxide as an oxidant for ethane oxydehydrogenation. *Appl. Catal., B* **64**, 201–208 (2006).
92. Kowalska-Kus, J., Held, A., and Nowinska, K. Oxydehydrogenation of $C_{2-}C_4$ hydrocarbons over Fe-ZSM-5 zeolites with $N_2O$ as an oxidant. *Catal. Sci. Tech.* **3**, 508–518 (2013).

93. Katerina, N., Roman, B., and Blanka, W. On the effect of cobalt loading on catalytic activity of Co-BEA zeolites in ethane oxidative dehydrogenation and ammoxidation. *Scientific papers of the University of Pardubice. Series A. Faculty of Chemical Technology* **9**, 111–128 (2003).
94. Woods, M. P., Mirkelamoglu, B., and Ozkan, U. S. Oxygen and nitrous oxide as oxidants: Implications for ethane oxidative dehydrogenation over silica–titania-supported molybdenum. *J. Phys. Chem. C* **113**, 10112–10119 (2009).
95. Kaddouri, A. CH bond activation in the presence or absence of oxygen or nitrous oxide. *React. Kinet. Catal. Lett.* **82**, 401–409 (2004).
96. Ovsitser, O., and Kondratenko, E. V. Similarity and differences in the oxidative dehydrogenation of $C_2$–$C_4$ alkanes over nano-sized $VO_x$ species using $N_2O$ and $O_2$. *Catal. Today* **142**, 138–142 (2009).
97. Nakamura, M., Mitsuhashi, H., and Takezawa, N. Oxygen species formed on different surface sites of CaO by decomposition of $N_2O$ and the reactivity. *J. Catal.* **138**, 686–693 (1992).
98. Choudhary, V., Mondal, K., and Mulla, S. R. Non-catalytic pyrolysis of ethane to ethylene in the presence of $CO_2$ with or without limited $O_2$. *J Chem Sci* **118**, 261–267 (2006).
99. Bortolozzi, J. P., Banús, E. D., Terzaghi, D., Gutierrez, L. B., Milt, V. G., and Ulla, M. A. Novel catalytic ceramic papers applied to oxidative dehydrogenation of ethane. *Catal. Today* (in press).
100. Nguyen, T. T., Burel, L., Nguyen, D. L., Pham-Huu, C., and Millet, J. M. M. Catalytic performance of MoVTeNbO catalyst supported on SiC foam in oxidative dehydrogenation of ethane and ammoxidation of propane. *Appl. Catal., A* **433–434**, 41–48 (2012).
101. Shi, X., Ji, S., Wang, K., and Li, C. Oxidative dehydrogenation of ethane with $CO_2$ over novel Cr/SBA-15/$Al_2O_3$/FeCrAl monolithic catalysts. *Energy Fuels* **22**, 3631–3638 (2008).
102. Yang, B., Yuschak, T., Mazanec, T., Tonkovich, A. L., and Perry, S. Multi-scale modeling of microstructured reactors for the oxidative dehydrogenation of ethane to ethylene. *Chem. Eng. J.* **135,** S147-S152 (2008).
103. Lobera, M. P., Balaguer, M., Garcia-Fayos, J., and Serra, J. M. Rare earth-doped ceria catalysts for ODHE reaction in a catalytic modified MIEC membrane reactor. *ChemCatChem* **4**, 2102–2111 (2012).
104. Hamel, C., Wolff, T., and Seidel-Morgenstern, A. Compatibility of transport and reaction in membrane reactors used for the oxidative dehydrogenation of short-chain hydrocarbons. *Int. J. Chem. Reactor Eng.* **9**, 1542–6580 (2011).
105. Rihko-Struckmann, L., Munder, B., Chalakov, L., and Sundmacher, K. Solid electrolyte membrane reactors. In *Membrane Reactors*, 193–233 (2010).
106. Czuprat, O., Werth, S., Schirrmeister, S., Schiestel, T., and Caro, J. Olefin production by a multistep oxidative dehydrogenation in a perovskite hollow-fiber membrane reactor. *ChemCatChem* **1**, 401–405 (2009).
107. Rodriguez, M. a. L., Ardissone, D. E., Lemonidou, A. A., Heracleous, E., López, E., Pedernera, M. N., and Borio, D. O. Simulation of a membrane reactor for the catalytic oxidehydrogenation of ethane. *Ind. Eng. Chem. Res.* **48**, 1090–1095 (2008).

108. Wang, H., Tablet, C., Schiestel, T., and Caro, J. Hollow fiber membrane reactors for the oxidative activation of ethane. *Catal. Today* **118**, 98–103 (2006).
109. Hamel, C., Tóta, Á. , Klose, F., Tsotsas, E., and Seidel-Morgenstern, A. Analysis of single and multistage membrane reactors for the oxidation of short-chain alkanes—Simulation study and pilot scale experiments. *Chem. Eng. Res. Des.* **86**, 753–764 (2008).
110. F. Hkonsen, S., and Holmen, A. Oxidative dehydrogenation of ethane at short contact times. In *Stud. Surf. Sci. Catal.*, Elsevier: Volume 167, 337–342 (2007).
111. Håkonsen, S., Silberova, B., and Holmen, A. Oxidative dehydrogenation of ethane on Pt–Sn impregnated monoliths. *Top. Catal.* **45**, 61–67 (2007).
112. Wang, D., Xu, M., Shi, C., and Lunsford, J. Effect of carbon dioxide on the selectivities obtained during the partial oxidation of methane and ethane over Li+/MgO catalysts. *Catal Lett* **18**, 323–328 (1993).
113. Krylov, O. V., Mamedov, A. K., and Mirzabekova, S. R. The regularities in the interaction of alkanes with $CO_2$ on oxide catalysts. *Catal. Today* **24**, 371–375 (1995).
114. Krylov, O. V., Mamedov, A. K., and Mirzabekova, S. R. Oxidation of hydrocarbons and alcohols by carbon dioxide on oxide catalysts. *Ind. Eng. Chem. Res.* **34**, 474–482 (1995).
115. Hong, Y., Liwu, L., Qingxia, W., Longya, X., Sujuan, X., and Shenglin, L. Oxidative dehydrogenation of ethane with carbon dioxide to ethylene over Cr-loaded active carbon catalyst. In *Stud. Surf. Sci. Catal.*, Elsevier: Volume 136, 87–92 (2001).
116. Wang, S., Murata, K., Hayakawa, T., Hamakawa, S., and Suzuki, K. Dehydrogenation of ethane with carbon dioxide over supported chromium oxide catalysts. *Appl. Catal., A* **196**, 1–8 (2000).
117. Mimura, N., Takahara, I., Inaba, M., Okamoto, M., and Murata, K. High-performance Cr/H-ZSM-5 catalysts for oxidative dehydrogenation of ethane to ethylene with $CO_2$ as an oxidant. *Catal. Commun.* **3**, 257–262 (2002).
118. Wang, S., Murata, K., Hayakawa, T., Hamakawa, S., and Suzuki, K. Oxidative dehydrogenation of ethane by carbon dioxide over sulfate-modified $Cr_2O_3/SiO_2$ catalysts. *Catal Lett* **63**, 59–64 (1999).
119. Solymosi, F., and Németh, R. The oxidative dehydrogenation of ethane with $CO_2$ over $Mo_2C/SiO_2$ catalyst. *Catal Lett* **62**, 197–200 (1999).
120. Valenzuela, R. X., Bueno, G., Solbes, A., Sapiña, F., Martínez, E., and Cortés Corberán, V. Nanostructured ceria-based catalysts for oxydehydrogenation of ethane with $CO_2$. *Top. Catal.* **15**, 181–188 (2001).
121. Qiao, A., Kalevaru, V. N., Kumar, A. S. H., Lingaiah, N., Prasad, P. S. S., and Martin, A. Effect of $CO_2$ admixture on the catalytic performance of Ni-Nb-M-O catalysts in oxidative dehydrogenation of ethane to ethylene, Reducing the carbon footprint of fuels and petrochemicals, Berlin, DGMK, Hamburg: Berlin, 97–104 (2012).

## CHAPTER 9

# CATALYTIC ORGANIC SYNTHESIS: A NEW PARADIGM IN INDUSTRIAL PROCESS INTENSIFICATION

GANGAVARAM V. M. SHARMA, PALAKODETY R. KRISHNA, VENKATA R. DODDI, SUDHIR KASHYAP, and POST S. REDDY

## CONTENTS

## 9.1 INTRODUCTION

### 9.1.1 FOCUS

Novel scientific ideas in organic synthesis require unique mechanisms to transfer them into industrially important innovations. A detailed review on the past scientific advances reported under catalytic organic chemistry from the world-renowned academic research groups had indicated that highly successful industrial applications occurred whenever the opportunities presented by the industry formed the main themes of study. It is equally interesting to note that when Nobel Prize for chemistry was awarded in 2000 for the electrically conducting polymers, organic electronics found its place on the innovation radars of several companies in the concerned area (s).

In this chapter, we mainly discussed, selected industrial scale organic synthetic applications on C-C and C-O bond forming reactions involving transition metal and asymmetric catalysis. Its focus is on recent advancements of catalytic reactions developed by Noble Laureates viz., cross coupling reactions (Heck, Suzuki, and Negishi), asymmetric oxidations (Sharples) and metathesis (Grubbs) and two other noteworthy reactions viz., Jacobsen and Shi epoxidations. The industrial application areas include such diverse fields as pharmaceuticals, pesticides, polymers, fine and speciality chemicals including macromolecules and emergency management in oil spills. The predominance of process intensification and greener process is evident.

Catalysis has become increasingly important in application oriented organic synthesis; especially in pharmaceutical and polymer industries. Catalysis is the key to chemical transformations that facilitates cost-effective and environment friendly manufacturing processes.

In 1836, the term "catalysis" was introduced by Berzelius. Ostwald defined it in 1895 as "a catalyst accelerates a chemical reaction without affecting the position of the equilibrium" and is still a valid definition. Catalysis is a cyclic process in, which catalyst in one form is involved in chemical bonding with the reactants and the products are released from another form, thus catalyst regenerates to its initial state (Fig. 9.1) [1].

Catalysts not only accelerate reactions but also influence the selectivity of chemical reactions. Different catalysts give different products from the same reactant and this selectivity is often more crucial than the catalytic activity [2]. In contrast to stoichiometric and nonselective reactions, catalytic reactions generate less waste, while the use of catalysis provides many benefits to the industry, including cost reduction, time saving and reduction of overall carbon-footprint. Most industrial syntheses of pharmaceuticals, polymers, synthetic fibers, electronics, dyes, agro chemicals, resins, and pigments rely heavily on catalysis [3]. Nobel Prizes for the work on catalysis starting from Ostwald (1909) for his pioneering work on catalysis, chemical equilibria and reaction rates, Ziegler and Natta (1963) for catalysis in polymer synthesis, Wilkinson and Fischer (1973) for organometallic compounds, Sharpless, Knowles and Noyori (2001) for the catalytic asymmetric oxidation and hydrogenation, Chauvin, Grubbs and Schrock (2005) for metathesis to recent recognition of Heck, Negishi and Suzuki (2010) for their outstanding research on cross-coupling reactions illustrates the importance of catalysis in science and in industry [4].

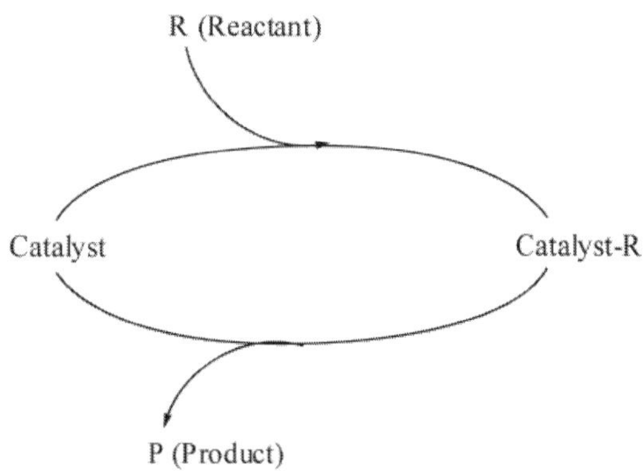


**FIGURE 9.1** Representation of catalytic cycle.

Although catalysis is being applied in chemical industry since 100 years, the importance in pharmaceutical industry is growing since last two decades. U.S. Food and Drug Administration (FDA) initiated "*quality by design*" (QbD) concept, which states that quality of medicinal products should be controlled not only after manufacturing (quality after design) but also before by controlling production processes.[5] Therefore, process chemists need to develop catalytic process with detailed mechanistic understanding that suit API manufacturing. A good catalytic process should possess the characteristics of a safe reaction profile, low catalyst loadings, high selectivity and robustness.

### 9.1.2 PROPERTIES OF CATALYST

A suitable catalytic industrial process must depend upon the following three properties of catalyst: 1) activity 2) selectivity and 3) stability. Each process will have prioritized importance of these properties, depending upon the demand of the industry [2].

**Activity:** Catalyst activity is nothing but a measure of how fast reaction proceeds in the presence of the catalyst. In terms of kinetics, activity can be defined formally in kinetic treatment as to measure reaction rates in terms of temperature and concentrations present in the reactor. Generally, in chemical industry, catalyst activity is measured in terms of turnover number (TON), turnover frequency (TOF) or turnover rate, which originated from the field of enzymatic catalysis.

***Turnover (TO):*** In general turnover can be defined as conversion of one equivalent of reactant to one equivalent of product (per equivalent of catalyst), nothing but one loop in catalyst cycle.

***Turnover number (TON):*** Turnover number is the number of reaction cycles occur at the reactive center through the catalytic cycle before catalyst deactivated [6]. TON is also defined as number of moles of reactant divided by number of moles of catalyst times the percentage of yield of product. For an ideal industrial application, the TON is in the range $10^6$–$10^7$. A high TON indicates a stable and long-lived catalyst.

In general, authors report the mol% of catalyst used in the reaction, which refers the amount of catalyst relative to the reactant. For example, 1 mole % of catalyst = 100 TON, 0.01% = 10,000 TON (Note: these numbers are for a theoretically maximum used catalyst).

***Turnover frequency (TOF) or Turnover rate:*** The turnover frequency TOF is the number of molecular reactions or catalytic cycles occurring at the center per unit time (typically sec, min or h). This number can be obtained by dividing the number of moles of product produced with number of moles of catalyst and time. Thus, the units are just $time^{-1}$. Most robust industrial applications have the TOF is in the range of $10^{-2}$–$10^2$ $s^{-1}$.

TOF and TON can be related as TON = TOF ′ Life time of a catalyst

**Selectivity:** The selectivity of a reaction is to determine the relation between desired and undesired products. Selectivity is of great importance in industrial catalysis, which includes chemoselectivity, regioselectivity, and enantioselectivity (wherever applicable). Chemoselectivity is the selective reactivity of the one functional group in the presence of others. Regioselectivity is the selective reactivity of the one region over the other region of the same functional group. Enantioselectivity can be defined as the selective formation of one enantiomer (e.g., *R* isomer) over the other enantiomer (e.g., *S* isomer).

***Enantiomeric excess (ee):*** Enantiomeric excess (ee) is generally used to measure a compound's enantiomeric purity or chiral purity. It is most often expressed as % of enantiomeric excess and can be formulated as shown in below formula in Fig. 9.2.

$$ee = \frac{|R - S|}{R + S} \text{ X } 100\%$$

**FIGURE 9.2** Formula for calculating enantiomeric excess.

Thus, a catalyst produces racemic mixture (an equal amount of *R* and *S* enantiomers) with 0% ee. In general, 85% or higher considered as good ee and it merely depends on the what other known catalysts can deliver relative to being reported.

**Stability:** Lifetime of a catalyst can be determined by its chemical, thermal, and mechanical stability in the industrial reactors. Catalyst stability is influenced by many factors like decomposition, leaching, fouling and poisoning. In the industrial catalysis, the total lifetime of a catalyst or a stable catalyst plays major role in the economics of a process. Usually, in industrial scale process, the catalytic process should be given the following order of priority: Selectivity > Stability > Activity.

In this chapter, we mainly discussed, selective industrial scale organic synthetic applications on C-C and C-O bond forming ractions involving transition metal catalysis and asymmetric catalysis. It focused on recent advancements of catalytic reactions developed by Nobel laureates, *viz.*, cross-coupling reactions (Heck, Suzuki and Negishi), asymmetric oxidation (Sharpless), metathesis (Grubbs') and two other noteworthy reactions, Jacobsen and Shi epoxidations.

## 9.2 CATALYTIC C-C BOND FORMATION

### *9.2.1 HECK OLEFINATION*

Heck olefination [7] is considered as an exceptionally resourceful process among the palladium catalyzed reactions and has been successfully employed as a powerful tool in industrial manufacturing of pharmaceuticals. In 1972, [8] Heck employed palladium catalyst in the preparation of stilbene from iodobeneze and industrial styrene. Over the years, advances were made on Heck reaction in a perspective with ligand systems, providing sterocontrol and practically adaptable route to quaternary carbon centers. The Heck reaction includes palladium catalyzed coupling of $sp^2$ hybridized centers, for example, unsaturated halides or triflates **1** ($R^1X$) with an activated alkene **2** in the presence of a base and ligand, resulting in a new olefinic system **3** (Scheme 9.1).

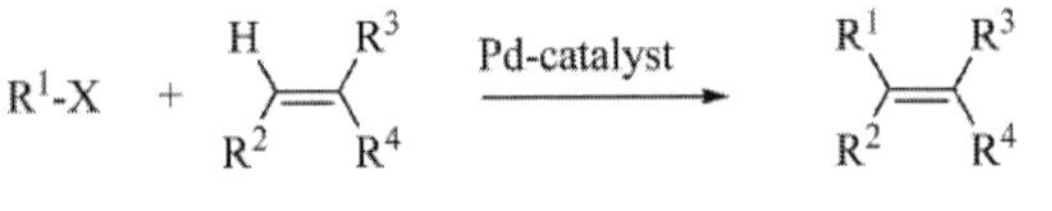


$R^1$ = aryl, alkenyl, benzyl, alkyl (no β-hydrogen)
$R^2$, $R^3$, $R^4$ = aryl, alkenyl, alkyl
X = Cl, Br, I, OTf, OTs, $N^+$

**1** **2** **3**

**SCHEME 9.1** General representation of Heck reaction.

### 9.2.1.1 *MECHANISM*

Since its discovery, details of various steps in Heck reaction [9] remained in focus for discussion; however, the accepted mechanistic Heck reaction can be simplified into five distinct steps. The first sequence comprises oxidative insertion of Pd(0) to alkyl halide bond (R-X). Initially, catalyst palladium(II) is reduced into Pd(0) by the use of phosphine ligands. Then, olefin coordinates to palladium *via* π bond formation to give a Pd-π complex. The next stereo-control step involves concurrent migration of organopalladium group (R) to the adjacent carbon (less hindered) of Pd-olefin species and effects *syn* addition of Pd and R groups to alkene. The resulting intermediate gets distorted into *trans* isomer to relieve the torsional strain and subsequent β-hydride elimination resulting in C-C coupled olefin. In the final step, decomposition of palladium (II) complex occurs, with stoichiometric amount of base, to regenerate Pd(0) *via* reductive elimination (Fig. 9.3).

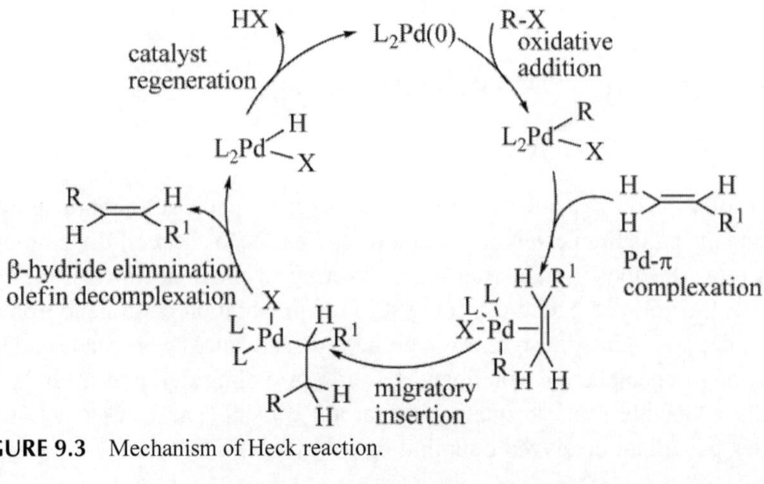


**FIGURE 9.3** Mechanism of Heck reaction.

### 9.2.1.2 INDUSTRIAL APPLICATIONS

The foremost industrial application of Heck coupling was described by Ciba-Geigy in 1994 in their synthetic route to sulfonyl urea herbicide, prosulfuron **8** [10]. In a single reactor, two successive steps, *viz.* Heck coupling of diazonium salt **4** using $Pd_2(dba)_3$ catalyst to generate an arylpalldium intermediate, that couples with trifluoropropene **5** to give **6**, and hydrogenation of **6** with activated charcoal, produced key intermediate **7** (Scheme 9.2).

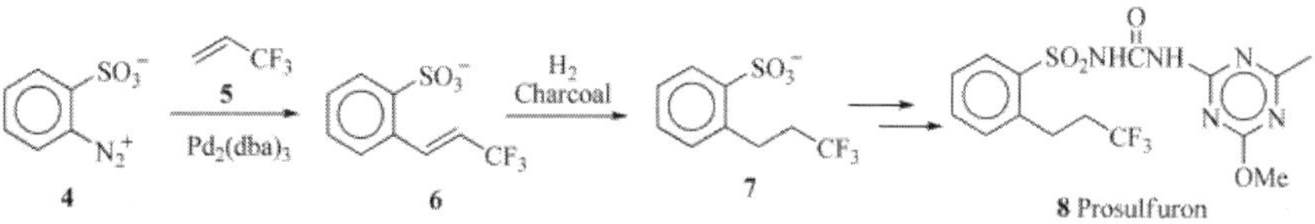


**SCHEME 9.2** The herbicide Prosulfuron.

Scientists at Merck explored the palladium catalyzed Heck coupling in their synthetic route to montelukast sodium **12** (Singulair), $LTD_4$ antagonist drug for asthma [11]. A ligand-free Heck reaction of methyl 2-iodobenzoate **10** with allylic alcohol **9** was affected by $Pd(OAc)_2$ as catalyst and $Et_3N$ as a base. Subsequent removal of residual Pd through Solka-Floc and crystallization gave key drug intermediate, ketone **11** in 83% yield (Scheme 9.3).

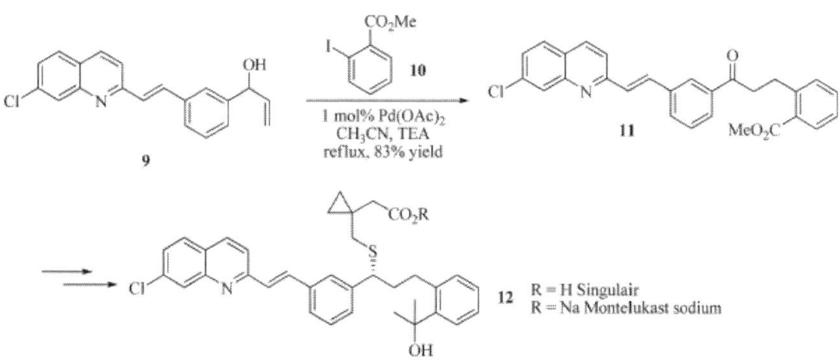


**SCHEME 9.3** Antiasthma agent: Singulair.

Industrial application of Heck reaction was employed at Albemarle for the synthesis of Naproxen **15**, a nonsteroidal antiinflammatory drug (NSAID) [12]. The process incorporated Heck coupling reaction with carbonylation (Scheme 9.4). Optimized reaction of bromonaphthalene **13** with ethylene (pressure 3kPa) in the presence of $PdCl_2$ and neomenthyldiphenylphosphine at 95° C gave key intermediate **14**, with a production over 500 tons/year.

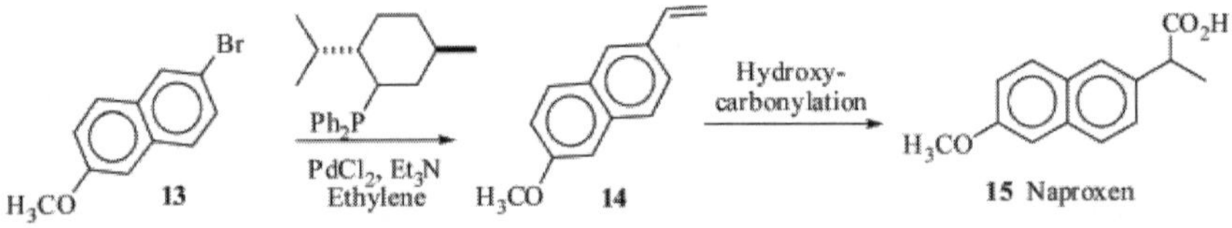


**SCHEME 9.4** Naproxen *via* Heck reaction and hydroxycarbonylation.

A double Heck reaction strategy was explored by Waite et al. [13] at Pfizer to assemble key intermediate **20** en route to **21**, an antagonist for the treatment of asthma. The first Heck reaction includes chemoselective coupling of benzophenone **16** (containing bromo and iodo substitution) and ethyl acrylate **17** using 5 mol% $Pd(OAc)_2$ and $Et_3N$ in MeCN. The resultant coupling product **18** underwent second Heck reaction with *N*-vinylphthalimide **19** using 5 mol% $Pd(OAc)_2$ and diisopropylamine in xylene to afford the intermediate **20** in 58% yield (Scheme 9.5).

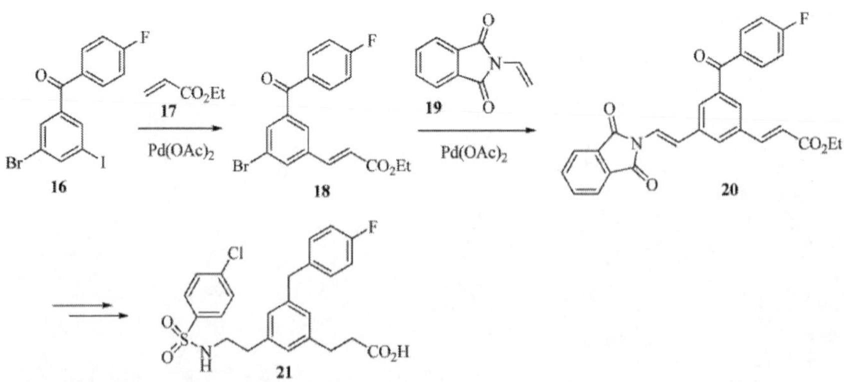


**SCHEME 9.5** Thromboxane receptor antagonist.

In 2004, a multikilogram-scale Heck coupling, using $Pd(OAc)_2$/P(*o*-tolyl)$_3$ catalytic system, was explored by Wallace et al. [14] at GlaxoSmithKline in their synthetic scheme of the vitronectin receptor antagonist **25**. The key intermediate, *trans*selective **24** was synthesized by Heck reaction between aryl bromide **22** and itaconic acid **23**, using 1 mol% $Pd(OAc)_2$, 3 mol% P(*o*-tolyl)$_3$, $Et_3N$ and *n*-$Bu_4NBr$ (Scheme 9.6).

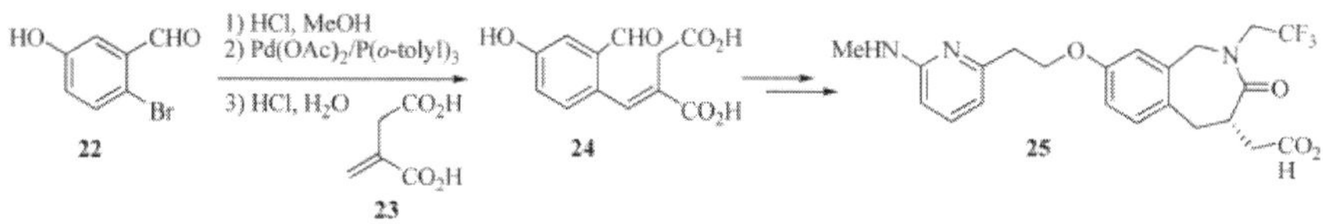

**SCHEME 9.6** Rvβ3 receptor antagonist.

Voight et al. [15] from GlaxoSmithKline developed a one-pot protocol comprising Heck reaction to assemble naphthyridinone **29** in their synthetic route to antibacterial drug **30**. On an industrial-scale, optimized Heck reaction between chloropyridine **26** and *n*-butyl acrylate (**27**) using $Pd(OAc)_2/P(t\text{-}Bu)_3{\bullet}HBF_4$ catalytic system and $(c\text{-}Hex)_2NMe$ as base, gave Heck coupling product **28**, which was cyclized into naphthyridinone **29** (Scheme 9.7).

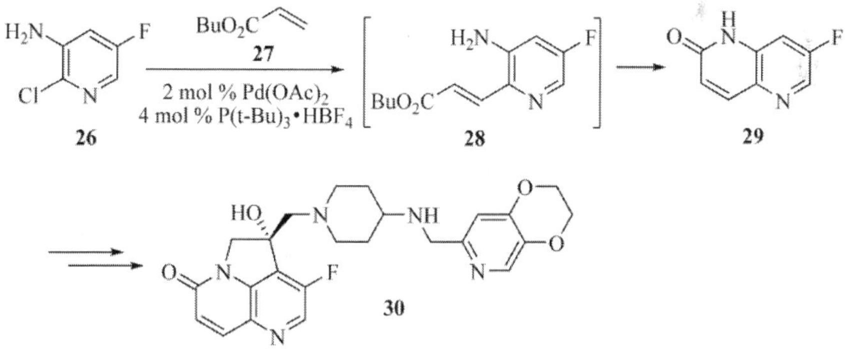

**SCHEME 9.7** Antibacterial drug.

Ripin et al. [16] from Pfizer developed a highly efficient Heck reaction of iodide **31** and bis-Boc allylimide **32** in their synthetic scheme for **33**, a selective ErbB2 angiogenesis inhibitor for the treatment of cancer. A Heck reaction between **31** and **32** in the presence of 1 mol% $Pd_2(dba)_3$, followed by treatment of the resulting product with HCl gave 90 Kg of **33** as a salt (Scheme 9.8).

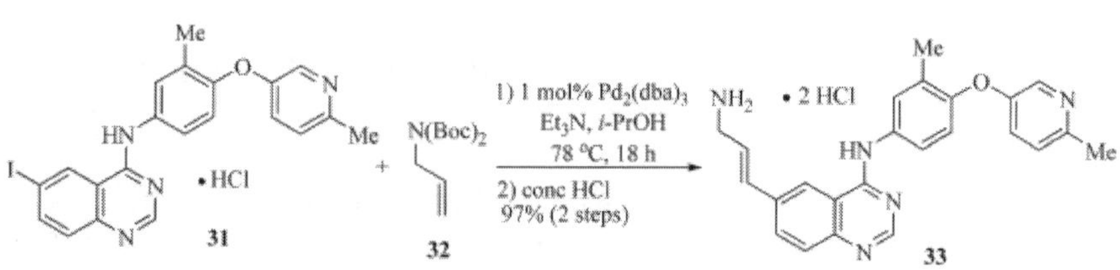

**SCHEME 9.8** ErbB2 angiogenesis inhibitor.

Johnson & Johnson accomplished a robust and multi kilogram scale (250 Kg) synthesis of intermediate **36**, en route to Rilpivirine **37**, a NNRTI [17]. It involves Heck coupling of iodoaniline **34** and acrylonitrile **35** under optimized conditions. Subsequent scavenging of metal residual and treatment with concentrated hydrochloric acid provides HCl salt **36** as an *E*:*Z* (98:2) mixture in 60–70% yield (Scheme 9.9).

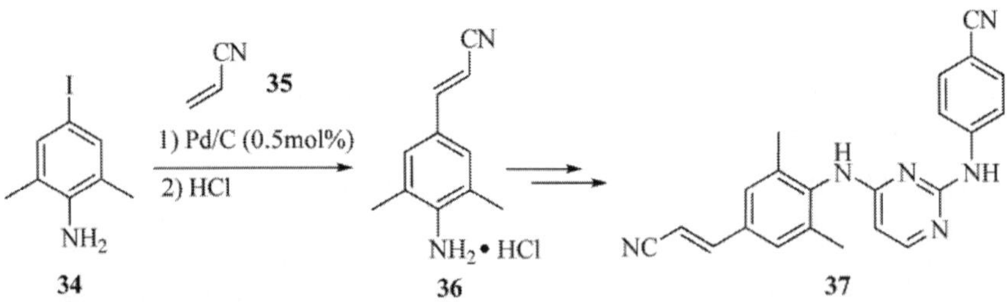


**SCHEME 9.9** Synthesis of Rilpivirine.

Zembower and co-workers explored the double Heck reaction in one-pot and completed the synthesis of DG-041–**41**, $EP_3$ receptor antagonist [18]. A practically reliable catalytic system $Pd(OAc)_2/(o\text{-tolyl})_3$ affected both the transformations, intramolecular Heck reaction of bromo intermediate **38** to form indole ring, as well as intermolecular Heck coupling of resulting indole intermediate with acrylic acid **39** to give **40**, en route to **41** (Scheme 9.10).

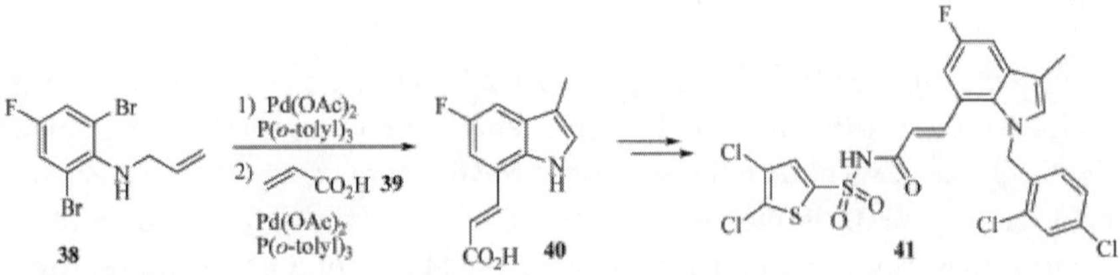


**SCHEME 9.10** Synthesis of DG-041.

Another industrial application of intramolecular Heck reaction was explored at GlaxoSmithKline in the synthetic approach for **44**, a glycine antagonist for the treatment of nicotine craving [19]. The synthesis of drug precursor **43** was accomplished *via* intramolecular Heck reaction of iodo alkene **42** using $PdCl_2/PPh_3$ catalytic system. Successive purging of Pd residual with trimercaptotriazine and crystallization from isooctane provides 245 Kg of **43** with > 99% ee in 74% yield (Scheme 9.11).

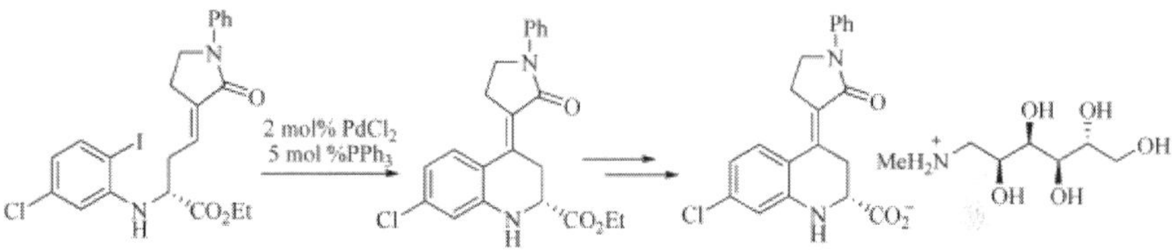


42 43 44

**SCHEME 9.11** Glycine antagonist.

Some more industrial applications of Heck reaction are presented in Appendix (Table 9.1).

### 9.2.2 SUZUKI REACTION

Suzuki was awarded the 2010 Nobel Prize in Chemistry for the part of his discovery and development of cross coupling reaction between organoboron ($R^1BR_2$; **45**) and organohalides, triflates, sulfonates or phosphates ($R^2X$; **46**), catalyzed by palladium(0) complexes [20,21]. Since its discovery in 1979 [22], Suzuki reaction has been extensively employed in pharmaceutical industry to assemble highly functionalized polyolefins, styrenes and biaryl drug targets ($R^1R^2$, **47**) on large-scale and established the most efficient method for biaryl C-C bond formation. Suzuki coupling, also referred to as Suzuki-Miyaura reaction, demonstrates a practically high yielding process, under mild reaction conditions with a wide range of functional group compatibility (Scheme 9.12).

$$\underset{\mathbf{45}}{R^1\text{-}BR^2} + \underset{\mathbf{46}}{R^2X} \xrightarrow[\text{base}]{\text{Palladium catalyst}} \underset{\mathbf{47}}{R^1\text{-}R^2}$$

$R^1,R^2$ = alkyl, aryl, alkenyl
$R_2$ = alkyl, OH, O-alkyl
X = Cl, Br, I, OTf

**SCHEME 9.12** A general representation of Suzuki cross-coupling reaction.

#### 9.2.2.1 MECHANISM

The mechanistic difference between Suzuki reaction and that of other similar couplings is that, the oxidative addition of R-X to Pd(0) follows metathesis of resulting organopalladium intermediate, giving another organopalladium species. On the other hand, activation of organoborane ($R^1$-$BR^2_2$) in presence of base ($R^3O$-$M^+$) produces borate. Activation of organoborane ligands results in the enhancement of polarization of boron-ate complex and facilitates the transmetalation with the organopalladium species to give a palladium complex with both R and $R^1$ groups on it. Finally, reductive elimination proceeds with C-C bond formation to give $R\text{-}R^1$ and restores the original Pd catalyst to start a new reaction (Fig. 9.4).

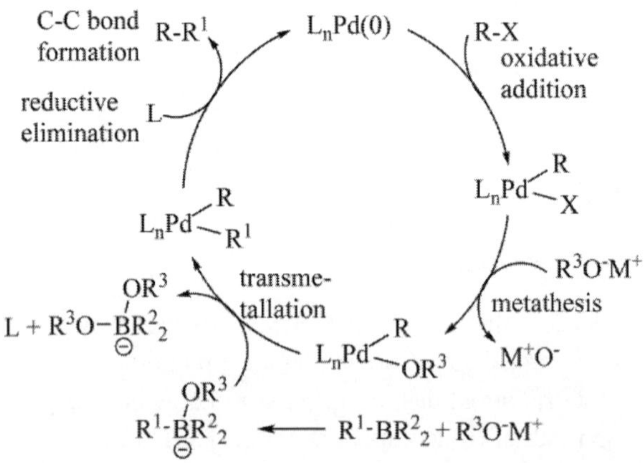


**FIGURE 9.4** Mechanistic representation of Suzuki cross-coupling reaction.

### *9.2.2.2 INDUSTRIAL APPLICATIONS*

It is beyond the scope of this chapter to include all the applications of Suzuki reaction that are of industrial importance. Hence, some important and landmark examples are cited here under.

In 1997, a patent from BASF, "Process for preparing nitrobiphenyls", described a commercially viable application of the Suzuki coupling [23]. On industrial-scale, biaryl intermediate **50** assembled *via* palladium catalyzed Suzuki coupling of **48** and *p*-chloro phenylboronic acid **49**, produced 1000 tons/year of fungicide, boscalid **51** (Scheme 9.13).

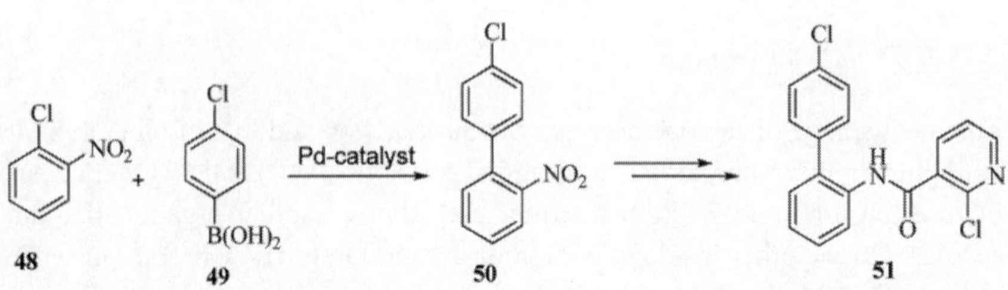


**SCHEME 9.13** Fungicide boscalid.

Chung and co-workers [24] at Merck reported a multikilogram synthesis of naphthyridinone **55**, a p38 MAP kinase inhibitor for the treatment of rheumatoid arthritis, Crohn's disease, and psoriasis. An improved biphasic Suzuki coupling of chloronaphthyridinone **52** and 2,4-difluorophenylboronic acid **53**, catalyzed by $Pd(OAc)_2$ gave drug precursor **54** in 91% yield (Scheme 9.14).

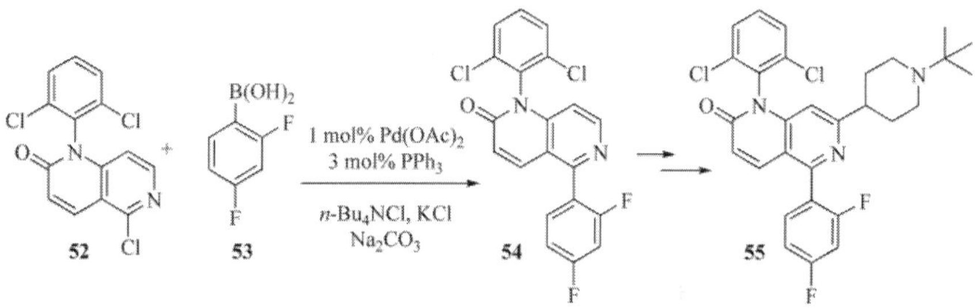


**SCHEME 9.14** p38 MAP kinase inhibitor.

The medicinal chemistry group at AstraZeneca (UK) implemented a pilot-plant scale Suzuki coupling for the synthesis **59**, a drug candidate for the treatment of asthma and rhinitis [25]. Under optimized conditions, the Suzuki coupling between bromothiophene **56** and boronic acid **57** in the presence of $Pd(OAc)_2$ and $P(p\text{-tolyl})_3$ afforded coupling product **58** in 99% yield (Scheme 9.15).

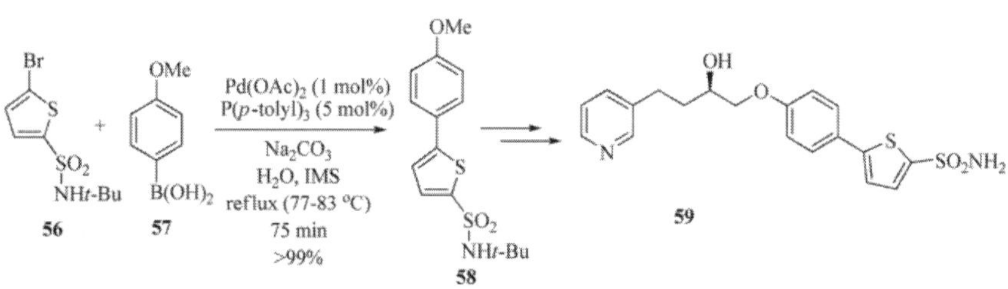


**SCHEME 9.15** A drug candidate for the treatment of asthma and rhinitis.

Scientists at SmithKline Beecham reported [26] Suzuki coupling reaction for the multikilogram preparation of **63**, a candidate for the treatment of depression *x*. The drug intermediate **62** was prepared *via* optimized Suzuki coupling of aryl bromide **60** and boronic acid **61** using Pd/C, in 91% yield (Scheme 9.16).

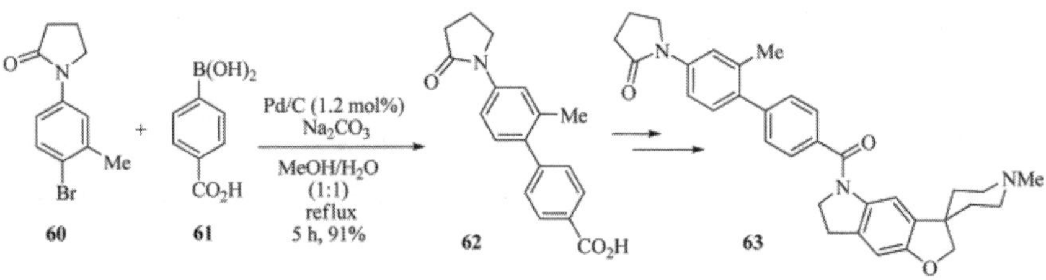


**SCH1P6**
**EME 9.16** Drug for treatment of depression.

Another industrial scale application of Pd/C in Suzuki reaction was explored by researchers at Merck for the preparation of **67**, a PDE4 inhibitor with potential treatment for asthma and chronic obstructive pulmonary disease [27]. Under the optimized Suzuki reaction, bromoquinoline **64** and boronic acid **65** in the presence of Pd/C and $K_2CO_3$ furnished drug intermediate **66** en route to **67** (Scheme 9.17).

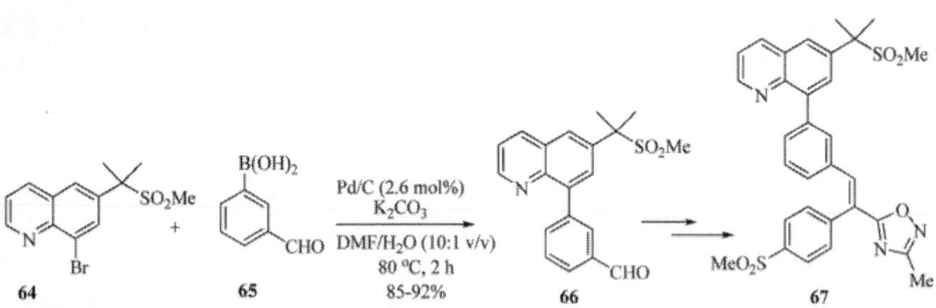


**SCHEME 9.17** PDE4 inhibitor.

The Suzuki coupling using $PdCl_2$(dppf) on commercial scale was explored by researchers at Merck [28], for the synthesis of **70**, a potent, selective and reversible inhibitor of cathepsin K. Thus, coupling of aryl bromide **68** and boronic acid **69** in the presence of $PdCl_2$(dppf)•$CH_2Cl_2$ and $K_2CO_3$ gave **70** (Scheme 9.18).

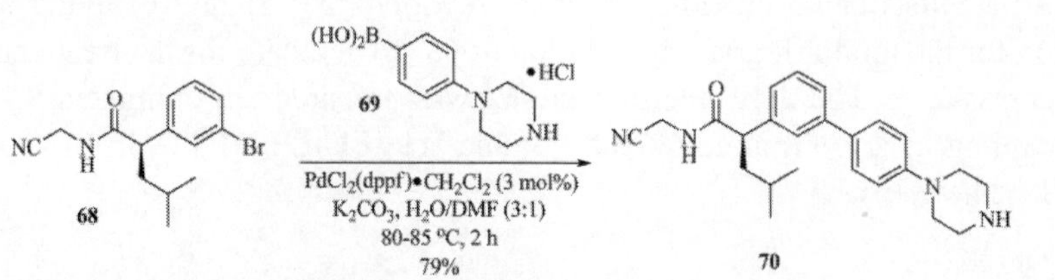


**SCHEME 9.18** An inhibitor of cathepsin K.

A large-scale Suzuki coupling, using Pd black, was published by Miller and co-workers [29] at Eli Lilly. The sulfonamide **75**, an AMPA potent inhibitor for the treatment of cognitive deficits associated with Alzheimer's disease, was assembled *via* double Suzuki coupling protocol. The first Suzuki coupling involves reaction between iodide **71** and pinacolborane **72**. Subsequently, hydrolysis of the coupled product with 48% HBr afforded aryl boronic acid **73**, which underwent second Suzuki reaction with aryl iodide **74**. The double Suzuki protocol afforded 335 g of the desired drug candidate **75**, using Pd black for both the steps (Scheme 9.19).

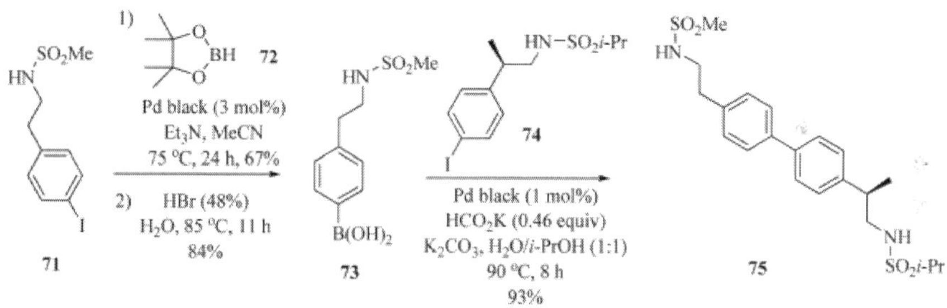


**SCHEME 9.19** Synthesis of AMPA potent inhibitor.

Richey and co-workers [30] at Eli Lilly revealed a cascade reaction with the perspective of Suzuki coupling and completed the synthesis of HCl salt **79**, an intermediate on the route to selective nuclear hormone receptor modulator. A Pd-catalyzed intramolecular carbometalation of alkyne **76** followed by slow coupling of resulting alkenylpalladium intermediate with *m*-nitroboronic acid **77** furnished the dibenzoxapine core **78**, a seven membered ring in 83% yield (Scheme 9.20).

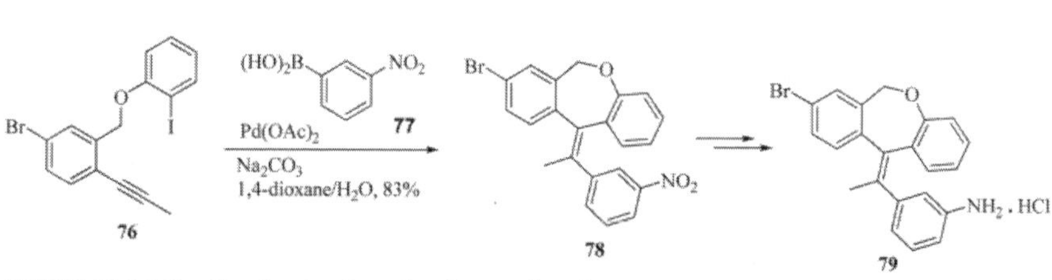


**SCHEME 9.20** Synthesis of a selective nuclear hormone receptor modulator.

A pilot plant-scale Suzuki coupling was employed by Sharp et al. [31] at GlaxoSmithKline in their synthetic route to **84**, a potent serotonin, noradrenaline and dopamine reuptake inhibitor. Initially, triflate **81**, prepared from ketone

**80** using triflic anhydride and DIPEA, on Suzuki reaction with boronic acid **82** in the presence of $Pd(OAc)_2$ and $PPh_3$, led to the biaryl intermediate **83**, a key drug moiety (Scheme 9.21).

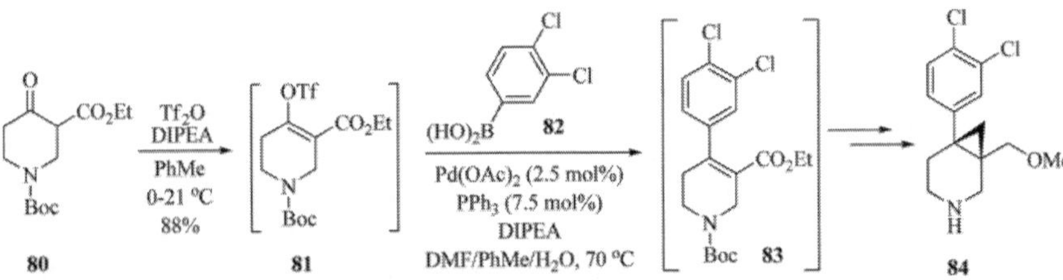


**SCHEME 9.21** Synthesis of Dopamine reuptake inhibitor.

A reliable route to Cl-1034–**89**, an endothelin receptor antagonist was evaluated by Jacks et al. [32] at Pfizer. The bromide **85** was converted into boronic acid **86** using *n*-BuLi and $B(Oi\text{-}Pr)_3$. Suzuki reaction of **86** with triflate **87** in the presence of $PdCl_2(PPh_3)_2$, followed by scavenging of Pd gave 80 Kg of drug intermediate **88** in 95% yield (Scheme 9.22).

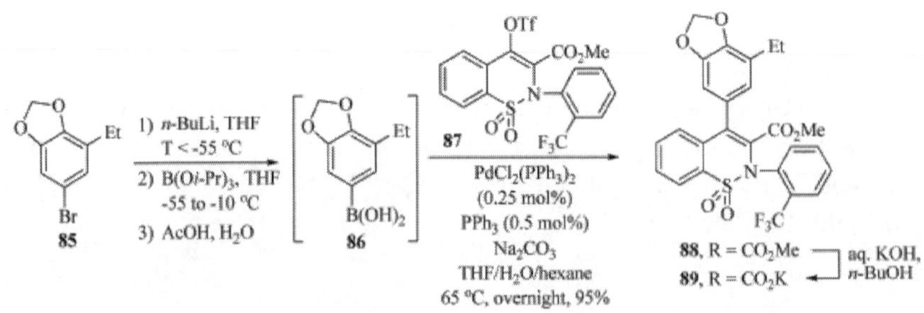


**SCHEME 9.22** Synthesis of endothelin receptor antagonist.

Some more industrial applications of Suzuki reaction are presented in Appendix (Table 9.2).

### *9.2.3 NEGISHI REACTION*

Negishi cross-coupling [33] has been moderately employed in pharmaceutical industry. In 1976, Negishi revealed organozirconium and organoaluminum reagents for cross coupling with organohalides **90** in the presence of a palladium catalyst. In a further improvement on coupling reaction, Negishi developed much milder and more selective organozinc compounds **91**. Due to greater ionic character of the C-Zn bond, organozinc reagents **91** behave more basic and nucleophilic. In addition, the organozinc reagent circumvented the drawbacks associated with other organometallic reagents. In 2006, researchers from different groups [34,35] developed methods for organozinc reagents in the presence of sensitive functional groups. Moreover, Pd(0) can be replaced with Ni-catalyst in Negishi cross-coupling reaction for C-C bond formation **92**. A general representation of Negishi reaction is shown below (Scheme 9.23).

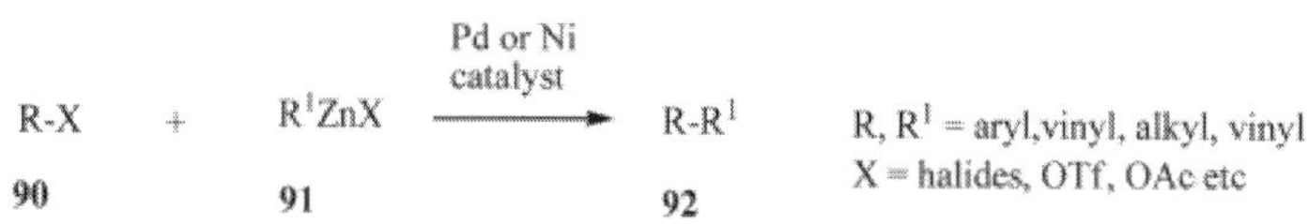


**SCHEME 9.23** A general representation of Negishi cross-coupling reaction.

### 9.2.3.1 MECHANISM

In general, mechanism of Negishi coupling is illustrated in three essential steps. The first step involves oxidative addition of R-X to Pd(0) to give an organopalladium intermediate and metalation of an oraganozinc precursor ($R^1$-X) to the corresponding organozinc intermediate ($R^1$-ZnX). In the second step, organopalladium intermediate exchanges its X with the $R^1$ group in $R^1$-ZnX, this transmetalation process (Zn-Pd exchange) brings the two organic groups (R and $R^1$) into close proximity on the same palladium atom. In the final step, reductive elimination regenerates the Pd(0) catalyst and C-C bond formation furnishes R-$R^1$ (Fig. 9.5).

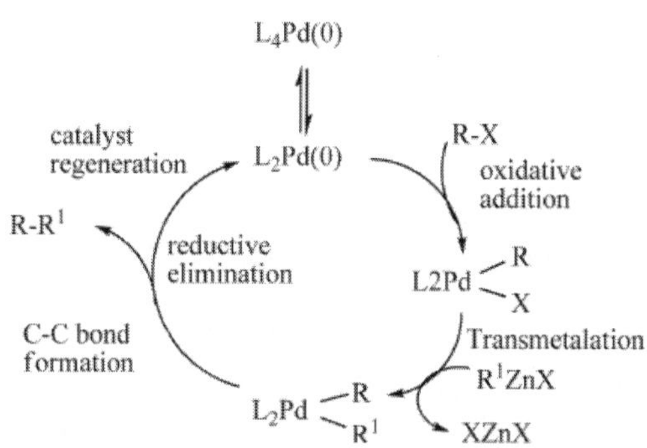


**FIGURE 9.5** Mechanistic representation of Negishi cross-coupling reaction.

### 9.2.3.2 INDUSTRIAL APPLICATIONS

Using stereospecific Negishi coupling as key step, Shang et al. [36] from Chiron Corporation designed a synthesis for selective NNRTI **97**. The reaction of anisole **93** with *n*-BuLi at −70° C followed by quenching with $ZnBr_2$ in THF solution gave aryl zinc **94**. The resulting solution of aryl zinc **94** on Negishi reaction with iodocyclopropane **95** in the presence of $Pd(OAc)_2$ and tris(2,4-di-t-butylphenyl)phosphate resulted in the desired coupling product **96** without epimerization (Scheme 9.24).

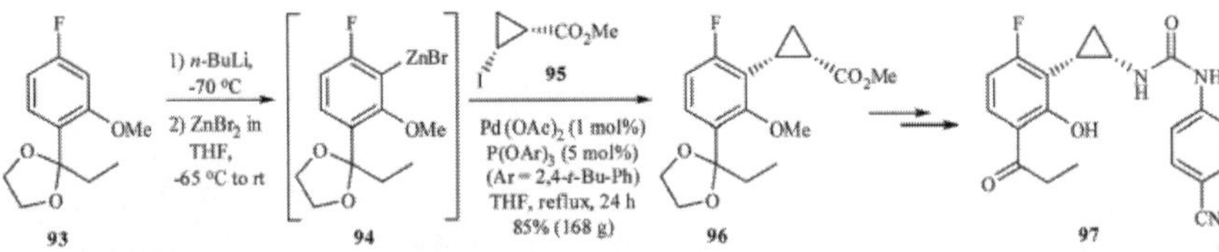


**SCHEME 9.24** B-Raf kinase inhibitor.

Another example of Negishi coupling using $Pd(OAc)_2$ on commercial scale was explored by Praveen et al. [37] from Dr. Reddy's Laboratories for the synthesis of an acyl-CoA cholesterol acyltransferase inhibitor, ezetimibe **102**. The carboxylic acid **98** was converted into acid chloride **99** and directly incorporated in a large-scale Negishi coupling with arylzinc complex **100** in the presence of $Pd(OAc)_2$ catalyst to afford the aryl ketone **101** (Scheme 9.25).

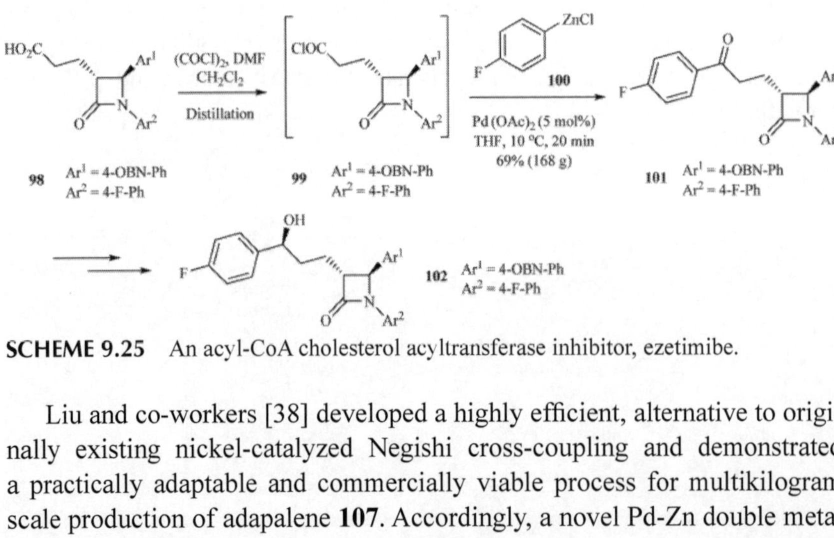


**SCHEME 9.25** An acyl-CoA cholesterol acyltransferase inhibitor, ezetimibe.

Liu and co-workers [38] developed a highly efficient, alternative to originally existing nickel-catalyzed Negishi cross-coupling and demonstrated a practically adaptable and commercially viable process for multikilogram scale production of adapalene **107**. Accordingly, a novel Pd-Zn double metal catalyzed cross-coupling reaction avoided the preparation of stoichiometric amounts of ArZnBr. Ester **106** was synthesized directly by the transmetalation of Grignard reagent **104**, generated from **103** and aryl bromide **105** in the presence of catalytic amount of $PdCl_2(PPh_3)_2$ and $ZnCl_2$ (5 mol%), en route to **107** (Scheme 9.26).

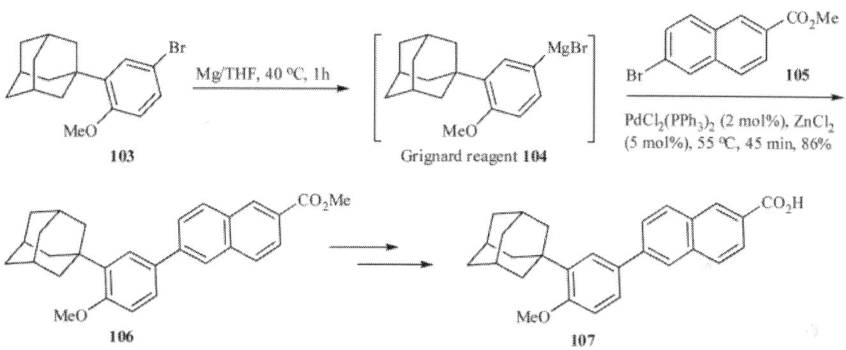


**SCHEME 9.26** Synthesis of adapelene.

Bänziger et al. [39] from Novartis used the Negishi coupling as key step for the preparation of drug intermediates in their route to B-Raf kinase inhibitor **112**. On large-scale, lithiation of bromoisoquinoline **108** using *n*-BuLi, followed by metal exchange with $ZnBr_2$ afforded arylzinc **109**. Negishi coupling of arylzinc **109** and triflate **110** in the presence of $Pd(PPh_3)_4$ gave crude product **111** (6 Kg) en route to **112** (Scheme 9.27).

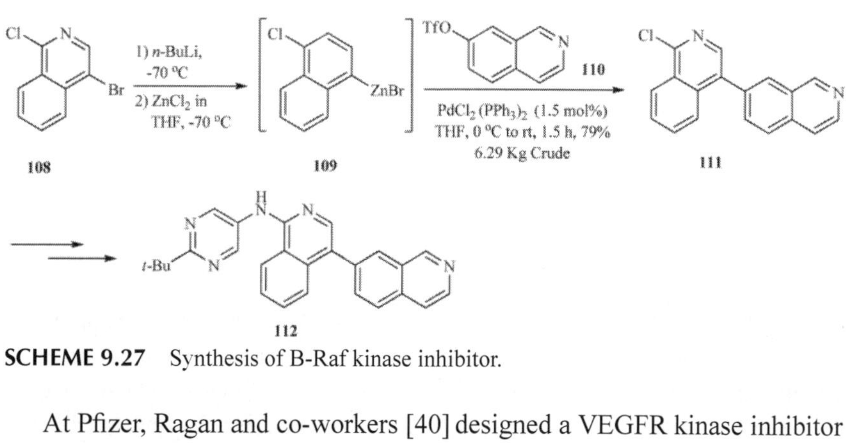


**SCHEME 9.27** Synthesis of B-Raf kinase inhibitor.

At Pfizer, Ragan and co-workers [40] designed a VEGFR kinase inhibitor **117**, using Negishi coupling to assemble key intermediate **116**. On a laboratory scale, lithiation of imidazole **113** with s-BuLi followed by addition of $ZnCl_2$ in THF gave imidazolylzinc **114**, which upon Negishi coupling with **115** in the presence of 10 mol% $PdCl_2(PPh_3)_2$ afforded drug precursor **116**. Thus, the Negishi coupling provided a better alternative to a more toxic Stille coupling, which involves a tributylstannylimidazole, en route to the synthesis of **117** (Scheme 9.28).

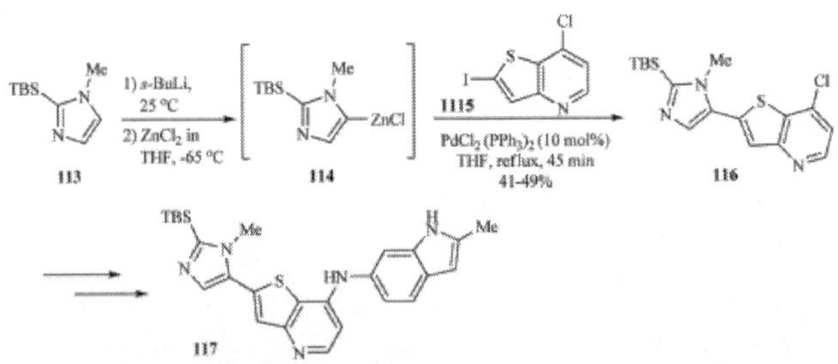


**SCHEME 9.28** Synthesis of a kinase inhibitor.

Perez-Balado et al. [41] at Johnson & Johnson, USA, developed a highly efficient Negishi coupling of pyridylzinc **118** and iodopyrimidine **119** in their synthetic route to **120**, an intermediate to a selective PDE-V inhibitor. A laboratory scale Negishi cross coupling of **118** and **119** in the presence of $Pd(PPh_3)_4$ afforded biaryl **120** in 65–70% yield. However, the yield on scale-up in the mini-plant fell to 40–48% and parallel six mini-plants were used for large-scale production of **120** (Scheme 9.29).

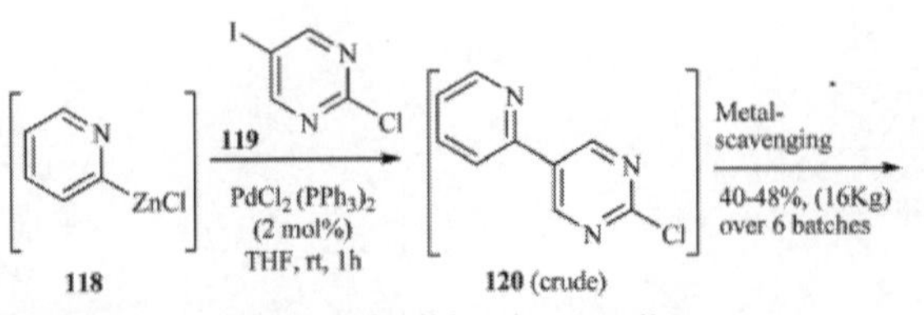


**SCHEME 9.29** A selective PDE-V inhibitor intermediate.

## 9.2.4 OLEFIN METATHESIS

The word 'metathesis' is derived from the Greek words *meta* (change) and *thesis* (position). Thus, olefin metathesis is a fundamental chemical reaction involving the rearrangement of carbon-carbon double bonds and can be used to couple, cleave, ring-close, ring-open or polymerize olefinic molecules (Scheme 9.30).

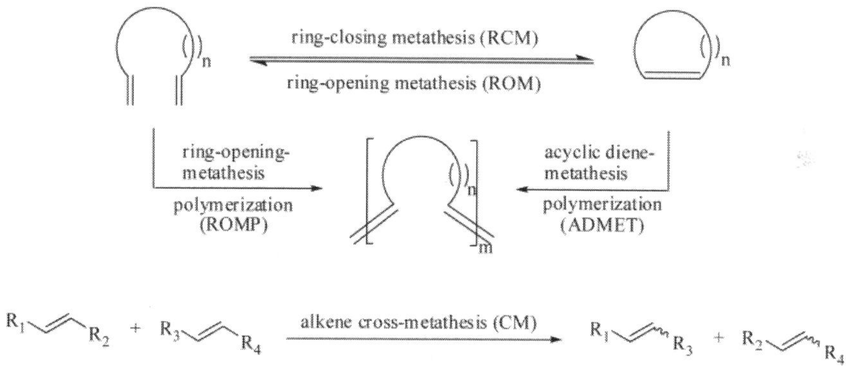


**SCHEME 9.30** The most commonly employed alkene-metathesis reactions in organic synthesis.

In the year 2005, Nobel Prize in Chemistry was awarded to Yves Chauvin, Robert H. Grubbs, and Richard R. Schrock "for the development of the metathesis method in organic synthesis" [41–43]. It is a general view that, this reaction revolutionized different fields of synthetic chemistry. Chauvin proposed the "carbene" mechanism to explain this reaction [44] and Schrock prepared the first well-defined highly active metathesis catalysts [45], while, Grubbs developed ruthenium based active and air stable catalysts, which are tolerant to various functional groups in the presence of olefins [46].

### 9.2.4.1 *MECHANISM*

Hérisson and Chauvin in 1971 [44], were the first to propose generally accepted mechanism, widely supported with key experimental evidences by Casey [47], Katz [48] and Grubbs groups [49] as shown in Scheme 9.31.

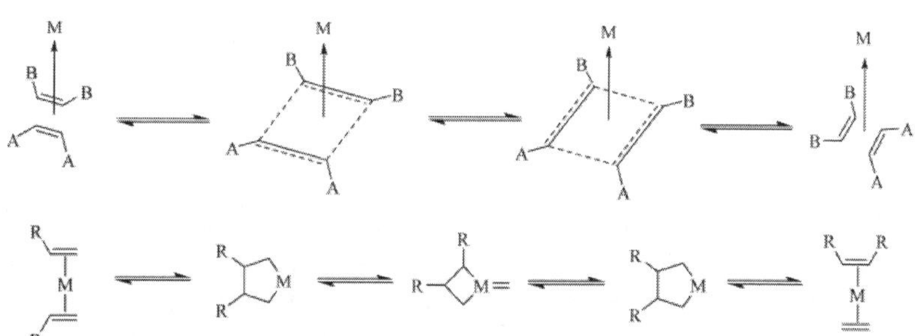

**SCHEME 9.31** Mechanisms for the olefin metathesis.

The work on single component catalyst for olefin metathesis reported in 1976, by Katz et al. [50], was the key milestone in the evolution of olefin metathesis. They demonstrated that single component tungsten carbenes could initiate olefin metathesis without added coactivators. As part of evolution of olefin metathesis, different groups Katz, Tebbe, Schrock and Grubbs developed different catalysts and their functional group tolerance is shown in Fig. 9.6 [51,52]. Grubbs catalysts are found to be relatively the best functional group tolerant and stable.

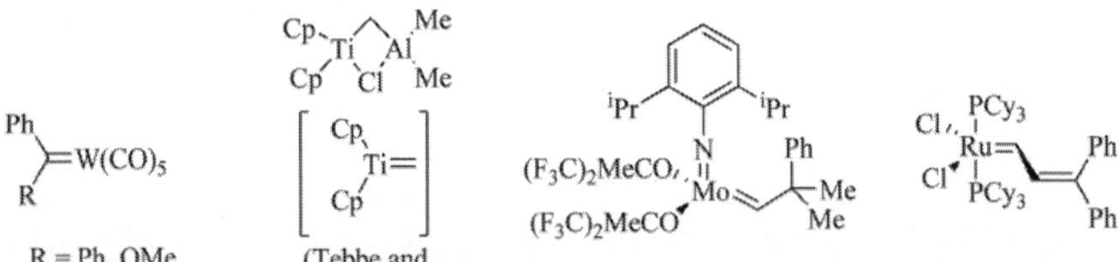


**FIGURE 9.6** The evolution of single-component olefin metathesis catalysts.

From 1992 onwards, ruthenium based catalysts gained importance in olefin metathesis and many groups developed variety of catalysts. Interestingly, Grubbs (**121**,**123**), Hoveyda (**122**,**124**) and Grela (**125**) catalysts are well used in the synthesis of a wide variety of natural products and molecules of pharmaceutical interest [53].

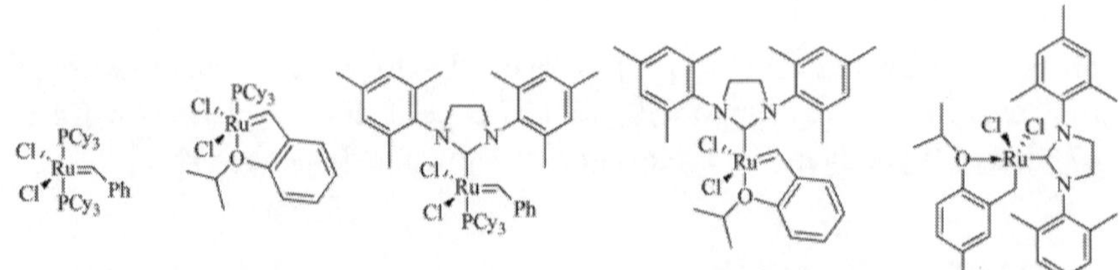


**FIGURE 9.7** Recent developments of olefin metathesis catalysts.

These catalysts are made up of ruthenium metal atom ligated by five ligands of, which two are neutral, two are anionic and one is alkylidene moiety. Trialkylphosphines, *N*-heterocyclic carbenes are generally employed as neutral ligands and halides are employed as monoanionic ligands. Phosphine ligands in the first generation catalysts are replaced with an *N*-heterocyclic carbene (NHC), which is characteristic of all the second-generation catalysts (Fig. 9.7). The second-generation catalysts are of same use in organic synthesis as the first generation catalysts, but generally with higher activity and more stability towards moisture and air, therefore easier to handle.

### 9.2.4.2 *INDUSTRIAL APPLICATIONS*

Alkene is the most abundant group present in many commercial products, or its precursors or starting materials. Therefore, olefin metathesis applications could be of enormous use in future, even though it is relatively new tool in pharmaceutical industry.

To date, commercial products range from a wooden baseball bat, treated with metathesis-based polymers to improve durability, to the preparation of highly functionalized pharmaceutical agents, many of, which are in the advanced stages of testing. In the 1960s, Philips Petroleum was the first to discover the metathesis reaction [54]. Royal Dutch Shell exploited the olefin metathesis in the *Shell's higher olefin process* (SHOP) for the production of fatty alcohols as ingredients of detergents [55].

The Phillips triolefin process [56] developed at Phillips Petroleum, used a heterogeneous $WO_3/SiO_2$ catalyst in metathesis reaction to convert propene **127** into mixture of ethene **125** and 2-butene **126**. As it is a reversible reaction (Scheme 9.32) and the price of propene rose high, the reverse reaction of Philips process offered is now by using Lummus' technology to produce propene known as 'Olefin Conversion Technology' (OCT).

$$\underset{\mathbf{125}}{CH_2{=}CH_2} + \underset{\mathbf{126}}{CH_3CH_2{=}CH_2CH_3} \longleftrightarrow \underset{\mathbf{127}}{2CH_3CH_2{=}CH_2}$$

**SCHEME 9.32** Olefin Conversion Technology.

In 1976, polynorbornene **129** under the trade name Norsorex® was the first commercial metathesis polymer product launched in market by CdF-Chimie in France. Later, in 1978, it was also marketed in the USA and Japan [57]. This polymer is produced by Ring-opening metathesis polymerization (ROMP) of 2-norbornene **128** (Scheme 9.33). This reaction was catalyzed by $RuCl_3$/HCl in butanol, which operates in air and produces a useful elastomer, used for oil spill recovery, for damping or as a sound barrier. It is presently produced in ASTROTECH Co. under the trade name Norsorex® APX, to use as a miscellaneous oil spill control agent by emergency management team in USA.

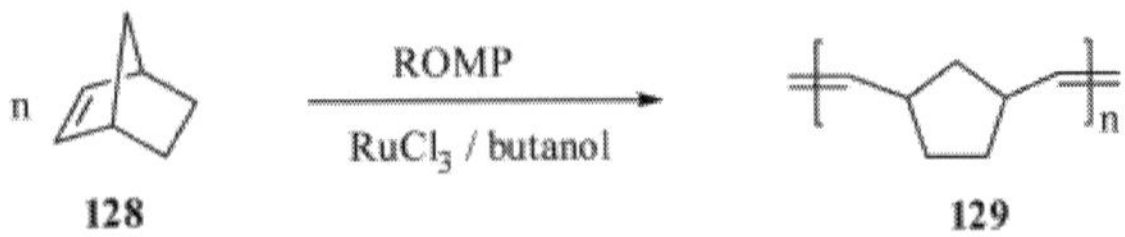


**SCHEME 9.33** Synthesis of Norsorex using ring opening Metathesis polymerization.

Another ruthenium catalyzed ROMP produced polydicyclopentadiene **131** (Scheme 9.34) from dicyclopentadiene **130** [58]. This resulted in a wide variety of polymer products, such as the composite baseball bats, bathroom fixtures, ballistic panels and large equipment body parts.

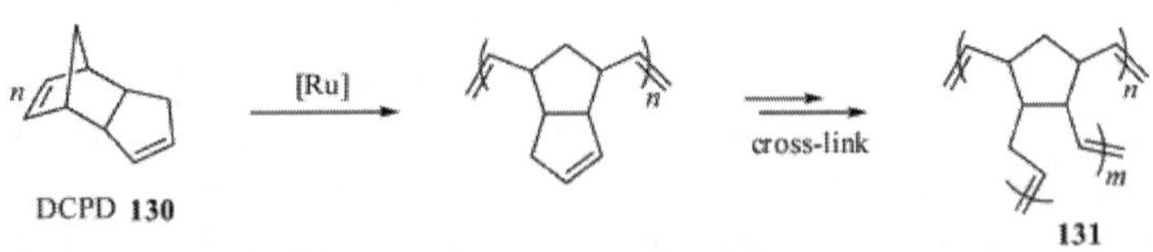


**SCHEME 9.34** ROMP for the preparation of dicyclopentadiene (DCPD) polymers.

### 9.2.4.3 *ENVIRONMENT FRIENDLY PRODUCTS*

Materia Inc. is using ruthenium-catalyst technology for the manufacture of environment friendly products such as the insect pheromones of the peach twig borer and the omnivorous leaf roller. 1-Hexene **132** and 5-hexenyl acetate **133** are the symmetrical internal olefins as starting materials, used for successive cross metathesis reactions to produce their higher carbon analog 5-decenyl acetate **134** (Scheme 9.35) as a major component of pheromone peach twig borer [59]. About 0.2 mol% of the Grubbs 2nd generation catalyst **123** (Fig. 9.7) generated an equilibration of the mixture. The reaction was run in neat and the product was separated by distillation and the starting olefins were recycled.

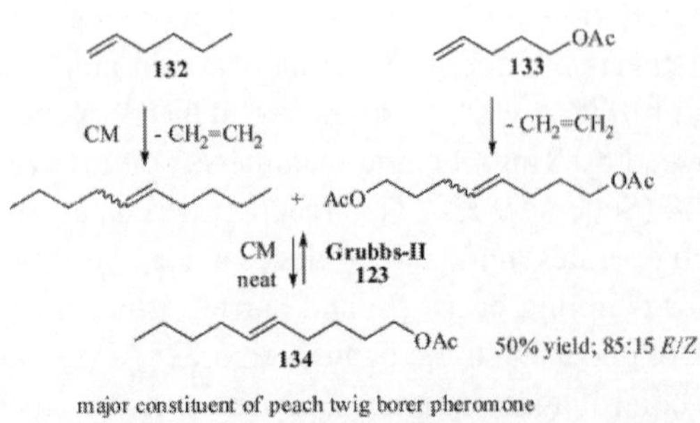


**SCHEME 9.35** Cross-metathesis in the production of "green" compounds.

### 9.2.4.4 INDUSTRIAL APPLICATIONS

Ring closing metathesis reaction has got its significance especially in organic synthesis to construct molecules of pharmaceutical interest. Although ring-closing metathesis (RCM) reaction is a popular reaction, it was not used much in large-scale chemical industry unlike other metathesis reactions, because of four main reasons: 1) high dilutions are generally needed to avoid dimerized products or oligomers, 2) relatively more catalyst loadings (3–7 mol%) are required to complete the reaction, 3) high cost of catalyst and low availability in bulk and 4) lesser stability of some of the active RCM catalysts at higher temperature s. However, due to its amazing ring-closure technology, pharmaceutical industry may still like to use RCM reaction, despite the drawbacks as mentioned above. In 2005, Boehringer Ingelheim started its large-scale application of RCM to construct a drug intermediate. Later, GSK and Merk used RCM reaction at large-scale to show that it is a gateway to generate a macrocycle on large-scale. Research towards more active catalysts and low cost technology will lead to many applications of RCM in pharmaceutical industry.

Among few disclosed large-scale applications, notable example is the preparation of antiviral drug **139** [60], a hepatitis C virus NS3 protease inhibitor by researchers at Boehringer Ingelheim. The group has widely screened diene substrates and catalysts and carried out RCM on diene **135** with **122–124** [61–64] (Fig. 9.7). They were successful to get macrolide **137** with **122** in >90% yield with no epimerization. Further investigation on the scale up studies with **122** revealed that it requires high catalyst loading (5–7 mol%), high dilution (0.01 M) and longer reaction times (24 h in $CH_2Cl_2$ at reflux) for the completion of reaction (Scheme 9.36).

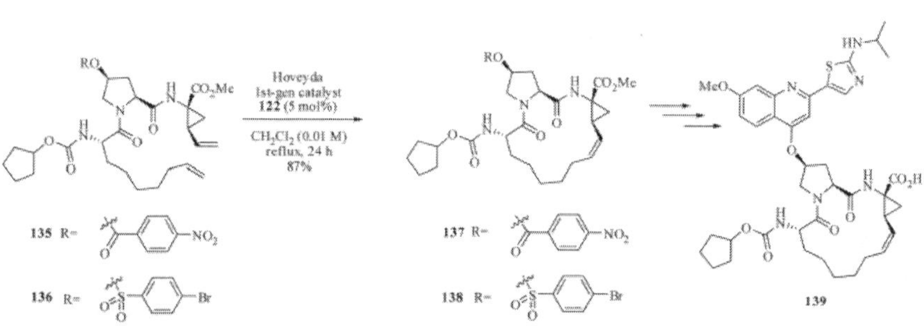


**SCHEME 9.36** Synthesis of antiviral drug 139.

Based on the above scale up studies, Boehringer Ingelheim performed RCM on **136** in toluene, as part of first scale up to produce 400 kg of **138** (Scheme 9.36) in two campaigns [65].

Further scale up studies, revealed Grela's catalyst **125** was more effective and provided higher TON (turnover number) and TOF (turnover frequency). Thus, the scale up was performed on protected diene **140** with Grela's catalyst **125** [66–68]. "Effective molarity" (EM) concept was adopted to reduce the high dilution and obtained satisfactory results in toluene (0.1 M) at reflux (Scheme 9.37). Finally, the optimized conditions gave the product in 95% yield in less than 30 min, with minimal oligomerization. Furthermore, product was obtained in a 27-fold increase in the EM with no epimerization [69].

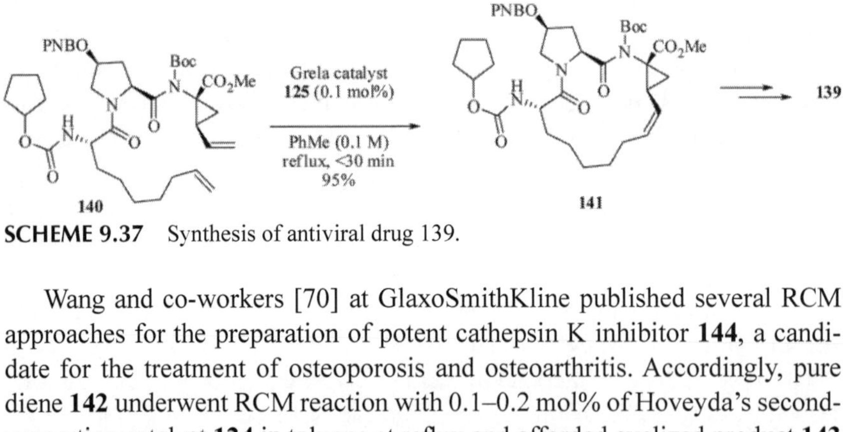


**SCHEME 9.37** Synthesis of antiviral drug 139.

Wang and co-workers [70] at GlaxoSmithKline published several RCM approaches for the preparation of potent cathepsin K inhibitor **144**, a candidate for the treatment of osteoporosis and osteoarthritis. Accordingly, pure diene **142** underwent RCM reaction with 0.1–0.2 mol% of Hoveyda's second-generation catalyst **124** in toluene at reflux and afforded cyclized product **143** (71.8 kg, 96% yield), without any byproducts (Scheme 9.38).

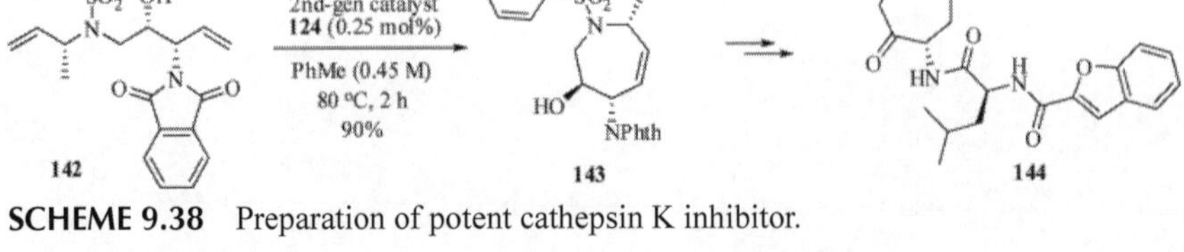


**SCHEME 9.38** Preparation of potent cathepsin K inhibitor.

The process research team at Merck developed a practical scalable route to the HCV protease inhibitor Vaniprevir (MK-7009) **149** [71]. RCM reaction of **145** with second-generation Hoveyda catalyst **124** formed the 20-membered macrocyclic core **147** in 91% yield. High dilution conditions were avoided by the simultaneous slow addition of catalyst and substrate to the reaction mixture, while 10 mol% 2,6-dichloroquinone additive suppressed the isomerization of the allylic diene and avoided side product (19-mr ring). The free acid **146**, a crystalline solid with high purity, under the above reaction conditions used for the ester **145**, gave the macrocyclic acid **148** in excellent yield (Scheme 9.39).

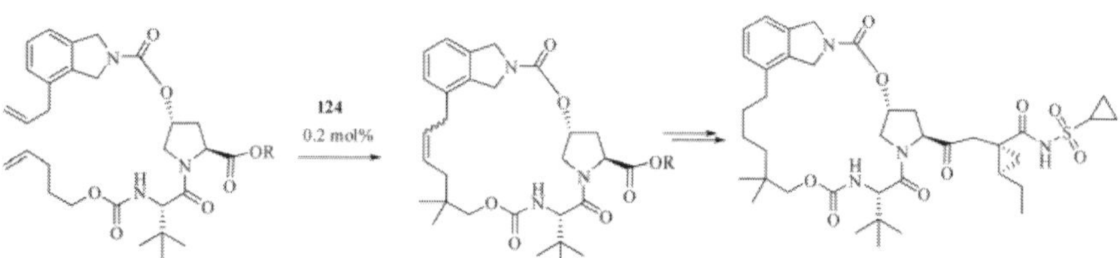


**SCHEME 9.39** Synthesis of HCV protease inhibitor Vaniprevir (MK-7009).

## 9.3 CATALYTIC C-O BOND FORMATION

In pharmaceutical industry, C-O bond formation is a very important reaction. Classical resolution is a very powerful and simple method for obtaining chiral alcohol/epoxide molecules, while, resolution is still being used along with enantioselective synthetic methods. Numerous asymmetric reactions are frequently employed for C-O bond formation in large-scale API synthesis. In 1980, Sharpless et al. [72] reported asymmetric titanium tartarate-catalyzed epoxidation of allylic alcohols, which was followed by the discovery of many asymmetric catalytic reactions. Yet, another important reaction on the asymmetric dihydroxylation was also reported by Sharpless et al. in 1988 [73].

Asymmetric manganese-salen-catalyzed epoxidation of unfunctionalized olefins was reported by Jacobsen et al. [74] in 1990, which allowed the enantioselective epoxidation of unfunctionalized olefins. In particular, the high enantioselectivities obtained for Jacobsen epoxidation on *cis*-olefins, nicely complement the Sharpless epoxidation and dihydroxylation protocols, which give reduced enantioselectivities for these substrates. The Sharpless and Jacobsen procedures are frequently used asymmetric oxidative reactions in API synthesis. More recently, organocatalytic procedures such as Shi epoxidations [75] were also employed to avoid toxic transition metal catalysts.

## 9.3.1 SHARPLESS ASYMMETRIC EPOXIDATION (SAE)

Sharpless' asymmetric epoxidation (SAE) is the epoxidation of allylic alcohols into asymmetric epoxides in high chiral purity (high enantioselectivity). Transition metal catalyst $Ti(OPr^i)_4$ with chiral additive, diethyl tartarate (DET), generates chiral catalyst (Scheme 9.40) which is responsible for the enantioselective outcome, while, *tert*-butyl hydroperoxide (TBHP) serves as an oxidant. Although, this catalytic system holds disadvantage of low turnover number (TON) with potential safety concerns for using concentrated solutions of peroxides, the reaction has nevertheless been extensively used in pharmaceutical industry [76].

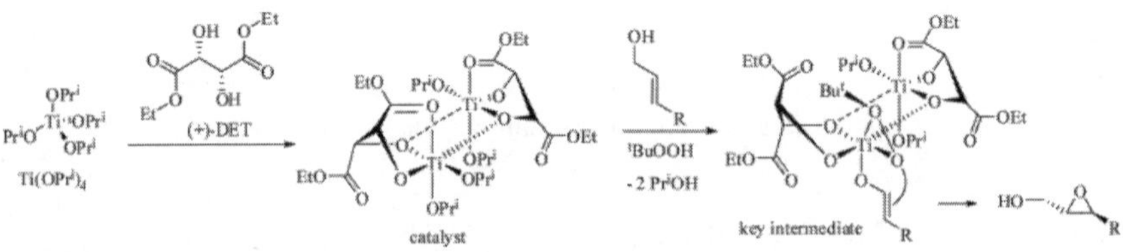


**SCHEME 9.40** Mechanism of asymmetric epoxidation.

### 9.3.1.1 INDUSTRIAL APPLICATIONS

Kimura et al. [77] at Eisai developed an enantioselective synthesis of (*S*)-noremopamil **153**, L-type calcium cshannel blocker, by using SAE as key step to introduce chirality (Scheme 9.41). Epoxidation of allylic alcohol **150** gave the crystalline epoxy alcohol **151** in 84% yield (> 99% ee). The epoxide **151** was elaborated into (*S*)-noremopamil as well as a library of related phenylalkylamines through the formation of intermediate **152** containing quaternary carbon.

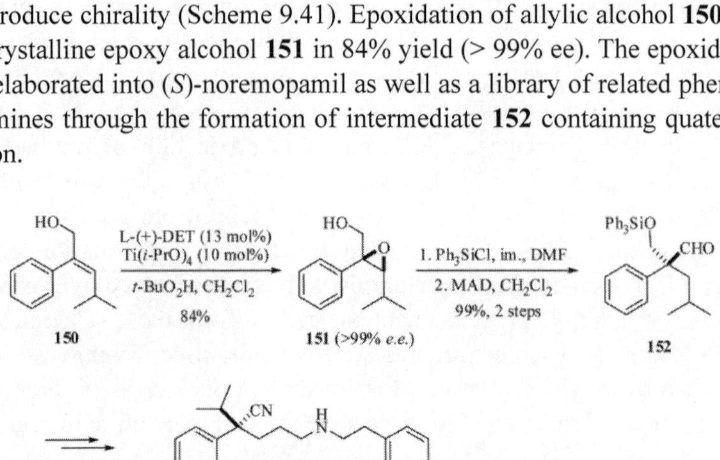


**SCHEME 9.41** Synthesis of (*S*)-noremopamil.

Henegar et al. [78] at Pfizer developed an efficient and greener synthesis of the (*S, S*)-reboxetine **157**, which is being evaluated for the treatment of neuropathic pain and a variety of other indications (Scheme 9.42). The reported chiral synthesis of **157** starts by SAE of cinnamyl alcohol **154** to give (*R, R*)-epoxide **155** in 89% yield (> 98% ee). Reaction of **155** with 2-ethoxyphenol gave crystallized product **156**. The overall yield of (*S, S*)-reboxetine succinate increased by 9%, compared to resolution method. The catalytic asymmetric process offers use of less solvent and reduces waste generation by approximately 50%, compared to the resolution route.

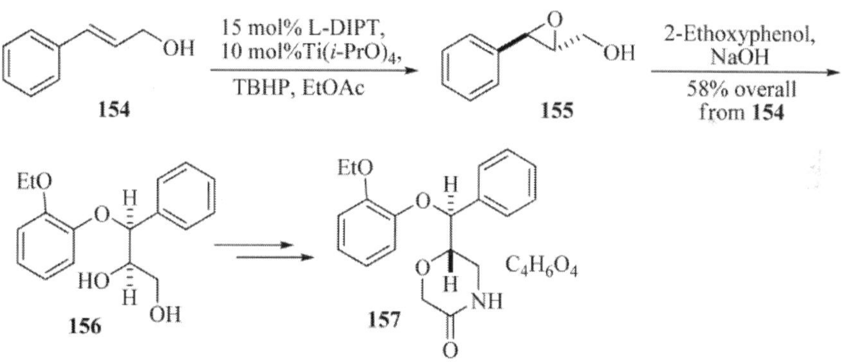


**SCHEME 9.42** Synthesis of (*S, S*)-Reboxetine succinate.

Sharpless asymmetric epoxidation is one of the most powerful tools in the field of asymmetric oxidation; it requires low temperature conditions to achieve proper epoxide yields and high enantiomeric excess (ee). Large operation costs associated with reaction conditions restricted the application to high scale production; however, Sharpless asymmetric dihydroxylation has much use in large-scale productions.

## 9.3.2 SHARPLESS ASYMMETRIC DIHYDROXYLATION (SAD)

The enantioselective 1,2-diol generation from unfunctionalized olefins can generally be achieved by Sharpless's asymmetric dihydroxylation reaction (SAD). "AD-mix $\alpha$" and "AD-mix $\beta$" are the catalysts employed along with methane sulfonamide to give the corresponding enantiomerically pure diols. The two letters AD, stand for asymmetric dihydroxylation. Both the reagents contain potassium osmate [$K_2OsO_2(OH)_4$], potassium ferricyanide [$K_3Fe(CN)_6$] as reoxidant in the catalytic cycle and potassium carbonate along with chiral ligand $(DHQ)_2$PHAL **158** in "AD-mix $\alpha$" and chiral ligand

$(DHQD)_2$PHAL **159** in "AD-mix $\beta$ " , being responsible for chiral induction (Fig. 9.8). These catalytic mixtures can be used as such in manufacturing.

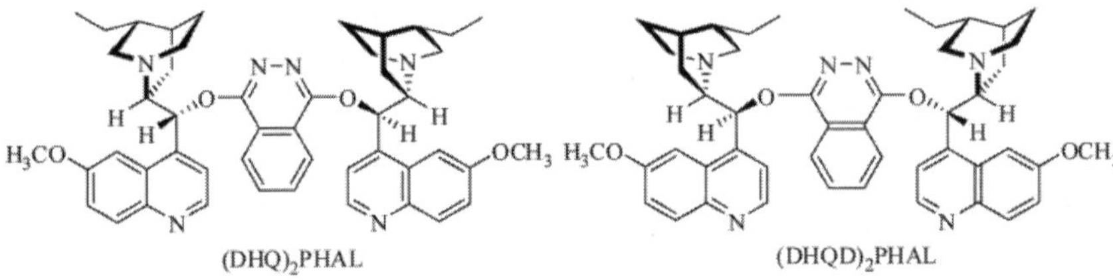


**FIGURE 9.8** Structures of chiral ligands responsible for asymmetric dihydroxylation.

### *9.3.2.1 INDUSTRIAL APPLICATIONS*

Prasad et al. [79] at BMS reported catalytic asymmetric synthesis of melatonin agonist **163** (Scheme 9.43), in, which the chirality of the molecule was introduced by SAD reaction on dihydrobenzofuran derivative **160**. Catalytic Sharpless dihydroxylation of **160** gave the diol **161** in 84–94% yield (98–99% ee). One-pot conversion of the diol **161** to epoxide **162** was developed using modified Sharpless-Kolb procedure. The two-step process was demonstrated on multi kilogram scale.

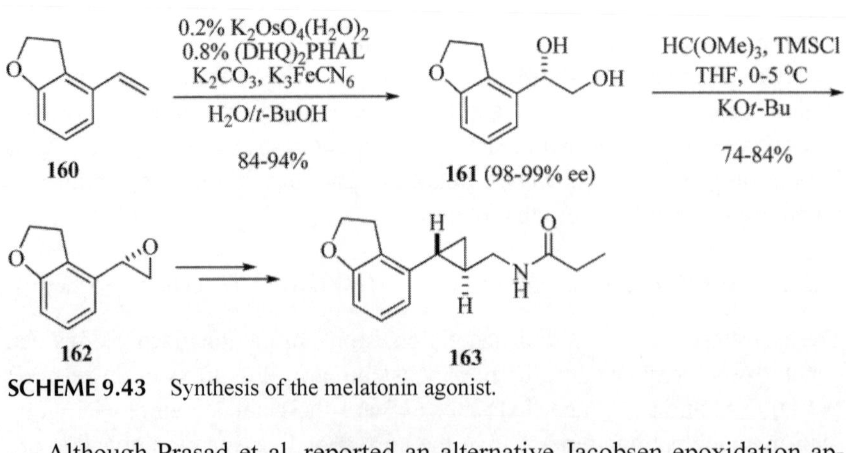


**SCHEME 9.43** Synthesis of the melatonin agonist.

Although Prasad et al. reported an alternative Jacobsen epoxidation approach for the preparation of **162**, the SAD procedure was chosen for further scale-up due to the higher overall yield in the synthesis of **163** (43% versus

22%). Simple wash with aqueous sodium sulfite during workup allowed them the easy removal of osmium metal to 6 ppm or less in the isolated diol.

Researchers at Merck reported SAD reaction of a styrene derivative **164** for the synthesis of the substance-P inhibitor **166** [80]. SAD on **164** using AD-mix α gave crude diol **165** with 87% ee optical purity, which was further upgraded to > 99% ee upon recrystallization (Scheme 9.44). Noteworthy, of the process is that the stereochemistry created in dihydroxylation reaction is responsible for the origin in installation of all the three chiral centers in the final drug substance.

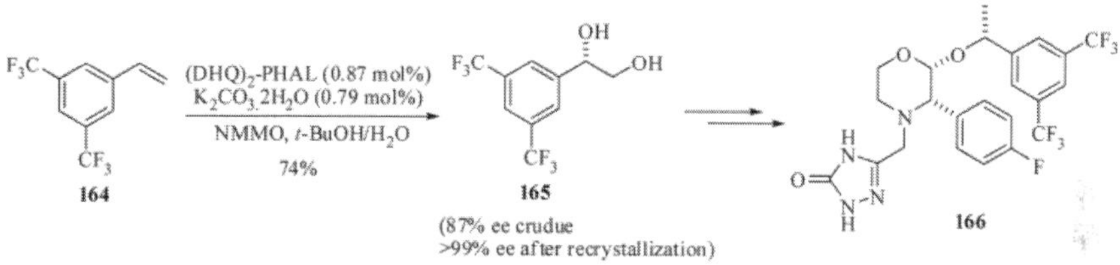


**SCHEME 9.44** Synthesis of the substance-P inhibitor.

### 9.3.3 JACOBSEN EPOXIDATION

Unlike the Sharpless epoxidation, which gives chiral epoxides from allylic alcohols, asymmetric epoxidation of unfunctionalized alkenes achieved by Jacobsen et al. [74], by using chiral salen-metal catalysts. Salen-Mn catalysts are preferred since manganese itself is relatively a low cost and nontoxic metal, primarily because of fewer side reactions over other metal complexes. Variety of simple oxidants, such as PhIO, NaClO, $H_2O_2$ and oxone are employed as reoxidants and best possible enantioselectivity for a given substrate could be achieved by choosing the proper metal-salen catalyst and reaction conditions [81]. The catalyst can be fine-tuned for required steric and electronic properties by making a variety of chiral salen ligands from various chiral diamines with salicylaldehyde derivates [82].

The general term Salen-type is used in literature to describe the class of [O, N, N, O] tetradentate bis-Schiff base ligands. Some of the salen ligands **167**, **168** and **169** and their metal complexes are illustrated in Fig. 9.9. Commercial synthesis of chiral Salen complex **171**, from chiral diamine salt **172** and salicylaldehyde derivative **173** is presented in Scheme 9.45 [83].

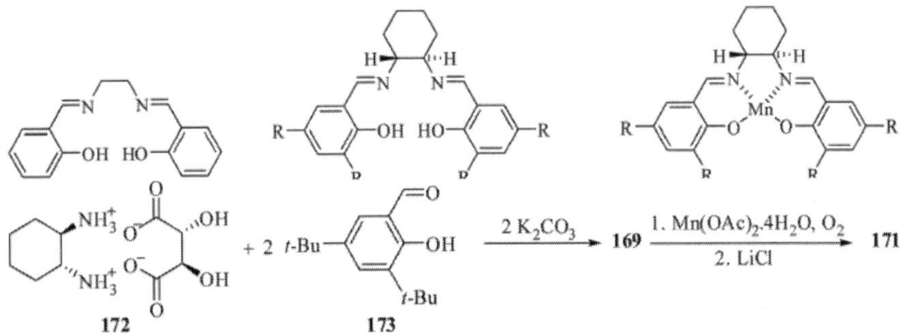


**SCHEME 9.45** Commercial synthesis of salen ligand.

### 9.3.3.1 *INDUSTRIAL APPLICATIONS*

Senanayake and co-workers [84–86] at Merck described a catalytic Jacobsen epoxidation of indene **174** using (*S, S*)-Mn(salen)Cl complex (enantiomer of **171**) **175** (0.7%) and 4-(3-phenylpropyl)pyridine *N*-oxide **176** (3%) to afford (1*S*,2*R*)-indene oxide **177** in 87% ee. The *N*-oxide **176** was employed to stabilize the catalyst and to accelerate the epoxidation. Crude **177** was subjected to crystallization with L-tartaric acid to provide enantiopure **178** (>99.5% ee) (Scheme 9.46). This two-step process was reported on 600 kg scale, which confirmed the efficiency and practicality of the catalytic chemistry.

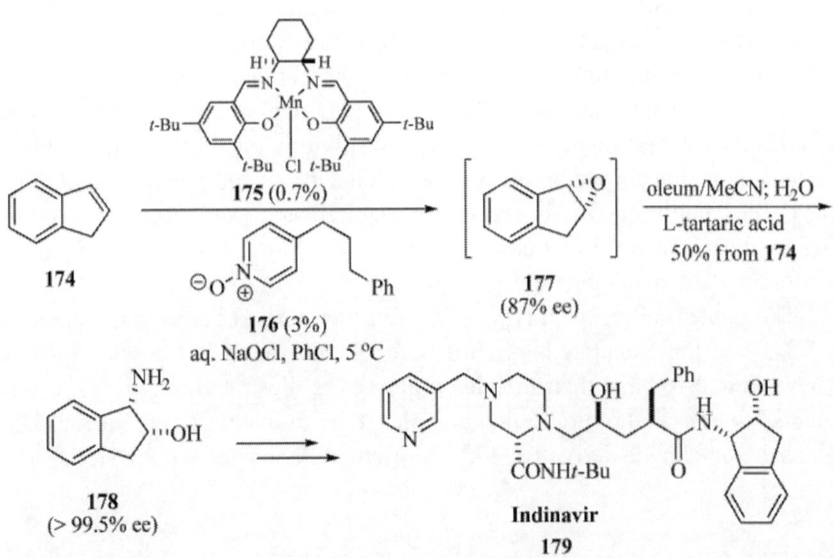


**SCHEME 9.46** Synthesis of HIV protease inhibitor crixivan (indinavir).

Lynch and colleagues [87] at Merck reported the synthesis of selective phosphodiesterase IV inhibitor CDP840–**182** (Scheme 9.47). The method used Jacobsen epoxidation of *Z*-olefin **180** to generate epoxide **181** with 89% ee with the catalyst (*S*, *S*)-Salen-Mn complex **175** (1%) and *N*-oxide **176**. Reductive opening of the epoxide in high diastereoselectivity (66:1) and deoxygenation of the secondary alcohol gave CDP840–**182**.

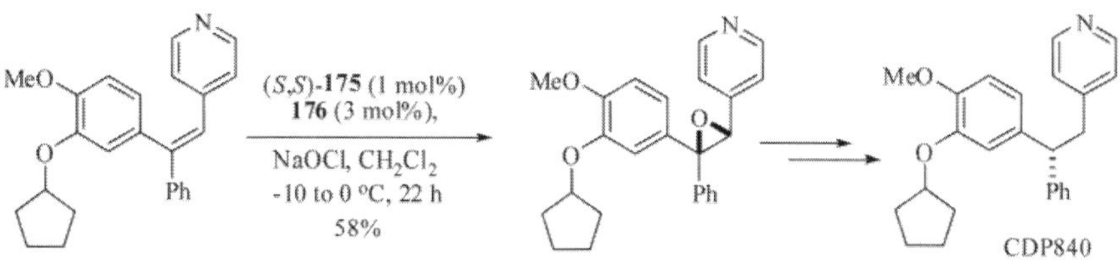


**180** **181** (89% ee) **182**

**SCHEME 9.47** Synthesis of phosphodiesterase IV inhibitor CDP840.

## *9.3.4 SHI EPOXIDATION*

In 1996, Shi et al. [75] developed a fructose-derived ketone (Epoxone®) **183** as a highly effective asymmetric epoxidation catalyst. Shi's epoxidation is known to be the best for the asymmetric epoxidation of *trans*olefins and trisubstituted olefins. Shi's ketone is readily available and an efficient and selective oxidant that requires mild conditions. Ketone **183** could be synthesized [88] from inexpensive chiral starting material D-fructose, by ketalization and oxidation (Scheme 9.48). The enantiomer of **183** can be synthesized from L-fructose, which in turn could be obtained from commercially available L-sorbose. Chemists at DSM developed a scalable process for the preparation of Epoxone® **183** in large quanities.

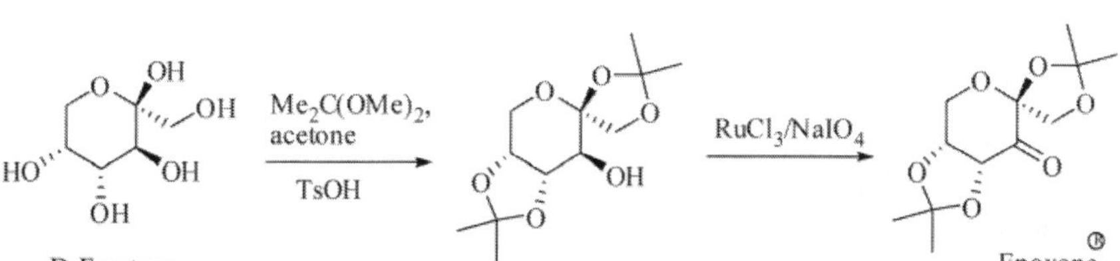


D-Fructose Epoxone® **183**

**SCHEME 9.48** Synthesis of Epoxone®.

Dioxirane, a powerful epoxidation reagent, is believed to be active species in this reaction, while dioxiranes are less stable, they can be generated by in situ oxidation of a ketone with potassium peroxymonosulfate (Oxone). As the fructose derivative is regenerated, catalytic amounts of it are sufficient for the completion of reaction (Fig. 9.10).

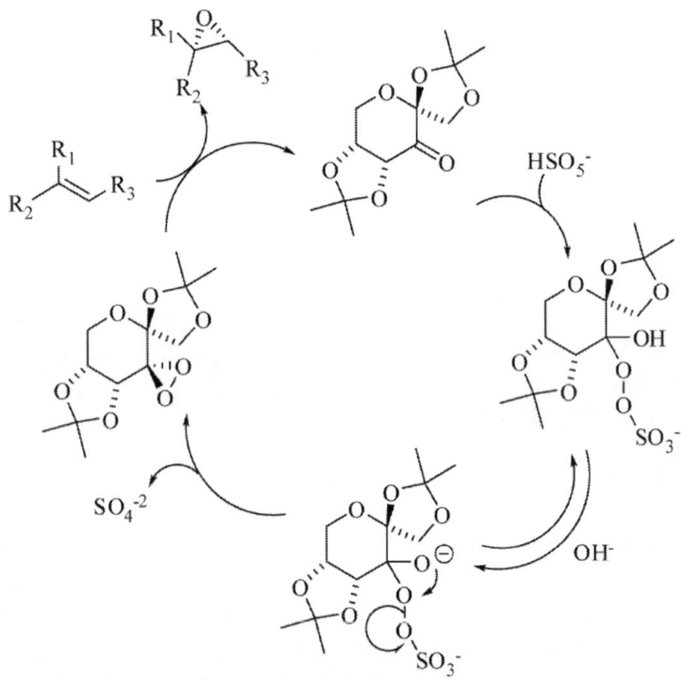


**FIGURE 9.10** Mechanism of Shi epoxidation.

### 9.3.4.1 *INDUSTRIAL APPLICATIONS*

Ager and co-workers [89] at DSM used organo catalytic asymmetric Shi's epoxidation as key step in the large-scale synthesis of **187**, which is an anti-HIV agent and blood-pressure modulator under development. Epoxidation of olefin **184** was performed with 30 mol% of catalyst Epoxone® **183**. Oxone® and KOH solutions were introduced simultaneously into the reactor at a rate to maintain the pH at 10–11 to give the epoxide **185**, which was subjected to intramolecular ring opening reaction to give the ~ 50-kg of desired lactone **186** (Scheme 9.49) in 63% yield (88% ee).

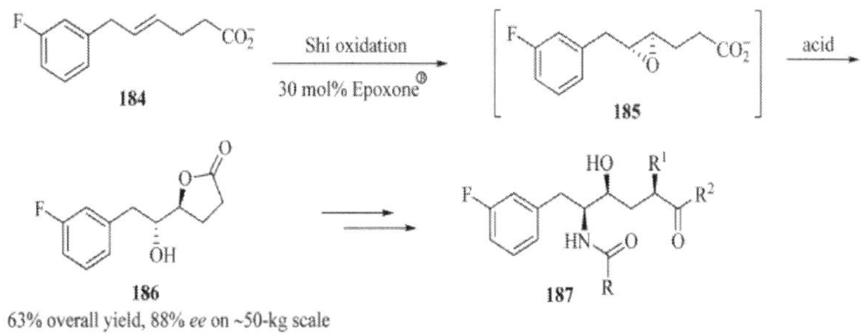


**SCHEME 9.49** Synthesis of antiHIV agent.

## 9.4 CONCLUSIONS AND PERSPECTIVES

This book chapter dwelt upon the pivotal role of catalysts in organic synthesis, specially focusing on its large-scale applications in the formation of C-C and C-O bonds. The predominant application of catalysis in the fine chemical and pharmaceutical industrial sectors was an account of their focus on achieving increased efficiency in terms of activity and selectivity enroute to environment friendly products and processes. Though, pharmaceutical industry rarely reports the catalyst's activity in terms of TON and TOF, these terminology was employed to have a better understanding on the catalyst performance under industrial environment.

Through the focus of our discussions centered around on C-C and C-O bond forming reactions with transition metal catalysis, the presented analysis highlights the process intensification efforts made to achieve improved yields and selectivities through various options, the notable among them being integration of more than one reaction, single pot synthesis, environment friendly product development, metal exchange in catalysts, TON/TOF improvement in catalysts, suppression of undesirable side reactions, less toxic process routing, scale up, less solvent usage, greener precursors and C-C bond rearrangement. They represent the challenges of synthetic organic chemistry in other fields as well. Each example exemplifies the academic excellence behind the concerned process intensification achievement. The early involvement of knowledge leaders has not only led to acquisition of new insights into the adopted pathways but also resulted in the evolution of very unique academic solutions to meet the industrial level challenges.

## APPENDIX

### TABLE 9.1

| Sr. No. | Industrial Applications of Heck Reaction | References |
|---|---|---|
| 1. | 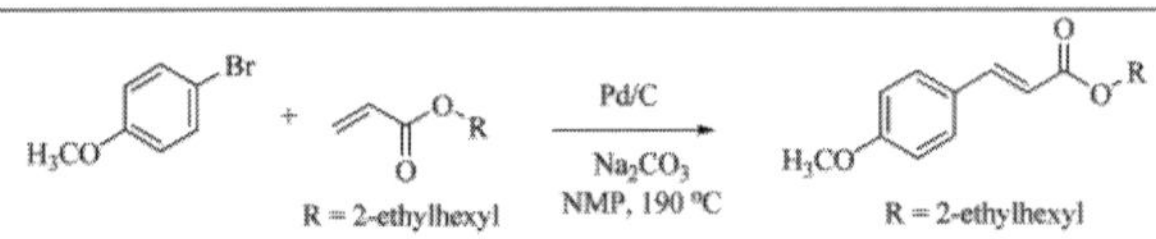 Sunscreen agent | Eisenstadt *et al.* *Chem. Ind.* **1998**, *75*, 415. |
| 2. | 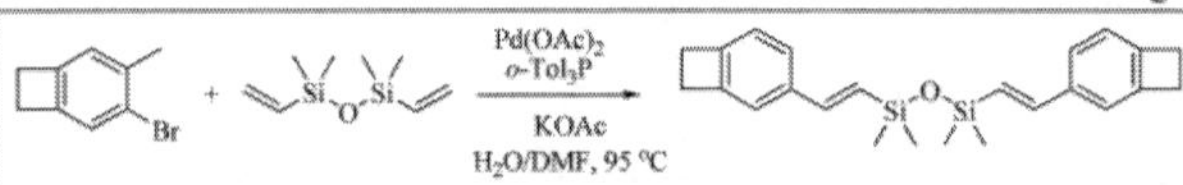 Monomers for coatings of electronic components: Cyclotene | DeVries *et al.* *Chem. Ind.* **1998**, *75*, 467. |
| 3. | 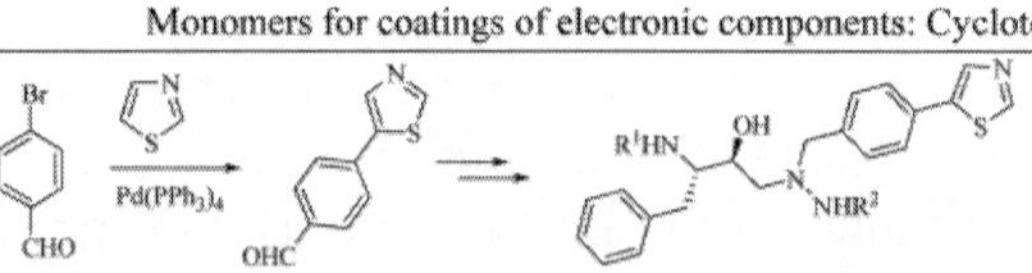 HIV-1 protease inhibitor | Bold *et al.* *J. Med. Chem.* **1998**, *41*, 3387-3401. |
| 4. | 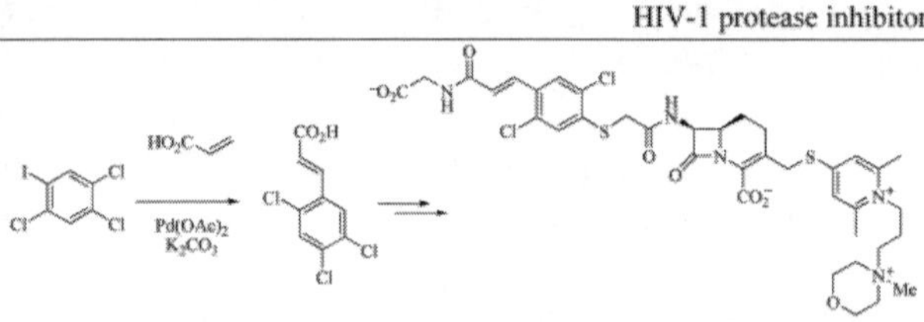 Antimethicillin resistant *Staphylococcus aureus* cephalosporin | Singh *et al.* *Org. Process Res. Dev.* **2000**, *4*, 488-497. |
| 5. | 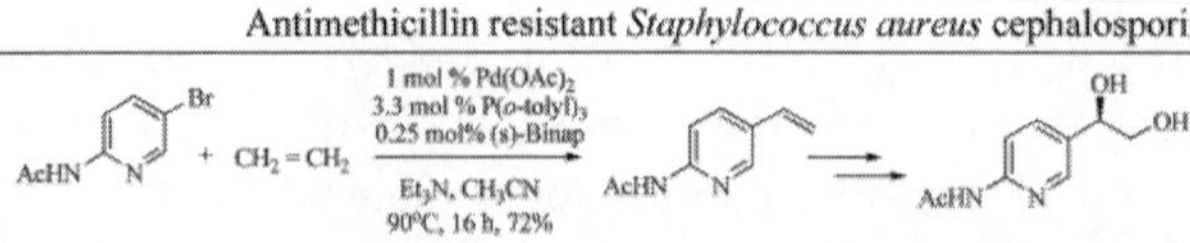 Vinylpyridine: A key drug intermediate | Raggon *et al.* *Org. Process Res. Dev.* **2002**, *6*, 67-69. |
| 6. | 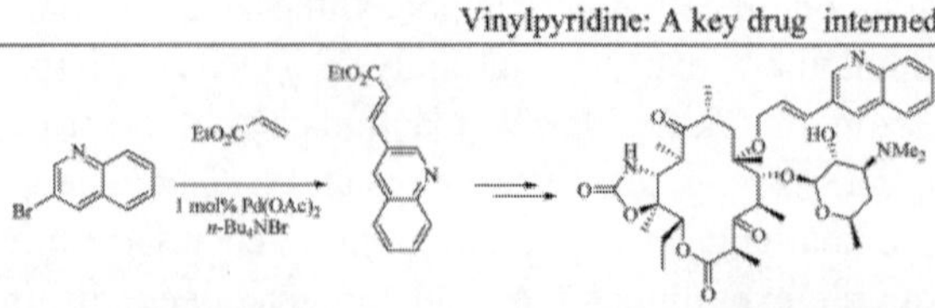 Antibiotic Cethromycin | Plata *et al.* *Tetrahedron* **2004**, *60*, 10171-10180 |

| | | |
|---|---|---|
| 7. | 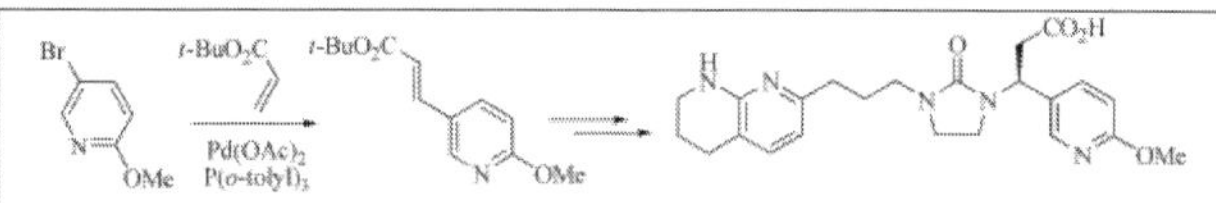 An $\alpha_v\beta_3$ integrin antagonist | Yasuda *et al.* *J. Org. Chem.* **2004**, *69*, 1959-1966. |
| 8. | 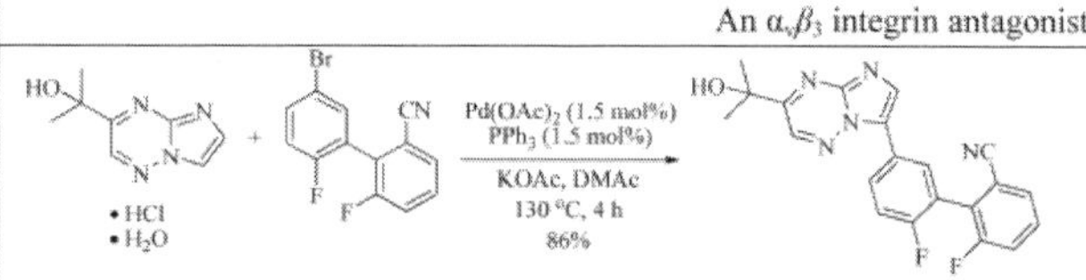 $GABA_A$ $\alpha_{2/3}$-selective allosteric modulator | Gauthier *et al.* *J. Org. Chem.* **2005**, *70*, 5938-5945. |
| 9. | 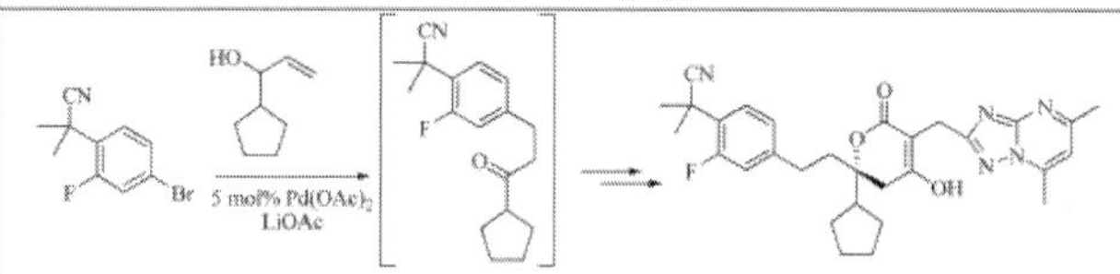 Hepatitis C viral polymerase inhibitor | Camp *et al.* *Org. Process Res. Dev.* **2006**, *10*, 814-821. |
| 10. | 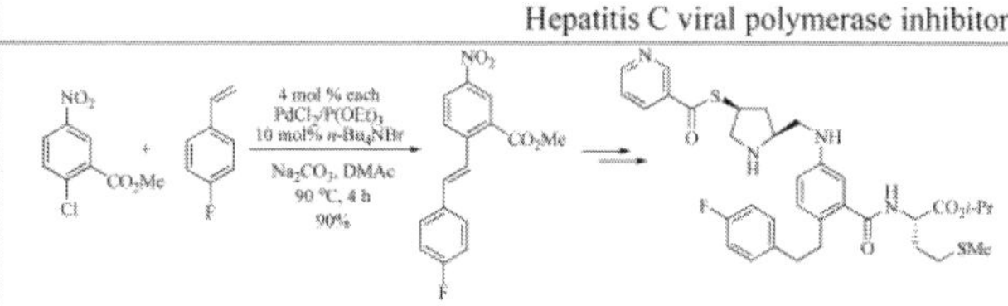 Antiproliferative agent for the treatment of cancer | Robinson *et al.* *Org. Process Res. Dev.* **2008**, *12*, 202-212. |
| 11. | 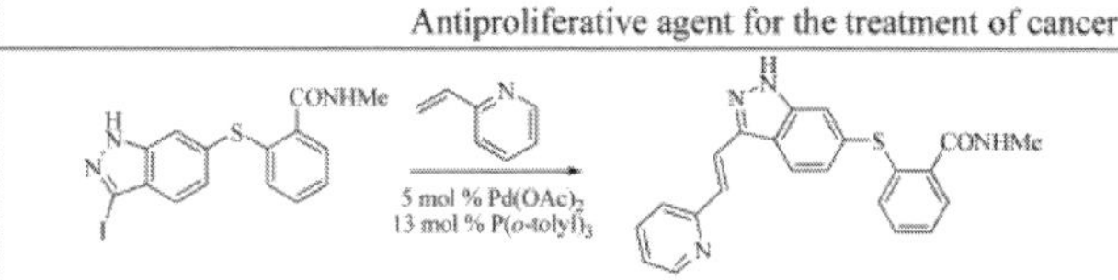 Axitinib, an inhibitor of vascular endothelial growth factor | Flahive *et al.* *Org. Process Res. Dev.* **2008**, *12*, 637-645. |

**TABLE 9.2**

| Sr. No. | Industrial Applications of Suzuki Reaction | References |
|---|---|---|
| 1. | 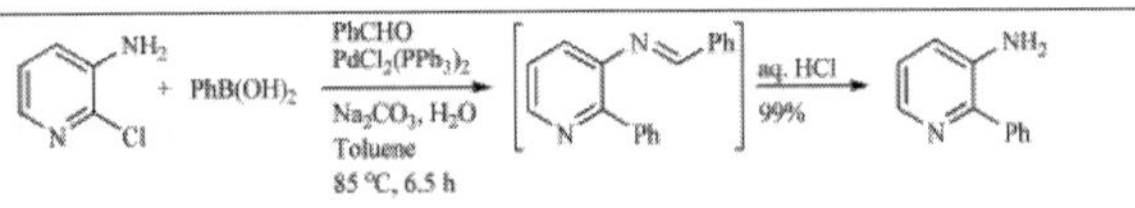 Nonpeptidic NK1 receptor antagonists | Caron *et al.* *Org. Process Res. Dev.* **2001**, 5, 254-256. |
| 2. | 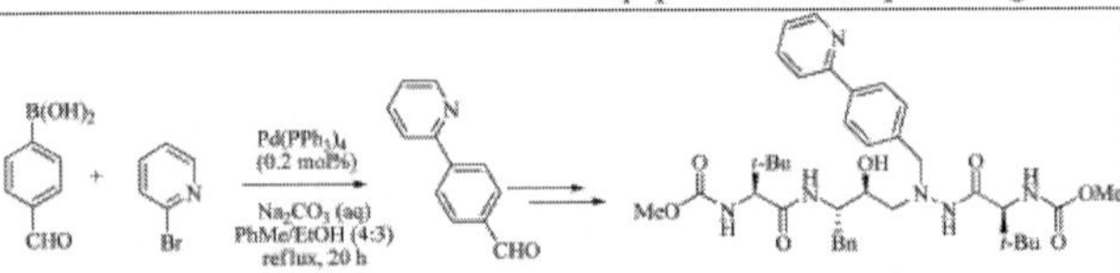 A potent HIV protease inhibitor | Xu *et al.* *Org. Process Res. Dev.* **2002**, 6, 323-328. |
| 3. | 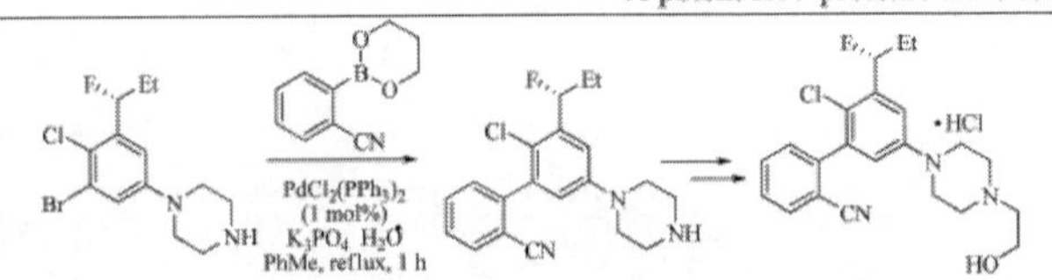 A potent antagonist of D3/D2/5-HT2 receptors | Urawa *et al.* *Org. Process Res. Dev.* **2003**, 7, 191-195. |
| 4. | 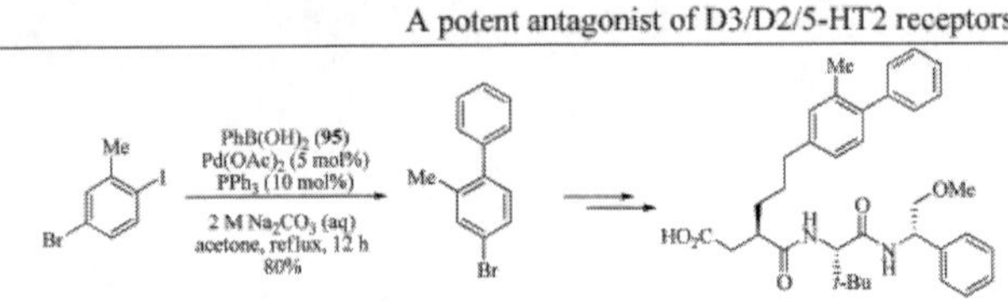 MMP-3 inhibitor | Thomson *et al.* *Org. Process Res. Dev.* **2003**, 7, 362-368. |
| 5. | 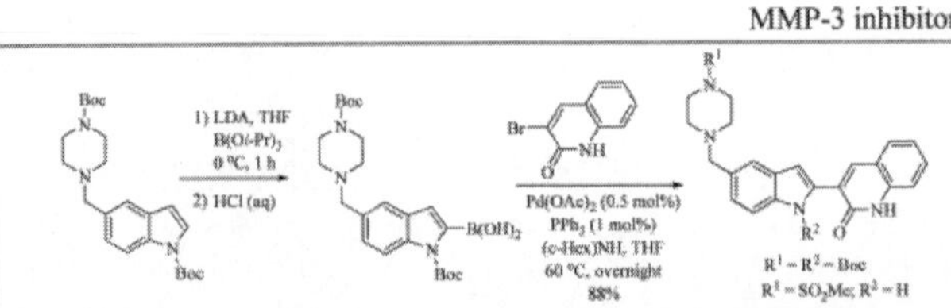 An antiangiogenic tyrosine kinase inhibitor | Payack *et al.* *J. Org. Chem.* **2005**, *70*, 175-178. |
| 6. | 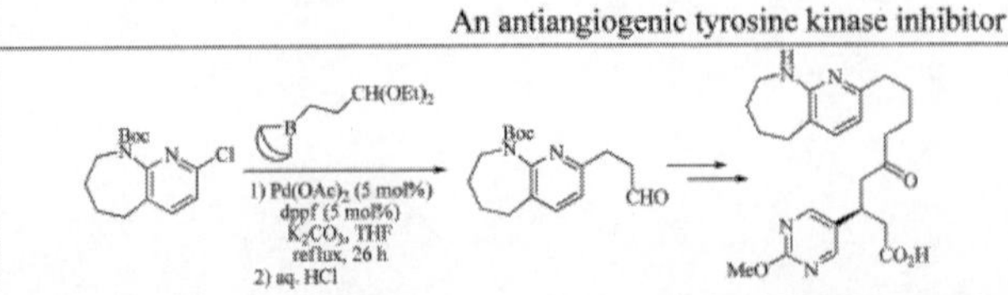 A nonpeptidic $\alpha_v\beta_3$ antagonist | Keen *et al.* *J. Org. Chem.* **2005**, *70*, 1771-1779. |

| | | |
|---|---|---|
| 7. | 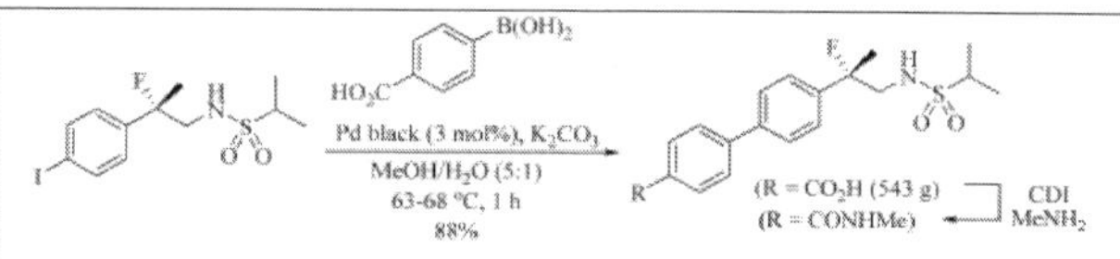 LY503430, A Parkinson's disease drug | Magnus *et al.* *Org. Process Res. Dev.* **2005**, *9*, 621-628. |
| 8. | 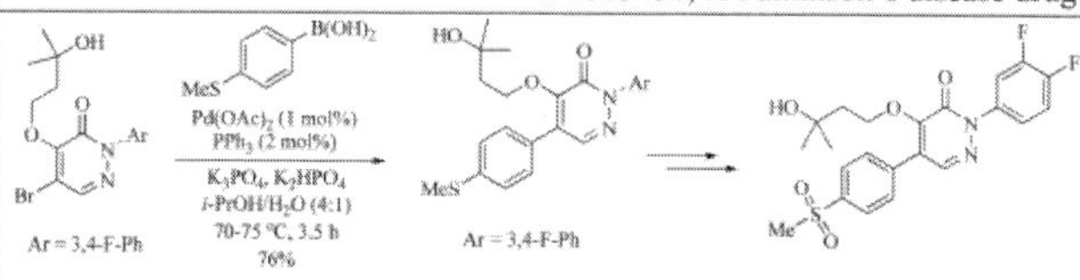 ABT-963, COX-2 inhibitor | Kerdesky *et al.* *Org. Process Res. Dev.* **2006**, *10*, 512-517. |
| 9. | 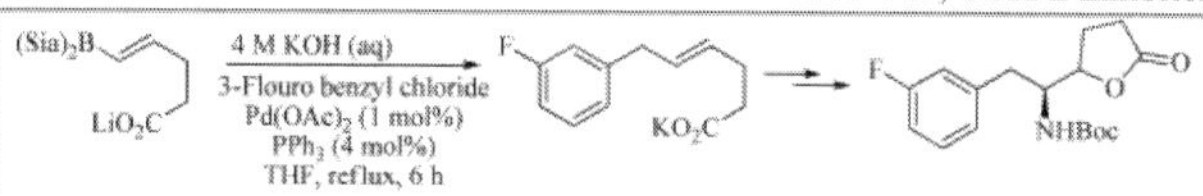 An intermediate for HIV drug | Ager *et al.* *Org. Process Res. Dev.* **2007**, *11*, 44-51. |
| 10. | 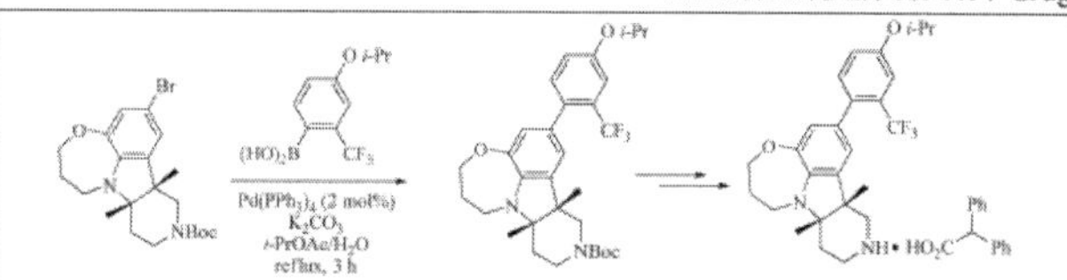 A $5HT_{2C}$ receptor for the treatment of obesity | Hobson *et al.* *Org. Process Res. Dev.* **2007**, *11*, 985-995. |
| 11. | 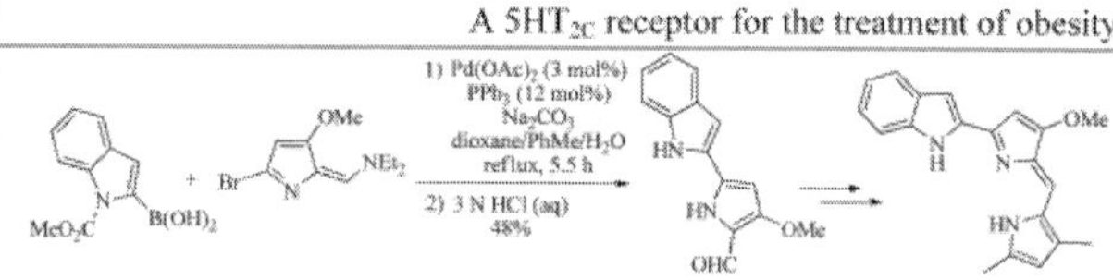 Obatoclax, an anticancer activity drug | Daïri *et al.* *Org. Process Res. Dev.* **2007**, *11*, 1051-1054. |
| 12. | 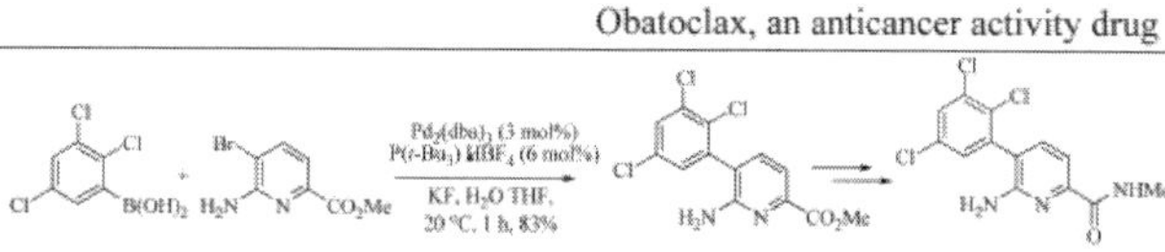 $Na_V1.8$ sodium channel modulator | Fray *et al.* *Org. Process Res. Dev.* **2010**, *14*, 263-271 |
| 13. | 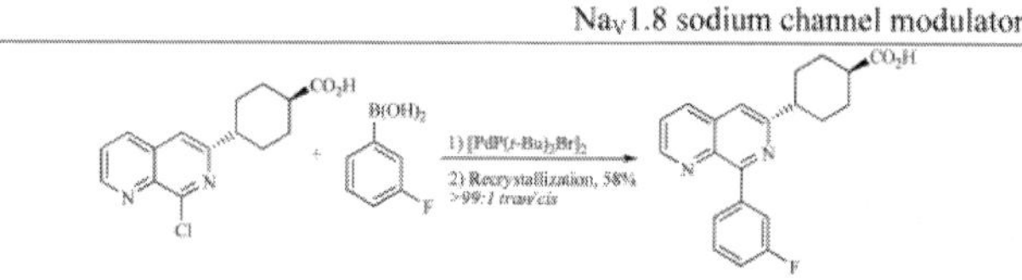 The phosphodiesterase-4 inhibitor | Jiang *et al.* *Org. Process Res. Dev.* **2010**, *14*, 883-889. |

## ACKNOWLEDGMENT

We thank CSIR-IICT and DST-INSPIRE for providing the infrastructure and funding.

## KEY WORDS

- **Heck coupling**
- **Jacobsen epoxidation**
- **Negishi coupling**
- **Olefin metathesis**
- **Sharpless asymmetric dihydroxylation**
- **Sharpless asymmetric epoxidation**
- **Shi epoxidation**
- **Suzuki coupling**

## REFERENCES

1. Kirk-Othmer, *Homogeneous Catalysis: Encyclopedia of Chemical Technology*, J. Wiley, New York, Vol. **5**; p 324 (1992).
2. Hagen, J. *Chemische Reaktionstechnik: Eine Einführung mit* Übungen, VCH, Weinheim (1992).
3. Busacca, C. A., Fandrick, D. R., Song, J. J. and Senanayakea, C. H. *Adv Synth Catal* **353**, 1825–1864 (2011).
4. Nobel Prize Home Page. http://nobelprize.org/nobel_prizes/chemistry/laureates (accessed Mar 05, 2013).
5. Yu, L. X. Pharmaceutical Quality by Design: Product and Process Development, Understanding, and Control. *Pharm Res* **25**, 781–791 (2008).
6. Ertl, G., Knözinger, H. and Weitkamp, J. *Handbook of Heterogeneous Catalysis*. VCH, Weinheim (1997).
7. Heck, R. F. Acylation, Methylation, and Carboxyalkylation of Olefins by Group VIII Metal Derivatives. *J Am Chem Soc* **90**, 5518–5526 (1968).
8. Heck, R. F. and Nolley, J. P. Palladium-Catalyzed Vinylic Hydrogen Substitution Reactions With Aryl, Benzyl, and Styryl Halides. *J Org Chem* **37**, 2320–2322 (1972).
9. Heck, R. F. Mechanism of Arylation and Carbomethoxylation of Olefins with Organopalladium Compounds. *J Am Chem Soc* **91**, 6707–6714 (1969).
10. Baumeister, P., Meyer, W., Oertle, K., Seifert, G., Siegrist, U. and Steiner, H. In *Heterogeneous catalysis and fine chemicals IV*; Blaser, H. U., Baiker, A. and Prins, R., Ed.; Elsevier Science: Amsterdam, p. 37 (1997).
11. Shinkai, I., King, A. O. and Larsen, R. D. A Practical Asymmetric Synthesis of LTD4 Antagonist. *Pure Appl Chem* **66**, 1551–1556 (1994).

12. Harrington, P. J. and Lodewijk, E. Twenty Years of Naproxen Technology. *Org Process*
13. Waite, D. C. and Mason, C. P. A Scalable Synthesis Of The Thromboxane Receptor Antagonist3-{3-[2-(4-Chlorobenzenesulfonamido)Ethyl]-5-(4Fluorobenzyl)Phenyl} Propionic Acid *via* a Regioselective Heck Cross-Coupling Strategy. *Org Process Res Dev* **2**, 116–120 (1998).
14. Wallace, M. D., McGuire, M. A., Yu, M. S., Goldfinger, L., Liu, L., Dai, W. and Shilcrat, S. Multi-Kiloscale Enantioselective Synthesis of A Vitronectin Receptor Antagonist. *Org Process Res Dev* **8**, 738–743 (2004).
15. Voight, E. A., Yin, H., Downing, S. V., Calad, S. A., Matsuhashi, H., Giordano, I., Hennessy, A. J., Goodman, R. M. and Wood, J. L. Target-Directed Synthesis of Antibacterial Drug Candidate GSK966587. *Org Lett* **12**, 3422–3425 (2010).
16. Ripin, D. H. B.; Bourassa, D. E.; Brandt, T.; Castaldi, M. J.; Frost, H. N.; Hawkins, J.; Johnson, P. J., Massett, S. S., Neumann, K., Phillips, J., Raggon, J. W., Rose, P. R., Rutherford, J. L., Sitter, B., Stewart, A. M., III., Vetelino, M. G. and Wei, L. Evaluation Of Kilogram-Scale Sonagashira, Suzuki, And Heck Coupling Routes to Oncology Candidate CP-724,714. *Org Process Res Dev* **9**, 440–450 (2005).
17. Schils, D., Stappers, F., Solberghe, G., van Heck, R., Coppens, M., Van den Heuvel, D., Van der Donck, P., Callewaert, T., Meeussen, F., De Bie, E., Eersels, K. and Schouteden, E. Ligandless Heck Coupling between a Halogenated Aniline and Acrylonitrile Catalyzed by Pd/C: Development and Optimization of an Industrial-Scale Heck Process for the Production of a Pharmaceutical Intermediate. *Org Process Res Dev* **12**, 530–536 (2008).
18. Zegar, S., Tokar, C., Enache, L. A., Rajagopol, V., Zeller, W., ▪ Connell, M., Singh, J., Muellner, F. W. and Zembower, D. E. Development of a Scalable Process for DG-041, a Potent EP3 Receptor Antagonist, *via* Tandem Heck Reactions. *Org Process Res Dev* **11**, 747–753 (2007).
19. Banks, A., Breen, G. F., Caine, D., Carey, J. S., Drake, C., Forth, M. A., Gladwin, A., Guelfi, S., Hayes, J. F., Maragni, P., Morgan, D. O., Oxley, P., Perboni, A., Popkin, M. E., Rawlinson, F. and Roux, G. Process Development and Scale Up of a Glycine Antagonist. *Org Process Res Dev* **13**, 1130–1140 (2009).
20. Suzuki, A. Organoboranes in Organic Syntheses Including Suzuki Coupling Reaction. *Heterocycles* **80**, 15–43 (2010).
21. Kotha, S., Lahiri, K. and Kashinath, D. Recent applications of the Suzuki–Miyaura cross-coupling reaction in organic synthesis. *Tetrahedron* **58**, 9633–9695 (2002).
22. Miyaura, N., Yamada, K. and Suzuki, A. A new stereospecific cross-coupling by the palladium-catalyzed reaction of 1-alkenylboranes with 1-alkenyl or 1-alkynyl halides. *Tetrahedron Lett* **20**, 3437–3440 (1979).
23. Eicken, K., Rack, M., Wetterich, F., Ammermann, E., Lorenz, G. and Strathmann, S. New Bi:Phenyl-Amide Derivatives are Active Against Wide Range of Phytopathogenic Fungi. (BASF AG, Ludwigshafen), German Patent DE 19735224, February 18 (1999).
24. Chung, J. Y. L., Cvetovich, R. J., McLaughlin, M., Amato, J., Tsay, F.-R., Jensen, M., Weissman, S. and Zewge, D. Synthesis of a Naphthyridone p38 MAP Kinase Inhibitor. *J Org Chem* **71**, 8602–8609 (2006).

25. Allsop, G. L., Cole, A. J., Giles, M. E., Merifield, E., Noble, A. J., Pritchett, M. A., Purdie, L. A. and Singleton, J. T. A Convergent Synthesis of AR-C123196 Utilizing Reaction between a Cyclic Carbonate and a Phenol for Aryl Alkyl Ether Formation. *Org Process Res Dev* **13**, 751–759 (2009).
26. Ennis, D. S., McManus, J., Wood-Kaczmar, W., Richardson, J., Smith, G. E. and Carstairs, A. Multikilogram-Scale Synthesis of a Biphenyl Carboxylic Acid Derivative Using a Pd/C-Mediated Suzuki Coupling Approach *Org Process Res Dev* **3**, 248–252 (1999).
27. Conlon, D. A., Drahus-Paone, A., Ho, G.-J., Pipik, B., Helmy, R., McNamara, J. M., Shi, Y.-J., Williams, J. M., Macdonald, D., Deschěnes, D., Gallant, M., Mastracchio, A., Roy, B. and Scheigetz, J. Process Development and Large-Scale Synthesis of a PDE4 Inhibitor. *Org Process Res Dev* **10**, 36–45 (2006).
28. Chen, C.-*y*., Dagneau, P., Grabowski, E. J. J., Oballa, R., O'Shea, P., Prasit, P., Robichaud, J., Tillyer, R. and Wang, X. Practical Asymmetric Synthesis of a Potent Cathepsin K Inhibitor. Efficient Palladium Removal Following Suzuki Coupling. *J Org Chem* **68**, 2633–2638 (2003).
29. Miller, W. D., Fray, A. H., Quatroche, J. T. and Sturgill, C. D. Recent Advances in Developing Chemoenzymatic Processes for Active Pharmaceutical Ingredients. *Org Process Res Dev* **11**, 359–267 (2007).
30. Richey, R. N. and Yu, H. Development of an Efficient Palladium-Catalyzed Intramolecular Carbometalation Reaction for the Synthesis of a Dibenzoxapine Containing Tetra-substituted Exocyclic Alkene. *Org Process Res Dev* **13**, 315–320 (2009).
31. Elitzin, V. I., Harvey, K. A., Kim, H., Salmons, M., Sharp, M. J., Tabet, E. A. and Toczko, M. A. Development of a New Synthesis for the Large-Scale Preparation of Triple Reuptake Inhibitor (−)-GSK1360707. *Org Process Res Dev* **14**, 912–917 (2010).
32. Jacks, T. E., Belmont, D. T., Briggs, C. A., Horne, N. M., Kanter, G. D., Karrick, G. L., Krikke, J. J., McCabe, R. J., Mustakis, J. G., Nanninga, T. N., Risedorph, G. S., Seamans, R. E., Skeean, R., Winkle, D. D. and Zennie, T. M. Development of a Scalable Process for CI-1034, an Endothelin Antagonist. *Org Process Res Dev* **8**, 201–212 (2004).
33. Negishi, E., King, A. O. and Okukado, N. Selective carbon-carbon bond formation *via* transition metal catalysis. A highly selective synthesis of unsymmetrical biaryls and diarylmethanes by the nickel-or palladium-catalyzed reaction of aryl-and benzylzinc derivatives with aryl halides. *J Org Chem* **42**, 1821–1823 (1977).
34. Uchiyama, M., Furuyama, T., Kobayashi, M., Matsumoto, Y. and Tanaka, K. Toward a Protecting-Group-Free Halogen-Metal Exchange Reaction: Practical, Chemoselective Metalation of Functionalized Aromatic Halides Using Dianion-type Zincate, $^tBu_4Zn$-$Li_2$. *J Am Chem Soc* **128**, 8404–8405 (2006).
35. Furuyama, T., Yonehara, M., Arimoto, S., Kobayashi, M., Matsumoto, Y. and Uchiyama, M. Development of Highly Chemoselective Bulky Zincate Complex, $^tBu_4ZnLi_2$: Design, Structure, and Practical Applications in Small-/Macromolecular Synthesis. *Chem-Eur J* **14**, 10348–356 (2008).
36. Cai, S., Dimitroff, M., McKennon, T., Reider, M., Robarge, L., Ryckman, D., Shang, X. and Therrien, J. Process Development on an Efficient New Convergent Formal Synthesis of MIV-150. *Org Process Res Dev* **8**, 353–356 (2004).

37. Sasikala, C. H. V. A., Padi, P. R., Sunkara, V., Ramayya, P., Dubey, P. K., Uppala, V. B. R. and Praveen, C. An Improved and Scalable Process for the Synthesis of Ezetimibe: An Antihypercholesterolemia Drug. *Org Process Res Dev* **13**, 907–910 (2009).
38. Liu, Z. and Xiang, J. A High Yield and Pilot-Scale Process for the Preparation of Adapalene. *Org Process Res Dev* **10**, 285–288 (2006).
39. Denni-Dischert, D., Marterer, W., Bänziger, M., Yusuff, N., Batt, D., Ramsey, T., Geng, P., Michael, W., Wang, R.-M. B., Taplin, F., Jr., Versace, R., Cesarz, D. and Perez, L. B. The Synthesis of a Novel Inhibitor of B-Raf Kinase. *Org Process Res Dev* **10**, 70–77 (2006).
40. Ragan, J. A., Raggon, J. W., Hill, P.D., Jones, B. P., McDermott, R. E., Munchhof, M. J., Marx, M. A., Casavant, J. M., Cooper, B. A., Doty, J. L. and Lu, Y. Cross-Coupling Methods for the Large-Scale Preparation of an Imidazole−Thienopyridine: Synthesis of [2-(3-Methyl-3H-imidazol-4-yl)-thieno[3,2-b]pyridin-7-yl]—(2-methyl-1H-indol-5-yl)— amine. *Org Process Res Dev* **7**, 676–683 (2003).
41. Chauvin, Y. Olefin Metathesis: The Early Days (Nobel Lecture). *Angew Chem, Int Ed* **45**, 3740–3747 (2006).
42. Schrock, R. R. Multiple Metal-Carbon Bonds for Catalytic Metathesis Reactions (Nobel Lecture). *Angew Chem Int Ed* ***45***, 3748–3759 (2006).
43. Grubbs, R. H. Olefin-Metathesis Catalysts for the Preparation of Molecules and Materials (Nobel Lecture). *Angew Chem Int Ed* **45**, 3760–3765 (2006).
44. Hérisson, J.-L. and Chauvin, Y. Catalyze De transformation des oléfines par les complexes du tungstène. II. Télomérization des oléfines cycliques en présence d'oléfines acycliques. *Makromol Chem* **141**, 161–176 (1971).
45. Schrock, R. R., DePue, R. T., Feldman, J., Schaverien, C. J., Dewan, J. C. and Liu, A. H. Preparation and Reactivity of Several Alkylidene Complexes of the Type W(CHR')(N-2,6-C6H3-iso-Pr2)(OR)2 and Related Tungstacyclobutane Complexes. Controlling Metathesis Activity through the Choice of Alkoxide Ligand. *J Am Chem Soc* **110**, 1423-1435 (1988).
46. Schwab, P., Grubbs, R. H. and Ziller, J. W. Synthesis and Applications of $RuCl_2(=CHR')(PR_3)_2$: The Influence of the Alkylidene Moiety on Metathesis Activity. *J Am Chem Soc* **118**, 100–110 (1996).
47. Casey, C. P. and Burkhardt, T. J. Reactions of (diphenylcarbene)pentacarbonyltungsten(0) with Alkenes. Role of Metal-Carbene Complexes in Cyclopropanation and Olefin Metathesis Reactions. *J Am Chem Soc* **96**, 7808–7809 (1974).
48. Katz, T. J. and McGinnis, J. Mechanism of the Olefin Metathesis Reaction. *J Am Chem Soc* **97**, 1592–1594 (1975).
49. Grubbs, R. H., Burk, P. L. and Carr, D. D. Mechanism of the Olefin Metathesis Reaction. *J Am Chem Soc* **97**, 3265–3267 (1975).
50. McGinnis, J., Katz, T. J. and Hurwitz, S. Selectivity in the Olefin Metathesis of Unsymmetrically Substituted Ethylenes. *J Am Chem Soc* **98**, 605–606 (1976).
51. Tebbe, F. N., Parshall, G. W. and Overnall, D. W. Titanium-Catalyzed Olefin Metathesis. *J Am Chem Soc* **101**, 5074–5075 (1979).
52. Schrock, R. R., Murdzek, J. S., Bazan, G. C., Robbins, J., DiMare, M. and O'Regan, M. B. Synthesis of Molybdenum Imido Alkylidene Complexes and Some Reactions Involving Acyclic Olefins. *J Am Chem Soc* **112**, 3875–3886 (1990).

53. Vougioukalakis, G. C. and Grubbs, R. H. Ruthenium-Based Heterocyclic Carbene-Coordinated Olefin Metathesis Catalysts. *Chem Rev* **110**, 1746–1787 (2010).
54. Banks, R. L. and Bailey, G. C. Olefin Disproportionation. A New Catalytic Process. *Ind Eng Chem Prod Res Dev* **3**, 170–173 (1964).
55. Reuben, B. and Wlttcoff, H. The SHOP Process: An Example of Industrial Creativity. *J Chem Educ* **65**, 605–607 (1988).
56. Mol, J. C. Industrial applications of olefin metathesis. *J Mol Catal A Chem* **213**, 39–45 (2004).
57. Marbach, A. and Hupp, R. The evolution of polynorbornene. *Rubber World*, June 1989, p. 30 (1989).
58. Woodson, C. S. and Grubbs, R. H. Polymeric Composites Including Dicyclopentadiene and Related Monomers. U.S. Patent 6,310,121 B1, October 30 (2001).
59. Pederson, R. L. and Grubbs, R. H. Metathesis Syntheses of Pheromones or Their Components. U.S. Patent 6,696,597, February 24 (2004).
60. Yee, N. K., Farina, V., Houpis, I. N., Haddad, N., Frutos, R. P., Gallou, F., Wang, X.-j., Wei, X., Simpson, R. D., Feng, X., Fuchs, V., Xu, Y., Tan, J., Zhang, L., Xu, J., Smith-Keenan, L. L., Vitous, J., Ridges, M. D., Spinelli, E. M., Johnson, M., Donsbach, K., Nicola, T., Brenner, M., Winter, E., Kreye, P. and Samstag, W. Efficient Large-Scale Synthesis of BILN 2061, a Potent HCV Protease Inhibitor, by a Convergent Approach Based on Ring-Closing Metathesis. *J Org Chem* **71**, 7133–7145 (2006).
61. Kingsbury, J. S., Harrity, J. P. A., Bonitatebus, P. J., Jr. and Hoveyda, A. H. A Recyclable Ru-Based Metathesis Catalyst. *J Am Chem Soc* **121**, 791–799 (1999).
62. Schwab, P., France, M. B., Ziller, J. W. and Grubbs, R. H. A Series of Well-Defined Metathesis Catalysts–Synthesis of $[RuCl_2(=CHR')(PR_3)_2]$ and Its Reactions. *Angew Chem Int Ed* **34**, 2039–2041 (1995).
63. Scholl, M., Ding, S., Lee, C. W. and Grubbs, R. H. Synthesis and Activity of a New Generation of Ruthenium-Based Olefin Metathesis Catalysts Coordinated with 1,3-Dimesityl-4,5-dihydroimidazol-2-ylidene Ligands. *Org Lett* **1**, 953–956 (1999).
64. Garber, S. B., Kingsbury, J. S., Gray, B. L. and Hoveyda, A. H. Efficient and Recyclable Monomeric and Dendritic Ru-Based Metathesis Catalysts. *J Am Chem Soc* **122**, 8168-8179 (2000).
65. Nicola, T., Brenner, M., Donsbach, K. and Kreye, P. First Scale-Up to Production Scale of a Ring Closing Metathesis Reaction Forming a 15-Membered Macrocycle as a Precursor of an Active Pharmaceutical Ingredient. *Org Process Res Dev* **9**, 513–515 (2005).
66. Farina, V., Shu, C., Zeng, X., Wei, X., Han, Z., Yee, N. K. and Senanayake, C. H. Second-Generation Process for the HCV Protease Inhibitor BILN 2061: A Greener Approach to Ru-Catalyzed Ring-Closing Metathesis. *Org Process Res Dev* **13**, 250–254 (2009).
67. Michrowska, A., Bujok, R., Harutyunyan, S., Sashuk, V., Dolgonos, G. and Grela, K. Nitro-Substituted Hoveyda–Grubbs Ruthenium Carbenes: Enhancement of Catalyst Activity through Electronic Activation. *J Am Chem Soc* ***126***, 9318–9325 (2004).
68. Shu, C., Zeng, X., Hao, M.-H., Wei, X., Yee, N. K., Busacca, C. A., Han, Z., Farina, V. and Senanayake, C. H. RCM Macrocyclization Made Practical: An Efficient Synthesis of HCV Protease Inhibitor BILN 2061. *Org Lett* **10**, 1303–1306 (2008).
69. Sheldon, R. A. The E Factor: Fifteen Years On. *Green Chem* **9**, 1273–1283 (2007).

70. Wang, H., Matsuhashi, H., Doan, B. D., Goodman, S. N., Ouyang, X., Clark, W. M. and Jr. Large-Scale Synthesis of SB-462795, A Cathepsin K Inhibitor: The RCM-Based Approaches. *Tetrahedron* **65**, 6291–6303 (2009).
71. Kong, J., Chen, C.-*y*., Balsells-Padros, J., Cao, Y., Dunn, R. F., Dolman, S. J., Janey, J., Li, H. and Zacuto, M. J. Synthesis of the HCV Protease Inhibitor Vaniprevir (MK-7009) Using Ring-Closing Metathesis Strategy. *J Org Chem 77*, 3820–3828 (2012).
72. Katsuki, T. and Sharpless, K. B. The First Practical Method for Asymmetric Epoxidation. *J Am Chem Soc* **102**, 5974–5976 (1980).
73. Jacobsen, E. N., Marko, I., Mungall, W. S., Schroder, G. and Sharpless, K. B. Asymmetric Dihydroxylation via Ligand-Accelerated Catalysis. *J Am Chem Soc* **110**, 1968 (1988).
74. Zhang, W., Loebach, J. L., Wilson, S. R. and Jacobsen, E. N. Enantioselective Epoxidation of Unfunctionalized Olefins Catalyzed by Salen Manganese Complexes. *J Am Chem Soc* **112**, 2801–2803 (1990).
75. Shi, Y. *Modern Oxidation Methods*, 2nd edn. (Eds.: J.-E. B_ckvall), Wiley-VCH, Weinheim (2010).
76. Farina, V., Reeves, J. T., Senanayake, C. H. and Song, J. J. Asymmetric Synthesis of Active Pharmaceutical Ingredients. *Chem Rev **106***, 2734–2793 (2006).
77. Kimura, T., Yamamoto, N., Suzuki, Y., Kawano, K., Norimine, Y., Ito, K., Nagato, S., Iimura, Y. and Yonaga, M. Practical Synthesis of Chiral Emopamil Left Hand as a Bioactive Motif. *J Org Chem* **67**, 6228–6231 (2002).
78. Henegar, K. E. and Cebula, M. Process Development for (*S, S*)-Reboxetine Succinate via a Sharpless Asymmetric Epoxidation. *Org Process Res Dev* **11**, 354–358 (2007).
79. Prasad, J. S., Vu, T., Totleben, M. J., Crispino, G. A., Kacsur, D. J., Swaminathan, S., Thornton, J. E., Fritz, A. and Singh, A. K. Development of Jacobsen Asymmetric Epoxidation and Sharpless Asymmetric Dihydroxylation Methods for the Large-Scale Preparation of a Chiral Dihydrobenzofuran Epoxide. *Org Process Res Dev* **7**, 821–827 (2003).
80. Pye, P. J., Rossen, K., Weissman, S. A., Maliakal, A., Reamer, R. A., Ball, R., Tsou, N. N., Volante, R. P. and Reider, P. J. Crystallization-Induced Diastereoselection: Asymmetric Synthesis of Substance P Inhibitors. *Chem Eur J **8***, 1372–1376 (2002).
81. Irie, R., Noda, K., Ito, Y., Matsumoto, N. and Katsuki, T. Catalytic Asymmetric Epoxidation of Unfunctionalized Olefins. *Tetrahedron Lett **31***, 7345–7348 (1990).
82. Minutolo, F., Pini, D. and Salvadori, P. Polymer-Bound Chiral (Salen)Mn(III) Complex as Heterogeneous Catalyst in Rapid and Clean Enantioselective Epoxidation of Unfunctionalised Olefins. *Tetrahedron Lett **37***, 3375–3378 (1996).
83. Larrow, J. F., Jacobsen, E. N., Gao, Y., Hong Y., Nie X. and Zepp C. M. A Practical Method for the Large-Scale Preparation of [N, N'-Bis(3,5-di-tertbutylsalicylidene)-1,2-cyclohexanediaminato(2-)]manganese(III) chloride, a Highly Enantioselective Epoxidation Catalyst. *J Org Chem* **59**, 1939–1942 (1994).
84. Senanayake, C. H., Roberts, F. E., DiMichele, L. M., Ryan, K. M., Liu, J., Fredenburgh, L. E., Foster, B. S., Douglas, A. W., Larsen, R. D., Verhoeven, T. R. and Reider, P. J. The Behavior of Indene Oxide in the Ritter Reaction: A Simple Route to *cis*-Aminoindanol. *Tetrahedron Lett* **36**, 3993–3996 (1995).
85. Senanayake, C. H., Smith, G. B., Ryan, K. M., Fredenburgh, L. E., Liu, J., Roberts, F. E., Hughes, D. L., Larsen, R. D., Verhoeven, T. R. and Reider, P. J. The Role of

4-(3-phenylpropyl)pyridine *N*-Oxide ($P_3NO$) in the Manganese-Salen-Catalyzed Asymmetric Epoxidation of Indene. *Tetrahedron Lett* **37**, 3271–3274 (1996).
86. Hughes, D. L., Smith, G. B., Liu, J., Dezeny, G. C., Senanayake, C. H., Larsen, R. D., Verhoeven, T. R. and Reider, P. J. Mechanistic Study of the Jacobsen Asymmetric Epoxidation of Indene. *J Org Chem* **62**, 2222–2229 (1997).
87. Lynch, J. E., Choi, W.-B., Churchill, H. R. O., Volante, R. P., Reamer, R. A. and Ball, R. G. Asymmetric Synthesis of CDP840 by Jacobsen Epoxidation. An Unusual Syn Selective Reduction of an Epoxide. *J Org Chem* **62**, 9223–9228 (1997).
88. Wang, Z. X., Tu, Y., Frohn, M., Zhang, J. R. and Shi, Y. An Efficient Catalytic Asymmetric Epoxidation Method. *J Am Chem Soc* **119**, 11224–11235 (1997).
89. Ager, D. J., Anderson, K., Oblinger, E., Shi, Y. and VanderRoest, J. An Epoxidation Approach to a Chiral Lactone: Application of the Shi Epoxidation. *Org Process Res Dev* **11**, 44–51 (2007).

## FOR ADDITIONAL READING:

1. *Industrial catalysis: A Practical Approach*; Hagen, J., Ed.; Wiely: New York, 2006.
2. *Catalysis of Organic Reactions: Twenty-second Conference*; Prunier, M. L.; Ed.; Chemical Industries series; CRC Press: England, United Kingdom, 2008.

## CHAPTER 10

# MICROWAVE ASSISTED SYNTHESIS: A VERSATILE TOOL FOR PROCESS INTENSIFICATION

LANKELA H. REDDY, DAMMA DEVAIAH,
and BENJARAM M. REDDY

## CONTENTS

## 10.1 INTRODUCTION

The green synthesis of chemicals and materials has attracted worldwide research attention in recent years. The number of scientific publications and patents on green synthesis has increased 20 times during the past decade. The focus was on making the processes environmentally friendlier as well as material/energy utilization efficient. Green synthesis has used the unique opportunities provided by the microwave irradiation in material synthesis. Microwaves are a nonionizing radiation, lies in between the infrared and radio waves with a wide frequency range of 0.3 to 300 GHz corresponding to the wavelengths of 0.001–0.3 m (energy = $10^{-5}$–0.01 eV). The maximum portion of the microwave spectrum is allocated for the telecommunication and radar technology. In order to avoid the interference with the telecommunications and radar technology, there is a specific type of frequencies in the microwave region namely, 27.12 MHz, 915 MHz, and 2.45 GHz that are used for industrial, scientific, and medical applications, respectively, and they are collectively called as ISM bands. Among them, the frequency of 2.54 GHz is most commonly employed in the design of microwave ovens for heating and research applications. There are many alternatives to the conventional heating, the notable among them are mechanochemical, ultrasound, microwave, photochemical, and X-ray or gamma-radiation mediated methods. The microwave-assisted synthesis has received a special attention from the scientific community and there is extensive reported literature on this subject [1–16]. The divergent fields that have been benefited by the microwave irradiation are synthetic organic chemistry, polymer chemistry, material science, nanotechnology, biochemical processes, peptide synthesis, food processing and analytical applications.

Several organic reactions including nucleophilic substitution, esterification, acetilization, hydrolysis, isomerization, oxidation, decarboxylation and several well-known name reactions have been made more efficient and faster with microwave mediation. Polar organic molecules are influenced most due to the dielectric polarization phenomenon. The microwave effects have manifested in the form of rapid volumetric heating and solvent superheating, enhanced mass transport properties, more favorable molecular orientation, significant increase in Arrhenius preexponential factor, decrease in reaction activation energy and development of hot spots in the reaction media.

An attempt has been made in this chapter to trace the evolutionary process of microwave mediation in material/chemical synthesis, its mechanistic aspects, type of microwave contactors and typical microwave mediated synthesis of inorganic and organic chemicals, nanomaterials and polymers.

## 10.2 EVOLUTION OF MICROWAVE HEATING

Nowadays, most of the kitchens are installed with the microwave ovens for food processing and heating purposes. The first commercial microwave oven was used in 1947. The heating phenomenon with the microwaves was first observed by Percy Spencer in 1940s accidentally while he was working on radar technology the candy bar in his pocket was melted. He thought that it was due to the occurrence of interaction between microwave radiation emitted from the radar and the materials. Thus it was established that microwave radiation is responsible for heating of the material. Later, microwave radiation is slowly implemented for various applications as heating aid (Scheme 10.1). The first commercial application being used was the microwave oven in 1986, which was reported independently by Gedye and Giguere [17, 18].

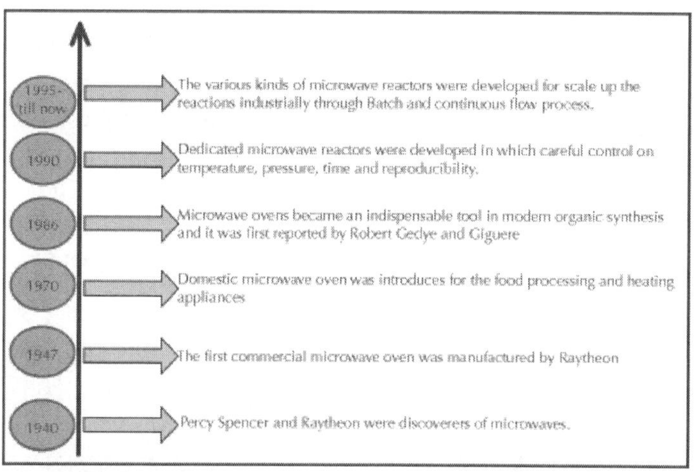


**SCHEME 10.1** Developments in microwave radiation as the source of heating.

The major advantage associated with the microwave heating is of its fast heating in comparison to the conventional heating. There is a direct interaction of microwaves with the materials and/or substrates, thereby, it can induce fast super heating of the materials, and it is independent on the thermal conductivity of the reaction vessel. The mechanistic aspects of microwave oven are outlined in the later paragraphs. In this method of heating there is no contact between the heating source and the reaction mixture. The only criterion for the molecule is to absorb the microwave radiation and it must have a dipolar nature and the vessels of the reaction mixture are made up as microwave transparent such as Teflon, glass, quartz, etc., and not as microwave reflectors or absorbers.

Generally, the conventional thermal processing methods such as hot plates, oil bath, heating mantle, etc. involve the conduction, convection, and radiation of heat from the surfaces of the material. In these methods, there is a sequential steps for the heating of the material, and involves heating of source (oil bath, heating mantle, hot plates, etc.) and then energy will be transferred from the heating source to the container of the reaction mixture, and next to the surface of the material. Finally, the bulk of the material will be heated. Unfortunately, it is a relatively slow and inefficient method. Furthermore, this process leads to the formation of thermal gradients in the material, which may lead to the overheating of the surface of the material. Particularly, this phenomenon is critically important when the reactions are handled industrially in the bulk scale as well as in the synthesis of nanomaterials where the thermal gradients have a strong impact on the nucleation and crystalline growth of the material. These two types of heating methods can be represented as shown in the Fig. 10.1 [19]. Here, the microwaves interact with the dipolar molecules, causing rotation and vibration of the molecular bonds, which result in heating of the mixture [19].

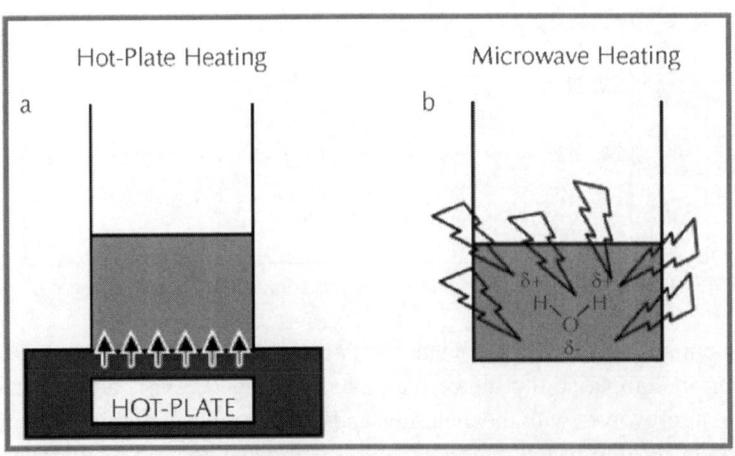


**FIGURE 10.1** Comparison of methods of heating: (a) Heat from the hot-plates heats the glass base of the beaker and is transferred by the convection. (b) The solution heats slowly as microwaves pass through the vessel walls directly into the polar liquid.

## 10.3 MICROWAVE HEATING MECHANISM

The energy of microwaves in comparison with various types of energies associated with the molecules such as chemical bonds, Brownian motion, and hydrogen bonds are shown in Table 10.1. As can be noticed that the microwave radiation has very less energy compared to others. Therefore, microwave energy could not break the bonds between the atoms and even the hydrogen bonds present in the molecules and thereby it cannot execute the chemical reactions and processes. But in reality it is established that the reactions are completed with much faster rate in a short span of time. There is specific mechanism acting in fast heating of the materials. This can be understood from following two main theories proposed for the microwave heating.

**TABLE 10.1** Comparison of energies of microwaves with other common bonds associated with chemical entities.

| | Microwave radiation | Chemical bonds | Brownian motion | Hydrogen bonds |
|---|---|---|---|---|
| Energy (eV) | $10^{-6}$ to $10^{-4}$ | 4.51 (C-H); 3.83 (C-C) | 0.026 (298K) | 0.04 to 0.44 |

### *10.3.1 DIPOLAR POLARIZATION*

The microwave radiation has two components, namely, electric and magnetic component. In general, when the dipolar molecules (polar molecules) are exposed to the microwave radiation, the dipoles are sensitive to the electric field of the microwave radiation and align them with the electric filed by rotation. As the applied field oscillates the dipoles, it tends to realign with the alternating filed. Overall, during this process, the energy is liberated in the form of heat through the molecular friction and it occurs very fast with the polar molecules. This can be best understood from the schematic representation as shown in Fig. 10.2. The changes to be undergone by simple water molecule under the influence of microwave heating are shown in Fig. 10.2 [12].

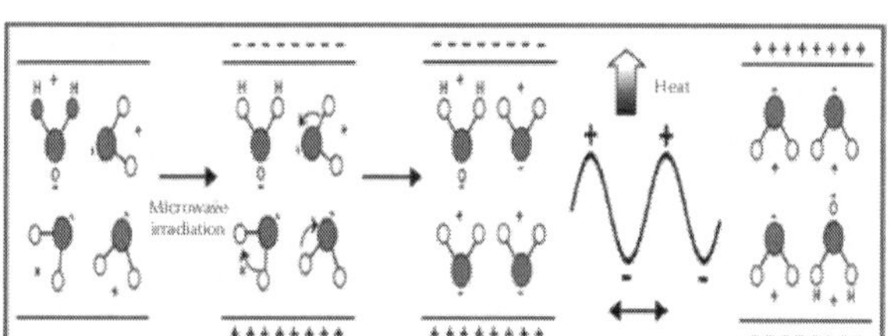


**FIGURE 10.2** Heating mechanism of $H_2O$ by microwave irradiation [12].

### 10.3.2 IONIC CONDUCTION

It is another mechanism proposed for the microwave heating, which involves the movement of charged particles by the back and forth motion present in the solution, finally the collisions with the surrounding molecules leads to the generation of heat [13]. The two mechanisms are diagrammatically represented in Fig. 10.3, and it explains how the polar molecules rotate very quickly with the alternating electric field [13].

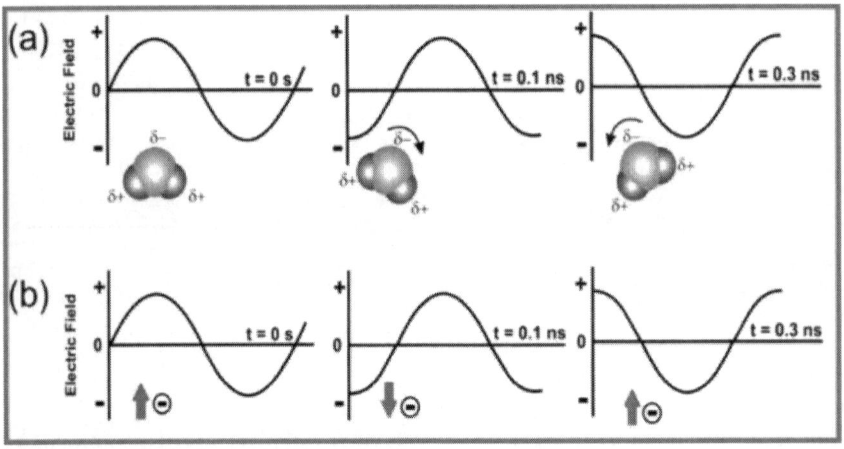


**FIGURE 10.3** Two main heating mechanisms under microwave irradiation: (a) dipolar polarization; (b) ionic conduction mechanism [13].

However, the rate of heating depends on the type of solvent employed and the individual chemical entities (substrates, catalysts, or reagents) for a particular reaction. The rate of heating by microwave radiation was studied by the term 'loss tangent' value, denoted by tan $\Delta$ and is obtained by $\varepsilon''/\varepsilon'$ where $\varepsilon'$ is the dielectric constant, and it represents the ability to be polarized by the electric component of the microwave radiation and the $\varepsilon''$ represents the dielectric loss, the efficiency of converting radiation into heat. The loss tangent values of some of the commonly employed solvents in the chemical processes are listed in Table 10.2. The higher the loss tangent value the higher is the microwave absorption power and thereby eventual rapid heating.

**TABLE 10.2** Loss tangent values of some of the commonly employed solvents [5].

| Solvent | tan δ | Solvent | tan δ |
|---|---|---|---|
| Ethylene glycol | 1.350 | DMF | 0.161 |
| Ethanol | 0.941 | 1,2-dichloroethane | 0.127 |
| DMSO | 0.825 | Water | 0.123 |
| 2-Propanol | 0.799 | Chlorobenzene | 0.101 |
| Formic acid | 0.722 | Chloroform | 0.091 |
| Methanol | 0.659 | Acetonitrile | 0.062 |
| Nitrobenzene | 0.589 | Ethyl acetate | 0.059 |
| 1-butanol | 0.571 | Acetone | 0.054 |
| 1,2-dichlorobenzene | 0.280 | Tetrahydrofuran | 0.047 |
| NMP | 0.275 | Toluene | 0.040 |
| Acetic acid | 0.174 | Hexane | 0.020 |

Though, some of the solvents such as toluene and tetrahydrofuran are poor absorbers of microwave radiation, they could be employed as solvents due to the polar nature of the substrates and the reagents or catalysts, since they can enhance the microwave absorbance power. In this context, ionic liquids, which exhibit high polar nature are widely employed as green solvents, or catalysts or structure directing agents in materials synthesis. Ionic liquids are used along with low microwave absorbing solvents to enhance the microwave absorption. However, the loss tangent values are found to vary with the frequency of used microwave irradiation as well as the temperature. Further, the values are useful in calculating the penetration depth as an important parameter for the scale-up of reactions. The loss tangent values of the materials such as Teflon and quartz are very less, 0.00028 and 0.00006, respectively. These observed values are less significant and there is no microwave absorption. Owing to this peculiar behavior, most of the reactors for microwave-assisted synthesis are made with Teflon or quartz or borosilicate materials [5, 13].

## 10.4 MICROWAVE OVEN AND LABORATORY GRADE MICROWAVE REACTORS

A typical microwave oven is constructed with three main parts namely, (i) magnetron, (ii) wave-guide, and (iii) applicator or cavity. The magnetron is the source for the generation of microwave radiation, and it converts the high electric voltage into the pulse of microwave radiation, thereby it is found to be the heart of the microwave oven. The generated microwaves travel through the waveguide, which is made with the metals having reflecting nature. The waveguide should reflect the microwave radiation. If it absorbs the radiation then the microwave radiation is disturbed. Finally, these waves travel to the cavity. Cavity is the empty place where the microwave radiation is exposed and the foodstuff and materials processing is done. A typical microwave oven is shown in Fig. 10.4.

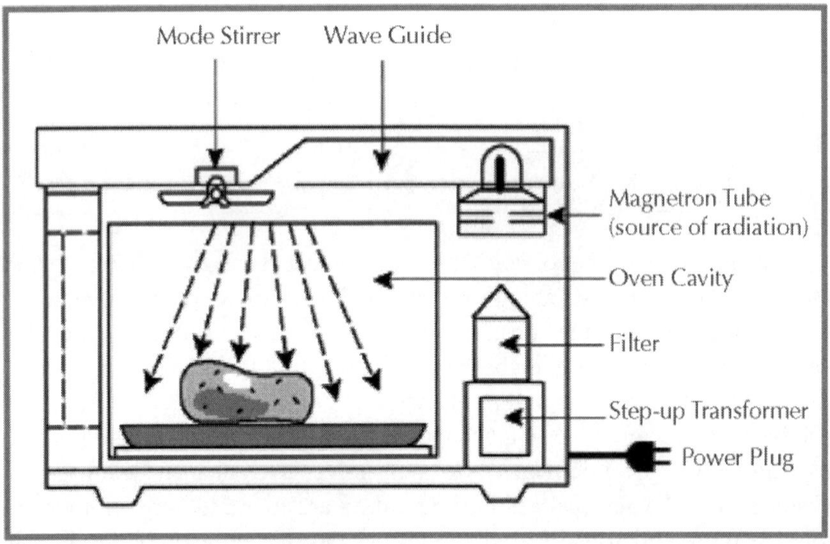


**FIGURE 10.4** A typical microwave oven.

Although, the first invented reactions were processed with the help of kitchen microwave ovens discussed above, later it was realized that there are many risks involved in their use of lack of built-in safety measures for reaction, which are explosive due to rapid heating of the organic solvents. They are also not reproducible. The precise temperature and pressure monitoring is difficult in them and there is no provision for controlling the stirring rate. All these demerits have been minimized in the specially designed microwave reactors for research purposes. The exact temperature and pressure developed during the chemical reactions could be monitored very easily. In addition, the stirring speed is controlled and forced airflow systems can be connected to the reactor facility to facilitate quick removal of moisture or gases released during the processing and for drying applications.

## 10.5 ADVANTAGES OF MICROWAVE-ASSISTED SYNTHESIS

The benefits associated with the microwave-assisted synthesis are ample compared to the conventional synthesis as listed below:

(i) Microwave method saves the precious time, which is a big benefit to perform number of reactions in different strategies in the development of drugs.
(ii) The important aspect in green technology is energy efficiency, which in turn is related to the economic issues.
(iii) Another green avenue is the use of environmentally benign solvents such as water and ionic liquids.
(iv) The synthetic methods are reproducible, reliable and often associated with the dedicated microwave reactors.
(v) Unlike conventional heating, it provides uniform heating to the material without superficial heating; thereby it is very crucial for nanomaterials synthesis.
(vi) Increases the percentage of yield.
(vii) Microwave assisted synthesis can be coupled with many other chemical approaches for materials synthesis such as microwave-assisted sol-gel, microwave assisted hydrothermal/solvothermal, and microwave assisted combustion.
(viii) Heating is achieved without contact between the source and sample.
(ix) Microwave heating is very quick in heating and cooling the materials when the microwave power is on and off, respectively.

## 10.6 MICROWAVE PROCESS EQUIPMENT AND SCALE UP ISSUES

The microwave heating has many advantages as discussed in the previous sections, and enabling it as one of the successful chemical tools for many research applications in academia and industry. There are many successful chemical reactions, which have been processed in milligram and/or gram scale. In this context, there is a growing demand in the scaling-up of chemical reactions by microwave heating owing to several advantages such as time saving, energy efficiency, and so on.

The scale up of the reactions could be carried out in two ways: 1) batch mode, and 2) continuous flow method. The batch mode uses large reactors or high capacity volume vessels for the scale up. However, the microwave radiation has severe limitations in heating the bulk samples. The most commonly used microwave frequency, that is, 2.45 GHz, which has a penetration depth of few centimeters for most of the solvents. Therefore, it will not induce the uniform heating as in the small scale. Moreover, it will heat the material conventionally from the surface and results thermal gradients in the sample. Therefore, most of the intriguing advantages of microwave heating are not replicated as the reactions are processed in small scale. In order to avoid this drawback, it needs employing a maximum number of reactors (parallel batch) with equal concentrations. In this context, it is important to know about the multi mode and mono or single mode microwave reactors. Although, most of initial chemical reactions have been processed by the domestic or modified microwave ovens, which are having the multimode microwave cavity, those reactions are not reproducible. This is due to the three-dimensional pattern stationary wave generated in the cavity and often it is non-uniform and the intensity is varied from place to place in the cavity. On the other hand, the single or mono mode microwaves are designed in such a way that the cavity of the microwave device is equal to the length of only one wave (single mode) and is focused towards the reactor situated in the middle of the cavity. These two types of microwave oven cavity are shown in Fig. 10.5. By this mode of heating, the chemical reactions are reproducible and these reactors are more helpful in optimizing the reaction conditions very efficiently and quickly. However, the single mode type of reactor has one limitation, which is only one vessel that can be irradiated. In contrast, multi mode cavity has a scope for inclusion of more number of reactors. However, the aforementioned mentioned parallel batch process uses only the multi mode reactors due to very large number of reactors and they are arranged in the cavity. In order to expose the microwave radiation on all these reactors, it is desirable to use multi mode cavity only. However, it needs the safety features associated with the pressure development due to overheating of the contents.

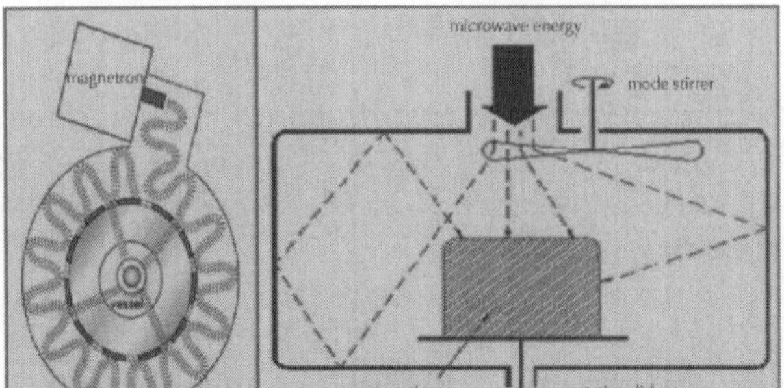


**FIGURE 10.5** Single mode heating apparatus (left side) and Multi-mode Heating Apparatus (right side) [20].

Microwave mediated processes can also be carried out in continuous mode. The typical continuous flow diagram is shown in Fig. 10.6. In this method, all the advantages of microwave heating as mentioned earlier are retained as in the small scale. Nevertheless, it has some drawbacks, such as difficulty to progress the reaction in case of solids, high viscous nature, and heterogeneous nature of the substrates.

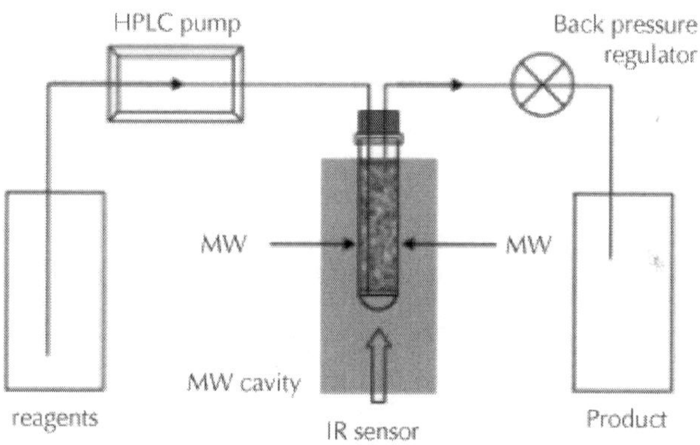


**FIGURE 10.6** Schematic diagram of the CF reactor [21].

Few prominent manufacturing companies, which can provide the dedicated microwave instruments for the academic laboratory and industry purposes are Milestone, Anton-Paar, CEM, SINEO Microwave, and Biotage as shown in Fig. 10.7 [22].

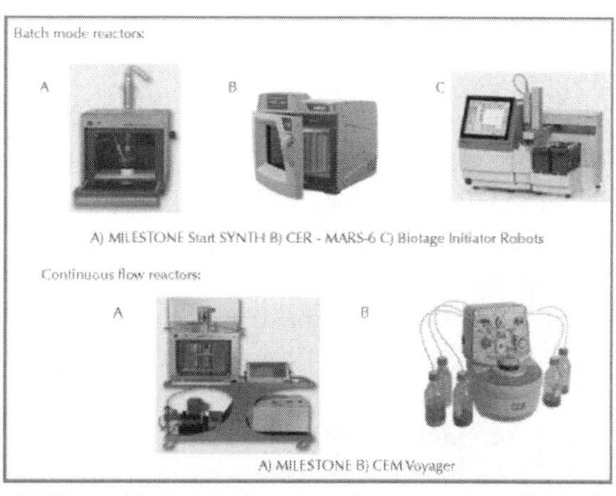


**FIGURE 10.7** Different kinds of microwave instruments available for academic and industry purpose [22].

## 10.7 APPLICATIONS

Microwave irradiation is well-known since 1960s as an alternative heating aid and the slow uptake of this technology for synthesis was due to the lack of fundamental knowledge and the safety issues concomitant with the small scale microwave ovens. However, with the availability of modified and dedicated microwave reactors as discussed in previous section, the use of this methodology both in the research laboratories and manufacturing industries is widely recognized. The microwave mediation has directly benefited the synthesis methodologies by various chemistry disciplines as indicated in Fig. 10.8. It has also made inroads into catalysis and nanomaterial synthesis, food processing, polymer science, biotechnology and analytical applications.

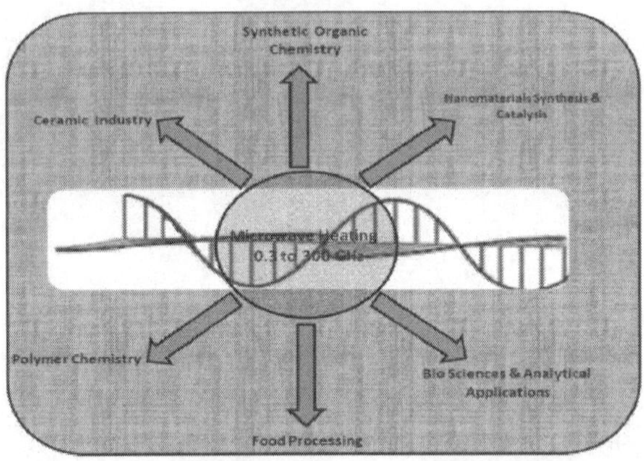


**FIGURE 10.8** Utilization of microwave heating in different fields.

## 10.8 ORGANIC AND PETROCHEMICAL SYNTHESIS

A wide spectrum of organic transformations (>3000 reactions) were reported in the literature by microwave heating. Many chemical reactions were reported to be completed within short reaction time, in more product yields, and high product selectivity. First synthetic application of microwave radiation was introduced independently by Giguere et al. [17] and Gedye et al. [18] using a domestic microwave oven in 1986. Thereafter, many of the reactions have been processed by this method of heating. Some of the organic reactions, which were greatly beneficiated by microwave heating, are schematically shown in Fig. 10.9. The microwave heating has a vivid role on the organic transformations and it is specifically named as Microwave Assisted Organic

Synthesis (MAOS). This topic is widely acknowledged in the literature with more number of the review articles and books [1–9].

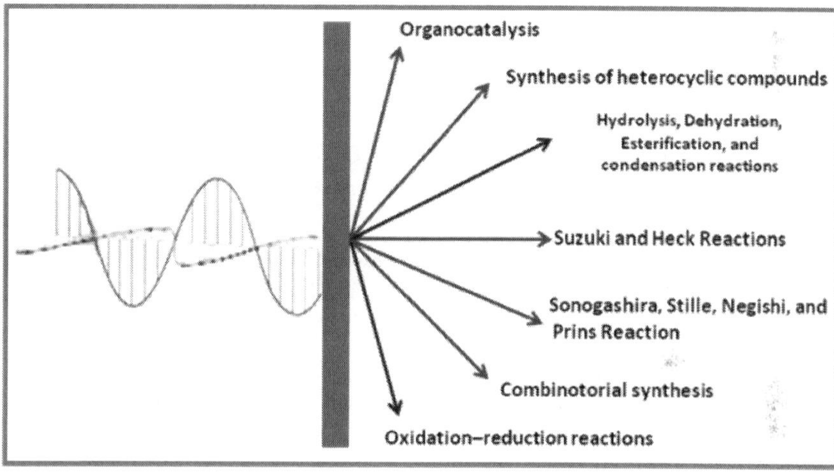


**FIGURE 10.9** Some of the famous organic transformations carried out by the microwave heating.

Microwave heating has a remarkable influence on the coupling reactions such as carbon–carbon coupling, aryl–nitrogen and aryl–oxygen coupling reactions [23, 24]. The carbon–carbon coupling reactions such as Suzuki, Heck, Songashira, etc. have been accomplished with the help of microwave heating and are very important steps for the synthesis of various organic molecules. Many of the coupling reactions need more time for the completion of the reaction by the conventional heating. For instance, the Suzuki reactions were performed by the microwave and conventional heating by keeping all parameters constant. The reactions were completed in 10 min by the microwave heating with 94% product yield. On the other hand, the conventional heating, the reaction was completed in 6 h with 92% of isolated product yield [25] (Scheme 10.2).

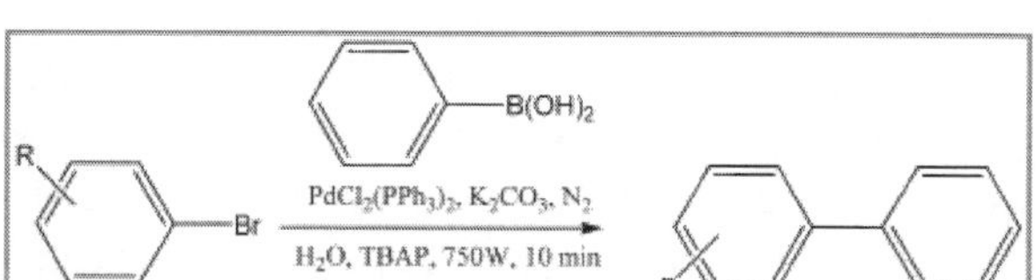


**SCHEME 10.2** Suzuki cross coupling between aryl bromide and phenylboronic acid under microwave irradiation [25].

In a similar fashion, many of the coupling reactions were greatly benefited by the microwave heating. Mehata et al. [23] and Nilsson et al. [24] have reviewed various carbon–carbon and carbon–hetero atom coupling reactions recently.

Microwave-assisted solid phase organic synthesis is another topic of most significant interest that is widely recognized in the field of drug discovery. Generally, solid phase reactions are heterogeneous and involve a high heating temperature and prolonged time. Consequently, the unwanted side products will be formed in large quantities. In this connection the microwave heating is one of the fast and reliable methods in heating and it has been identified as a versatile tool in the solid phase synthesis. The microwave irradiation was successfully employed as a heating aid for the peptide hydrolysis in 1990 [26]. Many have used the domestic microwave for the hydrolysis of polystyrene supported peptides and the hydrolysis was achieved in a short span of time (7 min). On the other hand, the conventional heating entails 24 h of reaction. Similarly, many of the solid phase chemical reactions have been reported in the literature. Continuously, the microwave heating has been slowly implemented for the solid–phase combinatorial library synthesis, where multi component reactants were involved to result a single product.

The microwave heating has a significant impact on many heterogeneous gas-phase catalytic reactions such as reforming of methane with carbon dioxide, hydrodesulfurization (HDS) of thiophene, oxidative coupling of methane, selective oxidation, decomposition of hydrogen sulfide, reduction of sulfur dioxide with methane, dehydrogenation, etc. The review articles published by Durka et al. [27], Zhang et al. [28] and Will et al. [29] independently emphasized the significance of microwave heating on various heterogeneous gas-phase catalytic reactions. Zhang et al. [30] employed the microwave irradiation for the reduction of sulfur dioxide with methane over 30 wt% $MoS_2$ supported on alumina that was prepared by mechanical mixing of the individual materials and compared the microwave heating with the conventional heating. The conversion of $SO_2$ and $CH_4$ as a function of temperature over the catalyst is shown in the Fig. 10.10. Both the conversion of $SO_2$ and $CH_4$ were highly dependent on the type of heating source employed. The microwave heating has a significant impact with nearly 200° C less temperature on the same activity in comparison to conventional method. This behavior was attributed to the nonhomogeneous heating of the catalyst, thereby generating hotspots in the catalyst at some regions. These results indicated that a high temperature was attained than the set temperature at some places on the catalyst. Similar phenomenon was explained for other reactions in the literature [27, 31, 32].

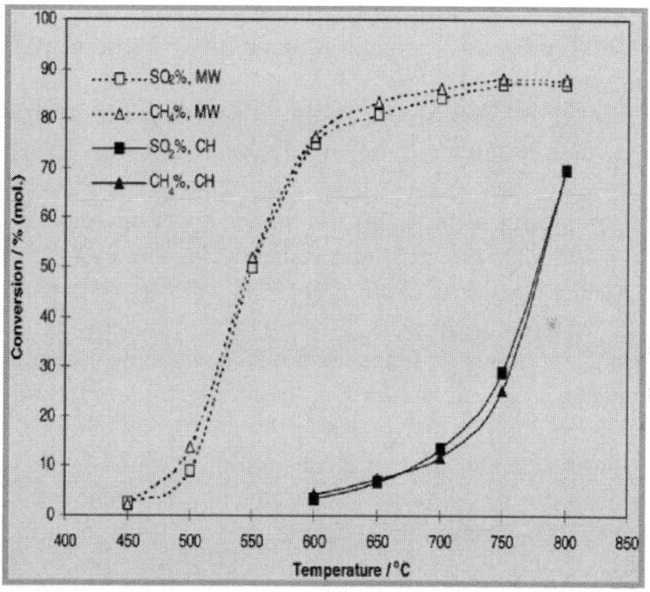


**FIGURE 10.10** $SO_2$ and $CH_4$ conversions as a function of temperature over catalyst [30].

Shekara et al. [33] examined the effect of conventional and microwave heating on the BEA zeolite catalyst for acylation reactions in the liquid-phase. There was a huge difference in the conversion of acids investigated, and it was found to be around 20% in the conventional method of heating and a 50–80% in the microwave heating. They characterized the spent catalysts after the reaction and it was observed that there is a large coke deposition on the catalyst in conventional processing than in microwave heating.

## 10.9 NANOMATERIAL SYNTHESIS

Various inorganic nanomaterials such as metals, metal oxides, and metal-organic frame works (MOFs) have been playing a crucial role in many applications including catalysis, gas absorption studies, organic transformations, etc. However, there should be a great control over the morphology and size of the materials, and these properties have a great impact for the type of reaction concerned. In this scenario, the uniform heating of the materials by microwave heating has a great control on the texture and size of the materials. This uniform heating can affect the nucleation and crystalline growth of the materials during the synthesis unlike conventional processing [10–13]. Here are some of the interesting examples illustrated by microwave processing of the materials.

Truong et al. [34] synthesized high purity $MnO_2$ with well-controlled morphology and crystallinty by microwave-assisted hydrothermal method. In brief, they employed an aqueous solution of hydrochloric acid (HCl) and potassium permanganate ($KMnO_4$) and treated under microwave at 150° C for 2, 5, 15, 30, 120, and 360 min, respectively. The morphology of the obtained products is shown in Fig. 10.11. It indicates that after 2 min of irradiation, the morphology was flower like with intersected nanosheets. When the heating time is prolonged to 5 min, a slight increase in the size of the flowers were observed and further, the purple color of the solution was completely disappeared, which indicated the completion of the reaction. Additionally, an increase in time of irradiation to 15 min, there is a change in the morphology from flower to wire type and complete transformation into nanowires has occurred at 30 min of heating. It can be attributed that there is phase change of the $MnO_2$ at this time. Further, annealing of these obtained nanowires were completely converted into nanotubes with open ended morphology as shown in the Fig. 10.11. Further, they have studied the prepared catalysts at various times of microwave treatment by XRD and found that the materials exhibited $\Delta$ $MnO_2$ phase at below 30 min of heating and the phase transition occurred at above 30 min to $\alpha$ $MnO_2$ for nanowires and tubes. Furthermore, at 15 min of heating both the phases were identified and it also concurred with SEM image at 15 min exhibiting both flower and nanowire type morphology. All these results concluded that the microwave heating has great control over the morphology and crystallinity, and they employed all the synthesized materials for the electrochemical process involved in lithium air batteries and among them the $\alpha$ $MnO_2$ phase with nantotube morphology showed better performance.

**FIGURE 10.11** SEM images of $MnO_2$ products obtained at different reaction times: (A) 2 (B) 5 (C) 15 (D) 30 (E) 120, and (F) 360 min [34].

For the first time, Vivek Polshettiwar et al. [35] synthesized hierarchical dendrite 3D nanostructures of various metal oxides such as iron oxide ($\alpha$ $Fe_2O_3$), manganese oxide ($Mn_2O_3$), cobalt oxide (CoO), chromium oxide ($Cr_2O_3$), and molybdenum oxide ($MoO_2$) in water by using $K_4Fe(CN)_6$, $K_3$ $Mn(CN)_6$, $K_3Co(CN)_6$, $K_3Cr(CN)_6$ $K_3$ $Mo(CN)_8$ as respective precursors under microwave hydrothermal treatment at 180° C for 3 h. The morphology of the obtained metal oxides was analyzed by SEM and TEM as shown in Fig. 10.12. It was observed that there is an attractive morphology for all the metal oxides and is different in each case. For instance, the nanostructures exhibited pine tree like morphology, triangular rod morphology, octahedral nanoparticles, dispersed spherical particles, and spherical nanoparticles for $Fe_2O_3$, $Mn_2O_3$, CoO, $Cr_2O_3$ and $MoO_2$, respectively. Further, all these materials were calcined at 700°**C** for 2 h in order to ascertain thermal stability of the materials and found that the morphology and crystallinty of the materials is retained. However, in the case of molybdenum and cobalt oxides even though the morphology is retained at higher temperature s, the crystalline phase was changed to $MoO_3$ and $Co_3O_4$.

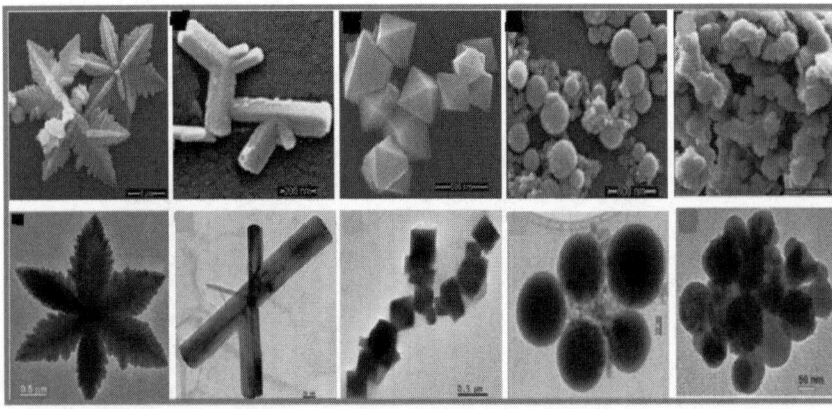

**FIGURE 10.12** SEM Images (upper part of the figure) of iron oxide, manganese oxide, cobalt oxide, chromium oxide, molybdenum oxide respectively (from left to right). The lower part of the Fig. 10.12 is corresponding TEM pictures [35].

## 10.10 CATALYST PREPARATION

They are more number of wet chemical methods available for the synthesis of various catalyst materials such as precipitation, sol–gel, combustion, hydrothermal/solvothermal synthesis, etc. In general, the heating is necessary for all these methods at a particular stage and consequently, the microwave heating was easily adopted and integrated with all these preparation methods. In this direction, Reddy et al. [31, 32] synthesized various mixed metal oxides by microwave-assisted solution combustion synthesis with a modified domestic microwave oven as shown in Fig. 10.13. They have synthesized the transition and rare earth metal-doped ceria-based oxides and copper promoted ceria-based mixed oxides for CO oxidation reaction. The synthetic procedures developed were very simple viz., dissolution of the metal nitrates of the respective oxide precursors in sufficient amount of water and addition of a calculated amount of urea and then exposure of the resultant solution to the microwave irradiation. Here the solvent is water and it is a good microwave absorber, besides the metal nitrates are also ionic in nature and hence a rapid heating was achieved by microwave heating. The combustion was initiated in 10 min and the combustion flame lasts for 1 min and ended up with the powder. The whole process was completed in 15 min for synthesizing the various materials.

**FIGURE 10.13** Modified domestic microwave oven for the synthesis of ceria based oxides by microwave assisted synthesis of combustion synthesis.

## 10.11 SINTERING OF INORGANIC OXIDES

Microwave heating has a prominent role in the sintering of materials such as ceramics, metals and metal oxides at reducing costs compared to conventional heating. Generally, the sintering of materials is carried out at higher temperature for longer timings, which leads to loss of theoretical density, loss of oxygen, and rapid grain growth of the materials. However, the microwave heating enables much faster in the reaction kinetics and sintering, which results finer and uniform microstructures in the product with better mechanical properties [36, 37].

Suresh Kumar and Mathews [38] studied the properties of sintered $Zr_{0.9}Ce_{0.1}O_2$ mixed oxides by microwave and conventional heating at 1100° C. The oxide materials exhibited a theoretical density of 92% and 86% in 1 and 24 h of duration by microwave and conventional processing, respectively. The prolonged heating of the mixed oxides resulted in fast grain growth and eventual loss of oxygen from the oxides. Apart from this, the phase segregation occurred at higher temperature for longer heating times. However, in the microwave sintering all these demerits were minimized.

Ferroelectric materials like $PbZr_{0.52}Ti_{0.48}O_3$ (PZT) are generally fabricated at higher temperature (1200° C) for several hours in the conventional process. However, the loss of lead oxide from the materials happened due to prolonged heating at higher temperature. The microwave heating was employed at 600° C and the material was formed with the required properties due to faster reaction kinetics and with minimal loss of the lead oxide and fine grain size [39]. Brosnan et al. [40] sintered a well-known ceramic material aluminum oxide by microwave and conventional heating. The density of the material obtained by both the processes is shown in Fig. 10.14 as a function of temperature. At 1400° C, the microwave method yielded the material with 100% density, on the other hand by the conventional process; it was only 52%.

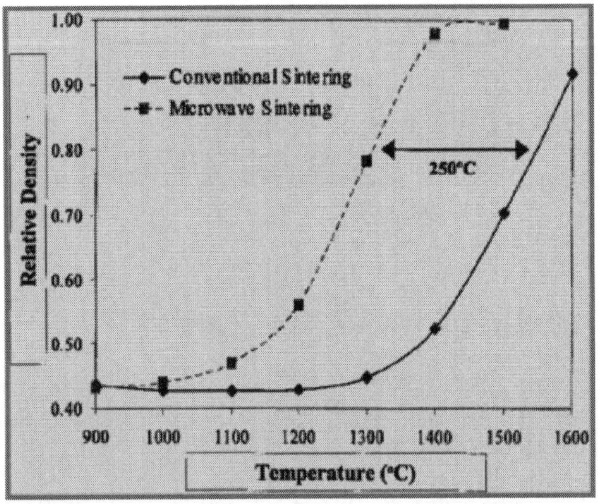


**FIGURE 10.14** Density of microwave-sintered and conventionally sintered alumina samples as a function of sintering temperature [40].

## 10.12 POLYMER CHEMISTRY

Microwave heating has a significant effect on the polymer processing. There are many polymer materials such as thermo-setters and thermoplastics that were successfully processed by the microwave heating [14, 15]. In the conventional method, the heating of the material depends upon the thermal conduction of the material. In that sense, the polymeric materials exhibit very low thermal conductivity and therefore, it is difficult in the polymers processing especially in curing. Whereas the microwave technology has a unique role in the polymers processing and it is quick and uniform in the bulk heating.

Gizdavic-Nikolaidis et al. [41] successfully synthesized the polyaniline nanofibers (PANI) by microwave heating with higher yields than conventional heating. The yield of the PANI under microwave heating was 76.2% after 5 min of irradiation at ambient conditions, on the other hand conventional synthesis resulted in 79.7% yield after 5 h. There is a remarkable influence of microwave heating on the synthesis of these materials. Despite the yield of products, both these methods produced nanofibrillar structures as analyzed by the SEM and TEM studies. Furthermore, they explored their studies on the yield of the materials at different power levels of microwaves such as 0, 3, 10,

40, and 70 W [42]. Increasing the power levels of the microwave radiation, the molecular weight of the polymer increased. For instance, the molecular weight of the polymer obtained at 3 W was 3900 g/mol and at 70 W it was 18300 g/mol at 5 min of irradiation. Despite this, there is no significant difference in the yield of the materials obtained at different microwave power levels. However, in general the heating affects the molecular weight of the resultant polymer in the conventional synthesis. As the temperature increases, the molecular weight of the polymer decreases, but in the microwave heating it was reversed, as the microwave power increases, the molecular weight increased.

## 10.13 ENHANCING MICROWAVE EFFECTS

### *10.13.1 IN CONNECTION WITH ULTRASOUND*

The microwave heating could be operated in conjunction with the ultrasound treatment. Ultrasound heating is of another green avenue for the chemical reactions. The mechanistic outlines for the interaction of ultrasound with the materials could be found in the literature [43]. In general, the heating of the chemical entities is due to the generation and explosion of millions of cavity microbubbles at low and high pressures, respectively through liquid medium. During the explosion, a high amount of energy is released and is responsible for heating the material. Based on this concept many successful applications were developed by ultrasound treatment. Very few reports exist on the combined use of microwave and ultrasound treatment for chemical processing despite their different mechanism of interaction with the molecules. The ultrasound treatment is greatly helpful for heterogeneous reactions. Pengab and Song [44] reported the Williamson synthesis by employing various types of phenols and benzyl chlorides and/or chloro benzene without the use of phase transfer catalysts by the assimilation of microwave heating with the ultrasound treatment. Generally, these reactions are carried out by employing the phase transfer catalysts and/or organic solvents with several hours of refluxing. For comparison purpose, these reactions were also performed under microwave, ultrasound and conventional heating alone. The best results were obtained with simultaneous microwave and ultrasound irradiation resulting in 50–86% yields in less time (50–150 s) compared to their counterparts. Further, they have extended the synthesis of hydrazides from the esters and hydrazine monohydrates, which are very useful for the synthesis of important chemicals [45] (Scheme 10.3).

$$RCO_2Me \xrightarrow[\text{SMUI/20-65 s}]{75\%N_2H_4.H_20} RCONHNH_2$$

**SCHEME 10.3** Simultaneous microwave and ultrasound irradiation.

The yield and time consumption for the reaction between methyl salicylate and hydrazine monohydrate are shown in Table 10.3 under different heat treatments. From Table 10.3, it is noted that this reaction signifies the utility of combined microwave and ultrasound treatment.

**TABLE 10.3** Hydrazinolysis of methyl salicylate using different methods.

| S. No. | Method of heating | Time | Yield (%) |
|---|---|---|---|
| 1. | Conventional | 9 h | 73 |
| 2. | Ultrasound | 1.5 h | 79 |
| 3. | Microwave | 18 min | 80 |
| 4. | Ultrasound/microwave | 40 s | 84 |

Additionally, they have examined the Knoevenagel–Doebner reaction in the aqueous media under simultaneous irradiation and observed to be greatly benefited [46]. Cravotto and Cintas have reviewed recently the combined use of ultrasound and microwave heating in the process chemistry [47].

The rate-determining step in some of the heterogeneous reactions is the mass transfer between the two interfaces. When the heterogeneous mixture is irradiated with the ultrasound, the liquid jet caused by cavitation propagates across the bubble with high speed and violent hit towards the interface and thereby by interchange of bubbles from both the phases. The two irradiations employed simultaneously enhance the heating and mass transfer over the conventional methods. This new joint endeavor is also explored for the synthesis of nanomaterials, food processing, etc. The randomly combined microwave-ultrasound and ultraviolet radiation instrument designed by Sineo microwave is shown in the Fig. 10.15.

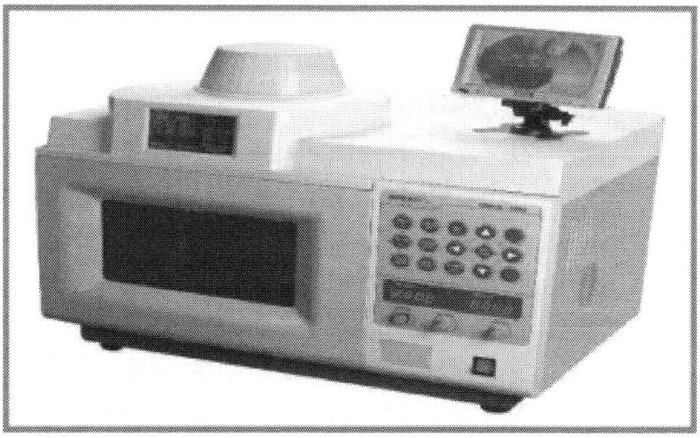

**FIGURE 10.15** Microwave-UV-US synthesis/extraction reactor [22].

### *10.13.2 IN CONJUNCTION WITH IONIC LIQUIDS*

Ionic liquids (IL) are environmentally benign green solvents, which are replacing the widely employed hazardous solvents for organic synthesis and are composed by the weakly bonding cation and anion arrangement. Furthermore, these ionic liquids exhibit interesting physiochemical properties such as high polarity, wide range of boiling points, inflammability, high thermal stability, better dissolution properties, and low interface tension, etc. [48]. All these properties of IL's make them as nourished or optimistic solvents under the microwave heating, in particular to the first three properties. As the molecules exhibit polar nature, they are good microwave absorbers, thereby facilitate fast heating. Further, over pressure development in the reactor with other solvents due to the rapid microwave heating can be avoided, whereas the ionic liquids are nonvolatile nature (high boiling point). Hence, under the microwave heating, the ionic liquids play an elegant role for the chemical reactions. Additionally, the ionic liquids act as structure directing agents for the synthesis of nanomaterials. Certainly, combining these two technologies will meet the green methodologies for the chemical processes [49].

There are many organic name reactions such as Pictet–Spengler, Pechmann, Stetter, Mannich condensation, Knoevenagel, Biginelli, Hantzsch and Biginelli, Tsuji–Trost Baylis–Hillman, Heck (phosphine-free), Morita–Baylis–Hillman, Diels–Alder Hetero Diels–Alder (inter and intramolecular), and Friedel–Craft acylation that were processed by microwave heating in conjunction with the ionic liquids [49, 50]. Apart from organic synthesis this approach is very helpful in the synthesis of nanomaterials and polymers. Xu et al. [51] have addressed the advantages of microwave heating over the conventional heating by ionic liquid mediated aluminophosphate molecular sieves synthesis. In brief, the reaction procedure involved mixing of the ionic liquid, that is, 1-ethyl-3-methylimidazolium bromide ([emim]Br) with $H_3PO_4$ in a round bottom flask and heating at 100° C for 10 min. Followed by the addition of an appropriate amount of $Al\{OCH(CH_3)_2\}_3$ and HF (40 wt% in water) separately to the above mixture and heating for 20 min. Finally, it involves treating the reaction mixtures at 150° C for different time intervals by both microwave and conventional heating. The morphology and relative crystallinity of the obtained products are shown in Fig. 10.16. Microwave irradiation enhanced the formation of crystalline materials within a short span of time. On the other hand in the conventional heating, a prolonged heating time was necessary to achieve full crystallinity. The morphology of these materials shows the formation of needle type and cuboid type crystals in conventional and microwave heating, respectively. This can be attributed to the fast heating of the materials by microwave irradiation induces the rapid crystallization, which facilitated the directional growth. On the other hand, in the conventional heating facilitates slow crystallization, which leads to the oriented attachment leading to formation of needle type morphology.

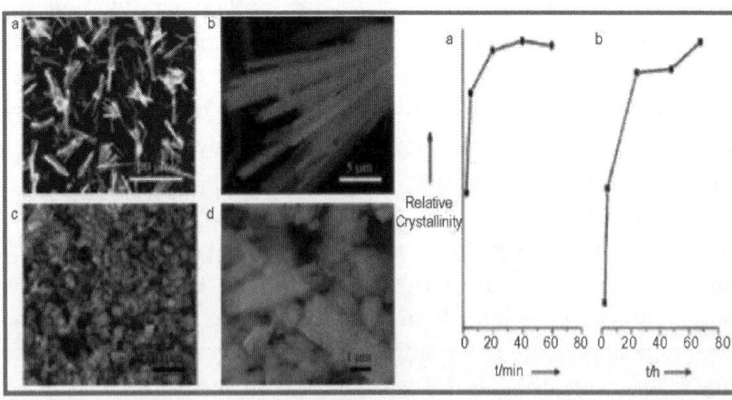


**FIGURE 10.16** SEM micrographs of aluminophosphate molecular sieves (AELtype) prepared by ionothermal synthesis: (a–b) samples after 68 h crystallization with conventional heating; (c–d) samples after 20 min crystallization with microwave heating. Relative crystallinities of aluminophosphate molecular sieves (AELtype) prepared by ionothermal synthesis as a function of time: (a) with microwave heating and (b) with conventional heating. Synthesis conditions: 150°C, ambient pressure, ionic liquid: [emim] Br [51].

Vollmer et al. [52] synthesized various nanoparticles such as Cr, Mo, W, Re, Mn, Fe, Ru, Os, Co, Rh, and Ir by the combined use of ionic liquids and microwave heating from their respective metal carbonyls. Zhu et al. [53] also used this combined technique for the synthesis of tellurium nanorods/nanowires with fast and high yields of the materials.

## 10.14 FOOD PROCESSING

One of the most successful applications of the microwave heating is in the food processing and now the microwave ovens are installed in every modern kitchen for cooking and reheating of the cooked food. For instance, in the drying of the food materials it has a significant role. In the conventional thermal processing the long time exposure to heating is needed consequently, there is a loss of nutrients and sensorial of the food. The other important areas where the microwave heating used were pasteurization, sterilization, and industrial blanching of various food materials [54]. The microwave radiation has an immense use in the biological transformations such as the synthesis of proteins, oligopeptides, peptoids, and carbohydrates processing especially, to understand the structure of proteomics. Furthermore, this method is frequently used for rapid hydrolysis of proteins [16, 26].

## 10.15 ENERGY EFFICIENCY

The green process of the chemical transformations should be economically viable and sustainable. In this connection, the microwave heating has a significant impact. The majority of the chemical reactions reported in the literature used microwave heating with the advantages of time saving as well as increased yield. Nevertheless, a few reports were concerned on the energy efficiency of the reactions. Therefore, there is a need to address on comparing both the conventional and microwave heating processes.

Razzaq and Kappe [55] studied the efficiency of microwave heating in four important reactions, namely, Diels–Alder Reaction of anthracene with maleic anhydride, hydrolysis of benzamide, cyclocondensation of diethyl malonate with N-methyl–aniline, and Suzuki coupling of 4-bromoanisole with phenylboronic acid. By keeping parameters such as concentration, temperature, and pressure constant, they studied energy utilization efficiency of microwave and continuous heating systems. They employed single mode microwave operators in sealed reactors. The contents were refluxed under microwave and conventional heating. All these reactions consumed a very less energy in microwave heating. Further, they have examined all these reactions in various processing conditions such as multi mode microwave systems, open vessel reactors and also with two distinct commercially available microwave instruments. It was found that when the reaction was processed in open vessel reactors by microwave heating and conventional refluxing the consumption of energy was more in the microwave processing. They further examined two types microwave exposures, that is, single-mode and multimode and found that more energy was consumed in multimode reactors. Table 10.4 compares the energy consumption for the benzamide hydrolysis under conventional and microwave heating.

**TABLE 10.4** Comparison of energy consumed by microwave and oil–bath heating for the hydrolysis of benzamide in 5% sulfuric acid.

| S. No. | Heating method | Scale (g) | $V_{solvent}$ [mL] | $T_R$ [°C] | $t_R$ [min][a] | Yield [%][b] | E[kWh][c] | $t_{\Delta total}$ [min][c] | E[kWh][c] |
|---|---|---|---|---|---|---|---|---|---|
| 1 | Oil bath | 1 | 10 | 100 | 720 | 91 | 0.605 | 735 | 80.10 |
| 2 | Oil bath | 5 | 50 | 100 | 720 | 94 | 0.709 | 750 | 18.26 |
| 3 | Isomantle | 5 | 50 | 100 | 720 | 95 | 0.755 | 745 | 19.24 |
| 4 | Sealed-MW-A[d] | 0.4 | 4 | 180 | 7 | 89 | 0.004 | 8.4 | 1.36 |
| 5 | Sealed-MW-B[e] | 0.4 | 4 | 180 | 7 | 94 | 0.026 | 8.5 | 8.38 |
| 6 | Sealed-MW-A[d] | 1 | 10 | 180 | 7 | 90 | 0.009 | 8.6 | 1.20 |
| 7 | Open-MW-B[e] | 5 | 50 | 100 | 120[f] | 76[g] | 1.793[f] | 720[f] | 48.23[j] |
| 8 | Open-MW-C[h] | 5 | 50 | 100 | 120[f] | 75[g] | 6.001[f] | 720[f] | 161.44[j] |
| 9 | Autoclave[i] | 0.4 | 4 | 180 | 7 | 91 | 0.094 | 27 | 31.30 |

[a] Reaction time ($T_R$) is defined as the time period the reaction mixture is kept at the specified maximum reaction temperature (fixed hold time in the case of microwave heating). [b] Isolated yield of product after filtration. [c] Energy consumption measured by the Wattameter for the total heating time ($t_{\Delta total}$) specified. For oil-bath/heating-mantle experiments, this included the time required for heating the device to the desired temperature. For microwave experiments, this includes a 30-s preheating period at room temperature, the ramp time and reaction time at the specified temperature. Additionally the energy consumption of the built-in magnetic stirrer and cooling fan (where appropriate) is also included. [d] Single-mode microwave reactor (Initiator Eight EXP 2.0, 400 W maximum magnetron output power). [e] Single-mode microwave reactor (Discover Labmate, 300 W maximum magnetron output power). [f] These reactions were run for 2 h, and the energy consumption was extrapolated for an 12 h run on the basis of the consumption between 60 and 120 min. [g] Isolated of the product after 2 h. [h] Multi-mode microwave reactor (MicroSYNTH, 1000 W maximum available magnetron output power). [j] Based on a product yield of 90%.

A similar conclusion was arrived by Clark et al. [56] on the energy efficiency of chemical reactions and there is a high reduction in the energy consumption when switching from conventional process to microwave heating.

## 10.16 CONCLUSIONS

In summary, this chapter addressed the definition of microwaves, difference between microwave and thermal heating as well as the mechanistic aspects of microwave heating. The advent of microwave heating over the other methods has been greatly acknowledged. This article focused on the distinct advantages of employing microwave mediation in solvent-free organic synthesis, solid-solid inorganic reactions, nanomaterial synthesis, derivatization of polymers and other transformations. Its thermal and nonthermal effects have greatly contributed to process intensification leading to greener synthesis. It is interesting to note the distinctive improvements to be gained in employing microwave irradiation in conjunction with ultrasound and structure directing ionic liquids. The formation of mono and multimode microwave cavatie endowed with entirely different microwave patterns within their microenvironment may prove advantages in achieving higher reaction specificity. However, the area of concern in microwave mediation is its restricted penetration depth into the reaction media. It has made process scale up follow modular rather than geometric approach.

## ACKNOWLEDGMENT

We wish to specially acknowledge all the researchers whose work is described in this chapter for their valuable contributions. L.H.R. and D.D. are the recipients of research fellowships of CSIR, New Delhi. Financial support was received from Department of Science and Technology, New Delhi, under SERC Scheme (SR/S1/PC-63/2008).

## KEYWORDS

- **Dipolar polarization**
- **Food processing**
- **Ionic conduction**
- **Ionic liquids**
- **Microwave heating**
- **Microwave heating mechanism**
- **Microwave oven**
- **Microwave reactors**
- **Microwave-assisted synthesis**
- **Nanomaterials synthesis**
- **Organic synthesis**
- **Sintering of inorganic oxides**

## REFERENCES


1. Hayes, B. L. Microwave synthesis: Chemistry at the speed of light. EM Publ: Matthews (2002).
2. Loupy, A. (ed.), Microwaves in organic synthesis; Wiley-VCH: Weinheim (2002).
3. Kappe, C. O. and Stadler, A. Microwaves in organic and medicinal chemistry; Wiley-VCH: Weinheim (2005).
4. Tierney, J. and Lidström, P. (eds), Microwave-assisted organic synthesis; Blackwell: Oxford (2005).
5. Kappe, C. O. Controlled microwave heating in modern organic synthesis. *Angew Chem Int Ed* **43,** 6250–6284 (2004).
6. Nasir Baig, R. B. and Varma, R. S. Alternative energy input: Mechanochemical, microwave and ultrasound–assisted organic synthesis. *Chem Soc Rev* **41,** 1559–1584 (2012).
7. Polshettiwar, V. and Varma, R. S. Aqueous microwave chemistry: A clean and green synthetic tool for rapid drug discovery. *Chem Soc Rev* **37,** 1546–1557 (2008).
8. Dallinger, D. and Kappe, C. O. Microwave-assisted synthesis in water as solvent. *Chem Rev* **107,** 2563–2591 (2007).
9. Strauss, C. R. and Varma, R. S. Microwaves in green and sustainable chemistry. *Top Curr Chem* **266,** 199–231 (2006).
10. Baghbanzadeh, M., Carbone, L., Cozzoli, P. D., and Kappe, C. O. Microwave-assisted synthesis of colloidal inorganic nanocrystals. *Angew Chem Int Ed* **50,** 11312–11359 (2011).
11. Klinowski, J., Almeida Paz, F. A., Silvab, P., and Rocha, J. Microwave-assisted synthesis of metal-organic frameworks. *Dalton Trans* **40,** 321–330 (2011).
12. Tsuji, M., Hashimoto, M., Nishizawa, Y., Kubokawa, M., and Tsuji, T. Microwave assisted synthesis of metallic nanostructures in solution. *Chem Eur J* **11,** 440–452 (2005).
13. Bilecka, I. and Niederberger, M. Microwave chemistry for inorganic nanomaterials synthesis. *Nanoscale* **2,** 1358–1374 (2010).
14. Sosnika, A., Gotelli, G., Abraham, G. A., Zhou, S., Sgriccia, N., and Hawley, M. C. Microwave assisted polymer synthesis (MAPS) as a tool in biomaterials science: How new and how powerful. *Prog Polym Sci* **36,** 1050–1078 (2011).
15. Zhong, L., Zhou, S., Sgriccia, N., Hawley, M. C., and Kempel, L. C. A Review of microwave-assisted polymer chemistry. *J Microwave Power EE* **38,** 49–74 (2003).
16. Collins, J. M. and Leadbeater, N. E. Microwave energy: A versatile tool for the biosciences. *Org Biomol Chem* **5,** 1141–1150 (2007).
17. Giguere, R. J., Bray, T. L., Duncan, S. M., and Majetich, G. Application of commercial microwave ovens to organic synthesis. *Tetrahedron Lett* **27,** 4945–4958 (1986).
18. Gedye, R., Smith, F., Westaway, K., Ali, H., Baldisera, L., Laberge, L., and Rousell, J. The use of microwave ovens for rapid organic synthesis. *Tetrahedron Lett* **27,** 279–282 (1986).
19. Cresswell, S. L. and Haswell, S. J. Microwave ovens–Out of the kitchen. *J Chem Educ* **78,** 900–904 (2001).

20. Charde, M., Shukla A., Bukhariya V., Mehta J., and Chakole, R. A review on: A significance of microwave assist techniques in green chemistry. *Int J Phytopharmacy* **2,** 39–50 (2012).
21. Bagley, M. C., Jenkins, R. L., Lubinu, M. C., Mason, C., and Wood. R. A simple continuous flow microwave reactor. *J Org Chem* **70,** 7003–7006 (2005).
22. (a) http://www.sineomicrowave.com/Products.asp?CPID=12. (b) http://www.milestonesrl.com/analytical/Products-Microwave-Synthesis/. (c) http://www.antonpaar.com/Microwave%20Synthesis/59_Corporate_en?productgroup _id=99&Microwave-Synthesis. (d) http://cem.com/. (e) http://www.biotage.com/DynPage.aspx?id=101913 (accessed on March 06, 2013).
23. Mehtaa, V. P. and Van der Eycken, E. V. Microwave-assisted C–C bond forming cross-coupling reactions: an overview. *Chem Soc Rev* **40,** 4925–4936 (2011).
24. Nilsson, P., Olofsson, K., and Larhed, M. Microwave-assisted and metal–catalyzed coupling reactions. *Top Curr Chem* **266,** 103–144 (2006).
25. Bai, L., Wang, J.-X., and Zhang, Y. Rapid microwave-promoted Suzuki cross coupling reaction in water. *Green Chem* **5,** 615–617 (2003).
26. a) Yu, H. M., Chen, S. T., Chiou, S. H., and Wang, K. T. Determination of amino acids on Merrifield resin by microwave hydrolysis. *J Chromatogr* **456,** 357–362 (1988). (b) Yu, H. M., Chen, S. T., and Wang, K. T. Enhanced coupling efficiency in solid–phase peptide synthesis by microwave irradiation. *J Org Chem* **57,** 4781–4784 (1992).
27. Durka, T., Gerven, T. V., and Stankiewicz, A. Microwaves in heterogeneous gas-phase catalysis: Experimental and numerical approaches. *Chem Eng Technol* **32,** 1301–1312 (2009).
28. Zhang, X. and Hayward, D. O. Applications of microwave dielectric heating in environment-related heterogeneous gas-phase catalytic systems. *Inorg Chim Acta* **359,** 3421–3433 (2006).
29. Will, H., Scholz, P., and Ondruschka, B. Microwave-assisted heterogeneous gas-phase catalysis. *Chem Eng Technol* **27,** 113–122 (2004).
30. Zhang, X., Hayward, D. O., Lee, C., and Mingos, D. M. P. Microwave assisted catalytic reduction of sulfur dioxide with methane over $MoS_2$ catalysts. *Appl Catal B* **33,** 137–148 (2001).
31. Reddy, G. K., Thrimurthulu, G., and Reddy, B. M. A rapid microwave-assisted solution combustion synthesis of ceria-based mixed oxides for catalytic applications. *Catal Surv Asia* **13,** 237–255 (2009).
32. Reddy, L. H., Reddy, G. K., Devaiah, D., and Reddy, B. M. A rapid microwave-assisted solution combustion synthesis of CuO promoted $CeO_2$–$M_XO_Y$ (M=Zr, La, Pr and Sm) catalysts for CO oxidation. *Appl Catal A* **445–446,** 297–305 (2012).
33. Chandra Shekara, B. M., Jai Prakash, B. S., and Bhat, Y. S. Microwave–induced deactivation-free catalytic activity of BEA zeolite in acylation reactions. *J Catal* **290,** 101–107 (2012).
34. Truong, T. T., Liu, Y., Ren, Y., Trahey, L., and Sun, Y. Morphological and crystalline evolution of nanostructured $MnO_2$ and its application in lithium–air batteries. *ACS Nano* **6,** 8067–8077 (2012).
35. Polshettiwar, V., Baruwati, B., and Varma, R. S. Self-assembly of metal oxides into three-dimensional nanostructures: Synthesis and application in catalysis. *ACS Nano* **3,** 728–736 (2009).

36. Agrawal, D. Latest global developments in microwave materials processing. *Mater Res Innov* **14,** 3–8 (2010).
37. Agrawal, D. K. Microwave processing of ceramics: A review. *Curr Opin Solid State Mater Sci* **3,** 480–486 (1998).
38. Kumar, K. S. and Mathews, T. Sol–gel synthesis and microwave assisted sintering of zirconia–ceria solid solution. *J Alloy Compd* **391,** 177–180 (2005).
39. Vaidhyanathan, B., Singh, A. P., Agrawal, D. K., Shrout, T. R., Roy, R., and Ganguly, S. Microwave effects in lead zirconium titanate synthesis: Enhanced kinetics and changed mechanisms. *J Am Ceram Soc* **84,** 1197–1202 (2001).
40. Brosnan, K. H., Messing, G. L., and Agrawal, D. K. Microwave sintering of alumina at 2.45 GHz. *J Am Ceram Soc* **84,** 1197–1202 (2001).
41. Gizdavic-Nikolaidis, M. R., Stanisavljev, D. R., Easteal, A. J., and Zujovic, Z. D. A Rapid and facile synthesis of nanofibrillar polyaniline using microwave radiation. *Macromol Rapid Commun* **31,** 657–661 (2010).
42. Gizdavic-Nikolaidis, M. R., Stanisavljev, D. R., Easteal, A. J., and Zujovic, Z. D. Enhanced microwave synthesis: Fine–tuning of polyaniline polymerization. *J Phys Chem C* **116,** 3235–3241 (2012).
43. Bang, J. H. and Suslick, K. S. Applications of ultrasound to the synthesis of nanostructured materials. *Adv Mater* **22,** 1039–1059 (2010).
44. Peng, Y. and Song, G. Combined microwave and ultrasound assisted Williamson ether synthesis in the absence of phase–transfer catalysts. *Green Chem* **4,** 349–351 (2002).
45. Peng, Y. and Song, G. Simultaneous microwave and ultrasound irradiation: A rapid synthesis of hydrazides. *Green Chem* **3,** 302–304 (2001).
46. Peng, Y. and Song, G. Combined microwave and ultrasound accelerated Knoevenagel–Doebner reaction in aqueous media: A green route to 3-aryl acrylic acids. *Green Chem* **5,** 704–706 (2003).
47. Cravotto, G. and Cintas, P. The combined use of microwaves and ultrasound: Improved tools in process chemistry and organic synthesis. *Chem Eur J* **13,** 1902–1909 (2007).
48. Wasserscheid, P. and Welton, T. Ionic liquids in synthesis.; Wiley-VCH, Weinheim: New York (2002).
49. Martínez-Palou, R. Ionic liquid and microwave-assisted organic synthesis: A “Green” and synergic couple. *J Mex Chem Soc* **51,** 252–264 (2007).
50. Martínez-Palou, R. Microwave-assisted synthesis using ionic liquids. *Mol Divers* **14,** 3–25 (2010) and references there in.
51. Xu, Y.-P., Tian, Z.-J., Wang, S.-J., Hu, Y., Wang, L., Wang, B.-C., Ma, Y.-C., Hou, L., Yu, J.-Y., and Lin, L.-W. Microwave–enhanced ionothermal synthesis of aluminophosphate molecular sieves. *Angew Chem Int Ed* **45,** 3965–3970 (2006).
52. Vollmer, C., Redel, E., Abu-Shandi, K., Thomann, R., Manyar, H., Hardacre, C., and Janiak. C. Microwave irradiation for the facile synthesis of transition–metal nanoparticles (NPs) in ionic Liquids (ILs) from metal–carbonyl precursors and Ru-, Rh-, and Ir-NP/IL dispersions as biphasic liquid–liquid hydrogenation nanocatalysts for cyclohexene. *Chem Eur J* **16,** 3849–3858 (2010).
53. Zhu, Y.-J., Wang, W.-W., Qi, R.-J., and Hu, X.-L. Microwave-assisted synthesis of single-crystalline tellurium nanorods and nanowires in ionic liquids. *Angew Chem Int Ed* **43,** 1410–1414 (2004).

54. Shaheen, M. S., El-Massry, K. F., El-Ghorab, A. H., and Anjum, F. M. Microwave applications in thermal food processing. In The development and application of microwave heating; Cao, W., Ed., InTech: (2012).
55. Razzaq, T. and Kappe, C. O. On the energy efficiency of microwave-assisted organic reactions. *ChemSusChem* **1,** 123–132 (2008).
56. Gronnow, M. J., White, R. J., Clark, J. H., and Macquarrie, D. J. Energy efficiency in chemical reactions: A comparative study of different reaction techniques. *Org Process Res Dev* **9,** 516–518 (2005).

# CHAPTER 11

# CATALYTIC PROCESSES FOR FLUOROCHEMICALS: SUSTAINABLE ALTERNATIVES

PAMULAPARTHY SHANTHAN RAO, BANDA NARSAIAH, YADLA RAMBABU, MADABHUSHI SRIDHAR, and KONDAPURAM VIJAYA RAGHAVAN

## CONTENTS

## 11.1 INTRODUCTION

Historically, fluorine is discovered by French Chemist Prof. Henri Moissan way back in 1886 by electrolysis method and became the only industrial production method. Its highest electro-negativity among all the known elements and small atomic radius give distinctive properties to many of this compounds. Fluorine is not considered as an essential mineral element for mammals and humans, in the sense of being necessary for life, however, its role in the natural and pharmacological prevention of dental caries is well established. Though fluorine was known long ago, its use in preparation of various compounds was not exploited. Due to unique properties of compounds with fluorine at an appropriate position, some of them function as refrigerant gases, fire suppressing agents, foam blowing agents and some are fluoropolymers having wide applications. This chapter deals with three important categories of fluorochemicals viz., hydrofluorocarbons, halon substitutes and fluoropolymers. Based on their molecular structure and possible atmospheric fate mechanisms, the fluorocarbons have been considered as ecofriendly molecules with negligible environmental risks. They are highly engineered chemicals to meet the technical requirements of several commercial applications.

The development of interim as well as long-term product alternatives in response to Montreal and Kyoto protocols has become the hallmark of fluorocarbon field since 1970s. This also necessitated alternative process technologies, which are sustainable and have a lower environment impact. The rapid phase out of CFCs in response to Montreal protocol had provided a grace period of 10 years to the world during, which to react to the potentially cataclysmic effects of ozone depletion and global warming. The interim alternatives like HCFCs are accordingly based on a concept that facilitates their reversing of their increased use to lead to a well-planned phase out. The new alternatives are selected based on their photochemical ozone creation potential, atmospheric lifetime, global warming potential and safety in terms of flammability and toxic exposure.

## 11.2 HYDROFLUOROCARBONS (HFCS) AS PROMISING REFRIGERANTS

Aliphatic chlorofluorocarbons known as CFCs have wide applications in refrigeration, air-conditioning, aerosols, foam blowing and others considered as wonder chemicals of twentieth century. They were discovered by Thomas Midgley, an American chemical engineer in the year 1930 and subsequently in 1936, CFCs entered into market as refrigerant gases. Some of the promising CFCs are CFC-11, CFC-12, CFC-113 and FC-14. After three and half decades of use, Prof. James E. Lovelock an environmental scientist from UK detected CFCs in atmosphere in 1970 for the first time and quantified 60 parts per trillion of CFC-11 over Ireland. Further developments led to the depletion of ozone layer by CFCs and an international symposium was organized in 1973 in Japan. In 1974, Sherwood Roland and M.J. Molina published results on ozone depletion, found that when CFCs are exposed to UV light, chlorine radical is generated and each chlorine radical destroys $10^5$ molecules of ozone by chain mechanism. Since then the efforts were focused to find the causes for depletion of ozone and ozone hole was found over Antarctica in 1985. There was an alarm world over on ozone depletion and a search for alternatives to CFCs had thus begun.

In 1987, big seven countries met in Toronto and passed a resolution called Montreal Protocol for phasewise reduction of CFCs. Initially hydrochlorofluorocarbons (HCFCs) were considered as transitional alternatives to CFCs with minimum ozone depletion but significant global warming potential. Some of the promising HCFCs are HCFC-21, HCFC-22, HCFC-122, HCFC-24, HCFC-252 and their ODP values are 0.04, 0.05, 0.08, 0.09, and 0.04, respectively. Research efforts were intensified to identify permanent alternatives to CFCs and found hydrofluorocarbons (HFCs) with minimum global warming and no ozone depletion potential. Some of the promising hydrofluorocarbons are HFC-134a, HFC-32, HFC-143a, and HFC-125. These compounds were used as refrigerant gases and also as blends. R&D work on development of efficient catalysts for fluorination reactions was initiated world over. The scientific groups led by Prof. Blanchard from France claimed that the reversibly oxidizable sites on the surface of the catalyst are responsible for effective fluorination reaction and Prof. Web & Winfield from UK and Prof. Kemnitz from Germany claim Lewis acidic sites on the surface of the catalyst as responsible. Based on the available literature, Imperial Chemical Industries (ICI) from UK developed chromia based catalysts for fluorination reactions and developed technologies for HFCs. M/s. Dupont from USA developed alumina-based catalysts with transition metal dopants. After achieving success with chromia and alumina-based catalysts, a combination of chromia and alumina were developed by coprecipitation method with transitional metals as dopants.

The technologies for HFCs were developed by multinational companies world over for commercial production. As CFCs were banned totally by 1996 in developed countries and by 2006 in developing countries, HFCs were made available as refrigerant gases world over. The Indian Institute of Chemical Technology (IICT), Hyderabad, India established an exclusive research division on fluorochemicals with excellent infrastructure facilities for catalytic vapor phase fluorination reactions. The IICT team developed technologies for HFCs employing novel catalysts based on coprecipitated chromia-alumina. Efforts are also being made to identify next generation refrigerant gases, which have no ODP and GWP.

### 11.2.1 HFC-134a (1,1,1,2-TETRAFLUOROETHANE)

HFC-134a has a boiling point of −27°C and is a highly recommended substitute for dichlorodifluoromethane (CFC-12) since it has zero ozone depletion potential, 13.8 years of life time and global warming potential of 1300. It is used in refrigeration, air-conditioning and aerosol products. HFC-134a is prepared [1–12] mainly from catalytic vapor phase fluorination of trichloroethylene (TCE) and 1,1,1-trifluoro-2-chloroethane (HCFC-133a) [13–21] employing anhydrous hydrogen fluoride. HFC-134a, HFC-125, HFC-143a were also prepared in single pot by catalytic vapor phase fluorination [30–31] of trichloroethylene, 1,1,1-trifluoro-2-chloroethane (HCFC-133a), [32] and chlorofluorination of TCE [33], respectively. A review on HFC-134a was published in 2001 [34]. The challenges of this process lie in the separation of organics by distillation and purification of HFC-134a from product mixture containing HF, HCl, un-reacted HCFC-133a, the starting material and other volatile products. The basic chemistry of HFC-134a is given by:

$$CCl_2 = CHCl + 3HF \longrightarrow CF_3CH_2Cl + 2HCl \quad (1)$$

$$CF_3 - CH_2Cl + HF \longrightarrow CF_3CH_2F + HCl \quad (2)$$

HFC-134a can also be synthesized [22–24] by catalytic vapor phase hydrogenation of 1,1,1,2-tetrafluoro-2,2-dichloroethane (HCFC-124) as per Eq. (3).

$$CF_3 - CFCl_2 + H_2 \longrightarrow CF_3CH_2F \quad (3)$$

It has been reported [25–29] that HFC-134a can also be prepared by liquid phase fluorination of trichloroethylene (TCE), 1,1-dichloro-2-fluoroethylene (HCFC-1121), 1,1,1-trifluoro-2-chloroethane (HCFC-133a) and 1,1,1-trifluoroethane (HFC-143a) in batch reactors.

$$CCl_2 = CHCl + 4HF \longrightarrow CF_3CH_2F + 3HCl \quad (4)$$

$$CCl_2 = CHF + 3HF \longrightarrow CF_3CH_2F + 2HCl \quad (5)$$

$$CF_3-CH_2Cl+HF \longrightarrow CF_3CH_2F \quad (6)$$

$$CF_3-CH_3+HF \longrightarrow CF_3CH_2F \quad (7)$$

The chlorine/fluorine exchange reaction, which is the main process for the synthesis of HFCs, is carried out in fixed bed reactors. The chromium-based catalyst, calcined in $N_2$ atmosphere, need to be preactivated in HF at elevated temperatures, in order to improve their activity. During the activation, the catalyst surface is partially fluorinated. The highly exothermic fluorination process require efficient cooling of the reaction zone in the fixed bed reactor to prevent overheating of the catalyst. The excess heat is also responsible for the production of undesirable by-products and reduced catalyst life time. The recovery of unreacted HF is difficult due to the formation of a low boiling azeotrope of HFC-134a and HF. The commercial fluorination plants are capital intensive and process economies are dictated by the efficiency with, which the undesirable coproducts are separated. A HFC-134a pilot plant was established at IICT, Hyderabad, India in 1999 as shown in Fig. 11.1.

**FIGURE 11.1** HFC-134a pilot plant.

It is a scaled-up version of the single tube fixed bed reactor facility established [35–37] by its own research team. It consists of inconel fixed bed reactors consisting of feed vaporization, preheating and two reaction sections. Figure 11.2 provides the process flow sheet. HF was continuously drawn in liquid form from pressurized tonners and vaporized before preheating. The precatalyst, after being charged into the reactor, was first activated with AHF at 375°C. The catalyst bed was maintained at 300°C and the fluorination of trichloroethylene was carried out in the reaction zone. The product gas stream is stripped of HCl before fractionated in a distillation column.

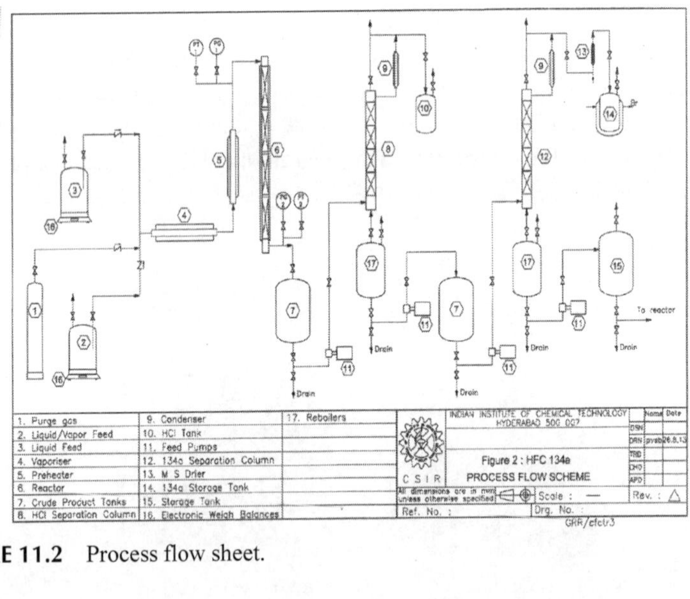


**FIGURE 11.2** Process flow sheet.

Figure 11.3 shows the reaction pathways involved in HFC-134a formation in a two-stage process by the fluorination of TCE. The direction of reactions depends on the concentration of the hydrogen halide. A coprecipitated $Cr_2O_3$/$Al_2O_3$ catalyst was employed.

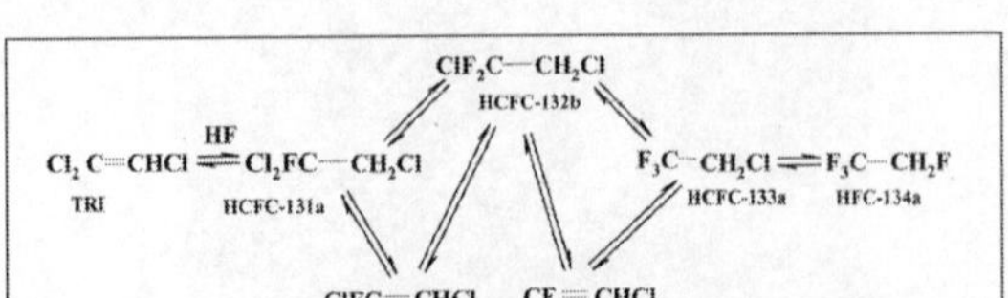


**FIGURE 11.3** Reaction pathway for formation of HFC 134a by fluorination of TRI.

The TCE conversion to HCFC-133a was 98–99% with 98% selectivity and the conversion of HCFC-133a to HFC-134a was between 25–30% with selectivity of 98%. The conversion of HCFC-133a to HFC-134a is thermodynamically restricted to not more than 30%.

### 11.2.2 HFC-32 (DIFLUOROMETHANE)

Difluoromethane (HFC-32) is a recommended refrigerant for use in household, industrial and commercial air-conditioning systems. It is a colorless gas (B.P. −51.7°C) flammable and available in the market as "Genetron®32." It is either used exclusively or blended with other components for commercial refrigeration. HFC-32 is practically nontoxic and nonozone depleting and environmentally friendly chemical with a global warming potential of 1800. It is prepared by catalytic fluorination or by reductive dehalogenation either by liquid or vapor phase process. The key raw material is anhydrous hydrogen fluoride and organic starting materials including methane, dichloromethane, or dichlorodifluoromethane. The catalyst will play a key role on the extent of conversion of raw material and selectivity towards the formation of HFC-32.

The HFC-32 is prepared by catalytic vapor phase fluorination of dichloromethane with anhydrous hydrogen fluoride using aluminum fluoride – chromium oxide [38], chromium oxide-zinc oxide [39, 40], bismuthoxide-manganese chloride doped aluminum fluoride [41], alumina doped with chromia aluminum fluoride [42], or metal halides like $SnCl_4$ [43], in the presence of chlorine [43a]. HFC-32 is also prepared by vapor phase fluorination of chlorofluoromethane using anhydrous hydrogen fluoride as fluorinating agent in presence of aluminum fluoride [44] as indicated in Fig. 11.4.

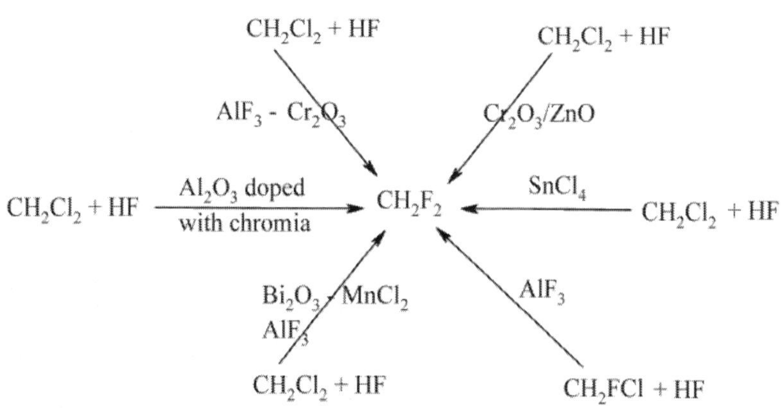


**FIGURE 11.4** Catalyst options for vapor phase synthesis of HFC-32.

HFC-32 can also be prepared by the reductive dehalogenations of dichlorodifluoromethane (CFC-12) performed in the presence of palladium on carbon [45, 46], palladium and zirconium on carbon [47], as follows.

$$CCl_2F_2 \xrightarrow[H_2]{Pd/carbon} CH_2F_2 \xleftarrow{Pd/Zr/carbon} CCl_2F_2 + H_2 \qquad (8)$$

Methane can be directly fluorinated with cobalt trifluoride [48] at 190–250°C as reaction temperature and pressure being 50–1000 psi.

$$CH_4 \xrightarrow{CoF_3} CH_2F_2 \qquad (9)$$

The liquid phase reactions for the production of HFC-32 involves the reaction of dichloromethane with hydrogen fluoride in presence of antimony pentachloride/tinchloride or titanium tetrachloride [49], antimony pentachloride, [50, 51], $SbCl_2F_3$ [52] and tin chloride [53] as indicated in Fig. 11.5. The reaction temperature are in the range of 50–150°C and pressure is in the range of 6–60 atmospheres.

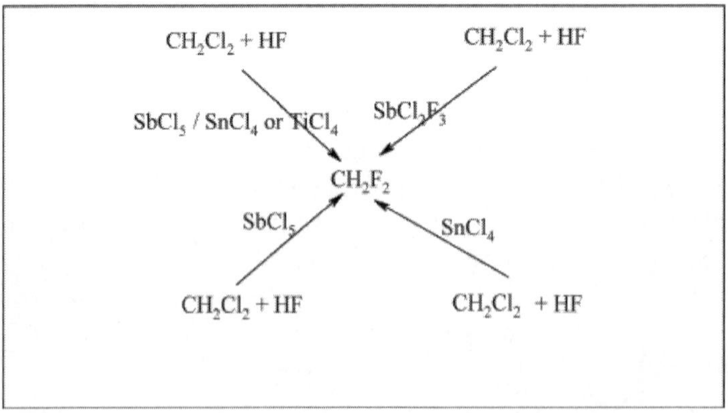


**FIGURE 11.5** Catalyst options for HFC-32 preparations in liquid phase.

### 11.2.3 HFC-143a (1,1,1-TRIFLUOROETHANE)

HFC-143a is an ideal commercial refrigerant mostly available as a blend with HFC-125 or HFC-134a for air-conditioning applications. HFC-143a is a colorless, nonozone depleting, nontoxic gas. It is obtained from vinylidene chloride [54, 55], 1,1,1-trichloroethane [56–58], or 1-chloro-1,1-difluoroethane (R-142b) [59, 60] by catalytic vapor phase fluorination using metal oxides as catalysts. Similarly, HFC-143a is also prepared by catalytic vapor phase fluorination using metal oxide or metal halide catalysts [61]. The source of fluorine is from anhydrous hydrogen fluoride. The reaction temperature varies in the range 50°–280°C depending on the type of catalyst employed. The feedstock and catalyst options for HFC-143a vapor phase synthesis are highlighted in Fig. 11.6.

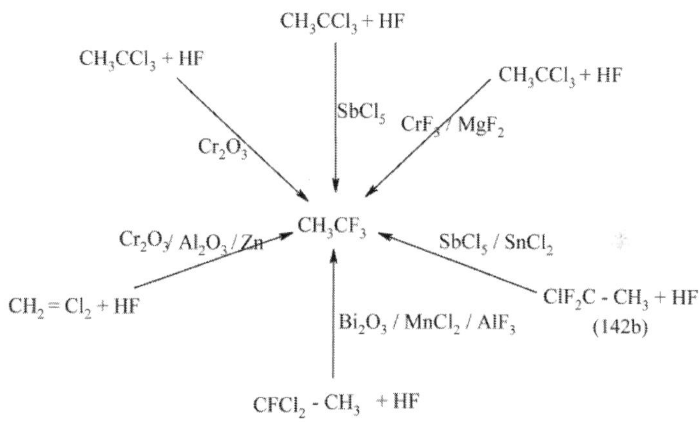


**FIGURE 11.6** Feed stock and catalyst options for HFC-143a vapor phase synthesis.

HFC-143a can also be synthesized in liquid phase medium starting from vinylidene chloride [62–64], vinylidene fluoride [65, 66], and 1,1-dichloro-1-fluoroethane [67] using catalysts like antimony pentachloride, titanium tetrachloride, tantalum pentachloride, tantalum pentafluoride, chloro/fluorosulfonic acid. The reaction temperatures are in the range of 30–120°C and reaction pressures are in the range of 1–20 bar. The feedstock and catalyst options for HFC-143a liquid phase synthesis are highlighted in Fig. 11.7.

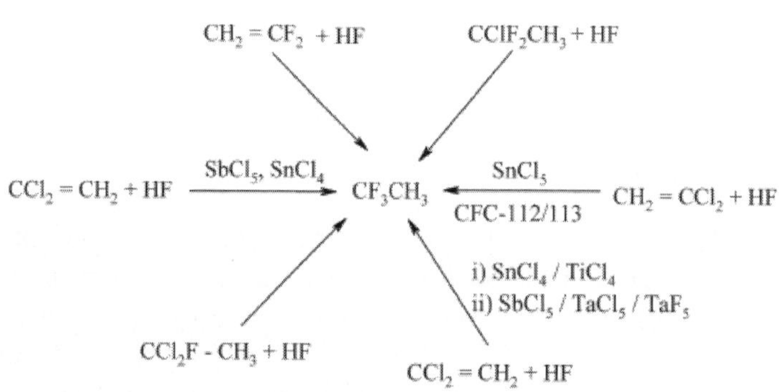


**FIGURE 11.7** Feed stock and catalyst options for HFC-143a liquid phase synthesis.

### 11.2.4 HFC-125 (1,1,1,2,2-PENTAFLUOROETHANE)

HFC-125 is a gas and is used in commercial refrigeration and air-conditioning applications. It is also recommended as fire suppression agent and considered as enviro-friendly chemical and a suitable replacement for Halon 1301. It can be synthesized from 1,1,1,2-tetrafluoroethane [68], 2-chloro-1,1,1,2-tetrafluoroethane (HCFC-124) [69, 70], or tetrachloroethylene [71] using hydrogen fluoride as fluorinating agent in presence of metaloxide catalyst. The details are given in Fig. 11.8.

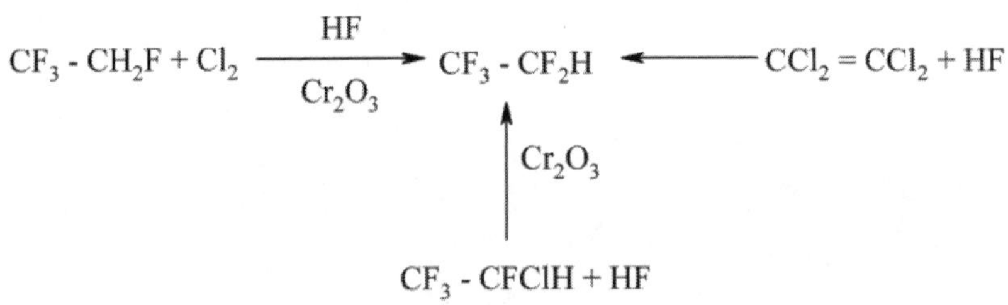


**FIGURE 11.8** Feedstock and catalyst options for HFC-125 synthesis.

The tetrachloroethylene is fluorinated with HF in vapor phase (320°C) in the presence of viz., chromium oxides, oxyfluorides or oxychlorides as catalyst to obtain HFC-125. It can also be obtained from dismutation of HCFC-124 and HCFC-123. The fluorination process has lower selectivity. In case of dismutation process, HCFC-124 has to be synthesized first by fluorination of tetrachloro ethylene. Fixed or flow bed reactors are used for the fluorination process. The catalyst can be regenerated by treatment with a mixture of air and HF followed by $H_2$ treatment.

## 11.3 HALONS AND THEIR SUBSTITUTES

### 11.3.1 HALONS

Halons are bromine containing alkanes used as efficient and convenient gaseous extinguishing agents in the fire and explosion suppression applications. The term, "halon" in the context of Clean Air Act of EPA (US) means a fire extinguishing agent essentially containing bromine, fluorine, and carbon atoms, and occasionally chlorine atom. Each halon is designated with a distinct four-digit number. The digits from left to right denote the number of carbon, fluorine, chlorine and bromine atoms (in increasing order of atomic weight). The most widely used halons in fire suppressing systems are:

(i) Halon 1301 (bromotrifluoromethane);
(ii) Halon 1211 (bromochlorodifluoromethane); and
(iii) Halon 2402 (dibromotetrafluoroethane).

Halons are fast-acting in controlling fires and safe to human exposure upto 10% concentration limit. They are electrically nonconducting and do not leave any residue on application. These properties made them as preferred fire extinguishing agents in occupied areas, computer/data processing systems, areas equipped with sensitive instruments and storage areas containing food and livestock.

The presence of bromine atoms and longer atmospheric lifetime of halons make them highly potent stratospheric ozone depleting agents. For instance, the ozone depletion potential (ODP) of halon 1301 is estimated to be 10 with respect to that of $CCl_3F$ (CFC-11) taken as 1. The ODP values for halon 2402 and halon 1211 are 6 and 3, respectively. The global warming potentials (GWP over 100 year time horizon, relative to $CO_2$ =1) of halon 1211, halon 1301 and halon 2402 are estimated as 1890, 7140 and 1640, respectively.

The absence of ozone molecules in any part of the stratospheric region is called ozone hole. The harmful UV radiation (wavelength between 295–320 nm), which is otherwise absorbed by ozone molecules reaches lower atmosphere through this ozone hole and affects all forms of life. Halons are very stable in troposphere and move along with air currents, eventually reaching the stratospheric region above poles, where they release bromine atoms after broken down by strong UV radiation. The bromine and chlorine atoms destroy several ozone molecules in a chain reaction causing ozone depletion in stratosphere. It is believed that one kilogram of halon released into the atmosphere is capable of destroying 240,000 to 800,000 kg of stratospheric ozone. Because of this reason, the production and use of these effective fire-extinguishing agents in noncore areas is prohibited worldwide under the Montreal protocol. As a result, several countries have started developing halon substitutes that have similar physical properties to fit into the halon-based fire extinguishing systems with minimum modifications.

### 11.3.2 HALON SUBSTITUTES

Traditional alternatives to halons include early detection and manual intervention, fine water spray/mist, carbon dioxide, inert gases, foam and dry powder.

Carbon dioxide is a known asphyxiant and also not suitable for controlling fires involving metals. Foam, water mist and dry powder applications are not used in sensitive areas. These traditional methods are not effective against running and spray fires.

HCFCs are allowed as stop-gap replacement for halons and CFCs. Ultimately, they will also be phased out of production due to their GWP and small ODP. Halocarbon gas systems are found to be more convenient in suppressing fires in place of halons. A number of fire extinguishing halocarbon gases with zero ozone depletion potential (ODP) have been developed. These include both HFCs such as pentafluoroethane, 2H-heptafluoropropane and 1,1,1,3,3,3-hexafluoropropane, and PFCs (perfluorocarbons) like perfluoropropane and perfluorobutane.

Ideal flame suppressing compounds are expected not to contain heavy halogen atoms such as chlorine, bromine, or iodine. Alternatively, they should have short atmospheric lifetime so that they are destroyed in the lower atmosphere and do not reach the stratosphere. They are also expected to be insoluble in water so that they don't end up in aquifers as pollutants. Perfluorocarbons satisfy the first condition but possess higher atmospheric life times causing global warming over longer period. Fluorocarbon compounds containing bromine or iodine molecules that are destroyed before they reach the stratosphere are of special interest as halon replacement agents. The most commonly used halon substitute, 2H-heptafluoropropane has zero ODP but due to its high stability contributes to global warming (GWP 2900).

### *11.3.3 PREPARATION OF HALOCARBON GASES*

The halocarbon gases such as pentafluoroethane (HFC-125), 1,1,1,3,3,3-hexafluoropropane (HFC-236fa) and 2H-heptafluoropropane (HFC-227ea) are manufactured by either vapor phase or liquid phase catalytic fluorination pathways in continuous mode of operation. The successful catalysts for vapor phase fluorination processes are usually chromium (III) oxide (preactivated with HF) and/or aluminum compounds doped with Lewis acids such as Co, Ni or Zn. Similarly, antimony (V) chlorofluoride ($SbCl_xF_y$)/HF is preferred as catalyst in liquid phase fluorination of chlorinated starting compounds. Some of the methods practiced for obtaining individual halocarbons are listed here.

As stated in Section 11.2.4, HFC-125 is industrially produced by heterogeneous continuous vapor phase fluorination of tetrachloroethylene over chromium oxyfluoride and allied catalysts at elevated temperature [72]. Chromium oxyfluoride is generated by treating chromium (III) oxide with AHF. Cr

(III) oxyfluoride transfers fluorine to the reactant and gets converted to inactive Cr (III) oxychloride. The active catalyst is continuously regenerated from Cr (III) oxychloride state by constant supply of AHF.

Pentafluoroethane is also prepared by contacting a steady stream of HFC-143a [73] chlorine and anhydrous HF with antimony (V) chlorofluoride in liquid phase at 100–130°C.

$$CF_3-CH_3+Cl_2+HF \xrightarrow[\text{HF, heat}]{SbCl_2F_3/SbCl_3F_2} CF_3-CHF_2+HCl \qquad (10)$$

### *11.3.2.1 HFC-236fa (1,1,1,3,3,3-HEXAFLUOROPROPANE)*

Hexafluoropropane is prepared by both heterogeneous and homogeneous catalytic processes starting from 1,1,1,3,3,3-hexachloroethane [74–76]. Both the process options are shown in Fig. 11.9.

$$CCl_3-CH_2-CCl_3 \xrightarrow[\text{HF, heat}]{SbF_3\ /\ SbF_5\ \text{or}\ SbF_5\ /\ HSO_3F} CF_3-CH_2-CF_3$$

$$CCl_3-CH_2-CCl_3 \xrightarrow[\text{HF, heat}]{\text{CrOF catalyst}} CF_3-CH_2-CF_3$$

**FIGURE 11.9** Liquid at vapor phase options for HFC-236fa synthesis.

### *11.3.2.2 HFC-227ea (2H-HEPTAFLUOROPROPANE)*

HFC-227ea is currently the most widely used halon substitute due to its efficiency, relatively lower cost and convenience in handling. This gas does not contaminate aquifers as its solubility in water is insignificant. It is manufactured in high selectivity and yield from hexafluoropropene by catalytic vapor phase hydrofluorination over chromium oxyfluoride/Aluminum fluoride [77,78] catalyst in presence of anhydrous HF. This compound is also produced by contacting hexafluoropropene along with HF with tertiaryammonium trihydrofluoride catalyst [79] in a tubular reactor. The reactions involved in both the processes are shown in Fig. 11.10.

$$CF_3 - CF = CF_2 + HF \text{ ------ } CF_3 - CHF\text{-}CF_3$$

**FIGURE 11.10** Liquid and vapor phase options for 2H-Heptafluoropropane.

### *11.3.2.3 PERFLUOROETHYL AND ISOPROPYL KETONE*

The search for a perfect fire suppression agent with zero ODP and GWP value of 1, which is nontoxic and highly efficient against different fires is still on. Recently, perfluoro ethyl isopropyl ketone with zero ODP and GWP of 1 due to its low atmospheric lifetime has been identified as one such molecule. Like HFC-227*ea*, the perfluoro ethyl isopropyl ketone is also prepared starting from hexafluoropropene. The sequence of reactions [80, 81] involved in its preparation as given below:

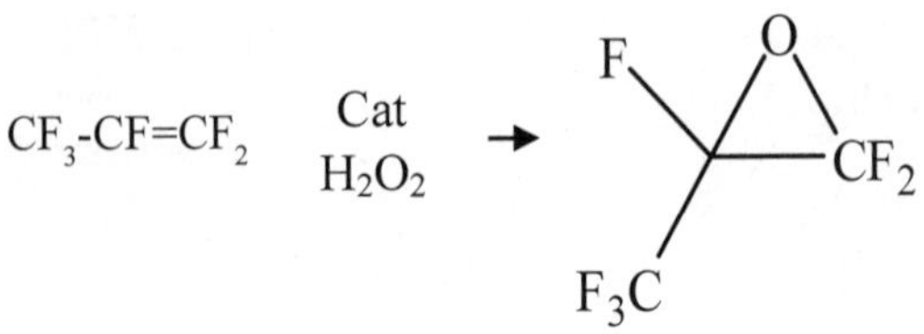


(11)

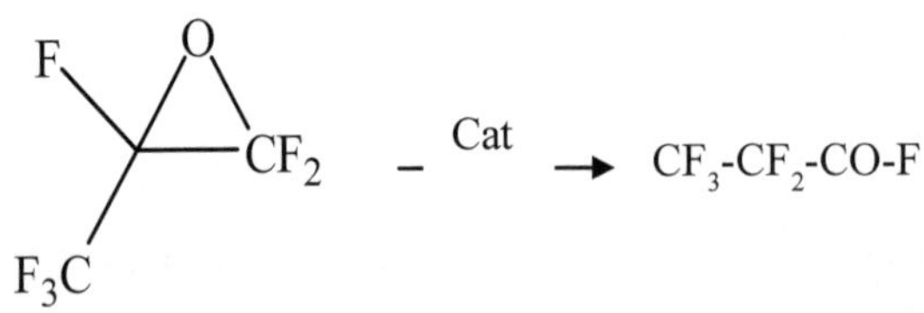


(12)

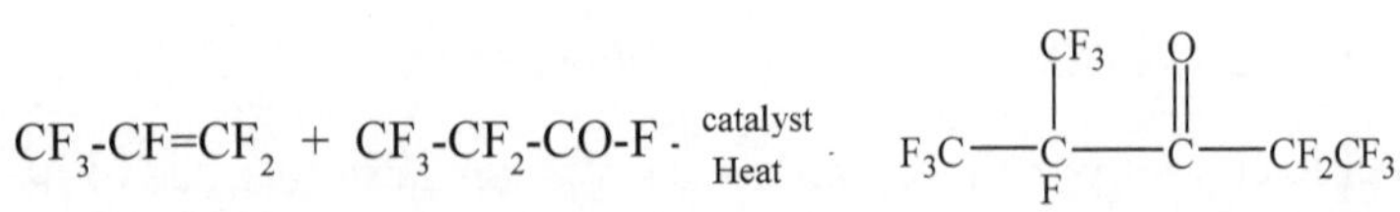


(13)

Perfluoro ethyl isopryl ketone or $C_6$-perfluoroketone with a boiling point of 49°C is a nontoxic volatile liquid at room temperature. It is easy to store and transport, and suitable for both streaming and total flooding applications during control of fires. It is also used as cover gas over molten metal for explosion suppression. However, the cost of production of this ecofriendly fire-extinguishing agent needs to be brought down considerably for wider acceptance.

## 11.4 FLUOROPOLYMERS

The serendipitous discovery of polytetrafluoroethylene (PTFE) in 1938 by Roy J. Plunkett [82] and his co-workers at DuPont research laboratories has spurred the development of a variety of fluorine containing polymers, which include fluorosilicones, fluorinated polyurethanes, fluorinated thermoplastic elastomers, etc., Many of the commercial fluoropolymers are suitable for melt processing *via* conventional injection molding, screw extrusion and blown film extrusion techniques.

Over the past three decades, the importance of fluoropolymers is ever increasing in chemical, automotive, electrical, electronics, communication, aviation, aerospace, architecture, construction, semiconductor fabrication and healthcare sectors. As a consequence, the worldwide consumption of fluoropolymers is seeing steady growth and it is at the rate of 5% over the past five years [83]. In 2011, the global demand of fluoropolymers was estimated to be worth 7.25 billion USD and it is expected to reach 9.4 billion USD by 2016 due to emergence of new markets from growing economies and also due to new developments of products, applications and processes [84].

The unique properties of fluoropolymers are due to the strong bond between carbon and fluorine and shielding of the carbon backbone by fluorine atoms. Fluoropolymers exhibit excellent physicochemical properties such as high thermal and oxidative stability, good weather stability, extreme resistance to chemical attack, low surface energy, low coefficient of friction and low dielectric constant. Fluoropolymers, with unrivaled combination of properties, showed solutions to several industrial problems and facilitated several technological developments in our way of life. This review addresses only popular industrial fluoropolymers with their chemistry, properties and general applications.

A good review on fluoropolymers was published in 2009 [85]. Fluoropolymers (olefinic) are classified as perfluorinated and partially fluorinated products. The former are homo and copolymers of TFE. The speciality fluorinated polymers include perfluoroethers, fluoroacrylates and fluoro silicones, which are used in significantly smaller volumes than olefinic fluoropolymers. The most important monomers are TFE, VDF, ethylene and CTFE. Figure 11.11 highlights fluoropolymers family tree.

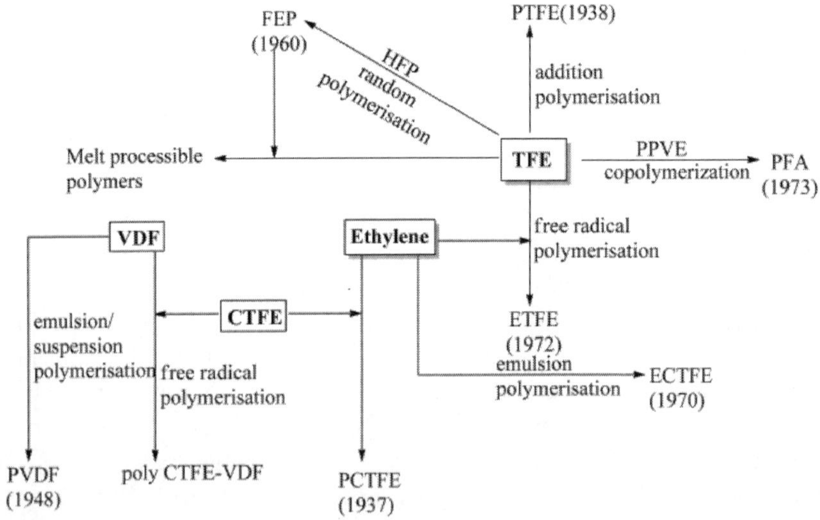


**FIGURE 11.11** A simplified fluoropolymers family tree.

Increasing the fluorine content of a polymer molecule increases its chemical, solvent and flame resistance and photo stability; imporves its electrical properties such as lower dielectric constant, lower coefficient of friction, raises melting point, increases its thermal stability and weakens its mechanical properties. Solubility of polymers in solvents usually decreases by increasing its fluorine content.

### *11.4.1 CRITICAL MONOMER SYNTHESIS*

TFE is prepared from pyrolysis of chlorodifluoromethane (CDFM). The pyrolysis products are cooled, scrubbed to remove HCl and dried. The resulting gas is compressed and distiled to recover the unreacted CDFM and highly pure TFE. The TFE polymerization requires high purity.

Several process options have been reported for VDF synthesis. The commercially relevant options are hydrofluorinatin of: (a) acetylene followed by chlorination [86]; or (b) trichloroethane [87]; or (c) vinylidene chloride [88, 89]. In all cases, VDF is stripped of HCl.

The manufacture of CTFE is quite straightforward. Dechlorination of TCTFE through vapor phase pyrolysis at 500–600°C is a well-practiced operation. The TCTFE is obtained through catalytic dechlorination of hexachloroethane employing HF. CTFE is purified on removal of methyl chloride, dimethyl ether and water through sulfuric acid scrubbing followed by passage through alumina packed column and condensation to liquid.

### 11.4.2 PTFE (POLYTETRAFLUOROETHYLENE)

PTFE is homopolymer of tetrafluoroethylene (TFE) and it is industrially manufactured from chloroform by a three-step process as shown below:

$$CHCl_3 + 2HF \xrightarrow[\text{catalyst}]{\text{anhy. HF}} CHClF_2 + 2HCl \quad (14)$$

$$2CHClF_2 \xrightarrow{\text{pyrolysis}} F_2C{=}CF_2 + 2HCl \quad (15)$$

$$F_2C{=}CF_2 \xrightarrow[\text{initiator}]{\text{polymerisation}} [\text{-}F_2C\text{-}CF_2\text{-}]_n \quad (16)$$

In the first step, chloroform is converted into chlorodifluoromethane by catalytic vapor phase fluorination reaction with anhydrous hydrogen fluoride and in the second step, chlorodifluoromethane is subjected to a noncatalytic gas phase pyrolysis at temperature 590–900°C and at atmospheric or subatmospheric pressures to obtain tetrafluoroethylene in about 95% yield.

Tetrafluoroethylene is a colorless, odorless gas (m.p. −142.5°C; b.p. −76.3°C). It can violently disproportionate into carbon and carbon tetrafluoride in the presence of air and hence, its handling and transport is highly hazardous. In the presence of air, it burns and also forms explosive mixtures with oxygen. Hence, polymerization of TFE requires special equipment and technology [89]. Tetrafluoroethylene is generally stored in steel cylinders in the form of liquid and stabilizers such as α-pinene and d-limonene, which act scavengers of oxygen, are used to prevent polymerization [90].

PTFE is a linear polymer like polyethylene (PE). Structurally, polyethylene has a planar zigzag chain confirmation, while PTFE has a helical chain confirmation. Due to the large fluorine volume and rigid helical polymer chains, PTFE crystallizes easily due to its high melting temperature (327°C) and a high melt viscosity (~1011 Pa s). Because of these properties, though PTFE synthesis is highly difficult as compared to conventional methods. Methods are now available for processing PTFE into various kinds of structures and shapes.

Polymerization of TFE is generally carried out in gaseous phase with water-soluble free radical initiators, such as peroxydisulfates, organic peroxide and aqueous medium. TFE polymerization proceeds well at moderate temperature (40–80°C) and pressures (102–406 psi) and it is a highly exothermic reaction (ΔH= −41 kcal/mol). The characteristics of PTFE are strongly dependent on the processing conditions such as sintering time, temperature and processing pressure.

Commercially PTFE is available in granular and fine powder resin forms and aqueous dispersions. The granular form of PTFE resin is prepared by suspension polymerization process in an aqueous medium with little or no dispersing agent. These forms of PTFE resins are mainly used for compression molding and ram extrusion. The fine PTFE powder is prepared by controlled emulsion polymerization and it is useful for paste extrusion into tapes, tubes, pipe liners, insulation layer of wires, gas-liquid separation membranes and fibers.

### *11.4.3 PCTFE (POLYCHLOROTRIFLUROETHYLENE)*

PCTFE is a linear polymer of chlorotrifluoroethylene as shown below. It was first commercialized in 1953 by M. W. Kellog Company and it was marketed under the trade name Kel-F. PCTFE is commercially processed by bulk, solution, suspension, and emulsion methods [91].

$$\underset{\text{CTFE}}{F_2C{=}CFCl} \xrightarrow{\text{initiator}} \underset{\text{PCTFE}}{[-F_2C\text{-}CFCl-]_n} \tag{17}$$

CTFE is a colorless gas (m.p. −158. 2°C; b.p. −27.8°C) at room temperature and atmospheric pressure. Unlike tetrafluoroethylene, CTFE monomer is highly stable at ambient temperature. It can be stored and transported without an inhibitor. It is prepared in two steps from perchloroethane. In the first step, perchloroethane is converted into 1,1,2-trichloro, 1,2,2-trifluoroethane (CFC-113) by catalytic fluorination with anhyd. HF and in the second step, CFC-113 is dechlorinated by reaction with zinc to give CTFE [92].

$$CCl_3\text{-}CCl_3 \xrightarrow[\text{catalyst, 250-350 °C}]{\text{anhy. HF}} \underset{\text{CFC-113}}{CClF_2\text{-}CCl_2F} \tag{18}$$

$$CClF_2\text{-}CCl_2F \xrightarrow{\text{Zn dust}} \underset{\text{CTFE}}{F_2C{=}CFCl} \tag{19}$$

In CTFE, the presence of chlorine atom, which has greater atomic radius than that of fluorine, causes hindrance to close packing. As a result, PCTFE shows low crystallinity and low melting temperature when compared to PTFE. PCTFE shows good chemical resistance to most chemicals, strong oxidizing agents and has zero-moisture absorption It has thermal resistance up to 175°C and nonflammable. It has low coefficient of thermal expansion [93] and its glass transition temperature ($T_g$) is about 45°C.

PCTFE in thin film form is useful for protection against moisture and finds wide application in pharmaceutical blister packaging, electroluminescent lamps and liquid crystal displays. Due to its chemical stability, PCTFE is also used in chemical industry in the form of tubes, valves, chemical tank liners, O-rings, seals and gaskets. Because of excellent electrical resistance and water repulsion properties, PCTFE finds applications also in preparation of flexible printed circuit boards, insulation of wires and cables.

### *11.4.4 FEP (FLUORINATED ETHYLENE PROPYLENE POLYMER)*

FEP polymer is a linear and random copolymer of TFE and hexafluoropropylene (HFP) containing $-CF_2CF_2-$ and $-CF_2CF(CF_3)-$ units in its structure. FEP and PTFE polymers are structurally similar except that FEP polymer contains trifluoromethyl pendant groups on its main chain as shown below.

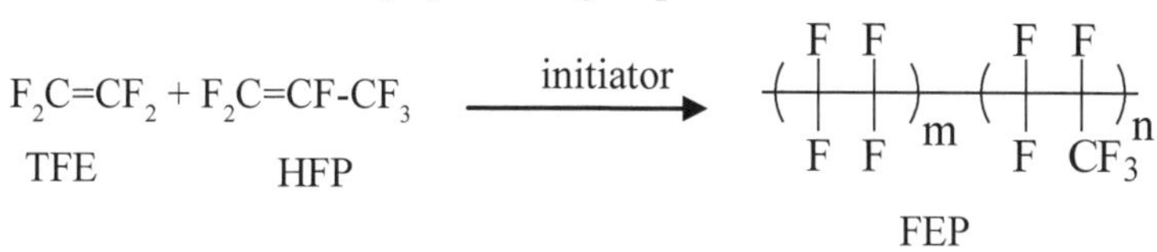


(20)

HFP monomer is a gas at normal temperature and pressure (m.p. −153°C; b.p. −28°C) and it does not homopolymerise easily [94]. However, it forms several industrially useful copolymers with other fluoromonomers. Oxidation of HFP gives industrially important intermediate hexafluoropropylene oxide, which is for preparation of number of perfluoroalkyl perfluorovinyl ethers [95]. Several methods were shown in literature for preparation of HFP. For example, HFP can be prepared in high yields by thermal cracking of TFE at 700–800°C under reduced pressure [96]. Another method for preparation of HFP good yields [97] is pyrolysis of PTFE under vacuum and at 860°C.

FEP polymer generally contains about 5 mol% of HFP. Because of the presence of HFP units, the crystallinity of FEP polymer is significantly lower than that of PTFE. Consequently, FEP polymer has relatively lower melting temperature (260°C) than PTFE (327°C). Several methods were described in literature for copolymerization of HFP and TFE using different catalysts. The commercial process routes for FEP are based on aqueous and nonaqueous dispersion polymerization techniques using conditions similar to those used in PTFE process. FEP has much lower melt viscosity and better processability than PTFE. Conventional polymer processing techniques such as injection molding, extrusion and film casting can be used for processing FEP, which has better impact strength, better wear resistance and less permeability for organic solvents compared to PTFE. FEP also exhibits chemical resistance, weather resistance, flame resistance, radiation resistance and the electrical properties similar to PTFE.

The major applications of FEP are electrical wire insulation, chemical resistance liners for pipes, fittings, tanks and reactors. FEP film products are used in preparation of solar panels and solar collectors.

### *11.4.5 POLY(CTFE-VDF)*

The copolymer of chlorotrifluoroethylene (CTFE) and vinylidene fluoride (VDF) is an industrially important thermoplastic polymer and it finds application as a binder for preparation of explosive composite materials such as polymer-bonded explosives. It was commercialized by Kellog Company in 1950s under the trade name Kel-F 800. The composition of this polymer contains about 35±5 mol% VDF and its structural formula is shown below:

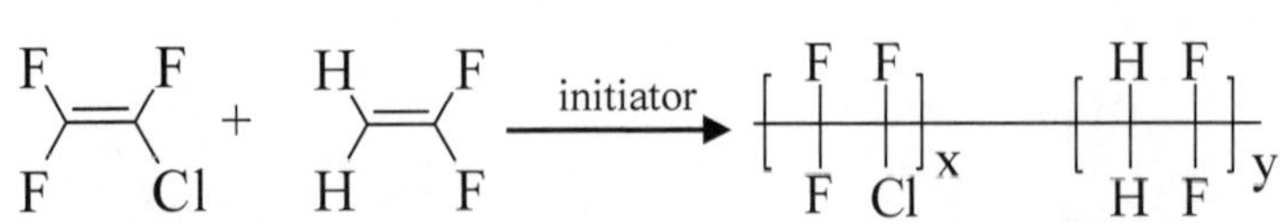

(21)

VDF is a colorless gas (m.p. −144°C; b.p. −83°C) at room temperature and pressure. It is flammable and can form explosive mixtures with oxygen. Industrially, VDF is produced in three steps starting from acetylene. In the first step, acetylene is reacted with HF in the presence of a Lewis acid such as $BF_3.Et_2O$ [98] to obtain 1,1-difluoroethane (HFC-152). In the second step, 1,1-difluoroethane is chlorinated to obtain 1-chloro-1,1-difluoroethane (CFC-142) [99]. In the final step, dehydrochlorination of 1-chloro-1,1-difluoroethane yields vinylidene fluoride [100].

$$HC{\equiv}CH \xrightarrow[BF_3.Et_2O]{HF} \underset{152}{CHF_2\text{-}CH_3} \tag{22}$$

$$CHF_2\text{-}CH_3 \longrightarrow CF_2Cl\text{-}CH_3 + HCl \quad (23)$$

152 142b

$$CF_2Cl\text{-}CH_3 \xrightarrow[\text{-HCl}]{700\text{-}900\ °C} CF_2{=}CH_2 \quad (24)$$

In another process, hydrofluorination of 1,1,1-trichloroethane gives 1-chloro,1,1-difluoroethane (CFC-142). In the next step, 2-chloro-1,1-difluoroethane is subjected to thermal or catalytic dehydrochlorination to obtain vinylidene fluoride (VDF).

The outstanding characteristic of poly(CTFE-VDF) is its chemical resistance to strong oxidizers, mineral acids, alkalis, peroxides and organic solvents. Poly(CTFE-VDF) is useful in manufacturing of special anticorrosive and heat-resistant (up to +200°C) varnishes and enamels as it exhibits 30 times more water resistant/water proof than polyethylene and it shows strong chemical resistance. CTFE-VDF polymer film is highly transparent in near-UV and near-infrared region of the spectrum and finds application in optical industry, infrared windows in radome covers and gage faces. This polymer is steam and gamma sterilizable, biochemically inert and hence, finds application in for food or pharmaceutical packaging.

### *11.4.6 VDF (VINYLIDENE FLUORIDE)*

Though several methods are available in literature on the preparation of vinylidene fluoride, IICT has adopted an elegant method starting from 2-Chloro-1,1,1-trifluoroethane (HCFC-133a). The chlorine of HCFC-133a was replaced by Iodine using Potassium Iodide and Iodine furnished 2-Iodo-1,1,1-trifluoroethane.The dehalogenation of Iodo compound in presence of zinc-methanol afforded vinylidene fluoride as depicted below:

$$CF_3CH_2Cl \xrightarrow[I_2]{KI} CF_3CH_2I \longrightarrow CF_2 = CH_2$$

(HCFC-133a) VDF

Figure 11.12 provides the flowsheet for VDF preparation as adopted by the IICT, Hyderabad.

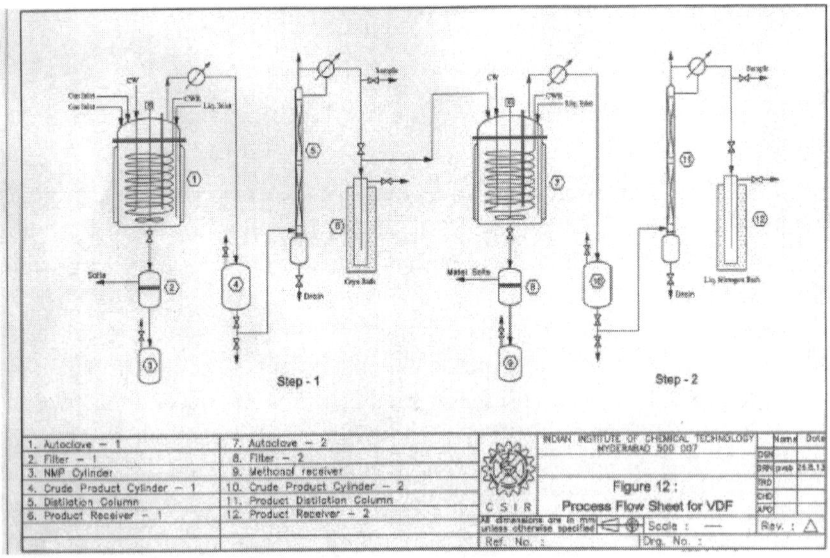


**FIGURE 11.12** The flowsheet for VDF preparation (as adopted by the IICT, Hyderabad).

## 11.5 CONCLUSIONS

Fluorochemicals have become part of our life in the form of drugs, refrigerants, aerosols, fire suppressing agents and has wide applications in strategic sectors. The emphasis on sustainability via greener product and process options with continue to be an important driver for the fluorochemical sector. It has to carve out its future growth map based on S&T developments that have low risk of being challenged. Considerable uncertainty persists about the climate change responses to greenhouse gas emissions due to the current uncertainties on atmospheric sources and risks. The future impact of climate change and likely mitigations strategies are expected to have significant impact on the future choice of fluorochemicals and their processes of manufacture.

## KEYWORDS


- **1,1,1,2-Tetrafluoroethane**
- **Chlorofluorocarbons**
- **Difluoromethane**
- **Ethylene chlorotrifluoro ethylene polymer**
- **Ethylene tetrafluoroethylene polymer**
- **Fluorinated ethylene propylene polymer**
- **Hydro chloro fluoro carbons**
- **Hydro fluoro carbon**
- **Ozone depleting potential**
- **Perfluoroalkoxy polymers**
- **Perfluorocarbon**
- **Polychlorotrifluoroethylene**
- **Polytetrafluoroethylene**
- **Polyvinylidene fluoride**
- **Tetrafluoroethylene**


## REFERENCES


1. Cuzzato, Paolo; Peron, Sergio. Two step continuous process for the preparation of 1,1,1,2-tetrafluoroethane (HFC-134a) from trichloroethylene (TCE). *IT 1324077 B1–20041028* (2004).
2. Lu, Jian; Zhang, Wei; Shi, Lei; Kou, Liangang; Pang, Guochuan; Xu, Qiang. Method for preparation of 1,1,1,2-tetrafluoroethane. *Faming Zhuanli Shenqing Gongkai Shuomingshu. CN 101117305 A 20080206* (2008).
3. Rao, Jampani Madhusudana; Pamulaparthy, Shanthan Rao; Attaluri, Siva Prasad; Narsaiah, Banda; Sripathi, Narayan Reddy; Kuppusamy, Radhakrishnan; Veeramachaneni, Vijayakumar; Sukhadeorao, Patil Kamalakar; Yadla, Rambabu; Srinivas, Koosampally; et al. A process for the preparation of 1,1,1,2-tetrafluoroethane with novel catalyst. *Indian Pat. Appl.* (2005) *IN 1999DE00062 A 20050701*
4. Kim, Hong-Kon; Lee, Byung-Kwon; Ahn, Byung-Sung; Kwon, Young-Soo; Lee, Sang-Deuk. Korea. Method for maintaining fluorination catalyst activity by the removal of water in the manufacture of 1,1,1,2-tetrafluoroethane from hydrogen fluoride and trichloroethane. *KR 137097 B1–19980425* (1998).
5. Sripathi, Narayan Reddy; Rao, Jampani Madhusudana; Pamulaparthy, Shanthan Rao; Attaluri, Siva Prasad; Narasaiah, Banda; Kuppusamy, Radhakrishnan, Veeramachaneni, Vijayakumar; Sukhadeorao, Patil Kamalakar; Yadla, Rambabu; Srinivas, Koo-

sampally; et al. Preparation of 1,1,1,2-tetrafluoroethane and mixed oxide catalyst. *WO 2001074483 A1–20011011* (2001).
6. Ahn, Byung-sung; Kim, Hong-kon: Lee, Byung-Kwon; Kwon, Bun-ill; Park, Sung. Process for the preparation of 1,1,1,2-tetrafluoroethane from 1,1,1-trifluoro-2-chloroethane and hydrogen fluoride. *KR 9608640 B1–19960628* (1996).
7. Oono, Hiromoto; Arai, Tatsuharu; Muramaki, Kazuo; Ooi, Toshio; Nakayama, Hidetoshi. Preparation of 1,1,1,2-tetrafluoroethane. *JP 06040961 A 19940215* (1994).
8. Oono, Hiromoto; Katamura, Koichi: Muramaki, Kazuo; Ooi, Toshio; Nakayama, Hidetoshi. Preparation of 1,1,1,2-tetrafluoroethane (F134a). *JP 06040963 A 19940215* (1994).
9. Oono, Hiromoto; Katamura, Koichi; Arai, Tatsuharu, Kobayashi, Hiroshi; Ooi, Toshio. Preparation of 1,1,1,2-tetrafluoroethane. *JP 06040962 A 19940215* (1994).
10. Scott, John David; Steve, Rachael Anne. Process for the preparation of 1,1,1,2-tetrafluoroethane (HFC 134a). *EP 449617 A2–19911002* (1991).
11. Scott, John David, Steven, Rachael Anne. Process for the preparation of 1,1,1,2-tetrafluoroethane (HFC 134a). *EP 449614 A2–19911002* (1991).
12. Yu, Xiuyuan; Zhang, Wucheng; Li, Na; Zhou, Zhiyong. Iron and chromium-based catalyst used for preparation of 1,1,1,2-tetrafluoroethane. *CN 1820845 A 20060823* (2006).
13. Quan, H.-D.: Yang, H.-e: Li, Z.; Ren, J.-Z.; Li, H.-L. Preparation of 1,1,1,2-tetrafluoroethane by the vapor-phase catalytic reaction of 1,1,1-trifluoro-2-chloroethane with anhydrous hydrogen fluoride. *Journal of Fluorine Chemistry.* **111**(*2), 193–199* (2001).
14. Lee, Byung-Gon; Park, Gun-U.; Kim, Hun-Shik; Jung, Mun-Cho. Chemoselective fluorodechlorination process and catalyst for the preparation of 1,1,1,2-tetrafluoroethane from hydrogen fluoride and 1,1,1-trifluoro-2-chloroethane. *KR 9514215 B1–19951123* (1995).
15. Lee, Hyunjoo; Kim, Hoon Sik; Kim, Honggon; Jeong, Woo Sik; Seo, Inseok. Preparation of 1,1,1,2-tetrafluoroethane by catalytic fluorination of 1,1,1-trifluoro-2-chloroethane over $CrF_3/MgF_{2-}AlF_3$. *Journal of Molecular Catalysis A:Chemical 136(1), 85–89* (1998).
16. No inventor data available. Preparation of 1,1,1,2-tetrafluoroethane over magenesium chromate catalysts. *JP 08027045 A 19960130* (1996).
17. Quan, Heng-dao; Li, Zhong; Zhao, Zhi-xia; Yang, Hui-e; Lu, Jian; Ren, Jian-gang; Chen, Shu-kang; Li, Hui-fang; Li, Hui-li. Preparation of 1,1,1,2-tetrafluoroethane (HFC-134a) by vapor-phase catalytic fluorination. *Applied catalysis, B: Environmental.* **8***(2), 209–15* (1996).
18. Tsuji, Katsuyuki, Tomota, Seiichi; Hotsuta, Masatoshi; Nakajo, Tetsuo; Nakayama, Hidetoshi. Preparation of catalyst for preparation of 1,1,1,2-tetrafluoroethane via fluorination of 1-chloro-2,2,2-trifluoroethane. *JP 06228022 A 19940816* (1994).
19. Galland, Jean Michel; Guiraud, Emmanuel; Schirmann, Jean Pierre. Preparation of 1,1,1,2-tetrafluoroethane. *EP 554165 A1–19930804.*
20. Cheminal, Bernard; Lacroix, Eric; Lantz, Andre. Process for the preparation of 1,1,1,2-tetrafluoroethane. *EP 486333 A1–19920520* (1992).
21. Manzer, Leo E. Preparation of 1,1,1,2-tetrafluoroethane by fluorination of 1,1,1-trifluorochloroethane with hydrogen fluoride over metal catalysts. *BR 8900588 A 19891010* (1989).

22. Reddy, Sripathi Narayan; Rao, Pamulaparthy Shanthan; Anuradha, Gundamaraju; Narsaiah, Banda; Rambabu, Yadla; Srinivas, Punnamraju Venkata Satya Surya; Ravindranath, Kajjam. Dehydrochlorination process for the preparation of 1,1,1,2-tetrafluoroethane from 1,1-dichlorotetrtafluoroethane. *IN 2005DE00475 A 20070105* (2007).
23. Oshio, Hideki, Mishumi, Sadaharu; Yagii, Kiyoshi; Yoshikawa, Satoshi; Murata, Katsuyoshi. Process for the preparation of 1,1,1,2-tetrafluoroethane by a selective hydrogenation of 1,1-dichloro-1,2,2,2-tetrafluoroethane. *DE 3917573 A1–19891214* (1989).
24. Furutaka, Yasuhisa; Aoyama, Hiroichi; Homoto, Yukio. Preparation of 1,1,1,2-tetrafluoroethane. *JP 01093549 A 19890412* (1989).
25. Pennetreau, Pascal; Janssens, Francine, Braun, Max; Eicher, Johannes; Hausmann, Eckhard; Rudolph, Werner. Preparation of 1,1,1,2-tetrafluoroethane. *EP 574077 A1–19931215.* (1993).
26. Eicher, Johannes; Fazniewscy, Karl Heinz; Rieland, Matthias; Rudolph, Werner. Preparation of 1,1,1,2-tetrafluoroethane. *EP 537560 A1–19930421* (1993).
27. Wu, Yongming; Lin, Yongda. Preparation of 1,1,1,2-tetrafluoroethane by aq. phase fluorination. *CN 1308043 A 20010815* (2001).
28. Yoshizumi, Motohiko, Yamashita, Yukya. Preparation of 1,1,1,2-tetrafluoroethane by fluorination of 1,1,1-trifluoro-2-chloroethane in aprotic polar solvents. *JP 01228925 A 1989091.* (1989).
29. Hotate, Makoto, Myoshi, Yasuo; Baba, Tatsuo; Suzuki, Akira. Preparation of 1,1,1,2-tetrafluoroethane from 1,1,1-trifluoroethane. *JP 04005247 A 199201019* (1992).
30. Lee, Byung Gwon; Kim, Hoon Sik; Kim, Honggon; Lee, Sang Deuk; Chung, Moon Jo. Fluorination process and catalysts for the concurrent production of 1,1,1,2-tetrafluoroethane and pentafluoroethane from 2-chloro-1,1,1-trifluoroethane. *US 5723700 A 19980303* (1998).
31. Ono, Hiroki; chujo, Tetsuo; Miyamura, Akira; Oi, Toshio. Simultaneous preparation of 1,1,1,2-tetrafluoroethane and pentafluoroethane from trichloroethylene and tetrachloroethylene. *JP 10007603 A 19980113* (1998).
32. Kim, H.S.: Lee, B.K. Simultaneous preparation of 1,1,1,2-tetrafluoroethane, pentafluoroethane and 1,1,1-trifluoroethane by fluorination of 1,1,1-trifluoro-2-chloroethane. *JP 09067278 A 19970311* (1997).
33. Manzer, Leo Ernest; Rao, Velliyur Nott Mallikarjuna. Process for the preparation of 1,1,1,2-tetrafluoroethane. *EP 3660404 A1–19900502* (1990).
34. Chen, Qing-Yun. 1,1,1,2-Tetrafluoroethane e-EROS *Encyclopedia of Reagents for Organic Synthesis*. DOI: 10.1002/047084289 *x*.rn00526 (2001).
35. Rao, J.M. Sivaprasad, A., Shanthan Rao, P., Narsaiah, B., Reddy, S.N., Vijaya Kumar, V., Manorama, S.V., Leela Krishna, K., Srinivas, K., Radhakrishnan, K and Badrinarayanan, S. *Journal of Catal., 184, 105–111*(1999).
36. Rao, J.M., Shanthan Rao, P., Vijaya Kumar, V., Rao, V.V., Narsaiah, B., Reddy, S.N., Leela Krishna, K., Srinivas, K., Radhakrishnan, K. and Patil, K.S. *J. Fluorine Chem., 95, 177–180 (1999).*
37. Leela Krishna, K., Radhakrishnan, K., Narsaiah, B., Siva A. Prasad, Shanthan Rao, P., Rao, J.M., Reddy, S.N., Srinivas, K, Patil, K.S., Vijaya Kumar. J. Chem. Engg. Data process for the preparation of HFC-134a. *US Patent 7982074 B2* (2006).

38. Ying, Mingming; Hu, Dejun; Ying, Zhenzhou; Yang, Yi; Yu, Sijia. Method of preparation difluoromethane (R32) by multistep continuous fluorination. *CN 1676501 A 20051005* (2005).
39. Tsuji, Katsuyuki; Kaga, Kazuari; Tomota, Seiichi; Nakajo, Tetsuo; Nakayama, Hidetoshi. Preparation of difluoromethane from dichloromethane. (1994), JP 06211707 A 19940802.
40. Bonniface, David William; Scott, John David; Watson, Michael John. Preparation of difluoromethane. *WO 9421579 A1–19940929* (1994).
41. Guo, Xinzheng; Ye, Zhaohui; Ma, Xinming. Preparation of difluoromethane by gas-phase fluorination of dichloromethane. *CN 1076686 A 19930929* (1993).
42. Jung, Kun-yong; Park, Jong-min; Lee, Yun-yong; Lee, Yun-woo. Fluorination process and catalysts for the preparation of difluoromethane from anhydrous hydrofluoric acid and dichloromethane. *KR 9711701 B1–19970714* (1997).
43. Wilmet, Vincent; Janssens, Francine. Process and catalysts for the preparation of dilfuoromethane hydrogen fluoride and dichloromethane. *EP 732314 A1–19960918* (1996).
43 a. Garrait, Dominique; Guiraud, Emmanuel. Fluorination process for the preparation of difluoromethane from dichloromethane and hydrogen fluoride. *EP 805136 A1–19971105.* (1997).
44. Requieme, Benoit; Lacroix, Eric. Gas-phase process for the preparation of difluoromethane from chlorofluoromethane. *FR 2736049 A1–19970103* (1997).
45. Kim, Jae Hun; Ha, Jeong Myeong; Park, Jong Min; Ah, Byeong Seong; Kim, Dae U. Method for the preparation of difluoromethane by hydrodechlorination of chlorodifluoromethane. *KR 2013005888 A 20130116* (2013).
46. Wiersma, Andre; van de Sandt, Emilius Johannes Albertus Xaverius; van Bekkum, Herman; Makkee, Michiel; Moulijn, Jacob Adriaan. Preparation of difluoromethane by reductive dechlorination. *NL 9401574 A 19960501* (1996).
47. Murthy, Janmanchi Krishna; Shekar, Sridara Chandra; Rao, Kamaraju Seetha Rama; Kumar, Vasireddy Siva; Raju, Burri David; Rao, Paladugu Venkateswara; Rao, Panja Kanta; Raghavn, Kondapuram Vijaya. Process for synthesis of a zirconia-modified, active carbon-supported palladium catalyst useful for the preparation of difluoromethane. *IN 2002DE00153 A 20061222.* (2006).
48. Kostyaev, Rodion A.; Asovich, Viktor S.; Kornilov, Vadim V. Preparation of difluoromethane and/or trifluoromethane by the fluorination of methane with cobalt trifluoride. *RU 2063953 C1–19960720* (1996).
49. Kang, Yun-Hwan; Lee, Yun-Yong; Jung, Kun-Yong. Catalytic fluorination process for producing difluoromethane from hydrogen fluoride and dichloromethane. *KR 134544 B1–19980420* (1998).
50. Nam, Kyung-hee; Na, Doo-chan; Kim, Dae-soo. Preparation of difluoromethane by the dechlorofluorination of dichloromethane with hydrogen fluoride in the presence of an antimony catalyst. *KR 9710459 B1–19970626* (1997).
51. Nam, Kyung H.; Na, Doo C.; Kim, Dae S. Method for the preparation of difluoromethane. *US 5495057 A 19960227* (1996).
52. Tsuda, Takehide; Shibanuma, Takashi. Preparation of difluoromethane and 1,1,1-trifluoroethane. *JP 08217704 A 19960827* (1996).

53. Wilmet, Vincent; Janssens, Francine. Process and catalysts for the continuous preparation of difluoromethane by the fluorination of dichloromethane with hydrogen fluoride. *EP 829462 A1–19980318* (1998).
54. Narsaiah, Banda; Rao, Pamulaparthy Shanthan; Srinivas, Punnamraju Venkata Satya Surya; Thomas, Bandanadam Vijaya. Development of new catalyst for preparation of 1,1,1-trifluoroethane (HFC-143a). *IN 2010DE01396 A 20121109* (2012).
55. Tang, Yueming; Chen, kefeng; Chen, Hongfu; Zhang, Xueliang; Yan, Xiaozhong. Process for preparation of 1,1,1-trifluoroethane. *CN 1699314 A 2005–1123* (2005).
56. Zhang, Wei; Lu, Jian; Kou, Liangang; Shi, Lei; He, Fei; Li, Fengxian. Process for the preparation of 1,1,1-trifluoroethane. *CN 101028990 A 20070905* (2007).
57. Kim, Honggon; Kim, Hoon Sik; Lee, Byung Gwon; Kwon, Young Soo; Park, Kun You. Preparation of 1,1,1-trifluoroethane by catalytic fluorination of 1,1,1-trichloroethane and 1-fluoro-1,1-dichloroethane. *Journal of Industrial and Engineering Chemistry. **3**(3), 203–209* (1997).
58. Swain, Charles F. Process and catalysts for the preparation of 1,1,1-trifluoroethane from hydrogen fluoride and 1,1,1-trichloroethane. *WO 9605156 A1–19960222* (1996).
59. Zhang, Xigong; Song, Shuwei; Sun, Sen. Process for the preparation of 1,1,1-trifluoroethane from 1-chloro-1,1-difluoroethane. *CN 1962584 A 20070516* (2007).
60. Mathieu, Veronique; Mross, Stefan. Process for preparation of 1,1,1-trifluoroethane using fluorination catalyst. *FR 2852007 A1–20040910* (2004).
61. Guo, Xinzheng; Zhao, Xuan; Zheng, Chengwu. Preparation of 1,1,1-trifluoroethane by gas-phase fluorination of 1,1,1-dichlorofluoroethane. *CN 1106780 A 19950816* (1995).
62. You, Laifang. Method for preparation of 1,1,1-trifluoroethane and 1,1,1-chlorodifluoroethane. *CN 1634816 A 20050706* (2005).
63. You, Laifang. Process for preparation of 1,1,1-trifluoroethane and 1,1,1-difluorochloroethane. *CN 1569784 A 20050126* (2005).
64. Guo, Xinzheng; Wang, Min; Zhao, Xuan. Preparation of 1,1,1-trifluoroethane by liquid-phase fluorination of 1,1-dichloroethylene. *CN 1106781 A 19950816* (1995),
65. Balthasart, Dominique; Pennetreau, Pascal. Process for the liquid-phase preparation of 1,1,1-trifluoroetahne with 1,1-dichloro-1-fluoroethane and/or 1-chloro-1,1-difluoroethane by the reaction of hydrogen fluoride with vinylidene fluoride and vinylidene chloride and/or 1,1,1-trichloroethane. *EP 703203 A1–19960327* (1996).
66. Pennetreau, Pascal; Balthasart, Dominique. Liquid-phase process for the preparation of 1,1,1-trifluoroethane via the hydrofluorination of vinylidene fluoride by hydrogen fluoride. *EP 703204 A1–19960327* (1996).
67. Guo, Xinzheng; Wang, Min; Ye, Zhaohui. Preparation of 1,1,1-trifluoroethane by liquid-phase fluorination of 1,1,1-dichlorofluoroethane. *CN 1106779 A 19950816* (1995).
68. Kono, Satoshi; Shibanuma, Takashi. Preparation of 1,1,1,2,2-pentafluoroethane and 2-chloro-1,1,1,2-tetrafluoroethane. *JP 2000034236 A 20000202* (2000).
69. Rao, Velliyur Nott Mallikarjuna; Rosenfeld, H. David; Subramoney, Shekhar; Subramanian, Munirpallam A.; Sievert, Allen C. Fluorination process and catalyst for the preparation of 1,1,1,2,2-pentafluoroethane from haloethanes and/or haloethylenes and hydrogen fluoride. *WO 2004018396 A1–20040304* (2004).

70. Kono, Satoru; Shibanuma, Takashi; Kanemura, Takashi; Yoshimura, Toshikazu; Takahashi, Kazuhiro. Process for the preparation of 1,1,1,2,2-pentafluoroethane. *WO 9919285 A1–19990422* (1999).
71. Oono, Hiromoto; Nakajo, Tetsuo; Myamura, Akira; Ooi, Toshio. Preparation of 1,1,1,2,2-pentafluoroethane as chlorofluorohydrocarbon alternate. *JP 07324044 A 19951212* (1995).
72. John, C. M., Nicolos, D.Y. Process for the manufacture of chlorodifluoromethane. *Eur. Patent EP1868973* (2007).
73. Barabanov, V.G., Shelopin, G.G, Ismagilov, N.G., Mitina, I.E., Zirka, A. A., Isupova, L.A., Reshetnikov, S.I., Simonova, L.G., Trukshin, I.G. Method for activation of a catalyst for fluorocontaining hydrocarbons. *Russ. Patent RU2449832* (2012); *CAN156: 590709.*
74. Webster, J.L., Lerou, J.J. Preparation de methanes fluores. *PCT Intl. W09425418* (1994).
75. Beatrice, B., Michel, D. Process for the manufacture of pentafluoroethane. *US8058487* (2006).
76. (a) Si, L., Zhang, P., Gu, H. Preparation method for pentafluoroethane. *Chinese Patent CN102617270* (2012); *CAN157: 261321.*
(b) Manogue, W. H., Rao, V. N. M., Sievert, A. C., Swearingen, S. H. Process for the manufacture of 1,1,1,3,3-pentafluoropropane, 2-chloropentafluoropropane and compositions comprising saturated derivatives thereof. *PCT Intl. WO9837043* (1998).
77. Scott, C.J., Paul, R.R., Steven, H.S. Producing $CF_3CH_2CF_3$ and/or $CF_3CH{=}CF_2$ by conversion of alpha-hydro perfluoroisobutyric acid compounds. *PCT Intl. WO9632364* (1996).
78. Qian, J. C., Iikubo, Y. Process for the preparation of 1,1,1,3,3,3-Hexafluoropropane. *PCT Intl. WO9615085* (1996).
79. Singh, R., Kapoor, A. K., Rao, P. S., Reddy, S. N., Rambabu, Y., Narsaiah, B., Srinivas, P. V. S. S. An improved process for the preparation of 2H-Heptafluoropropane. *Indian Pat. Appl. IN 20090403* (2009); *CAN 152:290542.*
80. Fenichev, I. M., Babenko, Y. I., Bispen, T. A., Vasil'ev, A. S., Moldavsky, D. D. Improvement of the method for synthesis of perfluoroethylisopropylketone. *Fluorine Notes*, **79**, Fen 1–6 (2011); *CAN156: 560157.*
81. (a) Huang, Z., Zhang, Y., Zhao, C., Qin, J., Li, H., Xue, M., Liu, Y. Direct gas-phase epoxydation of hexafluoropropylene with molecular oxygen using Ag catalyst. *Applied Catalysis, A: General, **303(1)**, 18–22* (2006). (b) Peter, H., Wolf-Dietmar, K. Method for preparing heptafluoropropane. *US 6118033* (2000). (c) Miller, R. N., Nappa, M. J., Toton, D. J., Shipley, J. E., et al. Vapor phase production of 1,1,1,2,3,3,3-Heptafluoropropane from hydrogen floride and hexafluoropropylene. *PCT Int. appl. WO 2003037832* (2003).
82. Plunkett, R. J. The History of Polytetrafluoroethylene: Discovery and development. In *High Performance Polymers: Their Origin and Development, Proceed. Symp.*; Seymour, R. B. and Kirshenbaum, G. S. (Eds.); Hist. High Performance Polymers at the ACS Meeting in New York, April 1986; Elsevier: New York, NY, USA. (1987).
83. Fearn, J. E. in *Fluoropolymers,* Wall, L. A. ( Ed.), Wiley-Interscience, New York, p. 3 (1972).
84. Acmite Market Intelligence, Market Report: Global Fluoropolymer Market (2012).

85. Ebnesajjad, S. Introduction to fluoropolyimers. In Applied plastics and engineering hand book, Part 1: Plastics, elastomeric and nanocomposite materials, Kutz, M. (Ed.), Elsevier Inc. (2011), p.49–60.
86. Schultz, N., Martens, P. and Vahlensiek, H. T. *German Pat. 2,659,712* (1976).
87. McBee, E., Hass, H., Bittenbender, W., Weesner, W., Toland, Jr., W., Hausch, W. and Frost. L. *Ind. Eng. Chem.*, ***39***, *409–412* (1947).
88. Kaess, F. and Michand, H. *U.S. pat. 3,600,450* (1971).
89. (a) Plunkett, R. J. *U.S. Patent 2,230,654* (1941). (b) Gangal, S. V. Polytetrafluoroethylene, homopolymers of tetrafluoroethylene. In *Encyclopedia of Polymer Science and Engineering*, 2nd ed.; John Wiley & Sons: New York, NY, USA (1989).
90. Dietrich, M. A. and Joyce, R. M., *U.S. Patent 2,407, 405* (1946).
91. Brasure, D.E. and Ebnesajjad, S. Vinyl fluoride polymers. In *Encyclopedia of Polymer Science and Engineering*, 2nd ed.; John Wiley & Sons: New York (1989).
92. Booth, H. S., Burchfield, P. E., Bixby, E. M. and McKelvey, J. B. Fluorochloroethylenes. *J. Am. Chem. Soc.* ***55***, *2231–2235* (1933).
93. Christopher, C. I. *Thermo Plastic Materials-Properties, Manufacturing Methods and Applications*. CRC Press, pp. *491*(2011).
94. (a) Nersasian, A. Redox catalyst system for preparing vinylidene fluoride interpolymers. *U.S. Pat. 3,467,636* (1969). (b) Bailor, F. V. and Cooper, J. R. Process for isolating fluorine containing polymer. *U.S. Pat. 3,536,683* (1970). (c) Dohany, J. E. and Whiton, A. C.; U.S. Elastomeric fluorinated terpolymer having good thermal stability. *Pat. 3,790,540* (1974). (d) Robinson, D. N. Low-modulus thermoplastic ethylene-tetrafluoroethylene-hexafluoropropene terpolymers. *U.S. Pat. 3,817,951* (1974).
95. (a) Chesire, J. R. Treatment of polymeric shaped structures. U.S. Patent 3,063,882 (1962). (b) Gangal, S. V. Polytetrafluoroethylene. In *Encyclopedia of Chemical Technology*, Vol. 11, Grayson, M. (Ed.), John Wiley & Sons, New York, p. *30* (1978).
96. (a) Nelson, D. A. Tube connection for fluid handling apparatus. U.S. Pat. 3,758,138 (1956). (b) B. Atkinson and A. B. Trenwith, The thermal decomposition of tetrafluoroethylene. *J. Chem. Soc., Part II 2082* (1953).
97. Waddell, J. S. Pyrolysis process for preparing hexfluoropropene from tetrafluoroethyle polymer. *U.S. Pat. 2759983* (1956).
98. Burk, R. E., Hoffman, D. D. and Kalb, G. H., Preparation of difluoroethane. *U.S. Pat. 2,425,991* (1947).
99. Calfee, J. D. and Florio, P. A. Photochemical manufacture of 1,1,1-difluorochloroethane. *U.S. Pat. 2,499,129* (1950).
100. Feasley, C. F. and Storer, W. H. Pyrolysis of difluoromonochloroethane. *U.S. Pat. 2,627,529* (1953).

# CHAPTER 12

# BASE CATALYZED TRANSESTERIFICATION FOR BIODIESEL: NOVEL CATALYST OPTIONS

SHIVA SHANKER KAKI and R. B. N. PRASAD

## CONTENTS

## 12.1 INTRODUCTION

The concept of using vegetable oil as a fuel dates back to 1895 when Dr. Rudolf Diesel developed the first diesel engine to run on vegetable oil. Diesel demonstrated his engine at the World Exhibition in Paris in 1900 using peanut oil as fuel [1]. Biodiesel is the name for alkyl esters of fatty acids made from vegetable oils or animal fats via transesterification. Biodiesel is renewable fuel and exhibits advantages over petroleum derived diesel fuel in terms of emissions and it does not require any modification of the existing diesel engines. It does not contain sulfur and aromatic molecules and the level of toxic emissions is lower compared to that of diesel fuel [2]. The most traditional methodology for the preparation of biodiesel from vegetable oils is transesterification with methanol (Fig. 12.1). Transesterification was conducted as early as in 1853. One of the first uses of biodiesel (transesterified vegetable oil) was powering heavy vehicles in South Africa before World War II. Transesterification is the reaction of oil with an alcohol to form biodiesel and a by-product, glycerol. It is in principle the action of one alcohol displacing another from an ester. As the reaction is reversible, an excess of alcohol is usually used to force the equilibrium to the product side. The stoichiometry for the reaction is 3:1 alcohol to oil; however in practice this is usually increased to 6:1 or more to increase product yield.

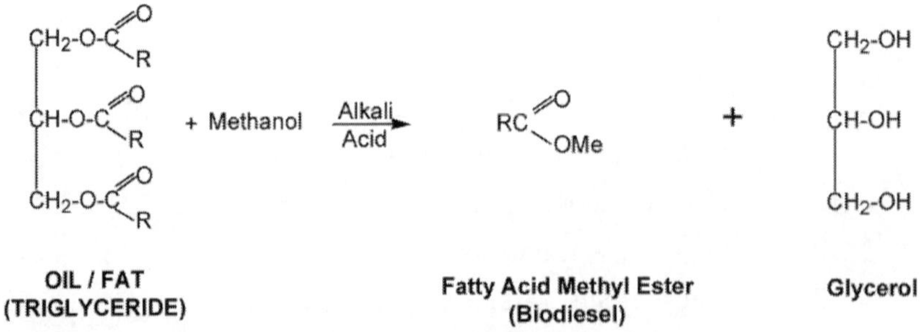


**FIGURE 12.1** Preparation of biodiesel employing transesterification reaction.

In case of high FFA oils, initially the fatty acid has to be esterified using acid catalyst like sulfuric acid (Fig. 12.2) followed by neutralization of the catalyst.

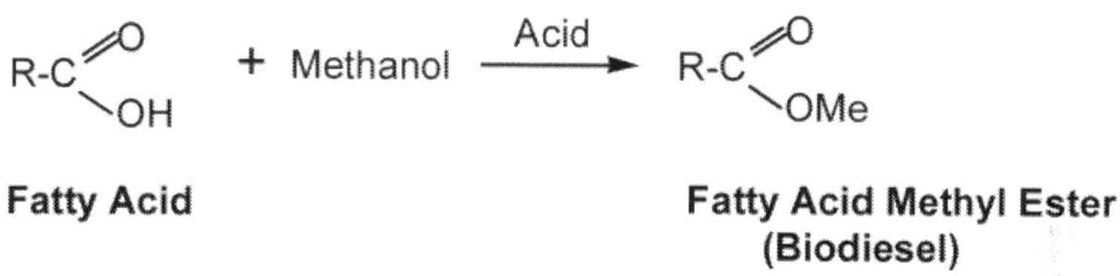


**FIGURE 12.2** Conversion of free fatty acid present in the oils to biodiesel.

The specifications for biodiesel are designed to ensure that consumers will not experience operational problems from the fuel's use. ASTM and European Union have standard specifications designed for B100 FAME used in biodiesel blending, which include GC (gas chromatographic analysis), elemental analysis, as well as physicochemical properties of the produced biodiesel [3, 4]. Quality fuel will provide the consumer with improved air quality and enhanced operability. Poor quality fuel will create operability problems and increased maintenance activity. It is really necessary to look for newer approaches for the production of biodiesel particularly for variable quality feed stock employing nonedible oils or used frying oils.

## 12.2 HOMOGENEOUS CATALYSIS

The crude vegetable oils contain several components like lecithin, free fatty acids, waxes, unsaponifiable matter and pigments. Depending on the content of lecithin the oil has to be degummed before converting it to biodiesel, otherwise the separation of glycerol from the product may create several problems due to the formation of strong emulsion. Bleaching may also be necessary when the oil contains large amounts of pigments. Free fatty acids have to be reduced either by converting them to fatty acid methyl ester or to remove them either by physical or chemical refining process. A typical flow diagram for the production of biodiesel is shown in Fig. 12.3, where the degumming of crude oil as described above is shown and the degummed or clean oil and the esters obtained are transesterified to obtain biodiesel and glycerol.

An acidic or basic catalyst is normally used to catalyze the reaction. The homogeneous alkali catalysts that are generally used include sodium hydroxide or potassium hydroxide or corresponding alkoxides such as sodium methoxide or ethoxide. Sodium hydroxide is the most common alkali used due to economical reasons and availability. Alkali-catalyzed reactions are used more often commercially than acid catalysts, as the reactions are faster [5].

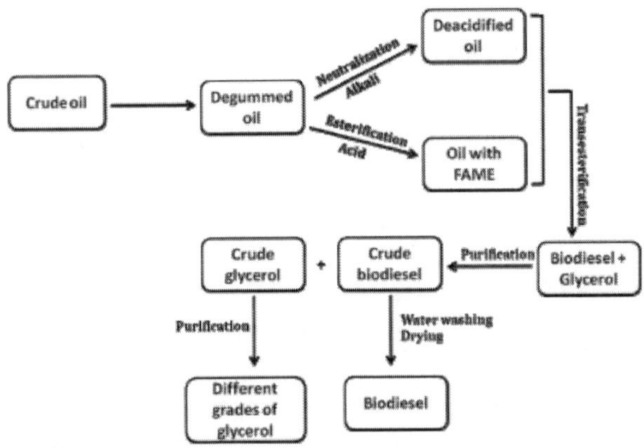


**FIGURE 12.3** Flow diagram of biodiesel production.

Technology for the preparation of biodiesel from low free fatty acid (FFA) containing high quality oils may not be much difficult as it involves only transesterification reaction. In principle, the chemistry of transesterification is very simple and only the efficient conversion rates of oil to fatty acid methyl esters are very important. The main challenge of the biodiesel technology is handling the multi feedstocks with high FFA and unsaponifiables. In the current commercial processes using crude feedstock, required alkali is added to remove all the FFA, in case FFA is less than 1 to 2%. For alkali-catalyzed transesterification, water and FFA are not favorable to the reaction; so anhydrous triglycerides and alcohol are necessary to minimize the production of soap. Soap production decreases the amount of esters and the separation of glycerol and esters becomes difficult.

Traditionally, biodiesel is prepared by the transesterification of triglycerides with methanol using an alkali catalyst. Homogeneous catalysts have the advantages of simple usage and require less time for conversion of oils to their respective esters and they are available at a lower price. Normally methanol and ethanol are the two mail alcohols, which are used. But among these two, methanol is more common because of its low cost and other physical properties. For transesterification with triglyceride, stoichiometrically 1:3 oil to methanol ratio is needed for complete transesterification. However, higher alcohol molar proportions (6 to 15 times) are often used to drive the equilibrium to get maximum yields of the ester [6]. The reaction is normally performed at 60–65° C. The glycerol and biodiesel are separated after the reaction, and glycerol will be removed by settling and the crude glycerol and biodiesel are then purified to match the required specifications.

The major problem of homogeneous acid catalyst like sulfuric acid is the formation of salts during neutralization of the un-reacted catalyst and also conversions are not very high like transesterification. The inefficient conversion rates of acid catalyst may result in the formation of biodiesel without matching the International Standards.

## 12.3 HETEROGENEOUS CATALYSIS

Production of biodiesel employing homogeneous catalysts namely acids for esterification and challenge for biodiesel industry is to handle higher amounts of effluent water and salts generated during homogeneous catalyst-based technologies. Biodiesel has to be thoroughly washed by water after neutralization of the alkali followed by drying to remove the resultant moisture. However, these catalysts are being used in large quantities, which cause various problems such as corrosion, catalyst separation, acid waste, salt formation and environmental pollution. Even though, some research groups are claiming that they have come out with a clean process using solid acid catalyst, it may take some more time to prove the efficacy of these innovative processes. Therefore, there is a need to replace the homogeneous catalysts with heterogeneous catalysts, which are easy to handle and environmental friendly. The by-product glycerol remains diluted with water using homogeneous catalysts and pose a problem while disposing off as a waste product. Glycerol will be almost free of water and its purification is very simple in case of solid acid catalyst-based transesterification. Especially, the pellet form of solid catalyst can be used in fixed-bed reaction processes. In addition, the solid catalysts can be very easily recycled for several times. Moreover, efficient solid catalysts will have excellent economical advantage because of productivity improvement resulting from its good catalytic activity and lowering the cost of equipment. Heterogeneous catalysts are the key to new developments in the production of biodiesel, combined with their application in a continuous process and easy separation. The most important requirement is preliminary catalyst trials to identity right heterogeneous candidates. Application of solid catalysts for both esterification and transesterification will have several advantages over homogeneous catalysts [7].

The esterification of carboxylic acids by heterogeneous acid catalysts is important considering that low-cost oil feedstock contains high concentrations of FFAs. Therefore, it is expected that a good solid acid catalyst must be able to carry out at least esterification so that mineral acid-based esterification can be avoided.

A number of bio-feedstocks like sunflower, soybean, palm, canola, rapeseed, cottonseed, non edible oils like jatropha and karanja, used frying oils, animal fats are currently being experimented for biodiesel production. Fatty acid composition of different vegetable oils and different variety of jatropha oils along with their physicochemical properties, which are important for feed stock selection were discussed in detail [8]. The production technologies for biodiesel from jatropha oils and the comparison of biodiesel properties from different raw materials are also discussed comprehensively. Apart from vegetable oil and animal fats as feedstocks for biodiesel production, algae have emerged as one of the alternative feed stock and are being projected as the most promising sources for biodiesel production. Algae can grow in places away from the farmlands and forests, thus minimizing the damages caused to the eco and food chain systems, in sewages and next to power-plant smokestacks where they digest the pollutants and give us oil. Oil productivity of many microalgae greatly exceeds the oil productivity of the best producing oil crops. The current oil crisis and fast depleting fossil oil reserves have made it more imperative for organizations and countries to invest more time and efforts into research on suitable renewable feedstock such as algae.

Excellent reviews have been published on the production of biodiesel employing different catalysts and articles are also published, which focus on the production technologies from alternative feedstocks [7, 9–13]. This chapter mainly aims at reviewing the status of heterogeneous catalysts for the production of biodiesel with emphasis on base catalyzed biodiesel production. Even though, none of these catalysts are ready for the commercialization of biodiesel proccss, this is a very important area to extend the research activities in a vigorous manner.

### *12.3.1 SINGLE METAL OXIDES*

Single component alkaline earth metal oxides such as Barium oxide (BaO), strontium oxide (SrO), calcium oxide (CaO) and magnesium oxide (MgO) are reported to possess higher basicity, lower solubility in alcohols and have potential for the production of biodiesel in higher yields [14]. Among the metal oxides studied, CaO is one of the widely used oxides for the transesterification of triglycerides for the production of biodiesel mainly due to easy availability and low cost.

Commercial CaO as heterogeneous catalyst was employed for biodiesel production by methanolysis of soybean/rapeseed oil and waste frying oils at 62° C and at atmospheric pressure. The yield of FAME was found to be about

97–98% when soybean/rapeseed oils were used whereas the yields were 86–87% when the waste frying oils were used [15]. The lower yields observed in the case of waste frying oils was due to the presence of free fatty acids, which neutralized the catalyst. The same authors reported the use of CaO for transesterification of semi refined oils, which produced 97% FAME. They also showed that there was a gradual decrease in the yield when oil acidified with oleic acid was used as substrate. The lower FAME yield was due to the neutralization of active sites of the catalyst [16]. In another study, similar results were obtained with transesterification of soybean oil with methanol where the yields obtained were 88% with about 4.5% of FFA in the feedstock and the lower yields are may be due to the presence of FFA in the oil [17].

One of the studies reported the initial activation of CaO by treatment with methanol at ambient temperature for a specific period of time and later addition of rapeseed oil for transesterification with activated CaO at 60° C produced 90% FAME with in 3 h [14]. Here the authors reported that the activation of CaO with methanol produced calcium methoxide, which acts as an initiating reagent. The activity and reusability of CaO in transesterification of sunflower oil with methanol was studied and it was found that CaO was deactivated mostly by carbon dioxide rather than by water. However, the catalyst can be reactivated and the carbon dioxide poisoning can be reverted by heating to >973K. The catalyst was used for eight times and yields of FAME for the first and eighth cycle were 94 and 81% respectively [18].

Nanopowder calcium oxide (Nano CaO) is another form of CaO, which has been tested for biodiesel production. Transesterification of soybean oil with methanol under microwave conditions was found to be more efficient than conventional heating for Nano CaO catalyzed biodiesel production [19] with conversion rates of over 96% with properties within the limits of standard biodiesel.

The reported order of activity of alkaline earth oxide catalysts was observed to be BaO>SrO>CaO>MgO. Among them, MgO was more active than CaO in the transesterification of rapeseed oil, but was also responsible for soap formation. In the transesterification of *Camelina sativa* oil, BaO and SrO were found to be effective catalysts for the synthesis of biodiesel at 100° C compared to CaO and MgO [20]. The authors suggested that SrO can be preferred over BaO due to the noxious nature of BaO. Similar results were also observed for the transesterification of palm oil using CaO, SrO and BaO as heterogeneous base catalyst under ultra sonic assisted preparation of biodiesel. BaO and SrO were found to be efficient catalysts than CaO for production biodiesel with a yield of 95%. With BaO, there was a drop in catalytic activity due to the dissolution of BaO catalyst in methanol [21].

BaO is reported to be harmful/poisonous catalyst and cannot be used as heterogeneous catalyst as it can be solubilized in methanol. Whereas, SrO on the other hand has stronger basic sites and it is insoluble in methanol. [22]. Transesterification of soybean oil using SrO (3%) as solid base catalyst at 65° C resulted in more than 95% of biodiesel in less than 30 min. It was observed that SrO had longer lifetime and activity was retained even after 10 cycles of biodiesel production [22]. In another study, transesterification of canola oil with MgO, CaO, SrO and BaO for biodiesel preparation, it was observed that BaO (1 mol%) showed >95% conversion in 1 h at 50° C [23].

Ba/CaO solid base catalyst was prepared from construction site marble waste and was employed in the transesterification of waste cooking oil to produce biodiesel. With 1:9 molar ratio of oil to methanol and 3% catalyst at 65° C, conversion of 88% was observed in 3 h reaction time. It was shown that the catalyst can be reused for over three cycles and the method is uses less energy and provides value addition to both waste cooking oil and construction waste [24]. Table 12.1 provides a brief description of the single metal oxides as base catalysts for the production of biodiesel.

**TABLE 12.1** Single metal oxides as heterogeneous base catalysts.

| S. No | Oil | Yield (%) | Catalyst | Characterization | Ref. |
|---|---|---|---|---|---|
| | Soybean/rapeseed | 97–98 | CaO | SEM, XRD, FT-IR | 15 |
| | Waste frying | 97 | CaO | XRD, SEM, Hammett indicator | 16 |
| | Soybean | 88.9 | CaO | — | 17 |
| | Rapeseed | 90 | CaO | XRD, Hammett indicators | 14 |
| | Sunflower | 80–94 | CaO | XRD, EGA-MS, XPS, FT-IR | 18 |
| | Soybean | 96 | Nanopowder CaO | BET, FT-IR | 19 |
| | *Camelina sativa* | 80 | BaO, SrO, MgO, CaO | SEM, XRD, BET | 20 |
| | Palm | 95 | CaO, SrO, BaO | Hammett indicators, SEM | 21 |
| | Soybean | 95 | SrO | BET | 22 |
| | Canola | 95 | MgO, CaO, SrO, BaO | XRD | 23 |
| | Waste cooking oil | 88 | Ba meliorated construction site waste marble | SEM, BET, EDX | 24 |

Among the single metal oxides, it is observed that CaO and SrO are widely used where as BaO is reported to be an efficient catalyst in terms of conversion. However, several authors suggested that BaO application is limited due to its poisonous nature and therefore, it appears that SrO could be a better choice for developing potential improved base catalyst for biodiesel production.

### 12.3.2 OXIDES WITH OTHER MATERIALS

Modified MgO with potassium hydroxide (20 wt% of KOH loaded on MgO) was reported to show high basicity and the transesterification of canola oil with methanol at reflux temperature with 3% of catalyst resulted in 95% of biodiesel yield [25]. In another study, CaO supported on silicas was prepared as a heterogeneous catalyst for biodiesel synthesis, which was claimed to be superior to other CaO catalyst in terms of leaching of CaO into the reaction medium. It was observed that conversion rates for sunflower and castor oils were about 95 and 68%, respectively [26].

Another oxide of calcium is calcium methoxide, which is reported to catalyze transesterification of soybean oil with methanol to produce biodiesel with 98% yield. It was shown that calcium methoxide had high basicity and activity and was active even after 20 times of recycling [27]. CaO when dipped in ammonium carbonate solution and further calcination at 900° C forms a base called solid super base, which is reported to have a basicity of 26.5. Transesterification of *Jatropha curcas* oil with 1.5% of solid super base of calcium at 70° C for 2.5 h with oil to methanol ration of 1:9 yielded 93% conversion. The authors also studied the purification of the produced biodiesel and showed that citric acid helped in decalcification [28]. CaO impregnated with 10% $KNO_3$ was examined as a heterogeneous catalyst for the transesterification of rapeseed oil with methanol. It was observed that 1% of this catalyst was able to convert 98% of triglyceride into biodiesel at 65° C with oil to methanol ratio of 1:6 with 3 h of reaction time. [29].

Zirconia ($ZrO_2$) loaded with potassium bitartrate ($C_4H_4O_6HK$) was prepared as a solid base nanocatalyst and was examined for transesterification of soybean oil to produce biodiesel. It was found that oil to methanol ratio of 1:16 and a temperature of 60C° with 6% catalyst was able to produce biodiesel with 98% yield [30]. This solid base nanocatalyst had a porous structure with particle sizes of 10–40 nm and was found to be effective even after 5 cycles.

Potassium fluoride (KF) loaded on to rare earth oxide like Europium Oxide ($Eu_2O_3$) was used as a super base catalyst for the transesterification of rapeseed oil to produce biodiesel. It was observed that 15% KF loaded $Eu_2O_3$ was an efficient catalyst for producing 92.5% of biodiesel at reflux temperature of methanol in 1 h and the oil to methanol ratio was 1:12 with 3% of catalyst [31].

Alkali metal ion (Li, Na and K) impregnated CaO catalysts were found to be effective in the transesterification of used vegetable oils with high moisture and free fatty acid content. It was observed that CaO impregnated with lithium carbonate, $Li_2CO_3$ (LiC-CaO-1.5) as a nanocatalyst was having the highest basic strength and the catalyst was able to convert cotton seed oil completely into biodiesel in 45 min at 65° C with oil to methanol ratio of 1:12 [32].

Alkaline earth metal-doped zinc oxide (ZnO) was examined as a catalyst for the transesterification of soybean oil to produce biodiesel. ZnO loaded with (strontium nitrate) Sr $(NO_3)_2$ followed by calcinations was found to be effective and a conversion of 94.7% was observed in the transesterification of soybean oil at reflux temperature of methanol with 5% of catalyst. The authors reported that when THF was used as a cosolvent, the conversion increased to 96.8%. However, a decrease in catalytic activity was observed when THF was used as a cosolvent [33].

It can be observed that metal oxides in combination with other materials such as silica, with other alkali and alkaline earth metal-doped type of catalysts are promising agents for higher conversion of triglycerides to biodiesel with lesser amounts of catalyst. All of these oxide combinations with other materials discussed above have been tabulated in the following Table 12.2.

**TABLE 12.2** Oxides in combination of other materials as heterogeneous catalysts.

| S. No. | Oil | Yield (%) | Catalyst | Characterization | Ref. |
|---|---|---|---|---|---|
| | Canola | 95 | MgO/KOH | XRD, SEM, Hammett indicator | 25 |
| | Sunflower, castor | 95 | CaO on mesoporous silica | XPS, SEM, DRX, FT-IR | 26 |
| | Soybean | 98 | Ca methoxide | SEM, BET, FT-IR, | 27 |
| | Jatropha | 93 | Super base of CaO | XRD, SEm, BET | 28 |
| | Rapeseed | 98.5 | KNO3/CaO | — | 29 |
| | Soybean | 98 | $ZrO_2/C_4H_4O_6HK$ | XRD, FT-iR, TEM, TGA, Hammett indicator | 30 |

**TABLE 12.2** *(Continued)*

| S. No. | Oil | Yield (%) | Catalyst | Characterization | Ref. |
|---|---|---|---|---|---|
| | Rapeseed | 92.5 | $KF/Eu_2O_3$ | XRD, FT-IR, TPD, Hammett titration | 31 |
| | Cotton seed | 99 | Li/CaO | XRD, BET, FESEM, TEM, FT-NMR | 32 |
| | Soybean | 94 | $ZnO/Sr(NO_3)_2$ | DTA-TG, IR, Hammett indicator | 33 |

## 12.3.3 MIXED OXIDE CATALYSTS

Transesterification of palm kernel oil with mixed oxides of Ca and Zn (10%) with 30 molar equivalents of methanol at 60° C was shown to produce 94% of FAME in 1 h. This catalyst was also employed for the transesterification of palmolein, soybean and sunflower oils and the reusability studies showed that the catalyst can be used for 3 batches without significant loss of activity [34]. When sunflower was used for transesterification with methanol using CaO supported on zinc oxide, biodiesel yield of 90% was obtained. The conditions for biodiesel production were, oil: methanol ratio was 12 with 1.3% of catalyst with a 16% of CaO at a temperature of 65° C and reaction time of 2 h was found to be optimum [35]. The authors observed that the method of preparation of ZnO-CaO was important as it had an influence on the physic-chemical characteristics of the catalyst and on the over process of transesterification. The binary oxide of CaO-MgO catalyst was tested for transesterification of *Jatropha curcas* oil to biodiesel. It was observed that a biodiesel yield of 90% was obtained with 3 wt% of catalyst at 120° C with 3 h of reaction time and 1:25 of oil to methanol ratio. This catalyst can be used for 4 times with a biodiesel yield of >80% [36].

MgAl and MgCa oxides were found to be efficient catalyst for transesterification of ethyl butyrate and methanol and for methanolysis of sunflower oil. With 1:12 of oil and methanol and 2.5% of catalyst at a temperature of 60° C, 92% of yield of FAME was obtained. The best ratio of Mg:Ca was found to be 3.8 and there was no leaching of catalyst into the reaction medium [37].

Thirteen kinds of metal oxides containing calcium, barium, magnesium and lanthanium were tested for transesterification of triglycerides. Among the tested catalysts, $CaZrO_3$ and $CaO\text{-}CeO_2$ heterogeneous base catalysts were effective catalyst for the transesterification of rapeseed oil at a temperature of 60° C and reaction time of 10 h to produce biodiesel in 90% yield [38].

CaO and $ZrO_2$ mixed oxides based heterogeneous catalyst was used for transesterification of waste cooking oil at 65° C with 2 h reaction time at 30:1 methanol to oil ratio yielded biodiesel in 92%. It was observed that the activity of catalyst increase with increasing Ca to Zr ratio but the stability of the catalyst decreased with increasing ratio of Ca to Zr and the optimum Ca to Zr molar ratio was found to be 0.5 [39].

Nanocomposite mixed oxides of (strontium-titanium) Sr-Ti prepared by sol-gel method was used a heterogeneous catalyst for the transesterification of soybean oil to produce biodiesel. It was found that 1% of the catalyst afforded biodiesel in 98% yield within 15 min at a temperature of 60° C and oil to methanol ratio of 1:15. It was found that Sr/Ti molar ratio of 0.8 and the calcination temperature of 965° C was the most efficient catalyst for transesterification of soybean oil to biodiesel [40].

Aluminium oxide modified Mg-Zn heterogeneous catalyst was tested for the transesterification of crude jatropha oil to produce biodiesel. It was found that the catalyst yielded 94% of methyl esters in 6 h with oil-methanol ratio of 1:11 at 182° C temperature with a catalyst amount of 8.68%. The study showed that catalyst was selective for the production of methyl esters from oil containing high free fatty acids and water in a single stage process, which is very useful when dealing with feed stocks with high free fatty acid content [41]. A brief summary of the mixed oxide catalysts for biodiesel production are given in Table 12.3.

**TABLE 12.3** Mixed oxide catalysts as heterogeneous base catalysts.

| S. No. | Oil | Yield (%) | Catalyst | Characterization | Ref. |
|---|---|---|---|---|---|
| | Palm kernel/palmolein/soybean | 94 | Ca/Zn mixed oxide | XRD, SEM, TGA | 34 |
| | Sunflower | 90 | CaO/ZnO | XRD, XPS, FT-IR, SEM | 35 |
| | Jatropha | 75–90 | Ca-MgO | XRD, BET, SEM, EDS | 36 |
| | Sunflower | 92 | MgAl/MgCA oxides | XRD, XPS, SEM, N2 adsorption | 37 |
| | Rape seed | 79–92 | $CaTiO_3$, $CaMnO_3$, $CaZrO_3$, Cao-$CeO_2$ | BET, XRD, Hammett indictor | 38 |
| | Waste cooking oil | 92.1 | Ca/Zr oxides | XRD, XRF, XPS, BET | 39 |

**TABLE 12.3** *(Continued)*

| S. No. | Oil | Yield (%) | Catalyst | Characterization | Ref. |
|---|---|---|---|---|---|
| | Soybean | 98 | Sr-Ti oxide nanocomposite | XRD, SEM, FT IR, TGA/DTA | 40 |
| | Jatropha | 94 | Al/Mg-Zn oxide | BET, N2 adsorption, XRD, SEM, DTA | 41 |

### 12.3.4 BASE CATALYSTS WITH ALUMINA ($Al_2O_3$)

Alumina loaded with potassium as solid base catalyst was prepared and tested for the transesterification of soybean oil. Potassium nitrate ($KNO_3$–35 wt%) on $Al_2O_3$ followed by calcinations at 773K for 5 h was found to be the active catalyst for transesterification at reflux temperature of methanol with oil to methanol ratio of 1:15, catalyst amount of 6.5% with 7 h of reaction time, which gave a conversion of 87% [42]. It was postulated that $KNO_3$ transformed to $K_2O$ or Al-O-K composites, which were proposed to be present in the active sites of the catalyst. In a similar type of study, $KNO_3$ (35 wt%) loaded on $Al_2O_3$ was used as catalyst for transesterification of jatropha oil gave 84% conversion with oil to methanol ratio of 1:12 at 70° C and the authors showed that the catalyst can be used for three times [43].

In a study conducted on the K/$Al_2O_3$ catalyst derived from $K_2CO_3$ precursor for triglyceride transesterification with methanol, the authors reported that the first run gave 99% FAME. With 5.5% of the prepared catalyst and at 60° C with oil to methanol ratio of 1:25, a near 100% yield was obtained within 1 h. However, the yield of FAME decreased during the subsequent reuse of the catalyst, which was due to the leaching of the catalyst into methanol. The authors also reported that the high yield in the first run was probably due to the leaching of potassium species into methanol, which served as homogeneous catalysis medium [44]. KF/$Al_2O_3$ solid-base catalyst for transesterification of palm oil was examined at 65° C with oil to methanol ratio of 1:12 with 3 h of reaction time. With 4 wt% of catalyst, FAME yield of 90% was observed with optimized reaction conditions [45]. The same authors reported the preparation of biodiesel from cotton-seed oil with KF/$Al_2O_3$ and they have observed a new $K_3AlF_6$ like phase in the calcined catalyst [46]. They also reported the transesterification of soybean oil using the similar catalyst with a yield of 99% with 2% of catalyst [47]. KI/$Al_2O_3$ prepared by impregnation of alumina with KI (35 wt%) followed by calcinations formed a solid base catalyst. This alumina loaded KI catalyst (2.5%) was used in the transesterification of soybean oil with methanol at a ratio of 1:15 at reflux temperature for 3 h resulted in 96% conversion [48].

K/KOH/$Al_2O_3$ as heterogeneous base catalyst was employed for the transesterification of rapeseed oil with methanol. Under optimized conditions of 1:9 of oil to methanol ratio at 60° C with 4% catalyst produced 84.5% yield of biodiesel in 1 h [49]. Though the catalyst was recovered by filtration, the reusability studies were not encouraging as the biodiesel yield with recycled catalyst decreased to 37.6%. The decrease in yield was observed due to the leaching of active species of catalyst into methanol. KOH loaded on $Al_2O_3$ and NaY zeolite support as heterogeneous was used for the transesterification of palm oil to biodiesel. With 25 wt% KOH/$Al_2O_3$ and 10% KOH/NaY, a biodiesel yield of 91.07% was achieved at 70° C in 3 h at 1:15 molar ratios of palm oil to methanol with 3–6% of catalyst. However, leaching of potassium species with both catalysts was observed, which decreases the biodiesel yield [50].

Me/$Al_2O_3$ (Me=Na, Ba, Ca and K) and monolithic K/$Al_2O_3$ catalysts were used for the transesterification of soybean oil and methanol 1:32 molar ratios at 120C° temperature with 1% of Me/Al2O3 and 0.5% of monolithic catalyst. It was observed that Na, Ca and K/$Al_2O_3$ produced very good biodiesel yields where as Ba/$Al_2O_3$ was not effective and monolithic catalyst gave 59.1% of biodiesel [51].

In another study, $Al_2O_3$ supported alkaline catalysts were examined for the transesterification of canola oil with methanol. Among the catalysts examined, it was observed that KOH/$Al_2O_3$ was found to be the best yielding 89.4% biodiesel with oil to methanol ratio of 1:6 at 60° C for 9 h with 3% catalyst [52].

Sodium aluminate was tested for the methanolysis of soybean oil at reflux temperature with 1:12 molar ratio of soybean oil to methanol for 1 h with 1.5% catalyst loading. It was observed that, 93.9% yield of biodiesel was obtained and the reusability studies showed a decrease in activity. It was also shown that, if the catalyst was used without calcinations before each batch, the biodiesel yields were very poor where as reuse after calcinations gave yields above 60%. The lower yields can be due to leaching of aluminate into methanol phase [53].

Na/NaOH/$Al_2O_3$ heterogeneous base catalyst for transesterification of vegetable oil to biodiesel was reported to show same activity compared to homogeneous NaOH catalyst. Under optimized conditions, with 1:9 of soybean oil to methanol and a temperature of 60° C with hexane as cosolvent, biodiesel production yield was observed to be 94% [54]. In another study, NaOH/$Al_2O_3$ catalyst was examined for the transesterification of sunflower oil with methanol in 1:6 molar proportions at 323K yielded 80% biodiesel. It was observed that about 5% leaching of sodium was observed after 9 h of test for leaching, which does not influence the reaction much [55].

$Al_2O_3$ supported CaO and MgO catalysts were studied for the transesterification of lipids of yellow green micro algae, *Nannochloropsis oculata* with methanol at 50° C. It was observed that pure CaO and MgO did not show any activity and among the tested catalysts, CaO/$Al_2O_3$ was found to be showing highest activity with a biodiesel yield of 80% with 2% of catalyst. When methanol molar ratio was increased to 30, the yield was found to be 97.5% [56]. It was also reported that activity of catalyst was not due to crystal size of CaO in $Al_2O_3$ but was because of basic sites and their density. It was observed that high basic site density and mild basic strength are required for higher biodiesel yield.

Alkali earth oxides such as CaO, BaO, MgO and $K_2CO_3$ supported on alumina catalysts were used for the transesterification of canola oil with methanol. Among the oxides, BaO was found to be effective yielding 85% of biodiesel but due to poisonous nature, it was not considered for further studies. It was found that $K_2CO_3$/$Al_2O_3$ was the best catalyst in terms of both ester yields and being environmentally friendly in nature. Based on response surface methodology, it was found that 1:11.5 ratio of oil to methanol with 3.16% catalyst at a reaction temperature of 60° C were found to be optimum conditions for obtaining 96.3% biodiesel in 2 h. [57].

$Eu_2O_3$/$Al_2O_3$ was used as a super base novel solid catalyst for the transesterification of soybean oil with methanol. However, a biodiesel yield of 63.2% was obtained with oil to MeOH ratio of 1:6 at 70° C for 8 h with 5% catalyst [58]. Table 12.4 describes the base catalysts prepared in combination of $Al_2O_3$ for biodiesel production from various oils.

Base catalysts in combination of alumina have been studied extensively for biodiesel production. However, most of the work has been reported on the modification of alumina with potassium and sodium containing base catalysts. Further research with other base combination with alumina could be interesting.

**TABLE 12.4** Base catalysts in combination with Alumina as heterogeneous catalysts.

| S. No. | Oil | Yield (%) | Catalyst | Characterization | Ref. |
|---|---|---|---|---|---|
| | Soybean | 87 | $K/Al_2O_3$ | XRD, IR, Hammett titration | 42 |
| | Jatropha | 84 | $KNO_3/Al_2O_3$ | Surface area analysis | 43 |
| | Sunflower | 99 | $K/Al_2O_3$ | ICP-AES, EGA-MS | 44 |
| | Palm | 90 | $KF/Al_2O_3$ | XRD, TGA | 45 |
| | Cotton seed | 90 | $KF/Al_2O_3$ | XRD, DSC-TGA | 46 |
| | Soybean | 99 | $KF/Al_2O_3$ | XRD, DSC-TGA | 47 |
| | Soybean | 96 | $KI/Al_2O_3$ | XRD, IR, SEM, Hammett indicator | 48 |
| | Rapeseed | 84.5 | $K/KOH/Al_2O_3$ | XRD, BET, IR, XPS | 49 |
| | Palm | 91 | $KOH/Al_2O_3$ and KOH/NaY | XRD, BET, XRF, $CO_2$-TPD | 50 |
| | Soybean | 94–97 | $Me/Al_2O_3$ and $K/Al_2O_3$ cordierite | XRD, BET, FT IR | 51 |
| | Canola | 89.4 | $Al_2O_3/KOH$ | Hammett indicators/ XRD, BET | 52 |
| | Soybean | 93.9 | Sodium aluminate | ICP-OES, TG/DTG, XRD, FTIR, Hammett indicator | 53 |
| | Soybean | 70 | $Na/NaOH/Al_2O_3$ | BET, XRD, TPD, TCD | 54 |
| | Sunflower | 80 | NaOH/alumina | XRD, BET | 55 |
| | Nannochloropsis oculata lipid | 97.5 | $Al_2O_3/CaO$ and MgO | XRD, $CO_2$-TPD | 56 |
| | Canola | 85 | CaO, BaO, $MgO/Al_2O_3$ | BET, ICP-MS, TPD, | 57 |
| | Soybean | 63.2 | $Eu_2O_3/Al_2O_3$ | BET, Hammett scale | 58 |

### 12.3.5 HYDROTALCITES

Mg-Al hydrotalcites (HTC) as heterogeneous base catalyst was studied for transesterification of soybean oil with methanol. It was found that oil to methanol ratio of 1:12 at a temperature of 60–65° C with 2.5% of catalyst and reaction time of 4 h were the optimized conditions. The characterization of the catalyst by (X-ray diffraction) XRD pattern showed that MgO was the main substance present in the structure [59]. Mg-Al hydrotalcite derived catalyst was examined for transesterification of poultry fat for the production of biodiesel. A temperature range of 60–120° C and a methanol ratio of 1:60 gave 94% yield of biodiesel in 6 h [60]. It is interesting to observe that contact of poultry fat with catalyst reduced the catalytic activity but recalcination of the catalyst restored the catalyst activity. This study shows that by-product can be converted to very important value added biofuel.

Mg-Al hydrotalcite catalysts were used for the transesterification of rapeseed oil with oil to methanol ratio of 1:6, at a temperature of 65° C with 1.5% catalyst in reaction time of 4 h. It was found that calcined hydrotalcite with Mg/Al molar ratio of 3.0 showed highest transesterification activity with a conversion of 90.5% and the catalyst could be separated easily [61]. Similar types of catalysts were employed for the transesterification of soybean oil with methanol. Oil to methanol ratio of 1:15 at reflux temperature for 9 h with 7.5% catalyst yielded 67% of biodiesel [62]. In a similar study for the transesterification of canola oil with methanol using Mg-Al hydrotalcite catalyst prepared by coprecipitation method, 71.9% conversion was observed. The conditions followed were 1:6 ratio of oil to MeOH 60° C for 9 h with 3% catalyst [63]. Cerium modified Mg-Al hydrotalcites were used for the transesterification of soybean oil with methanol. With a molar ratio of 1:9 of oil to methanol, at a temperature of 67° C and 5% of catalyst and reaction time of 3 h, a biodiesel yield of 90% was obtained. It was observed that catalysts with Ce/Mg atomic ratio of >0.3 was able to produce biodiesel of 90% yield [64]. Potassium loaded–calcined Mg-Al hydrotalcite was used as a base catalyst for the transesterification of palm oil with methanol. Biodiesel yield of 86.8% was achieved with 7% catalyst at 100° C for 6 h using 1:30 molar ratios of oil to methanol. The blend with prepared biodiesel was tested for its impact on elastomer properties [65].

Mg-Al-$CO_3$ hydrotalcite catalyst was used in the transesterification of refined and acidic cottonseed oil for the production of biodiesel. The catalyst was found to exhibit high activity for methanolysis and esterification in refined and acidic cottonseed oil. With oil to methanol ratio of 1:6 and 1% catalyst at a temperature of 180–200° C resulted in 80% of biodiesel yield [66].

Hydrotalcite-like compounds (MgAlFe) containing Mg2+, Fe3+ and Al3+ were used for the transesterification of soybean oil with methanol to produce biodiesel. Under optimized conditions of oil to methanol ratio of 1:21 and a temperature of 65° C with 3% catalyst and 4 h of reaction period resulted in 85% conversion [67]. In a similar study, hydrotalcite like compounds containing Mg2+, Ni2+ and Al3+ layered double hydroxide catalysts were used for transesterification with similar conditions resulted in 87% conversion with 0.3% catalyst [68].

Mg-Co-Al-La layered double hydroxide was examined for the transesterification of canola with ethanol. With 1:16 ratio of oil to ethanol and 2% of catalyst and at a temperature of 200° C or 473K in a reactor was observed to produce 97% of biodiesel in 4 h [69]. Calcined Li-Al, Mg-Al and Mg-Fe layered double hydroxides for the transesterification of both glyceryl tributyrate and soybean oil with methanol. With 1–3% of catalyst and oil to methanol ratio of 1:15 and at reflux temperature resulted in about 83% biodiesel yield [70].

KF/Ca-Al hydrotalcite solid base catalyst was used for the transesterification of palm oil with methanol. With oil to methanol ratio of 1:12 and temperature of 65° C with 5% catalyst produced biodiesel in 97.9% in 5 h. It was observed that loading of Ca-Al oxide with KF improved the activity of the catalyst [71]. KF/hydrotalcite solid base catalyst was tested for transesterification of palm oil with methanol to produce biodiesel. The optimum conditions were oil to methanol ratio of 1:12 at 338K temperature with 3% catalyst. The authors reported that 85% FAME yield was obtained at 3 h and the yield increased to 92% when the reaction time was increased to 5 h [72].

In another study, heterogeneous catalysts such as Mg/MCM-41, 10K/$ZrO_2$ and 20K/ $ZrO_2$ were tested for transesterification of rapeseed oil along with MgAl hydrotalcite catalyst. It was observed that among the heterogeneous MgAl hydrotalcite gave highest yields in both mechanical stirring mode and ultrasonication. With oil to alcohol ratio of 1:12 and temperature of 60° C the yield of 97% was obtained with mechanical stirring during 24 h of reaction time and 96% yield was obtained in 5 h of ultrasonication [73].

Both MgO and calcined hydrotalcites were used for transesterification of soybean oil with methanol. It was observed that hydrotalcite catalyst required 100° C where as other base catalysts required >180° C for transesterification reaction for obtaining good yields. With oil to methanol ratio of 1:10 with 5% catalyst and a reaction time of 1 h were found to be optimum conditions. It was observed that both hydrotalcite and MgO catalysts were unaffected by the presence of water, which is important while dealing with feedstocks containing moisture [74].

Magnesia-rich magnesium aluminate spinel ($MgO \cdot MgAl_2O_4$) were tested as solid base catalysts for the transesterification of soybean oil with methanol. With oil to MeOH ratio of 1:25 and 3% catalyst at 65° C, biodiesel with 60% yield was obtained in 10 h. The enhanced catalytic activity of the catalyst was suggested to be due to its higher basicity, pore volume and size and surface area compared to other catalyst of same Mg loading [75].

Poly (vinyl alcohol) membranes loaded with hydrotalcite was tested for the methanolysis of soybean oil to produce biodiesel. About 90% biodiesel was obtained with 1:12 ratio of oil to methanol at a temperature of 60° C with 1.3% catalyst for 24 h. It was observed that these catalysts were more active than the free hydrotalcite and can be reused for several runs [76].

Zn/Al complex oxides catalyst derived from a hydrotalcite-like precursor with a zinc/aluminum atom ratio of 3.74:1 was used for the transesterification of rapeseed oil with methanol. With oil to methanol ratio of 1:24 at 200° C and 2.5 MPa with 1.4% catalyst, conversion of 84% was observed in 1.5 h [77].

In one of the study transesterification of soybean oil with methanol over basic solid catalysts such as: MgO, ZnO, $Al_2O_3$, and mixed oxides derived from hydrotalcite like compounds (Mg/Al and Zn/Mg/Al) were investigated. With oil to MeOH ratio of 1:55 and a temperature of 130° C, a biodiesel yield of 60% was observed in 7 h [78].

Lipase (*Saccharomyces cerevisiae*) immobilized on Mg-Al hydrotalcite was examined for the transesterification of rapeseed oil with methanol. With oil to methanol ratio of 1:4–5 at 45° C with 1.5% of catalyst gave a conversion of 96% in 4.5 h. This study showed that Mg-Al hydrotalcite is a good support for lipase immobilization as the activity of the lipase was retained after immobilization [79].

Hydrotalcite catalysts or layered double hydroxides as catalysts are reported to give high conversion and yields in transesterification of triglycerides. Development of hydrotalcite materials for immobilization of biocatalysts for transesterification can overcome the need for higher temperature for hydrotalcite-catalyzed transesterification, which usually requires higher molar ratio of methanol. Table 12.5 summarizes the applications of hydrotalcite and hydrotalcite like compounds as heterogeneous base catalysts for the production of biodiesel.

**TABLE 12.5** Hydrotalcites as heterogeneous base catalysts.

| S. No. | Oil | Yield (%) | Catalyst | Characterization | Ref. |
|---|---|---|---|---|---|
| | Soybean | 97 | Mg-Al hydrotalcite | SEM, XRD, FT-IR, TG/DSC | 59 |
| | Poultry fat | 94 | Mg-Al hydrotalcite | BET, ICP, XDS, SEM-EDS | 60 |

**TABLE 12.5** *(Continued)*

| S. No. | Oil | Yield (%) | Catalyst | Characterization | Ref. |
|---|---|---|---|---|---|
| | Rapeseed | 90.5 | Mg-Al hydrotalcite | XRD, FT-IR, TGA-DTA, SEM | 61 |
| | Soybean | 67 | Mg-Al hydrotalcite | SEM, XRD, IR, DTA-TG, Hammett titration | 62 |
| | Canola | 71.9 | | BET | 63 |
| | Soybean | 90 | Ce/ Mg-Al hydrotalcite | SEM, XRD, XPS, FT-IR, TG-DTA-DSC, Hammett indicator | 64 |
| | Palm | 86.6 | K/ Mg-Al hydrotalcite | XRD, BET | 65 |
| | Cottonseed | 80 | Mg-Al-$CO_3$ hydrotalcite | — | 66 |
| | Soybean | 85 | HTC containing Mg/Fe/Al | XRD, TGA, FE-SEM, FT-IR | 67 |
| | Soybean | 87 | HTC containing Mg/Ni/Al | XRD, TGA | 68 |
| | Canola | 96–97 | Mg-Co-Al-La | XRD, XPS, particle analyzer | 69 |
| | Soybean | 77–83 | Li-Al, Mg-Al and Mg-Fe | ICP-OES, BET, XRD, XPS, SEM, Hammett indicators | 70 |
| | Palm | 97.9 | KF/Ca-Al | XRD, SEM | 71 |
| | Palm | 85–92 | KF/Hydrotalcite | XRD | 72 |
| | Rapeseed | 97 | Mg-Al and K/ Zirconia | XRD, SEM, EDS | 73 |
| | Soybean oil | 78–92 | MgO hydrotalcite | BET, XRD | 74 |
| | Soybean | 60 | MgO.Mg $Al_2O_4$ | XRD, TEM, HR-TEM, SEM, elemental analysis | 75 |
| | Soybean | 90 | PVA hydrotalcites | XRD | 76 |
| | Rapeseed | 84 | Zn/Al | XRD, XPS, FT-IR, TG-DTA, BET | 77 |
| | Soybean | 60 | Zn/Mg/Al HTC derived MgO, ZnO, $Al_2O_3$ | XRD, XRF, DTA, TGA | 78 |
| | Rape seed | 96 | Lipase immobilized on Mg-Al HTC | Lipase activity | 79 |

### 12.3.6 OTHER BASE CATALYSTS

NaX zeolite loaded with 10% KOH was used as a solid base catalyst for the transesterification of soybean oil with methanol. Best results were obtained with oil to MeOH ratio of 1:10 at reflux temperature for 1.5 with 3% catalyst, which resulted in 85.6% conversion. It was observed that the basic strength of parent NaX zeolite increased by the loading of KOH, which helped the transesterification [80].

In another study, neodymium oxide ($Nd_2O_3$) loaded with potassium hydroxide (30 wt%) was used as heterogeneous base catalyst for transesterification of soybean oil to produce biodiesel. With oil to MeOH ratio of 1:14 at 60° C with 6% catalyst, biodiesel yield of 92.41% was achieved in 1.5 h. It was showed that the catalyst was environmentally friendly and noncorrosive with longer lifetime and was used for five times without losing its activity [81].

Transesterification of oil from waste oilseed fruits with methanol was studied in both homogeneous and heterogeneous catalyzed systems. Potassium loaded on ITQ-6 (Delaminated zeolite ITQ-6) by ionic exchange was found to be an efficient base catalyst, which produced 87% biodiesel. The optimum conditions were found to be 1:20 ratio of oil to MeOH at 180° C with 5% catalyst with reaction time of 48 h. Though potassium leaching was observed, which led to deactivation of the catalyst, its regeneration and reusability were easy to perform [82]. The catalyst was also reported to be active in the conversion of oil containing free fatty acids (5.58%) without any soap formation and functions as acid/base catalyst.

1,1,3,3-Tetramethylguanidine (TMG) immobilized on mesoporous SBA-15 silica (SBA-15-pr-TMG) was examined for the transesterification of soybean oil with methanol.

With oil to MeOH ratio of 1:15 at reflux temperature for 12 h with 5% catalyst resulted in 91.7% conversion and it was reported that the catalyst can be recovered and reused without significant loss of activity [83].

A novel layered compound zinc hydroxide nitrate ($Zn_5(OH)_8(NO_3)_2.2H_2O$) heterogeneous catalyst was reported, which can be used for both esterification of free fatty acids and transesterification of vegetable oils. With oil/FFA to MeOH ratio of 1:6 and a temperature of140° C and 2 h reaction time, 95% biodiesel yield was reported with 2% catalyst in both the reactions [84]. It was shown that the catalyst is suitable when dealing with high FFA containing feed stocks and also the catalysts is nontoxic and recyclable.

KF loaded on to oxide supports containing Mg, Cu, Zn, Co, Al, Cr, Ni or Fe were studied for the transesterification of vegetable oil with methanol to produce biodiesel. With 1:6 of oil to MeOH ratio and temperature of 65° C with 3% catalyst, a biodiesel yield of 90% was observed with the catalyst containing KF whereas the oxides alone did not show transesterification [85].

Heterogeneous strong basic Mg/La catalyst was tested for the transesterification of edible and nonedible oils for producing biodiesel. With oil to MeOH ratio of 1:2.5 and catalyst of 20% complete conversion was observed within 0.3 h at 65° C and 2.2 h at room temperature [86]. The catalyst was shown to exhibit good activity for transesterification even in the presence of water and FFA at room temperature.

$Na_{0.1}Ca_{0.9}TiO_3$ nanorods were used as heterogeneous base catalysts for the transesterification of *Camelina sativa* oil to biodiesel. With oil to MeOH ratio of 1:36 at 60° C with 6% catalyst, a biodiesel yield of 93% was achieved in 8 h [87].

Modified dolomites ($CaMg(CO_3)_2$) were examined as base catalysts for the transesterification of palm kernel oil with methanol at 60° C. It was observed that 1:50 molar ratio of oil to MeOH with 10% catalyst and 3 h of reaction time was necessary for obtaining 96.1% of biodiesel yield [88].

*KOH* supported on *SBA* 15 was used as a heterogeneous base catalyst for the transesterification of palm oil with methanol. It was found that 1:11.6 ratio of oil to MeOH and a temperature of 70° C for 5 h with a catalyst amount of 3.91% resulted in 83.7% of biodiesel [89]. Table 12.6 gives brief information of base catalysts, which are not derived from metal oxides and hydrotalcites.

**TABLE 12.6** Base catalysts other than metal oxides and hydrotalcites.

| S. No. | Oil | Yield (%) | Catalyst | Characterization | Ref. |
|---|---|---|---|---|---|
| | Soybean | 85.6 | NaX zeolite with KOH | XRD, SEM, FT-IR, Hammett indicators | 80 |
| | Soybean | 92.4 | Nd2O3 | XRD, FT-IR, SEM, TGA, Hammett indicators | 81 |
| | Oil | 87 | K-ITQ-6 | XRD, ICP-MS, IR | 82 |
| | Soybean | 91.7 | TMG immobilized on mesoporous silica | XRD, FT-IR, SEM, TEM, Hammett titration | 83 |

**TABLE 12.6** *(Continued)*

| S. No. | Oil | Yield (%) | Catalyst | Characterization | Ref. |
|---|---|---|---|---|---|
| | Palm | 95 | Zinc hydroxide nitrate | XRD, FT-IR | 84 |
| | Oil | 90 | KF | XRD, FT-IR, BET, TGA, Hammett indicators | 85 |
| | Sun-flower | 100 | Mg/La | XRD, XPS, BET | 86 |
| | *Camelina sativa* | 93 | $Na_{0.1}Ca_{0.9}TiO_3$ nanorods | SEM, TEM, XRD, BET | 87 |
| | Palm kernel | 96.1 | $CaMg(CO_3)_2$ | XRD, XRF, TGA, DTA | 88 |
| | Palm | 87.3 | KOH/SBA 15 | — | 89 |

### 12.3.7 UNUSUAL SUPPORTS

Waste eggshell was calcined and tested for transesterification of soybean oil to produce biodiesel in economical route. With oil to MeOH ratio of 1:9 and temperature of 65° C and 3% catalyst, it was observed that 95% biodiesel was produced in 3 h. It was observed that the waste eggshell possess high activity and reusability and can make the biodiesel process more economical and environmental friendly [90]. In another study, heterogeneous catalyst derived from snail shell for the synthesis of biodiesel was investigated for the transesterification of waste frying oil. With 1:9.6 ratio of oil to MeOH at 60° C for 8 h with 2% catalyst gave 87.28% biodiesel yield [91]. Snail shell is a novel source for catalyst and this study showed that process address environmental and economic aspects for the production of biodiesel in a sustainable way.

South African Class F fly ash derived Na-X zeolite was used as heterogeneous base catalyst for the transesterification of sunflower oil to produce biodiesel. It was shown that a biodiesel yield of 83.53% was achieved with reaction conditions of oil to methanol ration of 1:6, catalyst amount of 3% and at a temperature of 65° C for a period of 8 h [92]. The study showed that the process offers value addition to the large amount of fly ash generated every year.

Cocoa pod husks derived solid base catalyst was examined for the transesterification of soybean oil to biodiesel was investigated where it was shown that, cocoa pod husk potash gave a biodiesel yield of 91.4% and cocoa pod husk/MgO catalyst gave 98.7% biodiesel yield. The optimum reaction conditions of oil to MeOH ratio of 1:6, catalyst amount of 1% and a temperature of 60° C for 1 h were employed. The study showed that cocoa pod husk based heterogeneous catalyst was a greener approach for commercial and sustainable production of biodiesel [93].

In another study, shrimp shell catalyst was developed by incomplete carbonization and loading of KF (30%) and subsequent activation of the material. The reaction conditions of oil to methanol ration of 1:9, with catalyst amount of 2.5% and reaction temperature of 65° C for 3 h gave 89% conversion for the transesterification of rapeseed oil with methanol. The study showed that the catalyst had a porous framework structure and has environmentally friendly properties and can be used as a heterogeneous base catalyst for biodiesel production [94].

Silica-supported guanidine as heterogeneous catalyst was employed for the transesterification of soybean oil with methanol for biodiesel production. The reaction conditions of oil to methanol ratio of 1:12, catalyst amount of 3.5% were used at a temperature of 80° C for 3 h, which gave 98% conversion [95].

Cinder supported CaO/KF was used as a heterogeneous base catalyst for the transesterification of raw soybean oil for producing biodiesel. The reaction conditions of oil to methanol ratio of 1:12, catalyst amount of 2.1% at a temperature of 65° C for 20 min gave 99% conversion. The study showed that the catalyst can be reused for several times with good conversions and it was reported that 96.6% conversion was observed in the fourth cycle [96]. Table 12.7 describes the use of natural materials as supports for heterogeneous base catalyst preparation for biodiesel preparation.

**TABLE 12.7** Base catalysts from unusual supports.

| S. No. | Oil | Yield (%) | Catalyst | Characterization | Ref. |
|---|---|---|---|---|---|
| | Soybean | 95 | Waste egg shell | XRD, TGA, SEM | 90 |
| | Soybean o | 87.2 | Snail shell | XRD, XRF, BET, TGA, Hammett indicators | 91 |
| | Sunflower | 83.5 | Fly ash based zeolite | XRD, XRF, SEM, BET, Hammett indicators | 92 |
| | Soybean | 91–98 | Cocoa pod husks | BET, AAS | 93 |
| | Rapeseed | 89 | Shrimp shell | XRD, TGA, DTA, Hammett indicators | 94 |
| | Soybean | 98 | Silica supported guanidine | TGA, BET | 95 |
| | Raw soybean oil | 99 | Cinder supported CaO/KF | XRD, SEM, EDS, BET | 96 |

### 12.3.8 CARBON-BASED CATALYSTS

In recent years carbon-based sulfonated catalysts (CBSC) have become very attractive catalysts with high stability and many strong protonic acid sites. The main attractive feature of CBSC is that it is prepared from several inexpensive biomasses like sugars, glycerol, oilseed cakes, etc. In one of the first communications in this area reported by Hara, the catalyst was prepared from D-glucose and sucrose by incomplete carbonization at low temperature to induce pyrolysis and the formation of polycyclic aromatic carbon rings. Later this material was reacted with sulfuric acid to introduce sulphonite groups ($-SO_3H$) for the esterification of fatty acids into biodiesel. The structural analysis of this catalyst was shown to consist of sheets of amorphous carbon bearing hydroxyl and carboxyl (-OH and –COOH) groups and high densities of $-SO_3H$ groups [97]. In another report, glycerol-based carbon catalyst was prepared by in situ partial carbonization and sulfonation of crude glycerol (by-product of biodiesel process), technical grade glycerol and glycerol pitch (waste from fat splitting industry) with sulfuric acid in a single pot reaction. The main application of this catalyst was focused on the esterification of fatty acids in place of sulfuric acid for the production of biodiesel [98]. This glycerol-based catalyst was further modified to base catalyst by treating with alkali in controlled conditions. The base catalyst was found to exhibit very good transesterification activity for the production of biodiesel from neat vegetable oils like sunflower oil, high oleic sunflower oil, soybean oil, palm oil, castor oil, rice bran oil and other refined clean oils. This catalyst was highly active, reusable, moisture tolerant and simplified the biodiesel production particularly for transesterification of vegetable oils replacing the traditional homogeneous alkaline catalysts [99]. Another carbon-based acid catalyst was prepared from de-oiled canola meal, which exhibited good activity for conversion of FFA present in the oils to biodiesel.. Partial carbonization of de-oiled canola meal was steam activated and further treatment with sulfuric acid produced the acid catalyst, which showed that the catalyst had acidic sites in the form of sulfonated groups, which catalyzed esterification [100]. However, all these carbon-based catalysts have been reported on a laboratory scale and extensive work has to be carried out in this area for commercialization of the catalyst.

### 12.3.9 BIOCATALYSTS

The disadvantages caused by chemical catalysts can be overcome by using biocatalysts particularly lipases as alternative catalysts for the preparation of biodiesel by employing greener approach. Biocatalysts are usually lipases; however conditions need to be well controlled to maintain the activity of the catalyst. Lipases are stable, do not require coenzymes and will often tolerate organic solvents. Recent patents and articles have shown that reaction yields and times are still unfavorable compared to base-catalyzed transesterification for commercial application. Several lipases were employed either in free from or in immobilized state for this purpose, however the reaction times are longer compared to the alkali-catalyzed processes. Immobilized *Candida Antarctica, C. rugosa Thermomyces lanuginosus, Rhizopus oryzae, Pseudomonas fluorescens, P. cepacia* were some of the lipases exploited for the preparation of biodiesel from different vegetable oils and fatty acids [101]. As lipases can simultaneously perform esterification of FFA and transesterification triacylglycerols, low quality oils like high FFA oils, acid oils, restaurant greases, oils isolated from spent bleaching earth were also attempted to convert into biodiesel in reasonably good yields. The main limitation of lipases is that the short chain alcohols like methanol and ethanol as acyl donors would lead to inactivation of lipases. Major by-product glycerol may also block the active sites of enzyme resulting in low lipase activity. Different strategies were initiated to regain the activity of lipases using short chain alcohols during the alcoholysis and acidolysis. It is interesting to observe that hydrotalcites are promising supports for lipase immobilization for biodiesel production. In addition, novel acyl acceptors like methyl acetate, ethyl acetate and propane-2-ol were tried so far successfully at laboratory scale to protect the activity of the lipases. A detailed overview on the lipase catalyzed preparation of biodiesel and the methodologies for stabilizing the lipases and the use of alternate acyl donors are reported in the literature. [102]. The cost of lipase production is still the main bottleneck to commercialize the lipase-catalyzed process for low value product like biodiesel. However, there is a lot scope for biotechnological revolution in this area.

## 12.4 CHARACTERIZATION OF THE CATALYSTS

The various characterization techniques employed for catalysts are Brunauer-Emmet-Teller (BET) for surface area characterization, Bopp-Jancso-Heinzinger (BJH) pore size measurements, N2 adsorption-desorption and temperature programmed desorption (TPD) for texture characteristics, X-ray powder diffraction (XRD) for crystalline structure, scanning electron microscopy (SEM) measurement for morphology characteristics. Energy dispersive X-ray spectrophotometer (EDX), X-ray fluorescence (XRF), ICP-MS, atomic absorption spectroscopic analysis is also used for additional characterization of the catalysts.

The basic strength or alkalinity of the catalyst ($H\_$) will be estimated using Hammet titration or Hammet indicators. Thermal decomposition of the catalyst is evaluated by thermo-gravimetric analysis (TGA) and differential thermal analysis (DTA). FT-IR analysis is also carried out to record the absorption peaks for further characterization of the catalysts. All these characterization techniques are normally used for base catalysts and the same has been reported in the literature, which is shown in the tables mentioned in the present chapter.

## 12.5 CONCLUSIONS

At present about 15 million tons of biodiesel is being produced globally employing homogeneous acid and base catalysts. A number of feedstocks like sunflower, soybean, palm, canola, rapeseed, cottonseed oils, nonedible oils like jatropha and karanja, used frying oils; animal fats are currently being examined for biodiesel production along with algae, which is being projected as promising source for biodiesel production. The main technology used in the industrial production of biodiesel is based on the transesterification of several refined vegetable oils with methanol using basic homogeneous catalysts like NaOH, KOH and their corresponding alkoxides. In case of FFA containing oils, initially the fatty acid is being esterified using acid catalyst like sulfuric acid followed by neutralization of acid catalyst before converting the triglycerides into biodiesel employing transesterification. The main challenge with homogeneous catalysis is handling high amounts of effluents and salts generated and the produced biodiesel must be washed with water. Also the catalysts cause corrosion and generate waste, which are responsible for environmental pollution. Hence there is a need to replace homogeneous catalysts with heterogeneous catalysts, which can address the issues mentioned above. Several heterogeneous base catalysts based on single metal oxides, mixed oxides, oxides in combination with other materials, alumina based catalysts, hydrotalcite-based catalysts, and other base catalysts and lipases have been reported for the conversion of vegetable oils, waste cooking oils and frying oils, animal fats to biodiesel.

Though homogeneous transesterification of vegetable oils for biodiesel production is a well reported technology, it suffers from disadvantages like high alcohol to oil requirement, adverse impact of impurities including FFA and unsaponifiables in the feed oils and glycerol separation problems. The literature reports on heterogeneous catalysis of above reactions indicate new process intensification options though none of them have been tested on commercial scale.

As most of the heterogeneous catalysts are metal based, the thrust may be given for carbon-based catalysts based on cheaper biomasses like carbohydrates, glycerol and oilseed cakes, etc. Also, catalysts derived from natural waste materials like shells of egg, snail and shrimp, cocoa pod husk and construction waste appear to be promising for biodiesel production. These types of studies are encouraging as they offer value addition to waste materials for producing high volume products like biodiesel. Future research directed towards development and modification of natural waste materials could be a potential platform for greener approach for sustainable production of biodiesel.

## KEYWORDS

- **Animal fats**
- **Biodiesel**
- **Fatty acid**
- **Fatty acid methyl ester**
- **Heterogeneous catalyst**
- **Homogeneous catalyst**
- **Transesterification**
- **Triglyceride**
- **Vegetable oils**

## REFERENCES

1. Tom Krawczyk, Biodiesel-Alternative Fuel Makes Inroads but Hurdles Remain. *INFORM* 7, 801–829 (1996).
2. Harrington, K. J. Chemical and physical properties of vegetable oil esters and their effect on diesel fuel performance. *Biomass* **9**, 1–7 (1986).
3. ASTM ASTM D 6751. Standard specification for biodiesel fuel blend stock (B100) for middle distilate fuels. ASTM International. West Conshohocken P
4. EN 14214. Automotive fuels—Fatty acid methyl esters (FAME) for diesel engines—Requirements and test methods.
5. Freedman, B., Pryde, E. H., and Mounts, T. L. Variables affecting the yield of fatty esters from transesterified vegetable oils. *J Amer Oil Chem Soc* **61**, 1638–1643 (1984)

6. Biodiesel production: a review. Fangrui Ma, Milford A. Hanna. Biores Technol **70**, 1–15 (1999).
7. Hara, M. Environmentally benign production of biodiesel using heterogeneous catalysts. *ChemSusChem* **2**, 129–135 (2009).
8. Chakrabarti, P. P., and Prasad, R. B. N. Biodiesel production from *Jatropha curcus* oil. N. Carels et al. (eds.), Jatropha, Challenges for a New Energy Crop: volume I: Farming, Economics and Biofuel, pp 463–490 (2012).
9. Helwani, Z., Othman, M.R., Aziz, N., Kim, J., and Fernando, W.J.N. Solid heterogeneous catalysts for transesterification of triglycerides with methanol: A review. *Applied Catal A: General* **363**, 1–10 (2009).
10. Sivasamy, A., Yoo Cheah, K., Fornasiero, P., Kemausuor, F., Zinoviev, S., and Miertus, A. Catalytic Applications in the Production of Biodiesel from Vegetable Oils. *ChemSusChem* **2**, 278–300 (2009).
11. Di Serio, M., Tesser, R., Pengmei, L., and Santacesaria, E. Heterogeneous Catalysts for Biodiesel Production. *Energy & Fuels* **22**, 207–217 (2008).
12. Sharma, Y. C., Singh, B., and Korstad, J. Latest developments on application of heterogeneous basic catalysts for an efficient and eco friendly synthesis of biodiesel: A review. *Fuel* **90**, 1309–1324 (2011).
13. Ashok Pandey. Handbook of plant-based biofuels, Boca Raton, FL. Taylor & Francis (2009).
14. Kawashima, A., Matsubara, K., and Honda, K. Acceleration of catalytic activity of calcium oxide for biodiesel production. *Biores Technol* **100**, 696–700 (2009).
15. Puna, J. F., Correia, M. N. J., Dias, A. P. S., Gomes, J., and Bordado, J. Biodiesel production from waste frying oils over lime catalysts. *Reac Kinet Mech Cat* DOI 10.1007/s11144–013–0557–2 (2013).
16. Dias, A. P. S., Puna, J., Correia, M. N. J., Nogueira, I., Gomes, J., and Bordado, J. Effect of the oil acidity on the methanolysis performances of lime catalyst biodiesel from waste frying oils (WFO). *Fuel Proces Technol* **116**, 94–100 (2013).
17. Ferdous, K., Rakib Uddin, M., Khan, M. R., and Islam, M. A. Preparation of biodiesel from soybean oil by using heterogeneous catalyst. *Int J Energy & Environ* **4**, 243–252 (2013).
18. Lopez Granados, M., Zafra Poves, M.D., Martın Alonso, D., Mariscal, R., Cabello Galisteo, F., Moreno-Tost, R., Santamarıa, J., and Fierro, J. L. G. Biodiesel from sunflower oil by using activated calcium oxide. *Applied Catal B: Environ* **73**, 317–326 (2007).
19. Hsiao, M. C., Lin, C. C., and Chang, Y. H. Microwave irradiation-assisted transesterification of soybean oil to biodiesel catalyzed by nanopowder calcium oxide. *Fuel* **90**, 1963–1967 (2011).
20. Patil, P. D., and Deng, S. Transesterification of Camelina sativa oil using heterogeneous metal oxide catalysts. *Energy Fuels* **23**, 4619–4624 (2009).
21. Mootabadi, H., Salamatinia, B., Bhatia, S., and Abdullah, A. Z. Ultrasonic-assisted biodiesel production process from palm oil using alkaline earth metal oxides as the heterogeneous catalysts. *Fuel* **89**, 1818–1825 (2010).
22. Liu, X., He, H., Wang, Y., and Zhu, S. Transesterification of soybean oil to biodiesel using SrO as a solid base catalyst. *Catal Comm* **8**, 1107–1111 (2007).

23. Gotcht, A. J., Reeder, A. J., and McCormick. A. Study of heterogeneous base catalysts for biodiesel production. *Journal of Undergraduate Chem Res* **8**, 22–26 (2009).
24. Balakrishnan, K., Olutoye, M.A., and Hameed, B.H. Synthesis of methyl esters from waste cooking oil using construction waste material as solid base catalyst. *Biores Technol* **128**, 788–791 (2013).
25. Ilgen, O., and Akin, A. N. Transesterification of canola oil to biodiesel using MgO loaded with KOH as a heterogeneous catalyst. *Energy & Fuels* **23**, 1786–1789 (2009).
26. Monica, C.G. A., Jimenez-Urbistondo, I., Santamarıa-Gonzalez, J., Merida-Robles, J. m., Moreno-Tost, R., Rodrıguez-Castellon, E., Jimenez-Lopez, A., Diana, C.S. A., Cavalcante Jr., C. L., and Maireles-Torres, P. CaO supported on mesoporous silicas as basic catalysts for transesterification reactions. *Applied Catal A: General* **334**, 35–43 (2008).
27. Liu, X., Piao, X., Wang, Y., Zhu, S., and He, H. Calcium methoxide as a solid base catalyst for the transesterification of soybean oil to biodiesel with methanol. *Fuel* **87**, 1076–1082 (2008).
28. Zhu, H., Wu, Z., Chen, Y., Zhang, P., Duan, S., Liu, X., and Mao, Z. Preparation of biodiesel catalyzed by solid super base of calcium oxide and its refining process. *Chin J Catal* **27**, 391–396 (2006).
29. Encinar, J. M., González, J. F., Pardal, A., and Martínez, G. Rape oil transesterification over heterogeneous catalysts. *Fuel Processing Technol* **91**, 1530–1536 (2010).
30. Qiu, F., Li, Y., Yang, D., Li, X., and Sun, P. Heterogeneous solid base nanocatalyst: Preparation, characterization and application in biodiesel production. *Biores Technol* **102**, 4150–4156 (2011).
31. Sun, H., Hu, K., Lou, H., and Zheng, H. Biodiesel Production from transesterification of rapeseed oil using KF/Eu2O3 as a catalyst. *Energy & Fuels* **22**, 2756–2760 *(*2008).
32. Kumar, D., and Ali, A. Nanocrystalline lithium ion impregnated calcium oxide as heterogeneous catalyst for transesterification of high moisture containing cotton seed oil. *Energy Fuels* **24**, 2091–2097 (2010).
33. Yang, Z., and Xie, W. Soybean oil transesterification over zinc oxide modified with alkali earth metals. *Fuel Processing Technol* **88**, 631–638 (2007).
34. Ngamcharussrivichai, C., Totarat, P., and Bunyakiat, K. Ca and Zn mixed oxide as a heterogeneous base catalyst for transesterification of palm kernel oil. *Appl Cata A: General* **341**, 77–85 (2008).
35. Alba-Rubio, A. C., Santamarıa-Gonzalez, J., Merida-Robles, J. M., Moreno-Tost, R., Martın-Alonso, D., Jimenez-Lopez, A., and Maireles-Torres, P. Heterogeneous transesterification processes by using CaO supported on zinc oxide as basic catalysts. *Catal Today* **149**, 281–287 (2010).
36. Taufiq-Yapa, Y.H., Lee, H.V., Yunus, R., and Juan, J.C. Transesterification of nonedible Jatropha curcas oil to biodiesel using binary. Ca–Mg mixed oxide catalyst: Effect of stoichiometric composition. *Chem Eng Journal* **178**, 342–347 (2011).
37. Albuquerque, M. C. G., Santamarıa-Gonzalez, J., Merida-Robles, J. M., Moreno-Tost, R., Rodrıguez-Castellon, E., Jimenez-Lopez, A., Azevedo, D. C. S., Cavalcante Jr., C. L., and Maireles-Torres, P. MgM (M = Al and Ca) oxides as basic catalysts in transesterification processes. *Applied Catal A: General* **347**, 162–168 (2008).
38. Kawashima, A., Matsubara, K., and Honda, K. Development of heterogeneous base catalysts for biodiesel production. *Biores Technol* **99**, 3439–3443 (2008).

39. Dehkordi, A. M., and Ghasemi, M. Transesterification of waste cooking oil to biodiesel using Ca and Zr mixed oxides as heterogeneous base catalysts. *Fuel Proces Technol* **97**, 45–51 (2012).
40. Rashtizadeh, E., and Farzaneh, F. Transesterification of soybean oil catalyzed by Sr–Ti mixed oxides nanocomposite. *Journal of the Taiwan Institute of Chemical Engineers* (2013) DOI: http://dx.doi.org/10.1016/j.jtice.2013.02.008.
41. Olutoye, M.A., and Hameed, B.H. Synthesis of fatty acid methyl ester from crude jatropha (Jatropha curcas Linnaeus) oil using aluminum oxide modified Mg–Zn heterogeneous catalyst. *Biores Technol* **102**, 6392–6398 (2011).
42. Xie, W., Peng, H., and Chen, L. Transesterification of soybean oil catalyzed by potassium loaded on alumina as a solid-base catalyst. *Applied Catal A: General* **300**, 67–74 (2006).
43. Vyas, A. P., Subrahmanyam, N., and Patel, P. A. Production of biodiesel through transesterification of Jatropha oil using KNO3/A12O3 solid catalyst. *Fuel* **88**–625–628 (2009).
44. Alonso, D. M., Mariscal, R., Moreno-Tost, R., Zafra Poves, M.D., Lopez Granados, M. Potassium leaching during triglyceride transesterification using K/A12O3 catalysts. *Catal Comm* **8**, 2074–2080 (2007).
45. Bo, X., Guomin, X., Lingfeng, C., Ruiping, W., and Lijing, G. Transesterification of palm oil with methanol to biodiesel over a KF/A12O3 heterogeneous base catalyst. *Energy & Fuels* **21**, 3109–3112 (2007).
46. Lingfeng, C., Guomin, X., Bo, X., and Guangyuan, T. Transesterification of cottonseed oil to biodiesel by using heterogeneous solid basic catalysts. *Energy & Fuels* **21**, 3740–3743 (2007).
47. Guangyuan, T., Gao, L., Guomin, X., and Liu, H. Transesterification of soybean oil to biodiesel over heterogeneous solid base catalyst. *Energy Fuels* **23**, 4630–4634 (2009).
48. Xie, W., and Li, H. Alumina-supported potassium iodide as a heterogeneous catalyst for biodiesel production from soybean oil. *J Mol Catal A: Chemical* **255**, 1–9 (2006).
49. Ma, H., Li, S., Wang, B., Wang, R., and Tian, S. Transesterification of rapeseed oil for synthesizing biodiesel by K/KOH/A12O3 as heterogeneous base catalyst. *J Am Oil Chem Soc* **85**, 263–270 (2008).
50. Noiroj, K., Intarapong, P., Luengnaruemitchai, A., and Jai-In, S. A comparative study of KOH/A12O3 and KOH/NaY catalysts for biodiesel production via transesterification from palm oil. *Renew Energy* **34**, 1145–1150 (2009).
51. Tonetto, G. M., and Marchetti, J. M. Transesterification of soybean oil over Me/A12O3 (Me 5 Na, Ba, Ca, and K) catalysts and monolith K/A12O3-cordierite. *Top Catal* **53**, 755– 762 (2010).
52. Ilgen, O., and Akin, A. N. Development of alumina supported alkaline catalysts used for biodiesel production. *Turk J Chem* **33**, 281–287 (2009).
53. Wan, T., Yu, P., Wang, S., and Luo, Y. Application of sodium aluminate as a heterogeneous base catalyst for biodiesel production from soybean oil. *Energy & Fuels*, **23**, 1089–1092 (2009).
54. Kim, H. J., Bo-Seung, K., Kim, M. J., Park, Y. M., Kim, D. K., Lee, J. S., and Lee, K. W. Transesterification of vegetable oil to biodiesel using heterogeneous base catalyst. *Catal Today* **93–95**, 315–320 (2004).

55. Arzamendi, G., Campoa, I., Arguinarena, E., Sanchez, M., Montes, M., Gandıa, L.M. Synthesis of biodiesel with heterogeneous NaOH/alumina catalysts: Comparison with homogeneous NaOH. *Chem Eng Journal* **134**, 123–130 (2007).
56. Umdu, E. S., Tuncer, M., and Seker, E. Transesterification of *Nannochloropsis oculata* microalga's lipid to biodiesel on Al2O3 supported CaO and MgO catalysts. *Biores Technol* **100**, 2828–2831 (2009).
57. D'Cruz, A., Kulkarni, M. G., Meher, L. C., and Dalai, A. K. Synthesis of biodiesel from canola oil using heterogeneous base catalyst. *J Am Oil Chem Soc* **84**, 937–943 (2007).
58. Li, X., Lu, G., Guo, Y., Guo, Y., Wang, Y., Zhang, Z., Liu, X., and Wang, Y. A novel solid superbase of Eu2O3/Al2O3 and its catalytic performance for the transesterification of soybean oil to biodiesel. *Catal Comm* **8**, 1969–1972 (2007).
59. Gomes, J. F. P., Puna, J F. B., Gonçalves, L. M., and Bordado, J. C. M. Study on the use of MgAl hydrotalcites as solid heterogeneous catalysts for biodiesel production. *Energy* **36**, 6770–6778 (2011).
60. Liu, Y., Lotero, E., Goodwin Jr., J. G., and Mo, X. Transesterification of poultry fat with methanol using Mg–Al hydrotalcite derived catalysts. *Applied Catal A: General* **331**, 138– 148 (2007).
61. Zeng, H. Y., Feng, Z., Deng, X., and Li, Y. G. Activation of Mg–Al hydrotalcite catalysts for transesterification of rape oil. *Fuel* **87**, 3071–3076 (2008).
62. Xie, W., Peng, H., and Chen, L. Calcined Mg–Al hydrotalcites as solid base catalysts for methanolysis of soybean oil. *J Mol Catal A: Chemical* **246**, 24–32 (2006).
63. Ilgen, O., Ismail., Yildiz, M., Alptekin, E., Boz, N., Canakci, M., Akin, A. N. Investigation of Biodiesel Production from Canola Oil using Mg-Al Hydrotalcite Catalysts. Turk *J Chem* **31**, 509–514 (2007).
64. Dias, A. P. S., Bernardo, J., Felizardo, P., and Correia, M. J. N. Biodiesel production over thermal activated cerium modified Mg-Al hydrotalcites. *Energy* **41**, 344–353 (2012).
65. Trakarnpruk, W., and Porntangjitlikit. S. Palm oil biodiesel synthesized with potassium loaded calcined hydrotalcite and effect of biodiesel blend on elastomer properties. *Ren Energy* **33**, 1558–1563 (2008).
66. Barakos, N., Pasias, S., and Papayannakos, N. Transesterification of triglycerides in high and low quality oil feeds over an HT2 hydrotalcite catalyst. *Biores Technol* **99**, 5037–5042 (2008).
67. Wang, S. H., Wang, Y. B., Dai, Y. M., and Jehng, J. M. Preparation and characterization of hydrotalcite-like compounds containing transition metal as a solid base catalyst for the transesterification *Applied Catal A: General* **439–440**, 135–141 (2012).
68. Wang, Y. B., and Jehng, J. M. Hydrotalcite-like compounds containing transition metals as solid base catalysts for transesterification. *Chem Eng Journal* **175**, 548–554 (2011).
69. Li, E., Xu, Z. P., and Rudolph, V. MgCoAl–LDH derived heterogeneous catalysts for the ethanol transesterification of canola oil to biodiesel. *Applied Catal B: Environ* **88**, 42–49 (2009).
70. Shumaker, J. L., Crofcheck, C., Tackett, S. A., Santillan-Jimenez, E., Morgan, T., Ji, Y., Crocker, M., and Toops, T. J. Biodiesel synthesis using calcined layered double hydroxide catalysts. *Applied Catal B: Environ* **82**, 120–130 (2008).

71. Gao, L., Teng, G., Xiao, G., and Wei, R. Biodiesel from palm oil via loading KF/Ca-Al hydrotalcite catalyst. *Biomass and bioenergy*, **34**, 1283–1288 (2010).
72. Gao, L., Xu, B., Xiao, G., and Lv, J. Transesterification of Palm Oil with Methanol to Biodiesel over a KF/Hydrotalcite Solid Catalyst. *Energy & Fuels* **22**, 3531–3535 (2008).
73. Georgogianni, K.G., Katsoulidis, A.K., Pomonis, P.J., Manos, G., and Kontominas, M.G. Transesterification of rapeseed oil for the production of biodiesel using homogeneous and heterogeneous catalysis. *Fuel Proces Technol* **90**, 1016–1022 (2009).
74. Di Serio, M., Ledda, M.., Cozzolino, M.., Minutillo, G., Tesser, R., and Santacesaria, E. Transesterification of soybean oil to biodiesel by using heterogeneous basic catalysts. *Ind Eng Chem Res* **45**, 3009–3014 (2006).
75. Wang, Y., Zhang, F., Xu, S., Yang, L., Li, D., Evans, D. G., and Duan, X. Preparation of macrospherical magnesia-rich magnesium aluminate spinel catalysts for methanolysis of soybean oil. *Chem Eng Sci* **63**, 4306–4312 (2008).
76. Guerreiro, L., Pereira, P.M., Fonseca, I.M., Martin-Aranda, R.M., Ramos, A.M., Dias, J.M.L., Oliveira, R., Vital, J. PVA embedded hydrotalcite membranes as basic catalysts for biodiesel synthesis by soybean oil methanolysis. *Catal Today* **156**, 191–197 (2010).
77. Jiang, W., Lu., H. F., Qi, T., Yan, S. L., and Liang, B. Preparation, application, and optimization of Zn/Al complex oxides for biodiesel production under subcritical conditions. *Biotechnol Adv* **28**, 620–627 (2010).
78. Antunes, W. M., Veloso, C. O., Henriques, C. A. Transesterification of soybean oil with methanol catalyzed by basic solids. *Catal Today* **133–135**, 548–554 (2008).
79. Zeng, H. Y., Liao, K. B., Deng, X., Jiang, H., and Zhang, F. Characterization of the lipase immobilized on Mg–Al hydrotalcite for biodiesel. *Process Biochem* **44**, 791–798 (2009).
80. Xie, W., Huang, X., and Li, H. Soybean oil methyl esters preparation using NaX zeolites loaded with KOH as a heterogeneous catalyst. *Biores Technol* **98**, 936–939 (2007).
81. Li, Y., Qiu, F., Yang, D., Li, X., and Sun, P. Preparation, characterization and application of heterogeneous solid base catalyst for biodiesel production from soybean oil. *Biomass and Bioenergy* **35**, 2787–2795 (2011).
82. Macario, A., Giordano, G., Onida, B., Cocina, D., Tagarelli, A., and Giuffrè, A. M. Biodiesel production process by homogeneous/heterogeneous catalytic system using an acid–base catalyst. *Applied Catal A: General* **378**, 160–168 (2010).
83. Xie, W., and Fan, M. Immobilization of tetramethylguanidine on mesoporous SBA-15 silica: A heterogeneous basic catalyst for transesterification of soybean oil. *Biores Technol* (2013). DOI: http://dx.doi.org/10.1016/j.biortech.2013.04.019.
84. Cordeiro, C. S., Arizaga, G. G. C., Ramos, L. P., and Wypych, F. A new zinc hydroxide nitrate heterogeneous catalyst for the esterification of free fatty acids and the transesterification of vegetable oils. *Catal Comm* **9**, 2140–2143 (2008).
85. Xu, C., and Liuab, Q. Catalytic performance and mechanism of KF-loaded catalysts for biodiesel synthesis. *Catal Sci Technol*, **1**, 1072–1082 (2011).
86. Seshu Babu, N., Rekha, S., Sai Prasad, P. S., and Lingaiah, N. Room-temperature transesterification of edible and nonedible oils using a heterogeneous strong basic Mg/La Catalyst. *Energy & Fuels* **22**, 1965–1971 (2008).

87. Man, L. F., Wong, W. T., and Yung, K. F. Alkali hydrothermal synthesis of Na0.1Ca0.9TiO3 nanorods as heterogeneous catalyst for transesterification of *Camelina Sativa* oil to biodiesel. *J Clust Sci* **23**, 873–884 (2012).
88. Ngamcharussrivichai, C., Wiwatnimit, W., and Wangnoi, S. Modified dolomites as catalysts for palm kernel oil transesterification. *J Mol Catal A: Chemical* **276**, 24–33 (2007).
89. Razali, N., Mootabadi, H., Salamatinia, B., Lee, K.T., and Abdullah, A.Z. Optimization of process parameters for alkaline-catalyzed transesterification of palm oil using response surface methodology. *Sains Malaysiana* **39**, 805–809 (2010).
90. Wei, Z., Xu, C., and Li, B. Application of waste eggshell as low-cost solid catalyst for biodiesel production. *Biores Technol* **100**, 2883–2885 (2009).
91. Birla, A., Singh, B., Upadhyay, S.N., and Sharma, Y.C. Kinetics studies of synthesis of biodiesel from waste frying oil using a heterogeneous catalyst derived from snail shell. *Biores Technol* **106**, 95–100 (2012).
92. Babajide, O., Musyoka, N., Petrik, L., and Ameer, F. Novel zeolite Na-X synthesized from fly ash as a heterogeneous catalyst in biodiesel production. *Catal Today* **190**, 54–60 (2012).
93. Ofori-Boateng, C., and Lee, K. T. The potential of using cocoa pod husks as green solid base catalysts for the transesterification of soybean oil into biodiesel: Effects of biodiesel on engine performance. *Chem Eng Journal* **220**, 395–401 (2013).
94. Yang, L., Zhang, A., and Zheng, X. Shrimp shell catalyst for biodiesel production. *Energy & Fuels* **23**, 3859–3865 (2009).
95. Balbino, J. M., W. de Menezes, E., Benvenutti, E. V., Cataluna, R., Ebeling, G., and Dupont, J. Silica-supported guanidine catalyst for continuous flow biodiesel production. *Green Chem* **13**, 3111–3116 (2011).
96. Liu, H., Su, L., Shao, Y., and Zou, L. Biodiesel production catalyzed by cinder supported CaO/KF particle catalyst. *Fuel* **97**, 651–657 (2012).
97. Toda, M., Takagaki, A., Okamura, M., Kondo, J. K., Hayashi, S., Domen, K., and Hara, M. Biodiesel made with sugar catalyst. *Nature* **435**, 178 (2005).
98. Prabhavathi Devi, B. L. A., Gangadhar, K. N., Sai Prasad, P. S., Jagannadh, B., and Prasad, R. B. N. A Glycerol-based Carbon Catalyst for the Preparation of Biodiesel. *ChemSusChem* **2**, 617–620 (2009)
99. Prabhavathi Devi, B. L. A., Vijayalakshmi, K., Sai Prasad, P. S., and Prasad, R. B. N. Novel glycerol-based heterogeneous carbon base catalyst useful for the transesterification of vegetable oils to biodiesel, a process and use thereof. Indian patent. Patent Appl No. 2291/DEL/2011 A.
100. Rao, B. V. S. K., Chandra Mouli, K., Rambabu, N., Dalai, A. K., and Prasad, R. B. N. Carbon-based solid acid catalyst from de-oiled canola meal for biodiesel production. *Catal Comm* **14**, 20–26 (2011).
101. Noureddini, H., Gao, X., and Philkana, R. S. Immobilized Pseudomonas cepacia lipase for biodiesel fuel production from soybean oil. *Biores Technol* **96**, 769–777 (2006).
102. Prasad, R. B. N., Rao, B. V. S. K. Lipase-catalyzed preparation of biodiesel. In Handbook of plant-based biofuels, pp 199–213, Boca Raton, FL. Taylor & Francis (2009).

# INDEX

## D



## E



## F

## G



## H

## T



## U



## V



## W



## X

## Z